AF411305

Sustainable Designed
Pavement Materials

Sustainable Designed Pavement Materials

Volume 1

Special Issue Editors

Sandra Erkens
Yue Xiao
Mingliang Li
Tao Ma
Xueyan Liu

MDPI • Basel • Beijing • Wuhan • Barcelona • Belgrade • Manchester • Tokyo • Cluj • Tianjin

Special Issue Editors
Sandra Erkens
Delft University of Technology (TUDelft)
The Netherlands

Yue Xiao
Wuhan University of Technology (WUT)
China

Mingliang Li
Research Institute of Highway Ministry of Transport (RIOH)
China

Tao Ma
Southeast University (SEU)
China

Xueyan Liu
Delft University of Technology (TUDelft)
The Netherlands

Editorial Office
MDPI
St. Alban-Anlage 66
4052 Basel, Switzerland

This is a reprint of articles from the Special Issue published online in the open access journal *Materials* (ISSN 1996-1944) (available at: https://www.mdpi.com/journal/materials/special_issues/sdpm).

For citation purposes, cite each article independently as indicated on the article page online and as indicated below:

LastName, A.A.; LastName, B.B.; LastName, C.C. Article Title. *Journal Name* **Year**, *Article Number*, Page Range.

Volume 1
ISBN 978-3-03928-985-1 (Hbk)
ISBN 978-3-03928-986-8 (PDF)

Volume 1-2
ISBN 978-3-03936-004-8 (Hbk)
ISBN 978-3-03936-005-5 (PDF)

Contents

About the Special Issue Editors

Sandra Erkens is the principal specialist in pavement materials and structures at Rijkswaterstaat, the Dutch highway authority. She is a full professor, holding the Chair of Pavement Engineering Practice, at Delft University of Technology. She is an internationally acknowledged expert in pavement materials and structures in general and asphalt concrete in particular. Prof. Erkens was a member of national and international groups involved in developing technical requirements for pavement materials and several (inter)national organizations for the dissemination of research. These include the ISAP technical committee on the Constitutive Modelling of Asphalt Concrete, the organization committee of the two-yearly Dutch conference on Infrastructure (CROW-infradagen) and the organizing committee of the 4th International Chinese European Workshop on Functional Pavement Design. She has been involved in road engineering research since 1997, has published over a hundred papers on her work and is a regular reviewer for conferences and journals.

Yue Xiao is a full-time research professor of the State Key Lab of Silicate Materials for Architectures in the Wuhan University of Technology. He was awarded the Fok Ying Tung Outstanding Young Teacher award by the Ministry of Education of China in 2018. He received the title of CHUTIAN Scholar in material science and engineering from the Hubei provincial department of education in 2014. Prof. Xiao received his Ph.D. degree in Road and Railway Engineering from Delft University of Technology, The Netherlands. He then joined Wuhan University of Technology in 2013. His research interests are in asphalt pavement surfacing, road and pavement materials, asphalt pavement design. Dr. Xiao is now conducting three innovative projects funded by the National Natural Science Foundation of China (NSFC), and projects supported by provincial departments as well. Since 2011, Dr. Xiao has published 62 SCI peer-reviewed journal papers.

Mingliang Li is a full-time associate researcher at the Road Research Center, Research Institute of Highway Ministry of Transport (RIOH) in China. Dr. Li received his Ph.D. degree in Road and Railway Engineering from Delft University of Technology, The Netherlands. His research interests are in functional pavement material and technology, pavement maintenance, asphalt materials, and recycling technology. He has participated in and completed more than 10 national and provincial scientific research projects, such as the National Key R & D plan of the Ministry of Science and Technology, research projects from the Ministry of Transport, etc. He was in charge of more than 20 design and consultant projects, including porous asphalt pavement, Sponge City permeable pavement and in-place recycling, etc. He has published one monograph, participated in the writing of six national and local standards, published more than 30 journal papers, and obtained 18 national patents.

Tao Ma is a full-time research professor and the vice dean of the School of Transportation at Southeast University. He is also the director of Road Engineering Department. His awards include the Young and Middle-Aged Leading Talent of Science And Technology Innovation In Transportation by the Ministry of Transport of China, Fok Ying Tung fund by the Ministry of Education of China, 333 High-Level Talent Project, and the Six Talent Summit Project by Jiangsu Province. He is the deputy director of the National Engineering Laboratory for Advanced Road Materials, deputy director of National Teaching Center of Road Traffic Virtual Simulation, and director of Jiangsu

Key Laboratory for Long-term Service and Safety of Road Infrastructure. His research interests are in asphalt pavement design, road and pavement materials, asphalt pavement maintenance and pavement sustainable development technologies. Dr. Ma is now conducting the science project for outstanding young people founded by the National Natural Science Foundation of China (NSFC), and other innovative projects supported by NSFC and provincial departments. Dr. Ma has published two books, more than 100 technical and journal papers, and has been awarded more than 20 national invention patents.

Xueyan Liu is currently an Associate Professor in the Section of Pavement Engineering of the Faculty of Civil Engineering & Geosciences of TU Delft. He works in the areas of constitutive modelling, numerical modelling, and material experimental characterization. Within the research program of the Section Pavement Engineering, his research topics mostly relate to the development and implementation of constitutive models for the simulation of the static and dynamic response of various pavement engineering materials like soils, asphalt concrete, liner and reinforcing systems, and sustainable development technologies, i.e., multiscale modelling of asphaltic materials, warm/cold asphalt concrete technology, durability of asphalt surfacings on orthotropic steel deck bridge, accelerated pavement test, and pavement continuous monitoring. Dr. Liu was granted his doctoral thesis in 2003. During the same period, Dr. Liu participated in the team that developed the ACRe model for Asphalt Concrete Response currently implemented in the 3D Computer Aided Pavement Analysis system (CAPA-3D). Dr. Liu has published more than 100 technical and journal papers on the mechanics and the finite element modelling of granular, concrete, and asphaltic materials. Dr. Liu is a member of the RILEM Technical Committee of Cracking in Asphalt Pavements WG3 and a member of the Delft Centre for Materials (DCMat). He is also a member of ISAP, AAPT, APSE, and IACMAG. Dr. Liu is an Editorial Board Member of Geomaterials (GM). Dr. Liu was appointed as a Board member of the International Association of Chinese Infrastructure Professionals (IACIP) and a member of the Academic Committee of the Key Laboratory of Road Structure and Materials Transportation Industry of the China Ministry of Transport. He is also actively involved in organizing international and national workshops and conferences and was invited as a Scientific/Technical committee member of several international conferences.

Editorial

Sustainable Designed Pavement Materials

Yue Xiao [1,2], Sandra Erkens [3], Mingliang Li [1,*], Tao Ma [4] and Xueyan Liu [3]

[1] Key Laboratory of Transport Industry of Road Structure and Material, Research Institute of Highway (RIOH), Ministry of Transport, Beijing 100088, China; xiaoy@whut.edu.cn
[2] State Key Laboratory of Silicate Materials for Architectures, Wuhan University of Technology, Wuhan 430070, China
[3] Faculty of Civil Engineering and Geosciences, Delft University of Technology, 2628 CN Delft, The Netherlands; s.m.j.g.erkens@tudelft.nl (S.E.); X.Liu@tudelft.nl (X.L.)
[4] School of Transportation, Southeast University, Nanjin 210096, China; matao@seu.edu.cn
* Correspondence: ml.li@rioh.cn; Tel.: +86-1581-033-9871

Received: 20 February 2020; Accepted: 25 March 2020; Published: 29 March 2020

Abstract: This Special Issue "Sustainable Designed Pavement Materials" has been proposed and organized as a means to present recent developments in the field of environmentally-friendly designed pavement materials. For this reason, articles included in this special issue relate to different aspects of pavement materials, from industry solid waste recycling to pavement materials recycling, from pavement materials modification to asphalt performance characterization, from pavement defect detection to pavement maintenance, and from asphalt pavement to cement concrete pavement, as highlighted in this editorial.

Keywords: pavement materials; sustainable designed pavement materials; recycling; recycled pavement materials; ageing resistance; modified asphalt materials; rejuvenator; skid resistance; pavement surfacing

---

This Special Issue "Sustainable Designed Pavement Materials" has been proposed and organized as a means to present recent developments in the field of environmentally-friendly designed pavement materials. It covers a wide range of selected topics on pavement materials. In total, 40 papers passed the peer-review and got published in this Special Issue. Universities and institutes considered as the most successful organizations, such as Wuhan University of Technology (10 papers), Southeast University (7 papers), Changsha University of Science & Technology (7 papers), Chang'an University (6 papers), Delft University of Technology (3 papers), Harbin Institute of Technology (2 papers), RWTH Aachen University (2 papers), Pennsylvania State University, Washington State University, Purdue University, and Michigan Technological University, have contributed a lot to this Special Issue. A brief summary of the articles is given in this editorial.

Research on solid waste recycling in pavement materials is considered as one of the most economic ways to achieve sustainable designed pavements. Kong et al. [1,2] studied the possibility of using oxygen furnace slag filler in asphalt mixture, and the BOF (Basic Oxygen Furnace) slag coarse aggregate was also presented in his research for making asphalt concrete. Three types of BOF slag fillers were concluded in their research. Ye et al. [3] investigated the effects of different cooling and treatment processes on the morphological features of BOF steel slag, and the effect of slag morphologies on the performance of asphalt mixtures. Another article from Qian et al. [4] also focused on slag pavement materials. Phosphorous slag was used as asphalt mixture aggregates on cement concrete deck to improve the interface bonding strength. Quarry fines were proposed by Zhang et al. [5,6] for pavement construction materials, by evaluating the properties of basic quarry fines and stabilized quarry fine specimens prepared using the gyratory compactor. Besides using slags as aggregates, crumb rubber was another widely used solid waste in pavement materials. The short-term aging of microwave activation crumb rubber was studied by Li et al. [7], the mixing and compaction temperatures of the crumb rubber modified asphalt mixture were discussed in the research from Li et al. [8], while Gu et al. [9] developed a meso-structure-based finite element model of rubber modified mixture to predict both the dynamic modulus and phase angle properties.

Reuse of pavement materials is another eco-efficient method for sustainable pavement design. Long-term ageing resistance and healing properties of pavement materials were discussed in detail. The ageing characteristics of asphalt materials during their service life were evaluated by Wang et al. [10], which made them differ from ageing research on the lab-aged materials. Rejuvenators were designed by Zhang et al. [11] and Su et al. [12] using petroleum technology and encapsulating rejuvenator fiber, and then added into recycled pavement material and lab-aged material. They concluded that rejuvenators can soften aged pavement materials and consequently recover the road performance. Furthermore, self-healing characteristics of pavement materials were reported by Wu et al. [13], Li et al. [14], and Cai et al. [15]. Wu et al. [13] found that UV irradiation will weaken the macro-structure and lower the failure strength and healing index. Li et al. [14] designed steel fiber modified asphalt concrete to promote the induction heating technology, while Cai et al. [15] used engineered cementitious composites mortar to prepare flexible pavement materials with certain healing property. Numerical simulation models of microwave heating of asphalt mixture, which can be used for pavement maintenance, recycling, and deicing, were developed with finite element software by Wang et al. [16].

Environmental conditions such as higher temperature, UV radiation, and moisture can introduce significant deteriorations of asphalt-based pavement materials. Materials modification technologies are thus widely used in pavement engineering to improve the long-term performance of pavement materials. For instance, ethylene bis stearamide based graphene [17], styrene-butadiene-styrene latex [18], styrene-butadiene rubber [19], aged lignin [20] and bio-based polyurethane [21] were used as modifier and detailed explained in this special issue. The viscos-elastic behavior, storage property, fracture energy, rutting resistance, and anti-cracking property were presented. Studies on high-viscosity modified asphalt binder [22] and fire-retardant asphalt [23] were also discussed in this special issue.

Characterization research on pavement materials is of important for this Special Issue. Performance studies on stress absorbing membrane interlayer and semi-flexible composite mixture were discussed by Yang et al. [24] and Zhang et al. [25]. The former article investigated the phase transition characteristics by dynamic mechanical analysis, while the second article presented the engineering properties by means of thermal cracking, fatigue, rutting resistance, and moisture resistance. Asphalt-based materials are composed of binder, filler, and aggregates, and the interaction between each different compound is the key to get a better understanding of pavement performance. The effect of aggregate meso-structure on the permanent deformation of asphalt mixture was discussed by Zhang et al. [26], with the three-dimensional discrete element model. Their model can capture the aggregate morphologies of angularity, orientation, and surface texture. Chen et al. [27] investigated the asphalt-filler/aggregate interaction on self-designed interface specimens with dynamic shear rheometer. They concluded that asphalt mortar could be the closest subscale in terms of performance to that of asphalt mixtures, making it a vital scale to bridge the gap between asphalt binder and asphalt mixtures in multiscale performance analysis. A unified strength model, which can be used to overcome the design deviation caused by the randomness of the laboratory strength test and improve the accuracy degree, was described by Xia et al. [28]. Different loading stresses were investigated to conclude the unified strength model, as well as to study the asphalt mixture moduli in the research presented by Fan et al. [29].

Field investigation is the principal requirement to ensure safe and well-accepted driving conditions in pavement maintenance. In the study by Pan et al. [30], piezo-ultrasonic wave technology was used for damage detection, including groove damage and cylinder cutting damage, in road engineering. Pan found that factors such as temperature, defect size, and ultrasonic velocity would affect the detection accuracy. Zhang et al. [31] reported the field investigation in full-depth asphalt pavement.

Other researches focused on the wet skid resistance [32] and gradation design [33] of pavement surfacings. Lyu et al. [34] introduced an outstanding durable road surface marking material, using persistent phosphors coated with silica-polymer hybrid shell. Cool coating materials for asphalt pavements were designed and discussed by Chen et al. [35]. These studies are critically important for pavement design and highway service. For instance, the cool coating can be widely used to solve the high-temperature-related defects in asphalt pavement.

Several other studies involved in this Special Issue look at cement concrete materials. Cement concrete and asphalt concrete are the two major materials in pavement engineering, so-called rigid pavement and flexible pavement. The micro-zone corrosion mechanism [36], cement mortar with super absorbent polymer [37], and exhaust-purifying cement [38] were discussed in detail. Polypropylene fiber was used in cement concrete and its performance was studied by Chen et al. [39] by dynamic compressive behavior analysis. Last, but not least, Yan et al. [40] presented their excellent work on anti-corrosion property of glass flake, which was designed for the reinforcement in chemically bonding phosphate ceramic coatings.

Author Contributions: Writing—original draft preparation, Y.X.; writing—review and editing, S.E., M.L., T.M. and X.L.; funding acquisition, Y.X. and T.M. All authors have read and agreed to the published version of the manuscript.

Funding: This research was funded by the National Natural Science Foundation of China (grant number No. 51922030, 51878164, and 51878526) and the 2019 Opening Funding Supported by the Key Laboratory of Transport Industry of Road Structure and Material in Research Institute of Highway, Ministry of Transport of China.

Acknowledgments: This Editorial was prepared by Yue Xiao at home, universities were closed due to the coronavirus, which has infected tens of thousands of people in China within these days. Revised by Sandra Erkens, Mingliang Li, Tao Ma, and Xueyan Liu. Many thanks to the doctors, nurses, and Chinese armies for their hard and unforgettable battle against the virus. They are the heroes who bring us hope and health. We wish China will receive the victory over novel coronavirus soon. Come on, Wuhan! Come on, China! We thank all the high-quality authors worldwide and reviewers for their contribution. Without their excellent work, this special issue would not get published successfully. We would also like to thank editors in MDPI for their assistance in managing this special issue.

References

1. Kong, D.; Wu, S.; Chen, M.; Zhao, M.; Shu, B. Characteristics of Different Types of Basic Oxygen Furnace Slag Filler and its Influence on Properties of Asphalt Mastic. *Materials* **2019**, *12*, 4034. [CrossRef]

2. Kong, D.; Chen, M.; Xie, J.; Zhao, M.; Yang, C. Geometric Characteristics of BOF Slag Coarse Aggregate and its Influence on Asphalt Concrete. *Materials* **2019**, *12*, 741. [CrossRef] [PubMed]

3. Ye, Y.; Wu, S.; Li, C.; Kong, D.; Shu, B. Morphological Discrepancy of Various Basic Oxygen Furnace Steel Slags and Road Performance of Corresponding Asphalt Mixtures. *Materials* **2019**, *12*, 2322. [CrossRef] [PubMed]

4. Qian, G.; Li, S.; Yu, H.; Gong, X. Interlaminar Bonding Properties on Cement Concrete Deck and Phosphorous Slag Asphalt Pavement. *Materials* **2019**, *12*, 1427. [CrossRef] [PubMed]

5. Zhang, Y.; Korkiala-Tanttu, L.K.; Borén, M. Assessment for Sustainable Use of Quarry Fines as Pavement Construction Materials: Part I—Description of Basic Quarry Fine Properties. *Materials* **2019**, *12*, 1209. [CrossRef]

6. Zhang, Y.; Korkiala-Tanttu, L.K.; Borén, M. Assessment for Sustainable Use of Quarry Fines as Pavement Construction Materials: Part II-Stabilization and Characterization of Quarry Fine Materials. *Materials* **2019**, *12*, 2450. [CrossRef]

7. Li, B.; Zhou, J.; Zhang, Z.; Yang, X.; Wu, Y. Effect of Short-Term Aging on Asphalt Modified Using Microwave Activation Crumb Rubber. *Materials* **2019**, *12*, 1039. [CrossRef]

8. Li, Y.; Lyu, Y.; Xu, M.; Fan, L.; Zhang, Y. Determination of Construction Temperatures of Crumb Rubber Modified Bitumen Mixture Based on CRMB Mastic. *Materials* **2019**, *12*, 3851. [CrossRef]

9. Gu, L.; Chen, L.; Zhang, W.; Ma, H.; Ma, T. Mesostructural Modeling of Dynamic Modulus and Phase Angle Master Curves of Rubber Modified Asphalt Mixture. *Materials* **2019**, *12*, 1667. [CrossRef]

10. Wang, X.; Guo, H.; Yang, B.; Chang, X.; Wan, C.; Wang, Z. Aging Characteristics of Bitumen from Different Bituminous Pavement Structures in Service. *Materials* **2019**, *12*, 530. [CrossRef]

11. Zhang, C.; Ren, Q.; Qian, Z.; Wang, X. Evaluating the Effects of High RAP Content and Rejuvenating Agents on Fatigue Performance of Fine Aggregate Matrix through DMA Flexural Bending Test. *Materials* **2019**, *12*, 1508. [CrossRef] [PubMed]

12. Shu, B.; Bao, S.; Wu, S.; Dong, L.; Li, C.; Yang, X.; Norambuena-Contreras, J.; Liu, Q.; Wang, Q. Synthesis and Effect of Encapsulating Rejuvenator Fiber on the Performance of Asphalt Mixture. *Materials* **2019**, *12*, 1266. [CrossRef] [PubMed]

13. Wu, S.; Ye, Y.; Li, Y.; Li, C.; Song, W.; Li, H.; Nie, S. The Effect of UV Irradiation on the Chemical Structure, Mechanical and Self-Healing Properties of Asphalt Mixture. *Materials* **2019**, *12*, 2424. [CrossRef] [PubMed]

14. Li, H.; Yu, J.; Wu, S.; Liu, Q.; Li, Y.; Wu, Y.; Xu, H. Investigation of the Effect of Induction Heating on Asphalt Binder Aging in Steel Fibers Modified Asphalt Concrete. *Materials* **2019**, *12*, 1067. [CrossRef]

15. Cai, X.; Huang, W.; Wu, K. Study of the Self-Healing Performance of Semi-Flexible Pavement Materials Grouted with Engineered Cementitious Composites Mortar based on a Non-Standard Test. *Materials* **2019**, *12*, 3488. [CrossRef]

16. Wang, H.; Zhang, Y.; Zhang, Y.; Feng, S.; Lu, G.; Cao, L. Laboratory and Numerical Investigation of Microwave Heating Properties of Asphalt Mixture. *Materials* **2019**, *12*, 146. [CrossRef]

17. Zhang, X.; He, J.X.; Huang, G.; Zhou, C.; Feng, M.M.; Li, Y. Preparation and Characteristics of Ethylene Bis(Stearamide)-Based Graphene-Modified Asphalt. *Materials* **2019**, *12*, 757. [CrossRef]

18. Ren, S.; Liu, X.; Fan, W.; Wang, H.; Erkens, S. Rheological Properties, Compatibility, and Storage Stability of SBS Latex-Modified Asphalt. *Materials* **2019**, *12*, 3683. [CrossRef]

19. Fan, X.; Lu, W.; Lv, S.; He, F. Improvement of Low-Temperature Performance of Buton Rock Asphalt Composite Modified Asphalt by Adding Styrene-Butadiene Rubber. *Materials* **2019**, *12*, 2358. [CrossRef]

20. Zhang, Y.; Liu, X.; Apostolidis, P.; Gard, W.; van de Ven, M.; Erkens, S.; Jing, R. Chemical and Rheological Evaluation of Aged Lignin-Modified Bitumen. *Materials* **2019**, *12*, 4176. [CrossRef]

21. Leng, C.; Lu, G.; Gao, J.; Liu, P.; Xie, X.; Wang, D. Sustainable Green Pavement Using Bio-Based Polyurethane Binder in Tunnel. *Materials* **2019**, *12*, 1990. [CrossRef] [PubMed]

22. Li, M.; Zeng, F.; Xu, R.; Cao, D.; Li, J. Study on Compatibility and Rheological Properties of High-Viscosity Modified Asphalt Prepared from Low-Grade Asphalt. *Materials* **2019**, *12*, 3776. [CrossRef] [PubMed]

23. Xu, G.; Chen, X.; Zhu, S.; Kong, L.; Huang, X.; Zhao, J.; Ma, T. Evaluation of Asphalt with Different Combinations of Fire Retardants. *Materials* **2019**, *12*, 1283. [CrossRef] [PubMed]

24. Yang, G.; Wang, X.; Zhou, X.; Wang, Y. Experimental Study on the Phase Transition Characteristics of Asphalt Mixture for Stress Absorbing Membrane Interlayer. *Materials* **2020**, *13*, 474. [CrossRef]

25. Zhang, W.; Shen, S.; Goodwin, R.D.; Wang, D.; Zhong, J. Performance Characterization of Semi-Flexible Composite Mixture. *Materials* **2020**, *13*, 342. [CrossRef]

26. Zhang, D.; Gu, L.; Zhu, J. Effects of Aggregate Mesostructure on Permanent Deformation of Asphalt Mixture Using Three-Dimensional Discrete Element Modeling. *Materials* **2019**, *12*, 3601. [CrossRef]

27. Chen, M.; Javilla, B.; Hong, W.; Pan, C.; Riara, M.; Mo, L.; Guo, M. Rheological and Interaction Analysis of Asphalt Binder, Mastic and Mortar. *Materials* **2019**, *12*, 128. [CrossRef]

28. Xia, C.; Lv, S.; You, L.; Chen, D.; Li, Y.; Zheng, J. Unified Strength Model of Asphalt Mixture under Various Loading Modes. *Materials* **2019**, *12*, 889. [CrossRef]

29. Fan, X.; Lv, S.; Zhang, N.; Xia, C.; Li, Y. Characterization of Asphalt Mixture Moduli under Different Stress States. *Materials* **2019**, *12*, 397. [CrossRef]

30. Pan, W.-H.; Sun, X.D.; Wu, L.M.; Yang, K.K.; Tang, N. Damage Detection of Asphalt Concrete Using Piezo-Ultrasonic Wave Technology. *Materials* **2019**, *12*, 443. [CrossRef]

31. Zhang, W.; Lee, J.; Ahn, H.J.; Le, Q.; Wu, M.; Zhu, H.; Zhang, J. Field Investigation of Clay Balls in Full-Depth Asphalt Pavement. *Materials* **2019**, *12*, 2879. [CrossRef] [PubMed]

32. Yan, B.; Mao, H.; Zhong, S.; Zhang, P.; Zhang, X. Experimental Study on Wet Skid Resistance of Asphalt Pavements in Icy Conditions. *Materials* **2019**, *12*, 1201. [CrossRef] [PubMed]

33. Cui, W.; Wu, K.; Cai, X.; Tang, H.; Huang, W. Optimizing Gradation Design for Ultra-Thin Wearing Course Asphalt. *Materials* **2020**, *13*, 189. [CrossRef] [PubMed]

34. Lyu, L.; Chen, Y.; Yu, L.; Li, R.; Zhang, L.; Pei, J. The Improvement of Moisture Resistance and Organic Compatibility of SrAl2O4: Eu2+, Dy3+ Persistent Phosphors Coated with Silica–Polymer Hybrid Shell. *Materials* **2020**, *13*, 426. [CrossRef]

35. Chen, Y.; Hu, K.; Cao, S. Thermal Performance of Novel Multilayer Cool Coatings for Asphalt Pavements. *Materials* **2019**, *12*, 1903. [CrossRef]
36. Wang, F.; Zhang, Z.; Wu, S.; Jiang, J.; Chu, H. Effect of Inhibitor on Adsorption Behavior and Mechanism of Micro-Zone Corrosion on Carbon Steel. *Materials* **2019**, *12*, 1901. [CrossRef]
37. Tan, Y.; Tan, Y.; Chen, H.; Wang, Z.; Xue, C.; He, R. Performances of Cement Mortar Incorporating Superabsorbent Polymer (SAP) Using Different Dosing Methods. *Materials* **2019**, *12*, 1619. [CrossRef]
38. He, R.; Huang, X.; Zhang, J.; Geng, Y.; Guo, H. Preparation and Evaluation of Exhaust-Purifying Cement Concrete Employing Titanium Dioxide. *Materials* **2019**, *12*, 2182. [CrossRef]
39. Chen, M.; Ren, C.; Liu, Y.; Yang, Y.; Wang, E.; Liang, X. Effects of Polypropylene Fibre and Strain Rate on Dynamic Compressive Behaviour of Concrete. *Materials* **2019**, *12*, 1797. [CrossRef]
40. Yan, G.; Wang, M.; Sun, T.; Li, X.; Wang, G.; Yin, W. Anti-Corrosion Property of Glass Flake Reinforced Chemically Bonded Phosphate Ceramic Coatings. *Materials* **2019**, *12*, 2082. [CrossRef]

Article

Characteristics of Different Types of Basic Oxygen Furnace Slag Filler and its Influence on Properties of Asphalt Mastic

Dezhi Kong [1], Shaopeng Wu [1], Meizhu Chen [1,*], Meiling Zhao [2] and Benan Shu [1]

[1] State Key Laboratory of Silicate Materials for Architectures, Wuhan University of Technology, Wuhan 430070, China; kongdz@whut.edu.cn (D.K.); wusp@whut.edu.cn (S.W.); shuba@whut.edu.cn (B.S.)

[2] Research Institute of Highway of Ministry of Transport, Beijing 100088, China; zhaoml@whut.edu.cn

* Correspondence: chenmzh@whut.edu.cn

Received: 24 October 2019; Accepted: 29 November 2019; Published: 4 December 2019

Abstract: The fillers of ordinary and pyrolytic basic oxygen furnace (BOF) slag were selected to investigate the properties of their asphalt mastic. XRF (X-Ray Fluorescence) was used to analyze chemical composition of fillers. Meanwhile, SEM (Scanning Electron Microscope) and AIMS (Aggregate Image Measurement System) were utilized to explore meso-morphology, angularity and sphericity. Penetration, softening point and viscosity of asphalt mastic were discussed, while the rheological properties of asphalt mastic were studied by means of DSR (dynamic shear rheometer) and BBR (bending beam rheometer) tests. The experimental results show that chemical composition of different types of BOF slag is similar. The grinding energy consumption of pyrolytic BOF slag is higher than that of limestone and ordinary BOF slag. It is not recommended that pyrolytic BOF slag filler is produced by grinding process. The micro-texture structure of ordinary BOF slag filler is more abundant and their angularity index is about 15% higher than that of limestone filler. The stiffness modulus and rutting factor of asphalt mastic with ordinary BOF slag filler is higher than that of limestone filler. Meanwhile the incorporation of BOF slag filler will further reduce the low-temperature flow performance of asphalt mastic. The effect of pyrolytic BOF slag filler on the performance of asphalt mastic is less than that of ordinary BOF slag. Ordinary BOF slag filler can effectively improve high temperature anti-rutting stability of asphalt mixture. Ordinary BOF slag has a useful application prospect as filler in asphalt mixture.

Keywords: BOF slag filler; asphalt mastic; morphological characteristics; rheological properties

---

1. Introduction

Steel slag is a common by-product of the steelmaking industry, and its output is about 10% to 15% of steel production in the world [1–3]. As a major type of steel slag, basic oxygen furnace (BOF) slag has strong alkalinity, rich angularity, tough surface characteristics and relatively good mechanical properties [4]. BOF slag is widely used as aggregate in asphalt mixtures in related research [5–7]. Pasetto et al. and Wu et al. [8–10] demonstrated that BOF slag aggregates improve performance of asphalt mixtures, such as moisture stability, high temperature deformation resistance, abrasion and skid resistance. BOF slag is an ideal substitute for natural aggregates in asphalt mixture.

Asphalt mixtures consist of asphalt binder, aggregate, and mineral filler. Asphalt mastic, which refers to mixture of asphalt binder and mineral filler, determines the final mechanical properties of asphalt mixture. Previous studies [11,12] have demonstrated that the properties of asphalt mastic are the factors influencing the rutting resistance and low-temperature crack resistance of asphalt mixtures, while the properties of filler are closely associated with the asphalt mastic. Many researchers [13–17] have proved that performances of asphalt mastic are affected by the volume content of fillers and performances of fillers like surface characteristics, alkalinity and size, as well as physical-chemical interaction between asphalt and fillers.

Xiao et al [18,19] discussed the feasibility of BOF slag as mineral filler in asphalt mixtures and the results show that the asphalt mastic with BOF slag filler has better high-temperature rheological properties than that of limestone filler. Qiu et al [20] examined the low-temperature fracture properties of asphalt mastic using steel slag powder. The results showed that the steel slag powder–asphalt system had higher fracture resistance than conventional systems and steel slag powder can diminish the severity of low-temperature reversible aging of modified asphalt. Song et al [21] demonstrated that the steel-making dust would be an alternative to the ordinary mineral filler to improve the performance of asphalt mortars and reduce the harm of the dust to the environment at the same time.

With the rapid development of highway construction projects and the consequential deterioration in high-quality mineral filler, there is an urgent need to broaden the source of the fillers that can be used in asphalt mixtures. Meanwhile, lots of numerous micro fillers are produced in the crushing process and magnetic separation of BOF slag [22]. The utilization of BOF slag as filler in asphalt mixture has attracted more concern. Furthermore, different types of BOF slag have different properties. Meanwhile, the difference in physicochemical properties of fillers and their influence on asphalt mastic is still unknown.

This study attempted to evaluate the feasibility of different types of BOF slag used as mineral filler to replace limestone filler (LF) in asphalt mixtures. All types of fillers were obtained by grinding 3–5 mm particle size range aggregates with the same processing conditions. It was found that 90% of all types of fillers could pass through 0.075 mm sieve. Figure 1 illustrates the research program on the characteristics of different types of BOF slag filler and their influence on properties of asphalt mastic. Firstly, chemical composition of different types of BOF slag filler was evaluated by XRF. Secondly, the geometrical properties of BOF slag filler were examined, such as meso-morphology and angularity and sphericity. Thirdly, the basic physical properties of asphalt mastic were discussed including penetration, softening point and viscosity. Finally, the rheological properties of asphalt mastic were studied at both low and high temperatures.

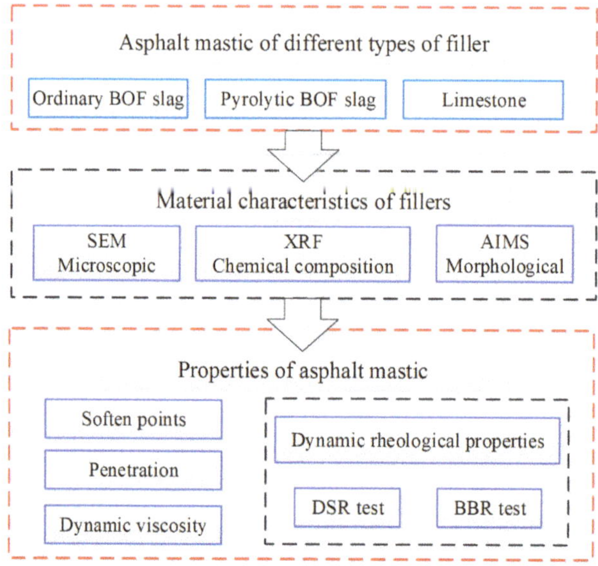

Figure 1. Finalized research program.

2. Materials and Methods

2.1. Raw Materials

Pen 60/80 bitumen binder, provided by Panjin Co., Ltd., Liaoning, Panjin, China, was used in this research. The basic properties of bitumen binder are shown in Table 1. They are all within the requirement specifications of China.

Table 1. Basic properties of the bitumen binder in this research.

Properties	Values	Specifications
Penetration [0.1 mm]	64	60–80
Penetration index	−0.6	−1.5–1.0
Softening point [°C]	46.9	≥46
Ductility, 5 cm/min, 15 °C [cm]	167	≥100
Dynamic viscosity (60 °C) [Pa·s]	172	≥160
Density [g/cm^3]	1.021	

Three types of BOF slag fillers and limestone filler were employed in this research. The limestone (L) was provided by Hubei province in China. Type A BOF slag (BS-A) was obtained from ironworks in Hubei province in China, while Type B BOF slag (BS-B) and pyrolytic BOF slag (PBS) were obtained from ironworks of Baotou in Inner Mongolia, China. BS-A and BS-B were basic oxygen furnace slag, and PBS was pyrolytic BOF slag which was treated with a hot stuffing process during the BOF slag cooling step.

All types of fillers in this research were obtained by grinding 3–5 mm particle size range aggregates with the same processing conditions. A ball mill was used and the time of the grinding was 60 min, 5,000 g for each sample and the rotating speed was 120 r/min. After grinding, more than 90% of all types of fillers could pass through the 0.075 mm sieve. The labels are as follows: limestone filler (LF), type A BOF slag filler (BSF-A), type B BOF slag filler (BSF-B), pyrolytic BOF slag filler (PBSF).

The basic properties of the four types of fillers are shown in Table 2. It can be deduced that the density of BOF slag is higher than that of LF. The density of BSF-A and BSF-B is almost equal, while they are about 18% higher than that of LF. Among them, the density of PBSF is the highest, which is attributed to the higher content of Fe_2O_3 and the hot stuffing process during BOF slag cooling step. The difference in hydrophilic coefficient of the four fillers is virtually negligible. However, the water absorption of BOF slag fillers is higher than that of LF, which is due to the particular pore structure of the surface of BOF slag.

Table 2. Basic properties of four types of fillers.

Property	Density (kg/m^3)	Hydrophilic Coefficient	Water Absorption (%)
LF	2725	0.64	0.53
BSF-A	3217	0.69	0.67
BSF-B	3244	0.73	0.63
PBSF	3478	0.65	0.64

2.2. Experimental Methods

2.2.1. Properties of Fillers

The chemical compositions of the four types of filler were evaluated by X-Ray Fluorescence (PANalytical. B.V., Zetium, Almelo, Netherlands). The surface characteristics of four types of fillers were evaluated using scanning electron microscope (JSM-IT300, SEM-JEOL, Tokyo, Japan). The aggregate image measurement system (AIMS) was used to analyze the morphological features of each type of filler. Meanwhile, the laser particle size analyzer (Mastersizer-2000, Malvern, England) was utilized to analyze the difference of particle size of four types of fillers.

2.2.2. Preparation of Asphalt Mastic

Specific amounts of each type of filler were incorporated into pen 60/80 asphalt binder to prepare the asphalt mastic. Firstly, asphalt binder was preheated to 150 °C for 30 min. Then, filler was gradually incorporated with a high shear instrument of the shear speed of 4000 rpm for 30 min. Homogeneous dispersion of the filler in the asphalt binder was required for further research. the filler–asphalt volume ratio of asphalt mastic used in this research was 0.4. The mass–volume conversions of four fillers were calculated by the density values in Table 2. After calculation, the filler-asphalt weight ratios of LF was 1.113, BSF-A was 1.314, BSF-B was 1.32, and PBSF was 1.42.

2.2.3. Properties of Asphalt Mastic

The penetration and softening point test were used to assess the basic properties of the asphalt mastic. In the preparation process of asphalt mixture, the workability is closely linked to the viscosity of asphalt mastic. A Brickfield viscometer was used to analyze the difference of four types of asphalt mastic. Test methods of penetration, softening point and dynamic viscosity refer to Chinese official standard JTG E20-2011 [23].

A dynamic shear rheometer (MCR101, DSR, Anton Paar, Graz, Austria) was utilized to evaluate the rheological properties of asphalt mastics. The DSR test was performed at a fixed frequency of 10 rad/s under variation of temperature from −10 °C to 80 °C with increments of 2 °C/min. In −10 °C to 3 °C, specimens were placed on a parallel plate geometry whose diameter was 8 mm, the thickness was 2 mm and strain level was 0.2%. In 30 °C to 80 °C, specimens were placed on a parallel plate geometry whose diameter was 25 mm, the thickness was 1 mm and strain level was 2.0%. The BBR (TE-BBR, Cannon, New York, NY, USA) was used to examine the rheological properties at a low temperature, which relates to the low-temperature cracking resistance. Preheated asphalt mastic was filled into an aluminum mold to prepare a mastic beam 102.0 ± 0.5 mm in length, 12.7 ± 0.5 mm in width, and 6.25 ± 0.5 mm in thickness. Drawing on the research of Xiao et al [18,19], tests were performed at a definite temperature (15 °C). Specimens were tested under a constant stress of 0.985 N for 250 s. Each test for different mastic included five duplicate specimens and the average value was adopted.

3. Results and Discussion

3.1. Material Characteristics of Fillers

3.1.1. Chemical Compositions of Fillers

The chemical compositions of the four types of filler from X-ray fluorescence analysis are shown in Table 3. Limestone is an alkali aggregate because the chemical composition of CaO was higher than fifty percent. The chemical composition of Fe_2O_3 in three types of BOF slags is more than 20%. Three types of BOF slag contain less SiO_2 and more than 30% CaO making them alkali aggregates.

Table 3. Chemical composition of the fillers in this research.

Composition [%]	SiO_2	CaO	MgO	Al_2O_3	Fe_2O_3	MnO	P_2O_5	Other	LOI
LF	0.86	51.2	2.36	0.85	0.12	0.7	1.02	0.19	42.7
BSF-A	19.2	42.7	5.19	3.25	23.9	1.77	1.41	0.22	2.36
BSF-B	17.7	39.7	5.56	2.91	24.4	4.55	1.68	0.09	3.41
PBSF	15.4	34.4	6.22	1.95	30.8	4.46	2.15	0.16	4.46

3.1.2. Microscopic Characteristics of Fillers

Figure 2 shows the diversity between limestone filler and three types of BOF slag fillers in SEM images. Compared with the micrographs of LF, BOF slag has different surface texture particularly due to the size and number of numerous tiny particles which are adhered to its surface s and its rough surface. The surface of LF is relatively smoother than that of BOF slag filler, where the latter is irregular shaped, rough, and bumpy, which might result in effective cohesion with asphalt binder and consequently lead to improved strength and water resistance. From practical viewpoint, the rough surface of BOF slag filler will cause an increase in the amount of asphalt required when BOF slag filler is used as filler in asphalt mixtures.

Figure 2. SEM images of investigated fillers: (**a**) limestone filler; (**b**) type A basic oxygen furnace (BOF) slag filler; (**c**) type B BOF slag filler; and (**d**) pyrolytic BOF slag filler.

3.1.3. Morphological Characteristics of Fillers

Morphological characteristics of fillers in this research were analyzed by AIMS, which is an integrated system comprising image acquisition hardware and a computer. Analysis of filler includes gradient angularity (the AIMS Angularity Index ranges from 1 to 10,000) and Form2D (AIMS Form2D Index ranges from 0 to 20).

Angularity is a description of edge sharpness of the boundary particles of aggregate. The angularity changes with filler granule boundary shape changes. The value of angularity is calculated based on the gradient on the particle boundary. Angularity is calculated with Equation (1) [24]. Its range is 0 to 10,000. The larger the value of angularity is, the boundary shape of filler is sharper.

$$Gradient\ Angularity = \frac{1}{\frac{n}{3}-1}\sum_{i=1}^{n-3}\left|\theta_i - \theta_{i+3}\right| \tag{1}$$

where θ is angle of orientation of the edge points, n is the total number of points, i is the ith point on the edge of the particle.

Figure 3 shows the comparison of distributions of gradient angularity indexes of four types of fillers. It can be clearly seen that LF has the lowest gradient angularity indexes while BSF-B has the highest gradient angularity indexes. The distribution range of the angularity index of LF is narrower than that of BSF-A, BSF-B and PBSF. Such differences illustrate that the distribution of gradient angularity index of LF is more uniform. By comparing three types of BOF slag filler, it can be found that the gradient angularity index of PBSF is smaller than that of ordinary BOF slag. The gradient angularity index of ordinary BOF slag is about 15% higher than that of LF. In summary, the particle shape of the BOF slag filler has more angular structure than that of limestone filler under the same grinding process conditions.

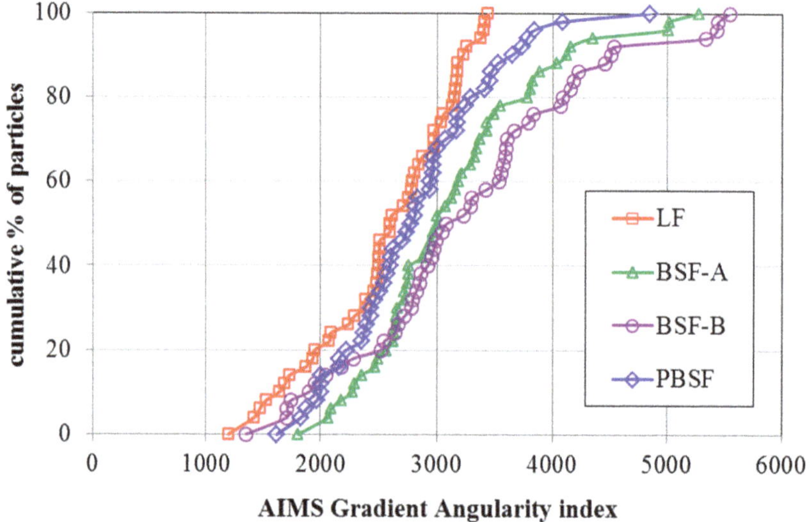

Figure 3. Comparison of distributions of angularity indexes of four types of fillers.

AIMS Form2D is applicable only to fine-sized aggregate and it quantifies the relative form from 2D images of aggregate particles. Form2D is calculated with Equation (2) [25] and its range is 0 to 20. A perfect circle has a Form2D value of 0. According to the definition of Form2D, a higher Form2D value would imply a relatively rougher surface, which would consequently indicate a positive contribution towards adhesive mechanical bonding of asphalt binder to filler.

$$Form\ 2D = \sum_{\theta=0}^{\theta=360-\Delta\theta} \left[\frac{R_{\theta+\Delta\theta} - R_\theta}{R_\theta} \right] \tag{2}$$

where R_θ is the radius of the particle at an angle of θ, $\Delta\theta$ is the incremental difference in the angle.

As shown in Figure 4, the Form2D values of four types of fillers have a wide distribution range of 5–12. For each type of BSF, nearly 80% of the particles have moderate and higher Form 2D values while about 40% of LF has lower Form 2D values. The Form2D values of LF is the lowest and the Form 2D of BOF slag is about 10% higher than that of LF. In summary, the surface of BOF slag filler was rougher and sharper than that of limestone filler. These results agree with the SEM results.

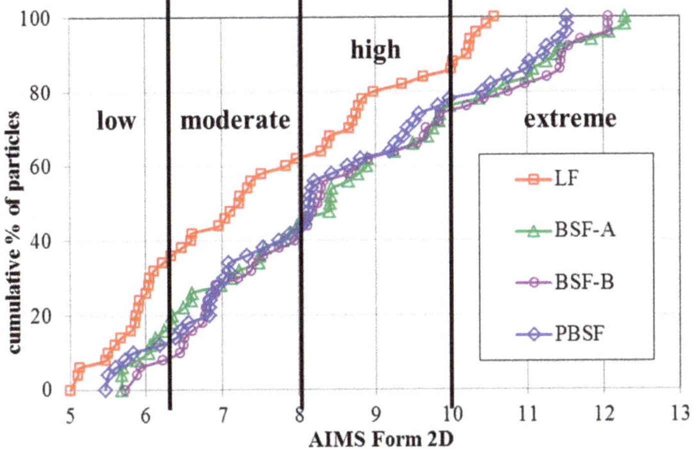

Figure 4. Distributions of Form2D values of four types of fillers.

3.1.4. Particle Size Analysis of Fillers

Figure 5 shows the particle size distribution of four types of filler. It reveals that the particle size of the four types of fillers is different under the same grinding process conditions. The particle size of LF is minimal, while the particle size of PBSF is maximal. Slight differences exist between LF and BSF-A in particle size. The order of particle size from small to large is LF, BSF-A, BSF-B and PBSF. The grinding efficiency of BOF filler is relatively lower than that of LF. The grinding efficiency of different kinds of BOF filler is also different. Pyrolytic BOF slag has the maximum particle size among the four types of fillers, indicating that pyrolytic BOF slag is more difficult to grind than ordinary BOF slag. In order to achieve the same particle size, the grinding process of pyrolytic BOF slag needs to consume more grinding time and more energy consumption.

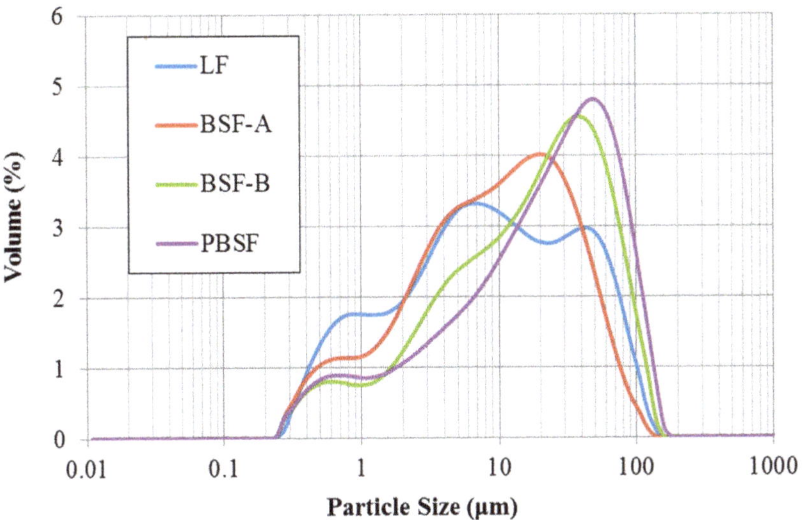

Figure 5. The particle size distribution of four types of filler.

3.2. Properties of Asphalt Mastic

Four types of asphalt mastics and basic bitumen binder were investigated in this research. Both basic properties and dynamic rheological properties were analyzed.

3.2.1. Basic Properties of Asphalt Mastic

The basic properties of asphalt mastic investigated in this research include soften points, penetration and dynamic viscosity.

The bitumen binder has a typical viscosity property. The smaller the penetration value is, the greater the viscosity of the asphalt material under low temperature conditions is, and the corresponding elastic deformation performance and recovery performance are also better. The softening point of the asphalt mastic is larger, indicating that the high temperature performance is better.

In Table 4 the values of soften points and penetration of asphalt mastic are presented. Compared with LF asphalt mastic, the softening point of asphalt mastic with BOF slag filler is lower, while its penetration is higher. BOF slag filler has a better viscosity-increasing effect on asphalt mastic than LF. The softening point and penetration of asphalt mastic with three types of BOF slag filler are similar. Although the PBSF has the largest particle size, its effect on softening point increase and penetration reduction is still effective. The particle size factor and morphological characteristics of BOF slag filler has little effect on the soften points and penetration of asphalt mastic.

Table 4. The soften points and penetration of asphalt mastic.

Property	Soften Points (°C)	15 °C Penetration (0.1 mm)	25 °C Penetration (0.1 mm)
LF	55.3	13.5	31.5
BSF-A	58.3	12.1	28.4
BSF-B	57.1	11.9	29.3
PBSF	58.5	12.5	29.8

The dynamic viscosity of asphalt is very important for asphalt mixture. The reasonable construction temperature is determined by the dynamic viscosity temperature range of 0.17 Pa s ± 0.02 Pa s, and construction rolling temperature is determined by the dynamic viscosity temperature range of 0.28 Pa s ± 0.03 Pa s [26]. The consistency of asphalt mastic in asphalt mixture construction is deeply influenced by the dynamic viscosity [27].

Viscosity–temperature value for bitumen binder and four types of asphalt mastic are shown in Table 5. In the case of limestone filler, the viscosity of asphalt mastic at different temperatures is approximately five times higher than that of bitumen binder. The addition of filler can effectively improve the viscosity of asphalt mastic. Compared with LF asphalt mastic, the viscosity of BSF asphalt mastic is higher, mainly because BOF slag filler has higher alkalinity and rougher micro-texture. The dynamic viscosity of the three types of asphalt mastic from large to small is BSF-A, BSF-B and PBSF. It is believed that the bigger particle size, more regular shape and smoother microscopic surface of PBSF result in a relatively lower dynamic viscosity of asphalt mastic.

Table 5. The dynamic viscosity test results of asphalt mastic.

Property	Dynamic Viscosity (Pa s)				
	90 °C	105 °C	120 °C	135 °C	150 °C
Bitumen binder	11.80	3.40	1.26	0.51	0.23
LF	52.00	14.25	4.95	1.97	0.94
BSF-A	78.00	22.25	7.25	3.00	1.40
BSF-B	77.00	21.05	7.15	2.80	1.38
PBSF	71.00	19.30	6.82	2.63	1.25

3.2.2. Dynamic Rheological Properties of Asphalt Mastic

As a typical viscoelastic material, the dynamic rheological properties of asphalt are closely related to its load and temperature conditions [28]. The viscoelastic characteristics of different types of asphalt and asphalt mastic are complex in different temperature conditions. Dynamic shear rheometer (DSR) tests were used to study dynamic rheological properties of asphalt mastic in this research. Complex shear modulus (G*) and phase angle (δ) are recorded and calculated by performing the DSR test at varying temperatures. G* can be decomposed into storage shear modulus (G' = G*cosδ) and loss shear modulus (G'' = G*sinδ), which is used to characterize the ability of asphalt mastic to resist deformation under repeated shear loads. The larger of the G* is, the higher of the resistance of asphalt mastic is in deformation. δ is the time lag of the applied stress and the resulting strain. The tangent value of δ is the ratio of the loss modulus to the storage modulus. A smaller δ indicates that there are more elastic components in the asphalt G*, and a larger δ indicates that there are more viscous components in the asphalt G*. For a fully elastic material, the phase angle δ is zero, at which point all deformations are recoverable. However, for viscous materials, the phase angle is close to 90°, at which point all deformation is permanent. G*/sinδ (the ratio of the complex modulus to phase angle sine) characterizes the ability of asphalt to resist high temperature rutting. G*/sinδ is called the rutting factor. Under the same temperature conditions, asphalt mastic with larger G*/sinδ has better rutting resistance [29].

Effect of different fillers on G* of asphalt mastic are shown in Figures 6 and 7. The G* of the asphalt mastic and asphalt decreases with the increase of temperature, which indicates that the rheological properties of the asphalt are greatly affected by the temperature. When the temperature rises, the volume of free asphalt increases, and the pitch changes from a high elastic state at a low temperature to a viscous state at a high temperature. As a result, the maximum shear stress and the maximum shear strain of the asphalt are increased when the shear force is applied, and therefore the G* is lowered.

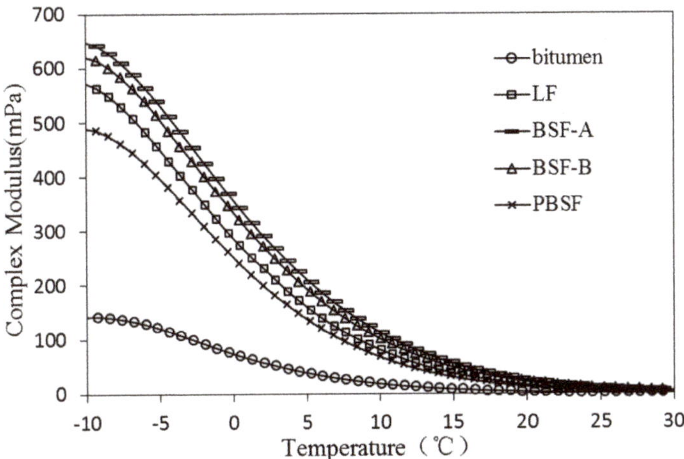

Figure 6. Effect of different fillers on G* of asphalt mastic (−10~30 °C).

In the temperature range of −10~30 °C, the higher the G* of the asphalt mastic is, the better the crack resistance of the asphalt under low temperature conditions. As shown in Figure 6, the values of complex modulus of BSFA are highest and PBSF has the worst effect on the increase of complex modulus. The values of PBSF are lower than that of LF, mainly due to a small alkalinity, a large particle size, a large specific surface area, less adsorbed asphalt, and relatively more free asphalt. Ordinary BOF slag filler has a good effect, improving the low temperature crack resistance of asphalt mastic.

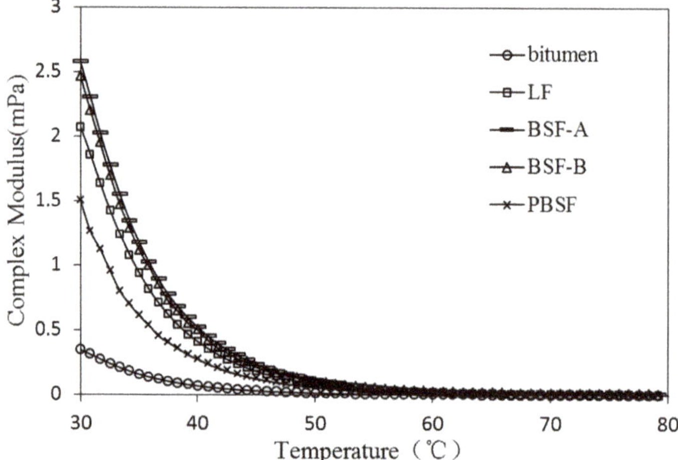

Figure 7. Effect of different fillers on G* of asphalt mastic (30~80 °C).

In Figure 7, the effect of different fillers on G* of asphalt mastic (30~80 °C) is presented. After adding BSF-A and BSF-B, the G* of asphalt mastic is higher than that of LF, which indicates that the high temperature stability of ordinary BOF slag filler asphalt mastic is better than that of LF.

In Figures 8 and 9 the values of different fillers on δ of asphalt mastic (−10~80 °C) are presented. The δ of asphalt mastic with different types of fillers increases with the increase of temperature, and the difference of δ values of asphalt mastic decreases with the increase of temperature. The sensitivity of the δ of the filler is continuously reduced with the temperature increasing. The phase angles of asphalt mastics with different types of fillers are arranged from large to small in order, pure asphalt, PBSF, LF, BSF-B and BSF-A. It is shown that ordinary BOF slag can effectively improve the elastic properties of asphalt mastic, and is beneficial, improving the anti-permanent deformation performance of asphalt mastic.

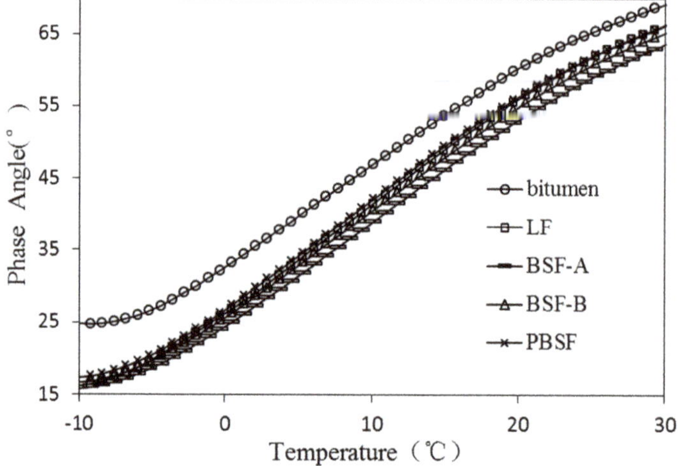

Figure 8. Effect of different fillers on δ of asphalt mastic (−10~30 °C).

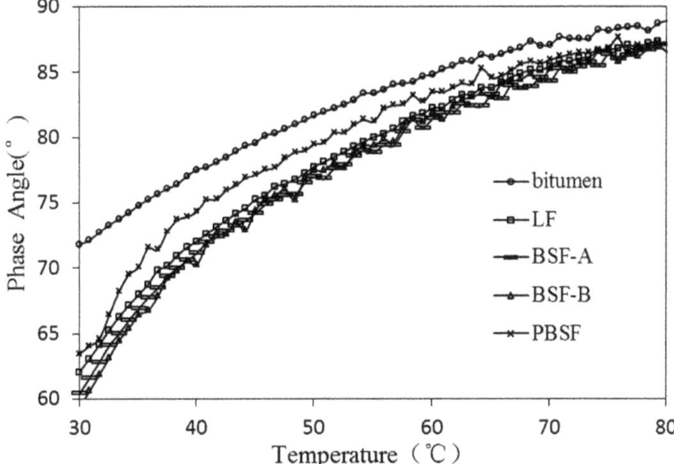

Figure 9. Effect of different fillers on δ of asphalt mastic (30~80 °C).

As shown in Figure 10, the rutting factor of the asphalt mastics with BSF-A and BSF-B are similar and larger than that of corresponding PBSF asphalt mastic. It demonstrates that all asphalt mastics containing ordinary BOF slag fillers have better high-temperature deformation resistance than ones with LF. BSF mastic presents the best deformation resistance. This was due to the chemical effect between alkaline components in ordinary BOF slag fillers and asphaltic acid in bitumen. The stiffness of BOF slag makes mastic structure more stable to resist permanent deformation.

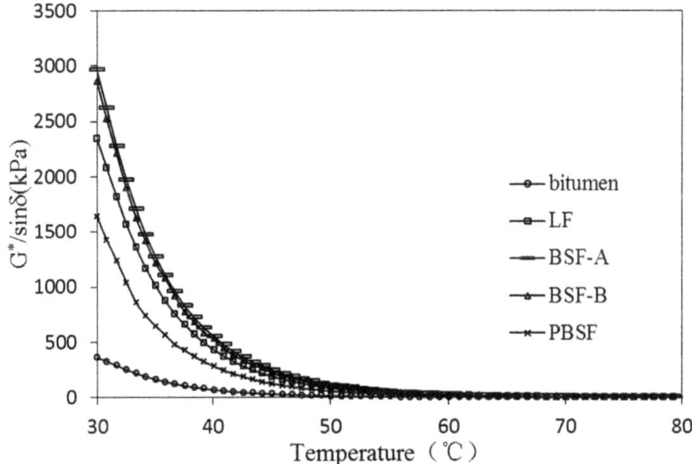

Figure 10. Effect of different fillers on G*/sinδ of asphalt mastic (30~80 °C).

3.2.3. Anti-Cracking Properties of Asphalt Mastic at Low Temperature

Bending beam rheometer (BBR) testing was used to investigate the low-temperature rheological properties of asphalt mastic in this research. Tests were performed at a fixed temperature (−15 °C) to discuss the different of low temperature performance of asphalt mastic with different filler, meanwhile the m-value and creep stiffness (S(t), MPa) were evaluated. The S(t) was calculated on the basis of Equation (3). The m-value signifies the rate that S(t) changes during loading time. Creep stiffness indicates the thermal stress of asphalt mastic under low temperature. Lower creep stiffness implies that the specimen has better rheological properties at lower temperatures. Scientifically, lower creep stiffness is positive because it corresponds to lower deformation stress. M-value reflects the stress relaxation property of asphalt mastic at low temperatures. A higher m-value is required since asphalt with a higher m-value has better ability to disperse deformation stress [30].

$$S(t) = \frac{PL^3}{4bh^3\Delta(t)}$$ (3)

where b, h and L are the width (mm), height (mm), length (mm) of specimen. P is the constant applied load. $\Delta(t)$ is the deflection of beam (mm) at different times (t).

The creep stiffness and m-value of asphalt mastic with different types of filler are shown in Figure 11. The change of m-value is similar to the stiffness. It can be clearly seen that the introduction of fillers increases the $S(t)$ and m-value. The $S(t)$ values of LF are about four times that of pure asphalt, so asphalt will become stiffer at a low temperature after mixing with fillers. Considering the value of stiffness, the stiffness of PBSF is smaller than that of LF, which indicates that pyrolytic BOF slag filler has a positive effect on the low-temperature crack resistance performance of asphalt mastic. Asphalt mastics with BSF-A and BSF-B are higher than ones with LF. It indicates that ordinary BOF slag filler has a certain increase in the stiffness of asphalt mastic, but has a negative influence on its low-temperature rheological properties. The incorporation of ordinary BOF slag filler reduces the low-temperature crack resistance performance of asphalt mastic compared with LF.

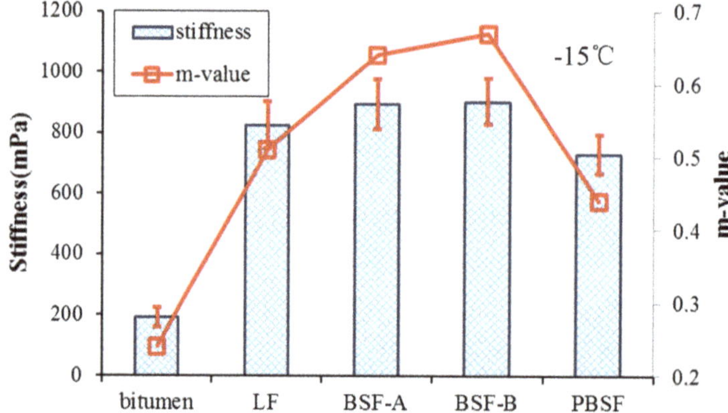

Figure 11. Bending beam rheometer (BBR) test results of four asphalt mastic and pure bitumen.

4. Conclusions

The microscopic characteristics, morphological characteristics (angularity and Form2D) and chemical properties of three types of BOF slag filler were investigated in the first part of this research. Then the basic physical properties and rheological properties of their asphalt mastic were studied. The overall conclusions are elaborated upon hereunder:

(1) The chemical composition of BOF slag is more complicated than that of limestone, which includes SiO_2, CaO, MgO, Al_2O_3, Fe_2O_3 and other components. The chemical composition of different types of BOF slag is similar. The micro-texture structure of BOF slag filler is more complex than that of LF. The angularity index of ordinary BOF slag is about 15% higher than that of LF. The angularity index, Form2D and micro texture of different types of BOF slag filler are also different.

(2) The asphalt mastic with BOF slag has higher soften points, lower penetration and higher dynamic viscosity than one with LF. The incorporation of BOF slag filler can significantly improve the high-temperature stability of asphalt mastic.

(3) Compared with rheological properties, asphalt mastic with BOF slag filler has higher stiffness modulus and rutting factor than that of LF asphalt mastic. The effect of pyrolytic BOF slag filler on the performance of asphalt mastic is less than that of ordinary BOF slag because of the bigger particle size, more regular shape and relatively clean surface. The incorporation of BOF slag filler will reduce the low temperature flow performance of asphalt.

(4) The grinding energy consumption of pyrolytic BOF slag is higher than that of limestone and ordinary BOF slag, meanwhile its chemical performance is relatively inactive. It is not recommended that pyrolytic BOF slag filler is produced by grinding technology. Ordinary BOF slag filler can effectively improve high temperature anti-rutting stability when used as filler in asphalt mixture. BOF slag filler has a good prospects for application as part of asphalt mixtures.

Author Contributions: D.K., M.Z., S.W. and B.S. conceived and designed the experiments; D.K., M.Z. and B.S. performed the experiments; D.K., M.C. and B.S. analyzed the data; M.C. and S.W. contributed reagents/materials/analysis tools; D.K., M.C. and M.Z. wrote the paper.

Funding: This work was financially supported by the Fundamental Research Funds for the Central Universities (WUT: 2017IVA083), Technological Innovation Major Project of Hubei Province (No. CXZ2016000024), Technological Innovation Major Project of Hubei Province (No. CXZ2019ACA147) and the Science and Technology Major Project of Inner Mongolia Autonomous Region (No. ZDZX2018029).

Acknowledgments: The authors gratefully acknowledge many important contributions from the researchers of all reports cited in our paper.

Conflicts of Interest: The authors declare no conflict of interest.

References

1. Han, F.; Zhang, Z.; Wang, D.; Yan, P. Hydration heat evolution and kinetics of blended cement containing steel slag at different temperatures. *Thermochim. Acta* **2015**, *605*, 43–51. [CrossRef]

2. Reddy, A.S.; Pradhan, R.K.; Chandra, S. Utilization of Basic Oxygen Furnace (BOF) slag in the production of a hydraulic cement binder. *Int. J. Miner. Process.* **2006**, *79*, 98–105. [CrossRef]

3. Shen, D.H.; Wu, C.M.; Du, J.C. Laboratory investigation of basic oxygen furnace slag for substitution of aggregate in porous asphalt mixture. *Constr. Build. Mater.* **2009**, *23*, 453–461. [CrossRef]

4. Chen, Z.; Wu, S.; Yue, X.; Zeng, W.; Yi, M.; Wan, J. Effect of hydration and silicone resin on Basic Oxygen Furnace slag and its asphalt mixture. *J. Clean. Prod.* **2015**, *112*, 392–400. [CrossRef]

5. Xue, Y.; Wu, S.; Hou, H.; Zha, J. Experimental investigation of basic oxygen furnace slag used as aggregate in asphalt mixture. *J. Hazard. Mater.* **2006**, *138*, 261–268. [CrossRef]

6. Iacobescu, R.I.; Angelopoulos, G.N.; Jones, P.T.; Blanpain, B.; Pontikes, Y. Ladle metallurgy stainless steel slag as a raw material in Ordinary Portland Cement production: A possibility for industrial symbiosis. *J. Clean. Prod.* **2016**, *112*, 872–881. [CrossRef]

7. Li, C.H.; Xiang, X.D.; Qin, X.X. Utilization of Steel Slag as Aggregates for SMA-13. *Appl. Mech. Mater.* **2015**, *768*, 402–405. [CrossRef]

8. Pasetto, M.; Baldo, N. Experimental evaluation of high performance base course and road base asphalt mixture with electric arc furnace steel slags. *J. Hazard. Mater.* **2010**, *181*, 938–948. [CrossRef]

9. Pasetto, M.; Baldo, N. Fatigue Performance of Asphalt Mixtures with RAP Aggregates and Steel Slags. *Rilem Bookseries* **2012**, *4*, 719–727.

10. Wu, S.; Xue, Y.; Ye, Q.; Chen, Y. Utilization of steel slag as aggregates for stone mastic asphalt (SMA) mixtures. *Build. Environ.* **2007**, *42*, 2580–2585. [CrossRef]

11. Antunes, V.; Freire, A.C.; Quaresma, L.; Micaelo, R. Effect of the chemical composition of fillers in the filler–bitumen interaction. *Constr. Build. Mater.* **2016**, *104*, 85–91. [CrossRef]

12. Wang, H.; Alqadi, I.L.; Faheem, A.F.; Bahia, H.U.; Yang, S.H.; Reinke, G.H. *Effect of Mineral Filler Characteristics on Asphalt Mastic and Mixture Rutting Potential*; Transportation Research Board: Washington, DC, USA, 2011.

13. Antunes, V.; Freire, A.C.; Quaresma, L.; Micaelo, R. Influence of the geometrical and physical properties of filler in the filler–bitumen interaction. *Constr. Build. Mater.* **2015**, *76*, 322–329. [CrossRef]

14. Faheem, A.F.; Wen, H.; Stephenson, L.; Bahia, H.U. Effect of Mineral Filler on Damage Resistance Characteristics of Asphalt Binders (With Discussion). *J. Assoc. Asph. Paving Technol.* **2008**, *1*, 77.

15. Geber, R.; Simon, A.; Kocserha, I.; Buzimov, A. In Microstructural and rheological analysis of fillers and asphalt mastics. *J. Phys. Conf. Ser.* **2017**, *790*, 012009. [CrossRef]
16. Grabowski, W.; Wilanowicz, J. The structure of mineral fillers and their stiffening properties in filler-bitumen mastics. *Mater. Struct.* **2008**, *41*, 793–804. [CrossRef]
17. Wang, D.; Zhou, G. Fatigue of asphalt binder, mastic and mixture at low temperature. *Front. Struct. Civ. Eng.* **2012**, *6*, 166–175.
18. Li, C.; Chen, Z.; Wu, S.; Li, B.; Xie, J.; Xiao, Y. Effects of steel slag fillers on the rheological properties of asphalt mastic. *Constr. Build. Mater.* **2017**, *145*, 383–391. [CrossRef]
19. Tao, G.; Xiao, Y.; Yang, L.; Cui, P.; Kong, D.; Xue, Y. Characteristics of steel slag filler and its influence on rheological properties of asphalt mortar. *Constr. Build. Mater.* **2019**, *201*, 439–446. [CrossRef]
20. Li, Q.; Qiu, Y.; Rahman, A. Application of Steel Slag Powder to Enhance the Low-temperature Fracture Properties of Asphalt Mastic and its Corresponding Mechanism. *J. Clean. Prod.* **2018**, *184*, 21–31. [CrossRef]
21. Song, L.; Wang, X.; Li, X.; Yang, Q.; Wang, P. Influence of the steel-making dust on high temperature and fatigue performance of asphalt mortars. *J. Wuhan Univ. Technol. Mater. Sci. Ed.* **2019**, *34*, 361–367. [CrossRef]
22. Topkaya, Y.; Sevinç, N.; Günaydın, A. Slag treatment at Kardemir integrated iron and steel works. *Int. J. Miner. Process.* **2004**, *74*, 31–39. [CrossRef]
23. Ministry of Transport. *Standard Test Methods of Bitumen and Bituminous Mixtures for Highway Engineering: JTGE20-2011*; China Communications Press: Beijing, China, 2011.
24. Kong, D.; Chen, M.; Xie, J.; Zhao, M.; Yang, C. Geometric Characteristics of BOF Slag Coarse Aggregate and its Influence on Asphalt Concrete. *Materials* **2019**, *12*, 741. [CrossRef]
25. Yue, X.; Feng, W.; Peide, C.; Lei, L.; Juntao, L.; Mingwei, Y. Evaluation of fine aggregate morphology by image method and its effect on skid-resistance of micro-surfacing. *Materials* **2018**, *11*, 920.
26. Geng, J.G.; Dai, J.L.; Sheng, Y.P. The Determination of Mixing and Compaction Temperature of Recycled Asphalt Mixtures. In Proceedings of the International Workshop on Energy & Environment in the Development of Sustainable Asphalt Pavements, Xi'an, China, 6–8 June 2010; pp. 44–49.
27. Tan, Y.; Meng, G. Study on the phase behavior of asphalt mastic. *Constr. Build. Mater.* **2013**, *47*, 311–317. [CrossRef]
28. Li, X.; Shan, L.; Tan, Y. Analysis of different indices for high- and low-temperature properties of asphalt binder. *Constr. Build. Mater.* **2015**, *83*, 70–76. [CrossRef]
29. Zu-Zhong, L.I.; Man, X.Y.; Chen, S.F. Study on Viscosity-temperature Characteristics of Asphalt Binder for Stress Absorbing Layers and Determination of Construction Temperature. *J. Wuhan Univ. Technol.* **2012**, *34*, 37–41.
30. Moon, K.H.; Falchetto, A.C.; Marasteanu, M.O. Investigation of limiting criteria for low temperature cracking of asphalt mixture. *KSCE J. Civ. Eng.* **2014**, *18*, 172–181. [CrossRef]

 *materials* 

Article

Geometric Characteristics of BOF Slag Coarse Aggregate and its Influence on Asphalt Concrete

Dezhi Kong [1], Meizhu Chen [1], Jun Xie [1,*], Meiling Zhao [2] and Chao Yang [1]

[1] State Key Laboratory of Silicate Materials for Architectures, Wuhan University of Technology, Wuhan 430070, China; kongdz@whut.edu.cn (D.K.); chenmzh@whut.edu.cn (M.C.); hbyangc@whut.edu.cn (C.Y.)

[2] Research Institute of Highway of Ministry of Transport, Beijing 100088, China; 197791@whut.edu.cn (M.Z.)

* Correspondence: xiejun3970@whut.edu.cn

Received: 31 January 2019; Accepted: 27 February 2019; Published: 4 March 2019

Abstract: In order to examine the geometric characteristics of BOF (blast oxygen furnace) slag coarse aggregate, the aggregate image measurement system (AIMS) was used to analyze the sphericity, gradient angularity and micro texture. Both volumetric and mechanical properties were studied to evaluate the influence of geometric characteristics of BOF slag coarse aggregate on asphalt concrete. The experimental results show that the BOF slag coarse aggregate has the characteristics of high sphericity, good angular performance and rough surface texture. The geometric characteristics of BOF slag has obvious influence on the volume performance of asphalt concrete. the higher sphericity of BOF slag causes an increase of the air voids of asphalt mixture. BOF slag coarse aggregate can effectively improve the road performances of asphalt concrete. BOF slag's higher sphericity and angularity improve the moisture damage resistance and rutting resistance of asphalt concrete. Results indicate that better angularity can slightly enhance the moisture resistance property of asphalt concrete, but excessively high angularity of BOF slag coarse aggregates reduces the anti-rutting properties of asphalt mixture.

Keywords: BOF slag; coarse aggregate; geometric characteristics; asphalt concrete; road performance

---

1. Introduction

With the increasing development of the transportation industry in the past few decades, the consumption of natural aggregate resource has been a close concern of scholars in this field. By the end of 2017, the total mileage of expressway has reached 0.137 million kilometers in China [1]. More than 90% of this is asphalt pavement, which has the advantages of low vibration, low noise, short construction time and convenient maintenance [2,3]. Expressway need lots of construction and maintenance, and more than 90% of the components of asphalt pavement are aggregates, which causes the depletion of natural resources [4,5]. Therefore, recycling solid waste and industrial smelting waste are becoming an effective option to relieve the supply pressure of natural aggregate resources [6–8].

Steel slag is the main solid waste in steel industry, accounting for more than 10% of crude steel production [9–11]. According to the different steelmaking processes, steel slag can be classified into three types: Electric arc furnace (EAF) slag, basic oxygen furnace (BOF) slag and ladle furnace (BF) slag. The large capacity of steel slag can meet the needs of aggregates for road construction, and it has good mechanical properties and high alkalinity [12,13]. The steel slag asphalt mixture has become an increasingly popular research topic in the field of environmental protection road materials in recent years [14,15].

Ahmedzade et al. [16] discussed the application of BOF slag coarse aggregate in AC-10 and AC-5 asphalt mixture. The research showed that the use of BOF slag coarse aggregate in asphalt mixture has better performance than that of limestone. The mechanical properties of steel slag improve the moisture damage resistance and rutting resistance performance of asphalt mixture. Behnood et al. [17,18] used steel slag coarse aggregate in stone mastic asphalt (SMA) mixture and found that steel slag applied in the stone mastic asphalt mixture is feasible. Using steel slag as coarse aggregate in SMA can effectively improve its water damage resistance and pavement durability. Pasetto et al. [19,20] evaluated high performance asphalt concrete with electric arc furnace steel slags. Their results showed that asphalt mixtures with EAF slags exhibited better mechanical characteristics than those of the asphalt mixtures with natural aggregates.

Wu et al. [21,22] studied the properties of BOF slag asphalt mixture from the viewpoint of physicochemical characteristics of basic oxygen furnace (BOF) slag. SEM, XRD, EPMA and other testing methods were used to analyze the physicochemical properties of BOF slag. The results showed that BOF slag has a rough surfaced micro texture, which is beneficial to the bonding properties of asphalt. The BOF slag has good angle properties and a rougher alkaline surface than natural aggregates. The asphalt mixture prepared by the BOF slag has better water damage resistance, and higher temperature rutting resistance, fatigue resistance and anti-skid resistance.

Although existing research has determined that the special physical properties of BOF slag can improve the performance of asphalt concrete, the geometric properties of BOF slag and its influence on asphalt mixture have not been effectively quantified. Nodes et al. [23] analyzed the performance of asphalt pavement with the effect of angularity on natural aggregate. The results showed that higher angularity of aggregates improved the interlocking interaction of coarse aggregates, which contributed to increased rutting resistance of asphalt pavement.

Digital image processing technology can solve this problem properly, as it has been used for accurately quantifying the sphericity, gradient angularity and micro texture of aggregates [24–26]. The aggregate image measurement system (AIMS) is a computer automation system that can accurately quantify the geometric properties of aggregates [27]. It uses a high resolution digital camera to quickly picture large numbers of aggregates, and then uses data processing software to calculate and process the shape and surface texture of each aggregate. Finally, quantitative properties, such as the angularity, sphericity and texture of the aggregates, can be obtained [28,29].

This research uses AIMS to quantify and analyze angularity, sphericity and texture of the BOF slag coarse aggregates, and then investigate the influence of geometric characteristics of coarse aggregates on asphalt concrete. The concept of effective density is used in this research to decrease the experimental error caused by the high density of BOF slag. Coarse aggregate of basalt, limestone and three types of BOF slag are used to prepare AC 13 (asphalt mixture with a maximum particle size of aggregate of 13 mm). Volumetric and mechanical properties are discussed to investigate the influence of geometric characteristics of BOF slag coarse aggregate on AC 13.

2. Materials and Methods

2.1. Aggregates

Two types of natural aggregates, basalt and limestone, were obtained from Hubei province in China for use in this research. Three types of BOF slag were used in this research: BOF slag #1 was obtained from Wuhan Iron and Steel Company in Hubei province, China; BOF slags #2 and #3 were obtained from Baotou Iron and Steel Company in Inner Mongolia. BOF slags 1 and 2 were basic oxygen furnace slag, while BOF slag #3 was pyrolytic BOF slag (basic oxygen furnace slag treated with hot stuffing process). Table 1 lists the chemical compositions based on X-ray fluorescence analysis.

Basalt and limestone are naturally available aggregates. As Table 1 shows, the chemical composition of SiO_2 in basalt is 47.9%, while chemical composition of CaO in limestone is more than 50%. Compared with naturally available aggregates, the chemical composition of Fe_2O_3 in the three types of BOF slag is more than 20%. The SiO_2 content in the three types of BOF slag is less than 30%.

The basic engineering properties of aggregates were tested according to ASTM standards [30–32] and the results are showed in Table 2. They all meet requirements.

Table 1. Chemical composition of the aggregates.

Aggregate Type	Composition (%)							
	SiO$_2$	CaO	MgO	Al$_2$O$_3$	Fe$_2$O$_3$	MnO	P$_2$O$_5$	LOI
Basalt	47.9	8.23	4.34	18.3	9.82	3.6	2.25	5.27
Limestone	0.86	51.2	2.36	0.85	0.12	0.7	1.02	42.7
1# BOF slag	19.2	42.7	5.19	3.25	23.9	1.77	1.41	2.36
2# BOF slag	17.7	39.7	5.56	2.91	24.4	4.55	1.68	3.41
3# BOF slag	15.4	34.4	6.22	1.95	30.8	4.46	2.15	4.46

Table 2. The basic engineering properties of the aggregates in this research.

Parameter Measured	Basalt	Limestone	BOF Slag			Requirements
			#1	#2	#3	
Los Angeles abrasion (%)	10.2	11.8	16.5	15.6	15.6	$\leq$28
Aggregate Crushing Value (%)	9.1	14.1	14.3	15.2	13.9	$\leq$26
Flakiness and elongation (%)	13.2	10.7	6.6	5.8	4.3	$\leq$18
Fine aggregate angularity (%)	42	46.3	NA	NA	NA	$\geq$30
Sand equivalent (%)	68.5	74.2	NA	NA	NA	$\geq$60

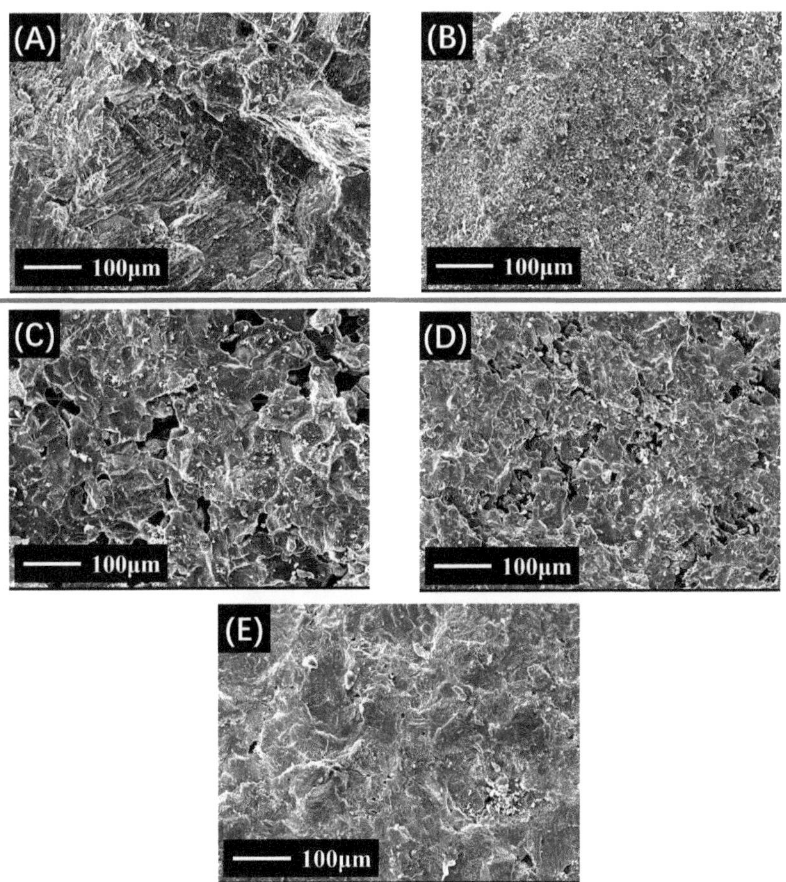

Figure 1. SEM images of investigated aggregates. **A**: Basalt; **B**: Limestone; **C**: #1 BOF slag, **D**: #2 BOF slag, **E**: #3 BOF slag).

Figure 1 shows the diversity between two types of naturally aggregates and three types of BOF slags in SEM images. According to the micrographs of basalt and limestone, the microscopic texture of basalt is relatively dense because it is a volcanic rock, while the microscopic surface of limestone consists of a clastic texture and tiny grain structure as it belongs to the carbonate sedimentary rocks. Comparing micrographs of BOF slags and natural aggregates, BOF slags show different surface textures, particularly on the size and number of surface microscopic pore structures. Micrographs of BOF slags #1 and #2 show that large numbers of pore structures with sizes of approximately 10–50 μm appear on the BOF slag surface. Micrographs of BOF slag #3 shows that the surface microscopic imaging of pyrolytic BOF slag is between basalt and BOF slag, which has a considerable number of pore structures of 1–10 μm in size. The greater the number of surface microscopic pore structures in BOF slag represents higher water absorption and asphalt absorption.

2.2. Asphalt Binder and Filler

Asphalt binder was used in this research and its optimum asphalt content is 4.7% in the mixture. Its characteristics are shown in Table 3. Table 4 presents the features of the used limestone filler in this research.

Table 3. Basic properties of the asphalt binder.

Properties		Values	Specifications
Penetration (0.1 mm)		63	60–80
Penetration index		−0.7	−1.5–1.0
Softening point (°C)		47.7	≥46
Ductility, 5 cm/min, 15 °C (cm]		>160	≥100
Dynamic viscosity (60 °C) (Pa·s)		179	≥160
Density (g/cm^3)		1.023	
After RTFOT (rotating thin film over test) ageing	Weight loss (%)	−0.05	−0.8–0.8
	Penetration ratio (%)	67.5	≥61
	Residual ductility 5 cm/min, 10 °C (cm)	11.5	≥6

Table 4. Basic properties of limestone filler in this research.

Properties		Values	Specifications
Apparent specific gravity		2.816	≥2.5
Sieves passing percentage (%)	<0.6 mm	100	100
	<0.15 mm	98.4	90–100
	<0.075 mm	92.2	75–100
Hydrophilic coefficient		0.4	<1.0

3. Experimental Details

3.1. Gravity Characteristics of Coarse Aggregate

Aggregate gravity is the first factor to be considered in asphalt mixture design. It includes apparent specific gravity, bulk specific gravity and water absorption.

The specific gravity of BOF slag and natural aggregates is quite different. The purpose of this research is to design an asphalt mixture with different types of coarse aggregates and fine aggregates. Insufficient consideration of aggregate specific gravity will lead to adverse influence on the volumetric and road performance of the asphalt mixture [33].

3.1.1. Aggregate Specific Gravity

Aggregate specific gravity is the ratio of the aggregate's density to water density at 23 °C. It is quite different between BOF slag and natural aggregate. The apparent specific gravity (Gsa) measures the volume of aggregate particle and impervious voids. The bulk specific gravity (Gsb) measures the volume of aggregate particle, impervious voids and water permeable voids. Water absorption refers to the amount of water absorbed by the void of the aggregate.

Aggregate with different grain size ranges have different aggregate specific gravity. Table 5 shows the Gsb and Gsa of aggregates at different grain size ranges according to the JTG42-2005 standard of China. Table 5 shows that the Gsb and Gsa of BOF slag is higher than that of basalt and limestone. Among three types of BOF slags, the Gsb and Gsa of BOF slag #1 and BOF slag #2 are similar. Due to the self-slaking process of BOF slag by pyrolytic and processing technology, BOF slag #3 (pyrolytic BOF slag) has the highest specific gravity. It is at least 15% higher than the basalt's specific gravity in each grain size. In the same types of aggregates, the Gsb increase with the increase of aggregate particle size range.

Table 5. Specific gravities of coarse aggregates at different size ranges.

Aggregate	Gravity	Grain Size (mm)			
		2.36–4.75	4.75–9.5	9.5–13.2	13.2–16.0
Basalt	Gsa	2.923	2.918	2.915	2.912
	Gsb	2.863	2.881	2.886	2.899
Limestone	Gsa	2.723	2.716	2.705	2.701
	Gsb	2.663	2.676	2.684	2.689
1#BOF slag	Gsa	3.308	3.302	3.293	3.306
	Gsb	3.092	3.102	3.125	3.153
2#BOF slag	Gsa	3.275	3.263	3.302	3.293
	Gsb	3.098	3.124	3.179	3.186
3#BOF slag	Gsa	3.543	3.587	3.563	3.582
	Gsb	3.379	3.426	3.453	3.517

Figure 2 shows the water absorption of the five types of aggregates at different grain size ranges. Immersion duration of aggregates is 24 h. Apparently, with the decrease of aggregate grain size, the water absorption becomes larger. Smaller grain size is logically related to bigger specific surface area, which will therefore decrease the ration of closed voids in aggregates and increase the ratio of open voids. Basalt and limestone have similar water absorption at different grain size ranges. Meanwhile, BOF slag has higher water absorption than natural aggregate. The order of their water absorption from large to small is: #1BOF slag, #2BOF slag and #3BOF slag.

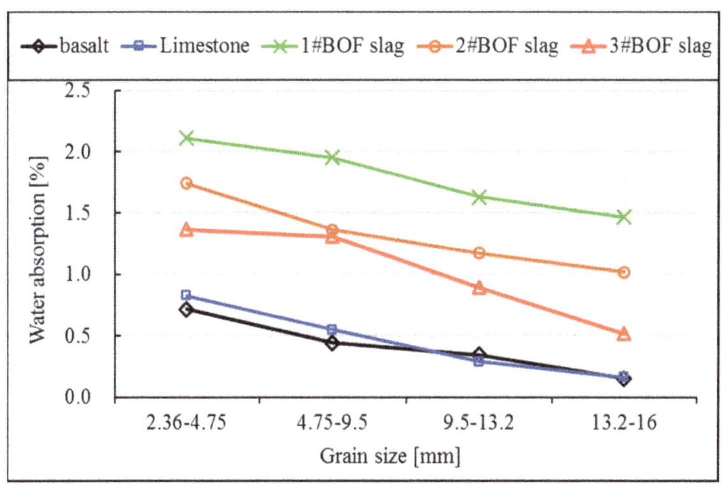

Figure 2. Water absorptions of aggregates.

3.1.2. Effective Specific Density

The mix ratio of asphalt mixture with single type of aggregate was based on aggregate weight. However, in the experimental design of this research, four types of aggregates were used to replace basalt coarse aggregate. The aggregate specific gravity results illustrate that big differences appear on specific gravities and water absorption between different aggregate types. Due to the difference in specific gravity, the gradation and optimum asphalt content in the design of asphalt mixture will change.

In the asphalt mixture, the asphalt can partly fill open voids on the surface of aggregate; therefore, the apparent density or bulk density of aggregates cannot effectively represent the density of aggregates in asphalt. It is more effective to use the specific density (Dse), taking into account the volume effect between aggregate types for designing asphalt mixture. The Dse of the different aggregates are shown in Table 6. Coarse aggregate was replaced by the ratio of effective specific density of different aggregates, aiming to reduce the inaccuracy of the test caused by aggregate density.

Table 6. Dse of aggregates at different grain size range.

Aggregate	Grain Size (mm)			
	2.36–4.75	4.75–9.5	9.5–13.2	13.2–16
Basalt	2.691	2.694	2.695	2.696
Limestone	2.896	2.905	2.902	2.904
#1BOF slag	3.272	3.264	3.217	3.186
#2BOF slag	3.214	3.197	3.245	3.153
#3BOF slag	3.474	3.513	3.498	3.421

3.2. Preparation of Asphalt Mixture

AC 13 asphalt mixture was selected to investigate the combination of two natural aggregates and three BOF slags. In order to reduce experimental errors, the aggregate gradation was kept constant. The aggregates distribution of basalt was used in the basic aggregate gradation, as shown in Table 7.

Table 7. Aggregate composite gradation of AC 13 asphalt mixture.

Coarse Aggregates (mm)		Weight Ratio (%)	Fine Aggregates (mm)		Weight Ratio (%)
9.5–16	13.2–16	4.1	0–2.36	1.18–2.36	6.3
	9.5–13.2	22.6		0.6–1.18	7.7
2.36–9.5	4.75–9.5	28.7		0.3–0.6	4.6
	2.36–4.75	13.3		0–0.3	8.7
	Filler				4%

The coarse aggregate and fine aggregate play different roles in the asphalt mixture. Coarse aggregates play a major role in supporting and reinforcing the construction of asphalt mixture, while fine aggregates are mainly used to fill the gaps between coarse aggregates. For the AC 13 asphalt mixture, the dividing sieve size between coarse aggregate and fine aggregate is 2.36 mm.

A coarse–fine composition method was proposed and used to keep fine aggregates such as basalt unchanged, while the coarse aggregates were fully replaced basing on effective specific density. Figure 3 explains the detail of coarse–fine composition.

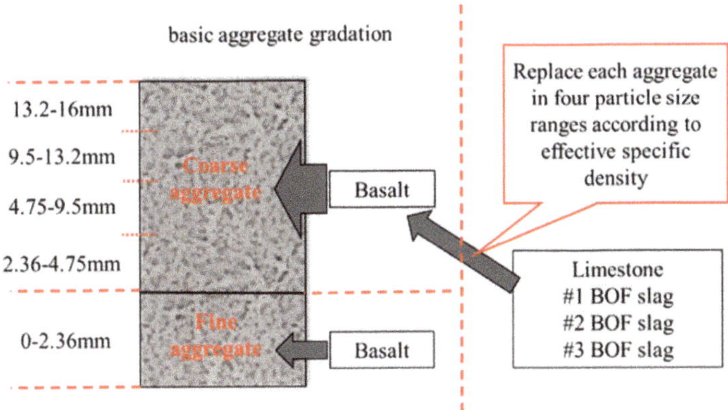

Figure 3. The coarse–fine composition method.

In asphalt concrete, the geometric characteristics of coarse aggregate, such as angularity, sphericity, surface texture, have a great influence on its performance. A marshall stability test was used to characterize the water resistance of asphalt mixture in 60 °C water; this test is still currently used in the road laboratories for the mechanical characterization of asphalt concretes. In order to change geometric characteristics of the coarse aggregate, a ball mill is used to process the coarse aggregate, which imitates the Los Angeles abrasion test. The time of the specific processing was 30 min, at a weight of 3000 g for each sample; the rotating speed was 120 rpm; and using abrasion with 8 steel balls (500 g each). In this research, #1 BOF slag was treated using the above method and named as processed #1BOF slag (SP1). AIMS (aggregate image measurement system) was used to analyze the angularity, sphericity and surface texture of each aggregate.

In this research, the coarse–fine composition of asphalt mixture method was applied, with 0% of coarse aggregates and 100% of fine aggregates passing through sieves of 2.36 mm. Table 8 lists the labels for every coarse–fine composition of asphalt mixture. Four specimens were tested in each and the average data were used for analysis.

Table 8. Labels for every coarse–fine composition of asphalt mixture.

Fine Aggregate	Coarse Aggregate	Label
	Basalt (B)	BB
	Limestone (L)	BL
Basalt	#1BOF slag (S1)	BS1
	#2BOF slag (S2)	BS2
	#3BOF slag (S3)	BS3
	processed #1BOF slag (SP1)	BSP1

3.3. Experimental Methods

3.3.1. Aggregates Morphology Test

Shape, angularity, and surface texture of aggregates have been shown to directly affect the engineering properties of highway construction such as HMA (hot mix asphalt concrete) and concrete. The aggregate image measurement system (AIMS) was used to analyze the angularity, sphericity and surface texture of coarse aggregates of each types of aggregate.

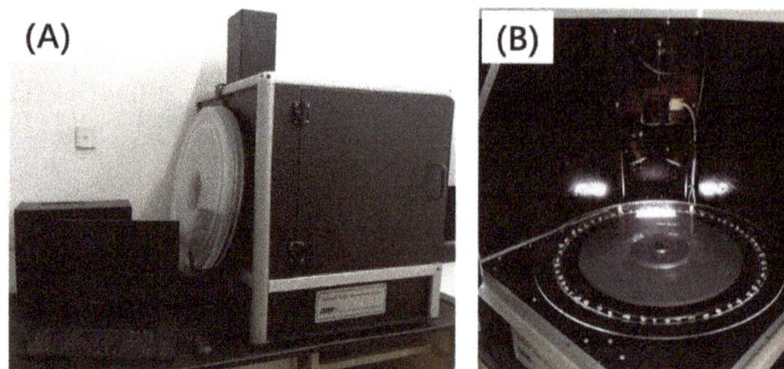

Figure 4. Aggregate image measurement system (**A**), and the tested coarse aggregate placed on tray (**B**).

As shown in Figure 4, the aggregate image measurement system is comprised of image acquisition hardware and a computer to run the system and process data. The image acquisition hardware used a high-resolution digital camera and a variable magnification microscope to collect digital images and measure aggregates. The system used a tray to place the aggregate under the view of the camera. The first scan used backlighting to describe a profile image of the aggregate particle, from which dimensions and angularity gradients of the edges were measured. The second scan utilized top-lighting and variable magnification to capture texture images and measure each aggregate particle's height. AIMS then used the computer to process the acquired images.

Angularity is a description of edge sharpness of the boundary particles of aggregate. The angularity changes with aggregate boundary shape changes. The value of angularity is calculated based on the gradient of the particle boundary. Angularity is calculated with Equation (1) and its range is from 0 to 1000. The lager the value of angularity, the sharper the boundary shape of the aggregate.

$$Angularity = \frac{1}{\frac{n}{3} - 1} \sum_{i=1}^{n-3} |\theta_i - \theta_{i+3}| \tag{1}$$

where θ is angle of orientation of the edge points, n is the total number of points, i is the ith point on the edge of the particle.

Sphericity is used to characterize the degree of similarity between the aggregate shape and the ideal sphere. Sphericity is calculated with Equation (2), its range is 0 to 1. If the value of sphericity is closer to 1, the shape of aggregate is more similar to the ideal sphere.

$$Sphericity = \sqrt[3]{\frac{d_S d_I}{d_L^2}} \tag{2}$$

where, d_S is particle shortest dimension, d_I is particle intermediate dimension, d_L is particle longest dimension.

3.3.2. Marshall Stability Test

The Marshall stability test is used to characterize the water resistance of asphalt mixture in 60 °C water. The compacted specimens were divided into two groups according to the numerical average of VV (percent air voids in asphalt mixtures). The first group was put in the water at 60 °C for 30 min. The second group was placed in water at 60 °C for 48 h. The specimens were then placed in a testing machine at a constant displacement load rate of 50 mm/min. We recorded the Marshall stability (MS, maximum load) and Marshall flow (the maximum load starts to decrease when the deformation is mm).

The retained Marshall stability (RMS) is defined as the ratio of the sample to Marshall stability after immersion in hot water for 48 h (MS1, kN) and 30 min (MS, kN). The larger the RMS, the better the water resistance performance of asphalt mixture.

3.3.3. Freeze-Thaw Splitting Test

The freeze-thaw splitting test can be used to measure the ability of moisture resistance of asphalt mixture at low temperatures. The main evaluation index is the splitting strength ratio TSR (%). The larger the value of TSR, the better the moisture resistance of asphalt mixture of the water at freezing and thawing environmental conditions.

Splitting strength ratio (TSR) is defined as the ratio between ITS after F-T circles and ITS of nor-freeze. Among them, the environmental conditions of F-T circles are freezing at $-20\,^{\circ}\mathrm{C}$ for 16 h.

3.3.4. Rutting Test

The rutting test was used to characterize the high temperature rutting resistance of asphalt mixture. The size of the rut specimen was $300 \times 300 \times 50$ mm, and the specimen density was controlled at 100 + 1.0% of the Marshall density. The rutting experiment temperature was 60 $^{\circ}\mathrm{C}$, the wheel pressure was 0.7 MPa, and the experimental wheel round-trip speed was 42 times/min. The dynamic stability (DS) is expressed by the rutting depth change rate of 45–60 min.

3.4. Research Program

Figure 5 illustrates the research program on the influence of geometric characteristics of BOF slag coarse aggregate on asphalt concrete. Firstly, the aggregate image measurement system (AIMS) was used to analyze the angularity, sphericity and surface texture of coarse aggregates. Then, the effective specific density was conducted during the asphalt mixture design. Basalt was used to create the AC 13 mixture, and the coarse aggregate was replaced by limestone and BOF slags by coarse-fine composition method, respectively. Volumetric and mechanical properties were studied to evaluate the influence of geometric characteristics of BOF slag coarse aggregate on asphalt concrete.

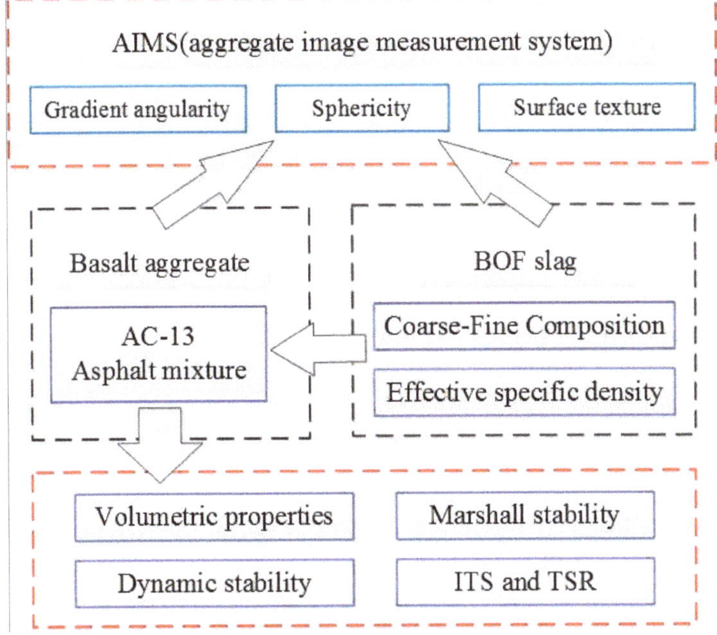

Figure 5. Finalized research program.

4. Results and Discussion

4.1. Geometric Characteristics of Coarse Aggregate

Coarse aggregates with sizes of 9.5–13.2 mm and 4.75–9.5 mm in total, account for 51.3% by the weight of the asphalt mixture. Therefore, the geometric characteristics of coarse aggregate were represented by the angularity, sphericity and surface texture of aggregates with those two sizes. Each test selected 200 aggregate samples, and the illustration of angularity, sphericity and texture is shown in Figure 6.

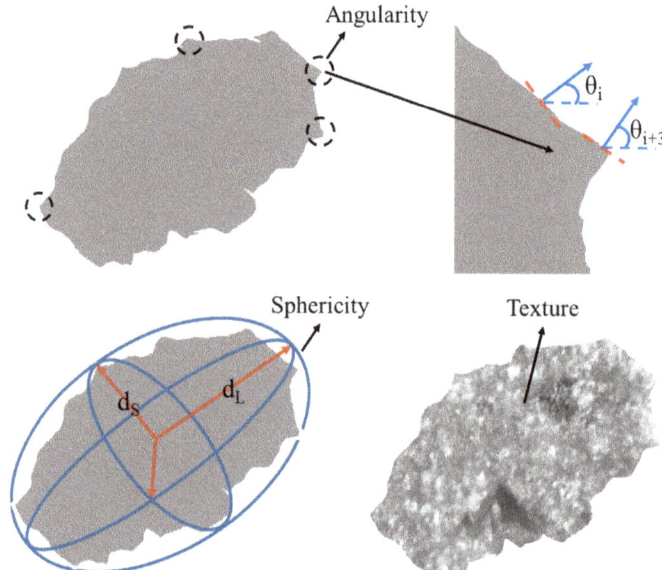

Figure 6. Illustration of angularity, sphericity and texture.

4.1.1. Angularity

Angularity is a description of edge sharpness of the boundary particles of aggregate. Profile images of different types of aggregate particle from the first scan of AIMS are shown in Figure 7.

The angularity of investigated coarse aggregates is shown in Table 9. This was obtained from the average values of 200 aggregate samples. The results show that the angularity of BOF slag is higher than that of natural aggregates. Different types of BOF slag have different angularity. Among them, in the same place of production, the angularity of basic oxygen furnace slag is higher than pyrolytic BOF slag. Meanwhile, limestone has the lowest angularity and #2 BOF slag has the highest angularity.

Table 9. Angularity of investigated coarse aggregates.

Grain Size (mm)	Basalt	Limestone	S1	S2	S3	SP1
9.5–13.2 mm	2965	2896	3064	3770	3180	2886
4.75–9.5 mm	2846	2733	3051	3283	2829	2765

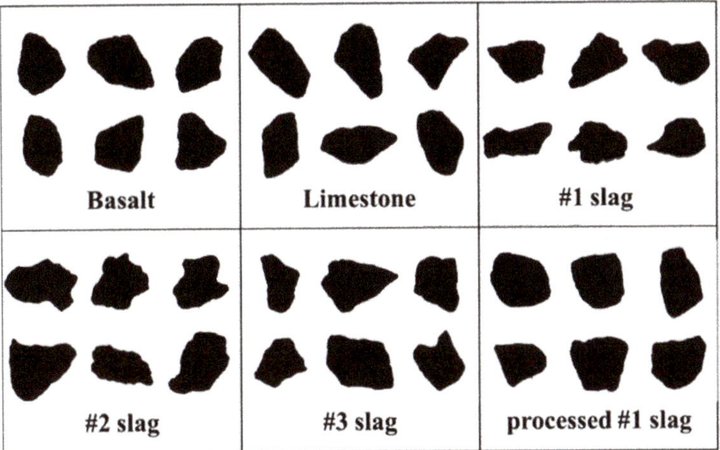

Figure 7. Parts of the profile image of different types aggregate particles.

4.1.2. Sphericity

The sphericity of the investigated coarse aggregates is shown in Table 10. The results were calculated from the 200 aggregate samples. It can be observed from Table 10 that the limestone had the lowest sphericity and #2 BOF slag had the highest sphericity. BOF slag had a higher degree of sphericity than natural aggregates.

Table 10. Sphericity of investigated coarse aggregate.

Grain Size (mm)	Basalt	Limestone	S1	S2	S3	SP1
9.5–13.2 mm	0.706	0.662	0.749	0.759	0.739	0.768
4.75–9.5 mm	0.639	0.609	0.693	0.718	0.726	0.729

4.1.3. Texture

Texture represents the relative roughness and smoothness of aggregate surfaces. The second scan utilized top-lighting and variable magnification to capture texture images of aggregates; these are shown in Figure 8.

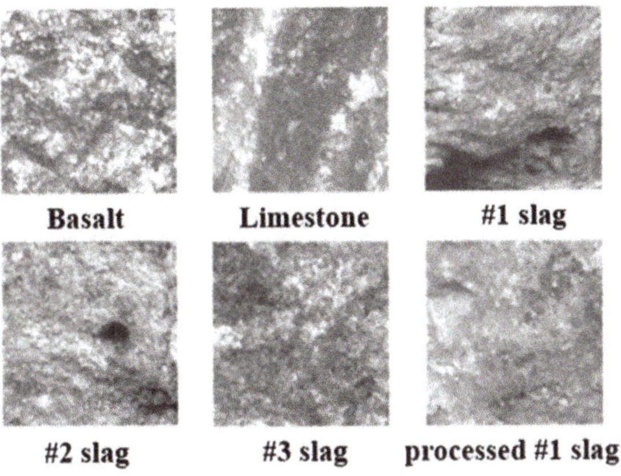

Figure 8. Parts of the surface texture images of aggregates.

The texture value of an ideal smooth surface is defined as zero. The AIMS texture analysis uses the wavelet method in the calculation and analysis of texture.

The texture of investigated coarse aggregate is shown in Table 11. From the texture values of six types of aggregate, the highest texture value is basalt. The texture value of BOF slag is smaller. As the surface of BOF slag has formed uniform color carbonized layers during the aging process, the photos taken by the digital camera of AIMS cannot represent the surface texture of the BOF slag. With reference to the water absorption of each aggregates in Table 8, #1 BOF slag is considered to be the most texture.

Table 11. Texture of investigated coarse aggregate.

Grain Size (mm)	Basalt	Limestone	S1	S2	S3	SP1
9.5–13.2 mm	708	450	349	356	407	419

4.2. Volumetric Properties of BOF Slag Coarse Aggregate Asphalt Concrete

In traditional asphalt mixture gradation designs the air voids in the asphalt mixture are the most important index of volumetric properties, which can also affect the performance of asphalt concrete. The skeleton structure of AC 13 asphalt mixture is a suspension-compact structure with 3–5% percent air voids in the asphalt mixture.

The percentage of air voids in the asphalt mixture (VV) is defined as the volume percentage of air voids in compacted asphalt mixtures. It is calculated with Equation (2).

$$VV = 100 \times \left[1 - \left(\frac{G_f}{G_t} \right) \right]$$

(3)

where G_t is the theoretical maximum specific gravity of bituminous mixture and G_f is bulk specific gravity of bituminous mixture.

Table 12 shows the volumetric properties of AC 13. Compare with group BB, the theoretical maximum density decrease and porosity decrease by about 40% in terms of limestone aggregate. The VV of BS2 is similar with BB, and the use of other groups of BOF slag can effectively improve the value of VV; the highest VV of BSP1 reached 5.83%.

Table 12. Volumetric properties with different coarse aggregates.

Aggregate Categories	Theoretical Maximum Density [g/cm^3]	Apparent Density [g/cm^3]	VV [%]
BB	2.689	2.585	3.873
BL	2.601	2.545	2.154
BS1	2.828	2.708	4.262
BS2	2.812	2.706	3.769
BS3	2.958	2.831	4.285
BSP1	2.827	2.662	5.832

When combining the volumetric properties analysis with geometric characteristics of aggregate, four types of BOF slags have higher sphericity than basalt. Among them, BSP1 has the highest sphericity and the lowest angularity and it has the largest value of VV. BS2 has the highest angularity when compared with BS1, but its VV is lower. This shows that the excessively high angularity reduces the value of VV.

The geometric characteristics of coarse aggregates have a great effect on the volumetric properties of asphalt mixture. Firstly, the rough surface texture and high water absorption of BOF slag coarse aggregate can absorb more free asphalt in the mixture and increase air voids of the asphalt mixture. Secondly, the high sphericity of the BOF slag contributes more skeleton support in the asphalt mix skeleton structure. In addition, BOF slag has a higher angularity than natural aggregate, so the larger sphericity of the BOF slag leads to a higher VV of the asphalt mixture. Lastly, if BOF slag has excessive angularity value, the edge angle of BOF slag would be destroyed during the compaction of the asphalt mixture, which would result in the decrease of the air voids of the asphalt mixture.

Using BOF slag coarse aggregates will result in a significant increase in the percentage of air voids in asphalt mixture. The rough surface texture and higher sphericity of BOF slag will cause the increase of the air voids of asphalt mixture. Besides, if the BOF slag has excessively high angularity, it will have an adverse effect on the increase of air voids.

4.3. Mechanical Properties of BOF Slag Coarse Aggregate Asphalt Concrete

The Marshall stability test, freeze-thaw (F-T) splitting test and rutting test were employed to evaluate the influence of geometric characteristics of BOF slag coarse aggregate on asphalt concrete.

4.3.1. Marshall Stability Test Results

As shown in Figure 9, the values of Marshall stability are presented. The RMS of the basalt-based asphalt mixture has the lowest value of 80.37%. Coarse aggregate of limestone and coarse aggregate of BOF slag can significantly enhance the moisture resistance property. BS1 has the highest water absorption and non-outstanding value of angularity and sphericity. The results of BS1 and BSP1 are similar, showing that improving the sphericity of BOF slag and reducing the angularity of BOF slag have little effect on the moisture resistance property of asphalt mixture.

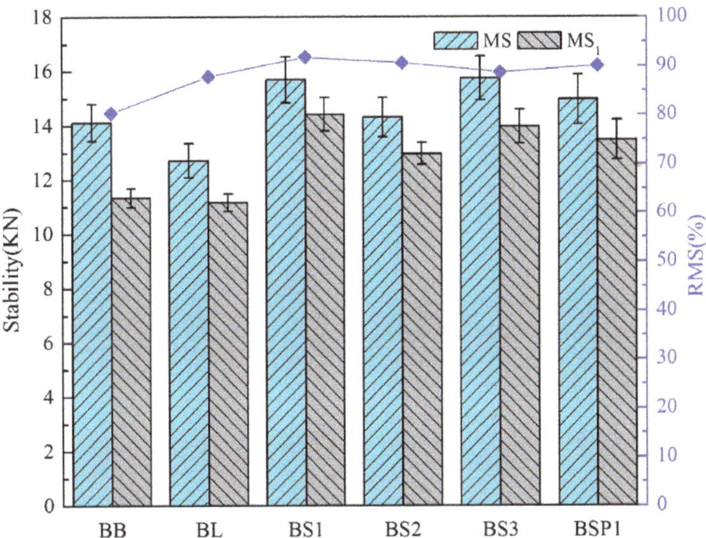

Figure 9. Marshall stability test results with different coarse aggregates.

As can be seen in Figure 10, there are close parabolic equation curve relationships between angularity or sphericity of BOF slag and Marshall stability test results. The R^2 of angularity and sphericity is 0.99 and 0.78, respectively. Both of the correlations of angularity and sphericity tend to increase first and then decrease. The optimal angularity and sphericity of Marshall results is 3267 and 0.755, respectively.

AC 13 has a suspension-compactness skeleton structure. The rough surface texture (high water absorption) of BOF slag and good adhesion with asphalt result in a good moisture resistance property of the asphalt mixture. Changing the degree of sphericity or angularity of BOF slag has little influence on the beneficial effect of moisture resistance property on the asphalt mixture.

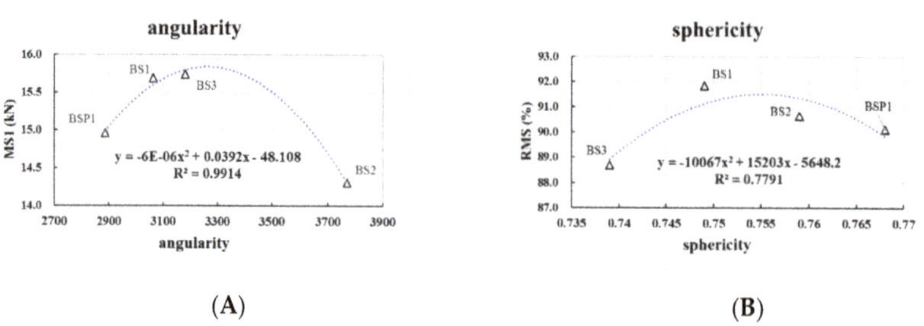

(A)   (B)

Figure 10. Correlation between Marshall stability test results and aggregate geometric characteristics. (**A**) correlation between MS1 and angularity; (**B**) correlation between RMS and sphericity.

4.3.2. Freeze-Thaw Splitting Test Results

Figure 11 presents the splitting rest results of composite asphalt mixtures. The tendency of freeze-thaw splitting test is similar to the Marshall stability test. The TSR of BB is the lowest. All types of BOF slags coarse aggregates composite asphalt mixtures possess higher TSR values. There is little difference between the TSR values of BS1 and BSP1. It shows that improving the sphericity of BOF slag and reducing the angularity of BOF slag have little effect on the moisture resistance property of BOF slag coarse aggregate asphalt mixture. TSR value of BS2 is the lowest among the 4 types of asphalt mixture with BOF slag.

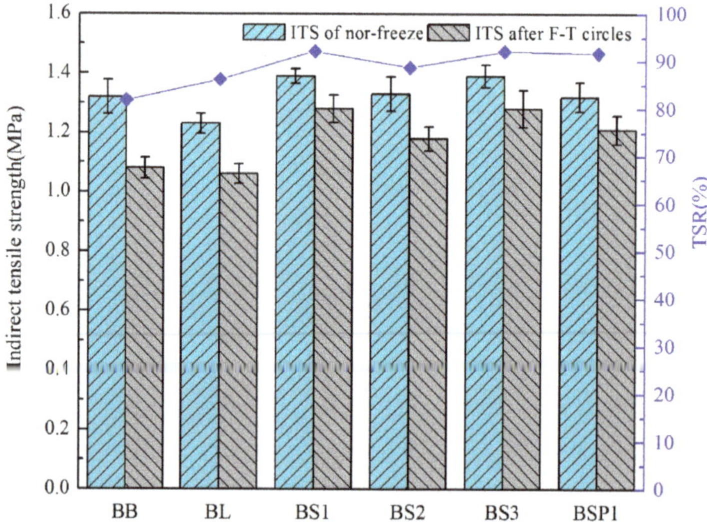

Figure 11. Splitting test results with different coarse aggregates.

As can be seen in Figure 12, there are close parabolic equation curves relationships between angularity of BOF slag and splitting test results, while R2 of angularity is 0.999. The correlations of angularity and TSR tend to increase first and then decrease. The optimal angularity of Marshall results is 3344. There is a weak linear relationship between sphericity and splitting test results.

BOF slag coarse aggregates have a rough surface and an alkaline surface, which enhances the moisture resistance under F-T circles of AC 13. Changing the degree of sphericity of BOF slag has little influence on the improving effect of low temperature moisture resistance property of asphalt mixture. The excessively high angularity of the aggregate can adversely affect the low temperature moisture resistance property of the asphalt mixture.

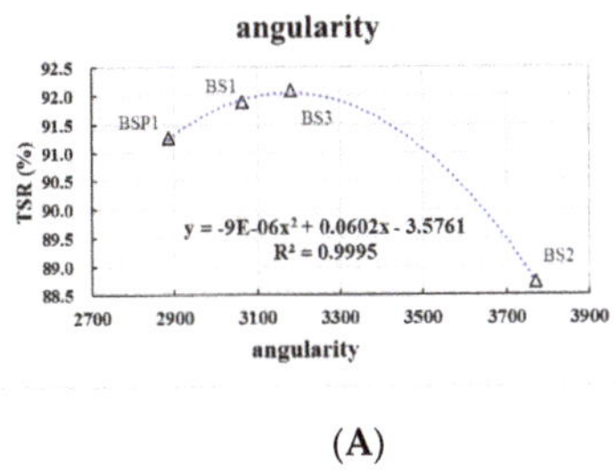

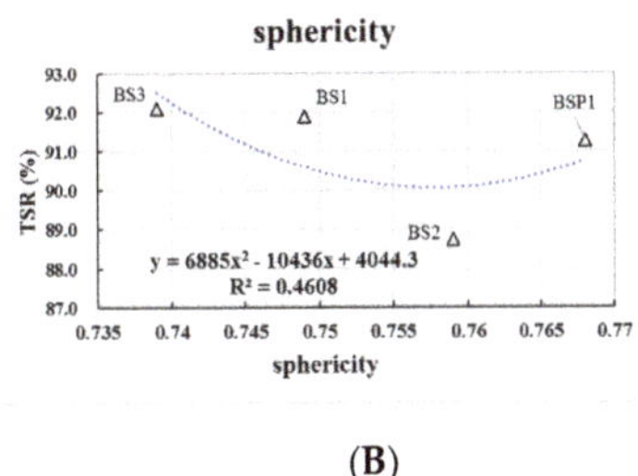

(**A**)

(**B**)

Figure 12. Correlation between splitting test results and aggregate geometric characteristics. (**A**) correlation between TSR and angularity; (**B**) correlation between TSR and sphericity.

4.3.3. Rutting Test Results

The Rutting test results are shown in Table 13. It can be clearly observed that the limestone coarse aggregate asphalt mixture has the lowest dynamic stability. Using four types of BOF slag coarse aggregates can improve the values of asphalt mixture dynamic stability in varying degrees, among which BS3 has the best high temperature rutting resistance of asphalt mixture. When compared with BS1, the dynamic stability of BSP1 and BS2 is decreased. This shows that the high sphericity and low angularity of BOF slag coarse aggregates will increase the dynamic stability of the asphalt mixture. Meanwhile, if the BOF slag has excessively high angularity, it will have an adverse effect on the increased dynamic stability.

Table 13. Rutting text results with different coarse aggregates.

Aggregate Categories	Rutting Depth (mm)	Dynamic Stability (mm)
BB	2.162	3056
BL	3.284	1863
BS1	1.952	4632
BS2	2.341	3517
BS3	1.842	4782
BSP1	2.080	4152

As can be seen in Figure 13, there are close parabolic equation curves relationships between angularity of BOF slag and dynamic stability, while R2 of angularity is 1.0. The curves of angularity tend to increase first and then decrease. The optimal angularity of dynamic stability is 3285. There is a weak linear relationship between sphericity and dynamic stability. With the rise of sphericity of BOF slag, dynamic stability has a slight reduction trend.

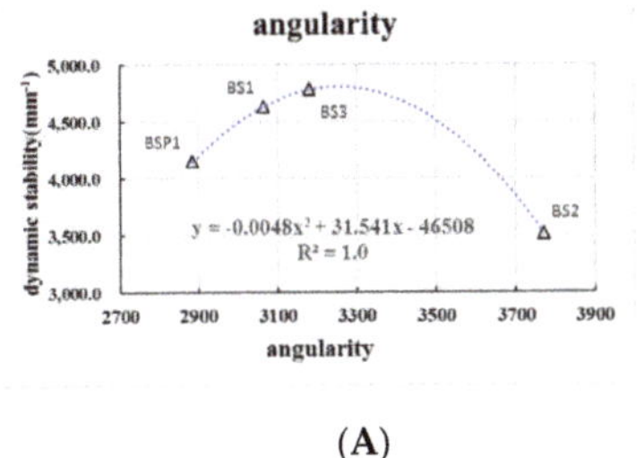

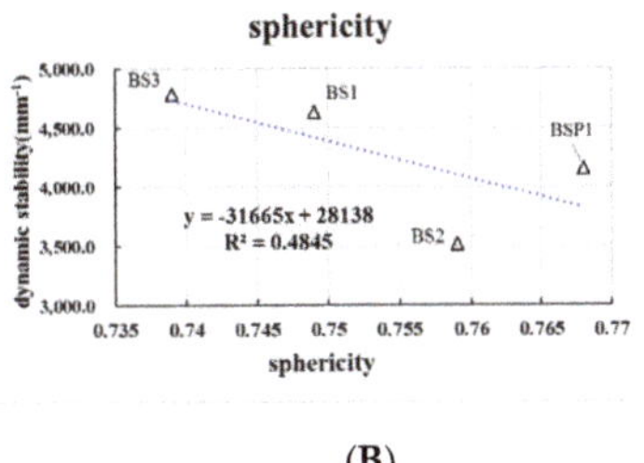

(**A**)

(**B**)

Figure 13. Correlation between rutting test results and aggregate geometric characteristics. (**A**) correlation between dynamic stability and angularity; (**B**) correlation between dynamic stability and sphericity.

5. Conclusions

This study investigated the geometric characteristics (sphericity, angularity and texture) of three types of BOF slag coarse aggregates by AIMS. A coarse-fine composition method and effective density were proposed and studied. Volumetric and mechanical properties of AC 13 were discussed to investigate the influence of geometric characteristics of BOF slag coarse aggregate. The following conclusions were obtained:

(1)   Compared with natural aggregates, the microscopic surface of BOF slag has a pitted and vesicular texture and a porous structure with particle range of 1–50 μm. BOF slag coarse aggregates have higher sphericity and angularity than natural aggregates. The BOF slag coarse aggregate has higher water absorption and higher effective density, among which the pyrolytic BOF slag has the largest effective density.

(2)   Using BOF slag coarse aggregates result in a significant increase on the VV of AC 13. The rough surface texture (higher water absorption) and higher sphericity of BOF slag causes the increase of the air voids in the asphalt mixture. The excessively high angularity of BOF slag can decrease the VV of AC 13 and the higher sphericity of BOF slag causes the increase of the air voids in the asphalt mixture.

(3)   According to the result of the Marshall stability, freeze-thaw splitting, and Rutting tests, using the coarse aggregate of three types of BOF slag and the processed BOF slag can enhance the properties of moisture damage resistance and rutting resistance of the asphalt mixture. A higher sphericity of BOF slag coarse aggregates leads to better rutting resistance. Excessively high angularity of BOF slag coarse aggregates reduces the anti-rutting properties of asphalt mixture. The optimal angularity of BOF slag is about 3300 based on the analysis of the result of Marshall stability test, freeze-thaw (F-T) splitting test and rutting test.

(4)   In the production of BOF slag coarse aggregates, the exorbitant angularity of coarse aggregates of BOF slag should be controlled. This will help to improve the service life of BOF slag asphalt pavement and the promotion of slag.

Author Contributions: Conceptualization, D.K.; data curation, D.K. and C.Y.; formal analysis, D.K., J.X., C.Y. and M.Z.; methodology, D.K., M.C. and C.Y.; project administration, J.X.; resources, M.C., J.X. and M.Z.; writing—review and editing, D.K., M.C. and J.X.

Funding: This research was funded by the National Key Research and Development Program of China grant number [2017YFE0111600] and the National Natural Science Foundation of China grant numbers [51708437 and 51778515].

Acknowledgments: The authors gratefully acknowledge many important contributions from the researchers of all reports cited in our paper.

Conflicts of Interest: The authors declare no conflict of interest.

References

1.   Ministry of Transport of the People's Republic of China. The Annual Statistics of Traffic and Transportation. Available online: http://zizhan.mot.gov.cn/zfxxgk/bnssj/zhghs/201803/t20180329_3005087.html (accessed on 2 July 2018).

2.   Skaf, M.; Manso, J.M.; Aragón, Á.; Fuente-Alonso, J.A.; Ortega-López, V. EAF slag in asphalt mixes: A brief review of its possible re-use. *Resour. Conserv. Recycl.* **2017**, *120*, 176–185. [CrossRef]

3.   Zheng, J. Design Guide for Semirigid Pavements in China Based on Critical State of Asphalt Mixture. *J. Mater. Civ. Eng.* **2013**, *25*, 899–906. [CrossRef]

4.   Ibrahim, A.; Faisal, S.; Jamil, N. Use of basalt in asphalt concrete mixes. *Constr. Build. Mater.* **2009**, *23*, 498–506. [CrossRef]

5.   Pan, P.; Wu, S.; Xiao, Y.; Liu, G. A review on hydronic asphalt pavement for energy harvesting and snow melting. *Renew. Sustain. Energy Rev.* **2015**, *48*, 624–634. [CrossRef]

6. Kambole, C.; Paige-Green, P.; Kupolati, W.K.; Ndambuki, J.M.; Adeboje, A.O. Basic oxygen furnace slag for road pavements: A review of material characteristics and performance for effective utilisation in southern Africa. *Constr. Build. Mater.* **2017**, *148*, 618–631. [CrossRef]

7. Jiang, Y.; Ling, T.C.; Shi, C.; Pan, S.Y. Characteristics of steel slags and their use in cement and concrete—A review. *Resour. Conserv. Recycl.* **2018**, *136*, 187–197. [CrossRef]

8. Poulikakos, L.D.; Papadaskalopoulou, C.; Hofko, B.; Gschösser, F.; Falchetto, A.C.; Bueno, M.; Arraigada, M.; Sousa, J.; Ruiz, R.; Petit, C. Harvesting the unexplored potential of European waste materials for road construction. *Resour. Conserv. Recycl.* **2017**, *116*, 32–44. [CrossRef]

9. Han, F.; Zhang, Z.; Wang, D.; Yan, P. Hydration heat evolution and kinetics of blended cement containing steel slag at different temperatures. *Thermochim. Acta* **2015**, *605*, 43–51. [CrossRef]

10. Reddy, A.S.; Pradhan, R.K.; Chandra, S. Utilization of Basic Oxygen Furnace (BOF) slag in the production of a hydraulic cement binder. *Int. J. Miner. Process.* **2006**, *79*, 98–105. [CrossRef]

11. Shen, D.H.; Wu, C.M.; Du, J.C. Laboratory investigation of basic oxygen furnace slag for substitution of aggregate in porous asphalt mixture. *Constr. Build. Mater.* **2009**, *23*, 453–461. [CrossRef]

12. Iacobescu, R.I.; Angelopoulos, G.N.; Jones, P.T.; Blanpain, B.; Pontikes, Y. Ladle metallurgy stainless steel slag as a raw material in Ordinary Portland Cement production: a possibility for industrial symbiosis. *J. Cleaner Prod.* **2016**, *112*, 872–881. [CrossRef]

13. Chao, L.I.; Chen, Z.; Xie, J.; Shaopeng, W.U.; Xiao, Y. A Technological and Applicational Review on Steel Slag Asphalt Mixture. *Mater. Rev.* **2017**.

14. Ahmedzade, P.; Sengoz, B. Evaluation of steel slag coarse aggregate in hot mix asphalt concrete. *J. Hazard. Mater.* **2009**, *165*, 300–305. [CrossRef] [PubMed]

15. Xie, J.; Wu, S.; Zhang, L.; Xiao, Y.; Ding, W. Evaluation the deleterious potential and heating characteristics of basic oxygen furnace slag based on laboratory and in-place investigation during large-scale reutilization. *J. Cleaner Prod.* **2016**, *133*, 78–87. [CrossRef]

16. Ameri, M.; Behnood, A. Laboratory studies to investigate the properties of CIR mixes containing steel slag as a substitute for virgin aggregates. *Constr. Build. Mater.* **2012**, *26*, 475–480. [CrossRef]

17. Behnood, A.; Ameri, M. Experimental investigation of stone matrix asphalt mixtures containing steel slag. *Scientia Iranica* **2012**, *19*, 1214–1219. [CrossRef]

18. Chen, Z.; Wu, S.; Xiao, Y.; Zhao, M.; Xie, J. Feasibility study of BOF slag containing honeycomb particles in asphalt mixture. *Constr. Build. Mater.* **2016**, *124*, 550–557. [CrossRef]

19. Pasetto, M.; Baldo, N. Experimental evaluation of high performance base course and road base asphalt concrete with electric arc furnace steel slags. *J. Hazard. Mater.* **2010**, *181*, 938–948. [CrossRef] [PubMed]

20. Pasetto, M.; Baldo, N. Fatigue Performance of Asphalt Concretes with RAP Aggregates and Steel Slags. *RILEM Bookseries* **2012**, *4*, 719–727.

21. Wu, S.; Xue, Y.; Ye, Q.; Chen, Y. Utilization of steel slag as aggregates for stone mastic asphalt (SMA) mixtures. *Build. Environ.* **2007**, *42*, 2580–2585. [CrossRef]

22. Xue, Y.; Wu, S.; Hou, H.; Zha, J. Experimental investigation of basic oxygen furnace slag used as aggregate in asphalt mixture. *J. Hazard. Mater.* **2006**, *138*, 261–268. [CrossRef] [PubMed]

23. Zhu, H.; Nodes, J.E. Contact based analysis of asphalt pavement with the effect of aggregate angularity. *Mech. Mater.* **2000**, *32*, 193–202. [CrossRef]

24. Amarasiri, S.; Gunaratne, M.; Sarkar, S. Use of Digital Image Modeling for Evaluation of Concrete Pavement Macrotexture and Wear. *J. Transp. Eng.* **2012**, *138*, 589–602. [CrossRef]

25. Wang, H.N.; Hao, P.W.; Pang, L.G.; Di, J.H. Investigation into grading characteristic of coarse aggregate via digital image processing technique. *J. S. Chin. Univ. Technol.* **2007**, *35*, 54–59.

26. Xiong, Q.; Wang, X.D.; Zhang, L. Research Summary of Digital Image Processing Technology on Coarse Aggregate Morphology Characteristics. *Subgrade Eng.* **2012**.

27. Araujo, V.M.C.; Bessa, I.S.; Branco, V.T.F.C. Measuring skid resistance of hot mix asphalt using the aggregate image measurement system (AIMS). *Constr. Build. Mater.* **2015**, *98*, 476–481. [CrossRef]

28. Bessa, I.S.; Soares, J.B. Evaluation of polishing and degradation resistance of natural aggregates and steel slag using the aggregate image measurement system. *Road Mater. Pavement Des.* **2014**, *15*, 385–405. [CrossRef]

29. Mahmoud, E.; Gates, L.; Masad, E.; Erdoğan, S.; Garboczi, E. Comprehensive Evaluation of AIMS Texture, Angularity, and Dimension Measurements. *J. Mater. Civ. Eng.* **2010**, *22*, 369–379. [CrossRef]

30. ASTM. Standard Test Method for Flat Particles, Elongated Particles, or Flat andElongated Particles in Coarse Aggregate. ASTM D4791. 2010.
31. ASTM. Standard Test Method for Resistance to Degradation of Small-sizeCoarse Aggregate by Abrasion and Impact in the Los Angeles Machine. ASTM C131. 2014.
32. ASTM. Standard Specification for Concrete Aggregates. ASTM C33. 2003.
33. Kong, D.; Xiao, Y.; Wu, S.; Tang, N.; Ling, J.; Wang, F. Comparative evaluation of designing asphalt treated base mixture with composite aggregate types. *Constr. Build. Mater.* **2017**, *156*, 819–827. [CrossRef]

Article

Morphological Discrepancy of Various Basic Oxygen Furnace Steel Slags and Road Performance of Corresponding Asphalt Mixtures

Yong Ye, Shaopeng Wu *, Chao Li *, Dezhi Kong and Benan Shu

State Key Laboratory of Silicate Materials for Architectures, Wuhan University of Technology, Wuhan 430070, China

* Correspondence: wusp@whut.edu.cn (S.W.); lic@whut.edu.cn (C.L.)

Received: 20 June 2019; Accepted: 19 July 2019; Published: 21 July 2019

Abstract: Due to the difference of cooling and treatment processes (rolling method, hot braised method, layer pouring method), basic oxygen furnace (BOF) steel slag can be mainly classified as roller steel slag (RSS), hot braised steel slag (HBSS) and layer pouring steel slag (LPSS). Treatment difference directly results in the performance variations of different BOF steel slag and corresponding asphalt mixtures. The primary purpose of this research was to examine the effects of different cooling and treatment processes on the morphological discrepancy of different BOF steel slag. Also, the road performances of corresponding asphalt mixtures, and mechanism between steel slag performance and road performance were studied. The results show that LPSS owns the largest variability of angular index and texture index, and RSS has the most balanced morphological parameters. The structure of RSS asphalt mixture is advantageous for improving the ability of the asphalt mixture to resist the deformation and enhancing the stability of structure. Higher content of CaO and lower content of SiO_2 make the acid-base reaction of RSS asphalt mixture most intense, which contribute to the best road performance of it.

Keywords: basic oxygen furnace steel slag; morphological discrepancy; road performance; mechanism research

---

1. Introduction

In recent years, rapid development of transportation infrastructure has aroused various obstacles [1], the lack of substantive natural aggregates which possess significant qualities is the most urgent one [2]. Substantive natural aggregates made asphalt mixtures inferior in moisture damage, crack resistance and rutting deformation in road construction [3,4]. The lack of them has made researchers to take other measures to alleviate the crisis, like the use of industrial waste [5].

Steel slag, one type of industrial wastes to replace superior natural aggregates, can be mainly classified as basic oxygen furnace (BOF) steel slag, electric arc furnace (EAF) steel slag, ladle refining (LF) steel slag, and casting residue according to the manufacturing types of steel production [2,6]. Among these, BOF steel slag has been widely utilized in producing numerous types of asphalt mixtures according to former researches [5,7,8]. Kambole reviewed common physical and mechanical characteristics of BOF slag and natural aggregates, and evaluated the influence of main aggregate performance on the properties of asphalt mixtures [9]. Results demonstrated that BOF steel slag owns very good technical and physical properties compared with natural stone aggregates, and proved its feasibility as a valuable resource for road pavements. In addition, mixtures of bitumen with BOF steel slag have yielded better resilient moduli, rutting resistance, bonding, moisture damage resistance and stripping resistance compared to mixes with natural aggregates [9–11]. López-Díaz used the BOF steel slag as coarse aggregate in asphalt concrete, and the results confirmed that it was feasible to use BOF steel slag to partially replace conventional aggregates in road paving [12]. Qazizadeh evaluated the effects of BOF steel slags on the fatigue behavior of asphalt mixes and found that the addition of slags in the mixtures considerably enhanced fatigue life of asphalt mixes [13]. Xue explored the feasibility of using BOF steel slag as aggregates in stone mastic asphalt (SMA) mixtures, results indicated that BOF steel slag improved the high-temperature stability and the low-temperature cracking resistance of SMA mixture when compared with basalt [14]. Shen studied the influence of porous asphalt mixture contained BOF steel slag on the mixture performance and sound absorption characteristic, results demonstrated that BOF steel slag enhanced the skid resistance, moisture susceptibility, rutting resistance, and sound absorption of porous asphalt mixtures [15].

Nevertheless, former researches neglected the influence of cooling and treatment techniques on the properties of BOF steel slag after being produced from furnace, and just placed different BOF steel slag in the same bracket indistinctly. Actually on the basis of different cooling and treatment techniques in China, the BOF steel slag can be mainly classified as roller steel slag (RSS), hot braised steel slag (HBSS) and layer pouring steel slag (LPSS) [16]. To obtain the RSS, firstly the liquid BOF steel slag is dumped into the rotating roller along the chute, steel balls are then added into the roller. By controlling the amount of water, finally steel slag can undergo heating, pulverizing, grinding, and cooling in the roller. The manufacture of HBSS is in accordance with the steps that liquid BOF steel slag is firstly dumped into the pit with the sprinkler and cover, and then smashed by mixing many saturated steams in the confined pit. Comparing to the former two treatments and cooling processes, the production of LPSS is much easier. The liquid BOF steel slag only needs to be pumped onto the slag bed (or inside the slag pit) firstly. Then, a proper amount of water is sprayed and LPSS is weathered in the natural environment.

Above treatment difference directly results in the performance variations of different BOF steel slag, such as the chemical composition, and morphology [17]. The chemical composition and morphology are directly related to the road performance of corresponding asphalt mixtures [18,19]. Therefore, in line with the properties aroused by different cooling and treatment techniques to select suitable application, this paper focused on the utilization of three BOF steel slags in asphalt mixtures, including RSS, HBSS, and LPSS, so that the BOF steel slag could have a broader utilization prospects in transportation infrastructure based on their difference. Firstly, both the surface texture of three BOF steel slags were investigated. Secondly, their chemical composition and morphological discrepancy were also included. Thirdly, three types of asphalt mixtures contained different steel slags were designed, and their road performances were tested. Finally, the mechanism between the BOF steel slag performance and road performance were also analyzed.

2. Materials and Methods

2.1. Raw Materials

The asphalt binder with penetration of 68.3 (0.1 mm at 25 °C), ductility of 151 cm (5 cm/min, 15 °C), and softening point of 47.5 °C, which supplied by Guochuang Co., Ltd., Wuhan, China, was used in this research [20]. Limestone aggregates with different particle size (0 mm–2.36 mm, 2.36 mm–4.75 mm, 4.75 mm–9.5 mm) and filler were supplied by Agoura Stone Processing Factory, Inner Mongolia and their properties were all met the standard specifications. Three types of basic oxygen furnace (BOF) steel slag, including roller steel slag (RSS), hot braised steel slag (HBSS), and layer pouring steel slag (LPSS), were supplied by Baotou Steel with particle size ranging from 9.5 mm to 19 mm, the basic properties of them were shown in Table 1. From the results of Los Angeles abrasion and crushing value, it can be seen that the mechanical properties of steel slag are better than natural aggregates. As these mechanical properties can reflect the hardness, wear resistance and anti-slip properties of aggregates closely related to the performance of asphalt mixtures [9]. Additionally, as shown in Figure 1, it can be seen that the three types of steel slag all have irregular shapes and porous structure. Among them, the surface texture of LPSS is the most abundant. This is because the aging process of LPSS occurs in natural conditions with no high temperature and high-pressure treatments. Therefore, the formation process of LPSS is uneven, slow and the pores are not fully filled with aged materials, which results in a topographical feature of the surface texture.

Table 1. Basic properties of used aggregates. RSS: roller steel slag; LPSS: layer pouring steel slag; HBSS: hot braised steel slag.

Aggregate	Properties [21]					
	Size	Water Absorption (%)	Apparent Specific Gravity	Log Angeles Abrasion (%)	Crushing Value (%)	f-CaO Content (%)
RSS	9.5–16 mm	1.16	3.220	12.9	18.5	0.40
LPSS	9.5–16 mm	1.44	3.260	14.5	13.9	1.17
HBSS	9.5–16 mm	1.75	3.100	13.2	16.8	0.81
Basalt [22]	9.5–16 mm	0.40	2.774	16.8	20.0	N/A
Limestone [23]	9.5–16 mm	1.10	2.650	20.4	N/A	N/A
Granite [22]	9.5–16 mm	0.60	2.723	21.6	21.9	N/A
Limestone	4.75–9.5 mm	1.30	2.726	N/A	9.3	N/A
Limestone	2.36–4.75 mm	1.50	2.822	N/A	9.3	N/A
Requirements [24]	N/A	≤3.00	≥2.500	≤30.0	≤26.0	≤2.00

Figure 1. Appearance of different basic oxygen furnace (BOF) steel slag: (**a**) RSS; (**b**) LPSS; (**c**) HBSS.

2.2. Asphalt Mixture Design

In this research, two types of asphalt mixtures were designed based on standard Marshall Method and Superpave method. Three types of AC-13 asphalt mixtures were mixed with same limestone aggregates, asphalt binder and different BOF steel slag, the gradation curve was shown in Figure 2. Additionally, three types of Superpave-13 asphalt mixtures were mixed the same materials as the AC-13 asphalt mixtures, and the gradation curve was shown in Figure 3.

2.3. Experimental Methods

2.3.1. Characterization

The surface textures of different steel slags were detected by a JSM-5610LV Scan Electronic Microscope (SEM) manufactured by JEOL, Tokyo, Japan [25]. The resolution of SEM in the high-vacuum and low-vacuum mode was 3.0 nm and 4.0 nm separately; the magnification of 1000× was adopted in this research. Chemical composition changes of different steel slags were determined using an AXIOS X-ray fluorescence spectrometer (XRF) manufactured by PANalytical B.V., Amsterdam, The Netherlands [26].

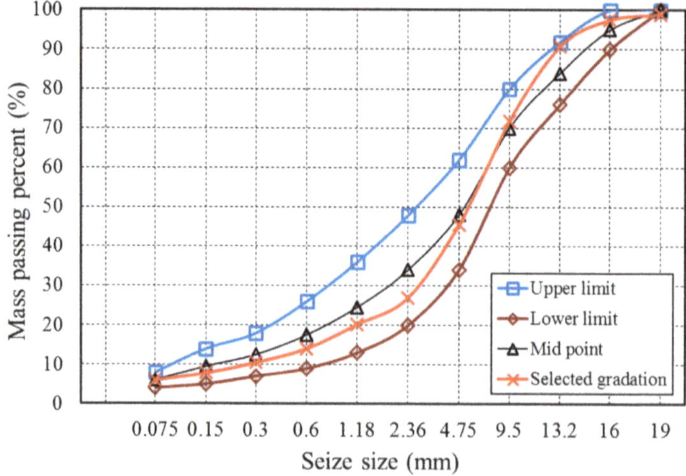

Figure 2. Gradation curve of AC-13 asphalt mixtures.

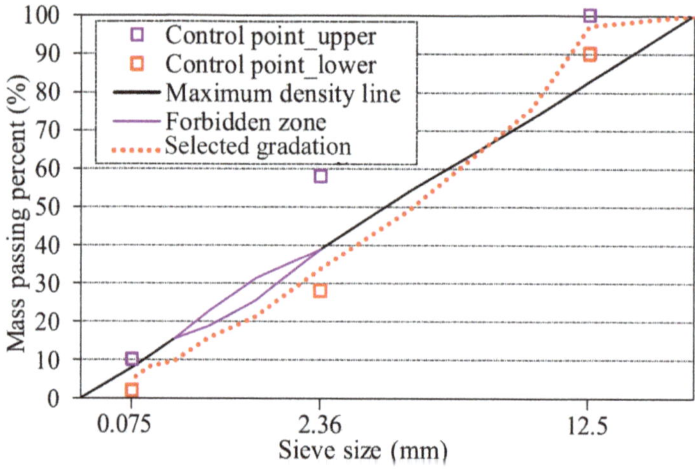

Figure 3. Gradation curve of Superpave-13 asphalt mixtures.

2.3.2. Morphological Discrepancy

AFA2 Aggregate Imaging System (AIMS) (PINE, New York, USA) was utilized to analyze the morphological properties of steel slag in this research, including angularity and texture. AIMS captures images of aggregates at different resolutions through a simple setup that consists of one camera and two different types of lighting schemes [27,28]. The image acquisition setup is configured to capture a typical image of 640 by 480 pixels at these resolutions in order to analyze the aggregates, and export the angularity index and texture index.

2.3.3. Road Performance

In this paper, high temperature stabilities of AC-13 asphalt mixtures were performed by using the wheel tracking device, which was normally used for HMA testing and the dimension of tested slab specimen was 300 mm × 300 mm × 50 mm [29]. A solid rubber wheel with a wheel pressure of 0.7 MPa is used to walk on the specimen at 60 °C. When measuring the deformation period of the test piece, the dynamic stability, which means the number of times it needs to walk for every 1 mm augment, was obtained to evaluate the high temperature stabilities [30].

Based on ASTM D1075 and AASHTO T283, the moisture resistance abilities of three steel slag asphalt mixtures were evaluated by the water immersion Marshall test and the freeze-thaw split test. Both AC-13 asphalt mixtures and Superpave-13 asphalt mixtures were evaluated [31].

To evaluate the low temperature cracking resistance abilities of AC-13 asphalt mixtures, specimen was loaded under a universal material testing machine (IPC, Sydney, Australia) based on the single axes compression test at 0 °C, and the stress–strain curves were obtained automatically by the data acquisition system. The maximum compressive strain energy density was used to evaluate the low temperature cracking resistance abilities and calculated according to the Equation (1) [14].

$$\frac{dW}{dV} = \int_0^{\varepsilon_0} \sigma_{ij} d\epsilon_{ij} \tag{1}$$

where *dW/dV* represents the compressive strain energy density function; σ_{ij} represents the stress fraction; ε_{ij} represents the strain fraction; and ε_0 represents the strain at maximum compressive stress.

3. Results and Discussions

3.1. Characterization

The SEM results of different types of steel slags are shown in Figure 4, it can be indicated that all steel slags have rough surface and numerous pores. This morphological property contributes to improve the bonding performance between aggregate and asphalt binder, resulting in the enhancement of asphalt concrete structural stability. The shape characteristics of RSS and HBSS are similar with little pits and gullies on the surface, which are relatively smoother than the surface texture of LPSS. LPSS not only contains numerous pits, but also has a very uneven surface contained crystal materials with regular shapes. This is because aging process of LPSS is in the natural environment, in which the free calcium oxide slowly forms a calcium carbonate with a crystal structure under the action of water and carbon dioxide. The formed calcium carbonate crystal gradually fills the original pores of LPSS, and due to the uneven distribution of the free material on the surface of LPSS, large amounts of pores still exist on the surface of LPSS. In contrast, under the condition of high temperature and high pressure, RSS and HBSS have been fully aged, the resulting aged products are fully fused with original components of steel slag and formed into a whole, so the surface textures of RSS and HBSS are relatively smoother than LPSS.

Figure 4. SEM results of different steel slags: (**a**) RSS; (**b**) LPSS; (**c**) HBSS.

The XRF results of different steel slags are shown in Figure 5, it can be inferred that the chemical composition of three steel slags mainly contains CaO, SiO_2 and Fe_3O_4, which accounts for more than 80% of the total composition. Compared with the main chemical composition of CaO obtained by the XRF test of traditional aggregate limestone, the steel slag contains a certain amount of silicate and iron-containing compounds due to the iron ore composition and the steel slag treatment process. Limestone is an excellent alkaline rock with good adhesion to asphalt binder. For steel slag, alkalinity which calculated by $M = w(CaO)/[w(SiO_2) + w(P_2O_5)]$, is used to evaluate the acidity and alkalinity of steel slag [32]. Wang Q divided BOF steel slags into three grades based on alkalinity: low alkalinity slag ($M < 1.8$), intermediate alkalinity slag ($1.8 < M < 2.5$) and high alkalinity slag ($M > 2.5$) [33]. Chen indicated that intermediate alkalinity slag and high alkalinity slag had better adhesion to asphalt binder [3]. Therefore, It can be seen from the XRF results, all three steel slags are medium-high alkalinity steel slags ($M > 1.8$) and has good adhesion to asphalt binder. In addition, M value of RSS is 26.84% higher than that of HBSS, and 9.72% higher than that of LPSS, which caused by the higher content of CaO and lower content of SiO_2 in RSS.

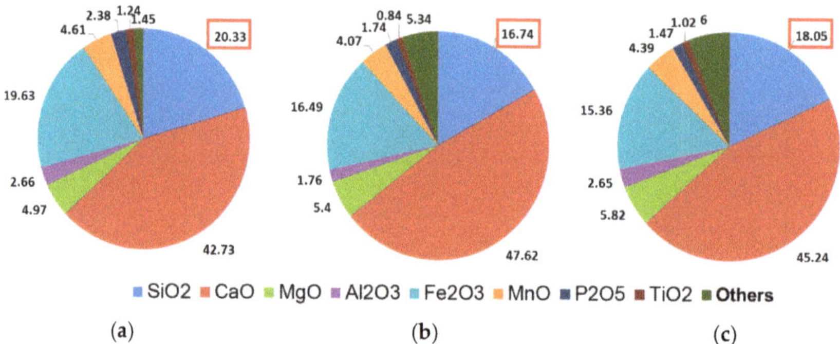

Figure 5. X-ray fluorescence spectrometer (XRF) results of different steel slags (**a**) LPSS; (**b**) RSS; (**c**) HBSS.

3.2. Morphological Discrepancy

Figure 6 shows the schematic diagram of morphological properties for coarse aggregates. Angularity is used to demonstrate variations at the corners, surface texture is used to describe the surface irregularity at a tiny scale to affect the overall shape. Because of different scales with respect to aggregate size, these two morphological properties can be distinguished and used to order them. Every property can be distinguished from other properties widely without necessarily affecting each other [28].

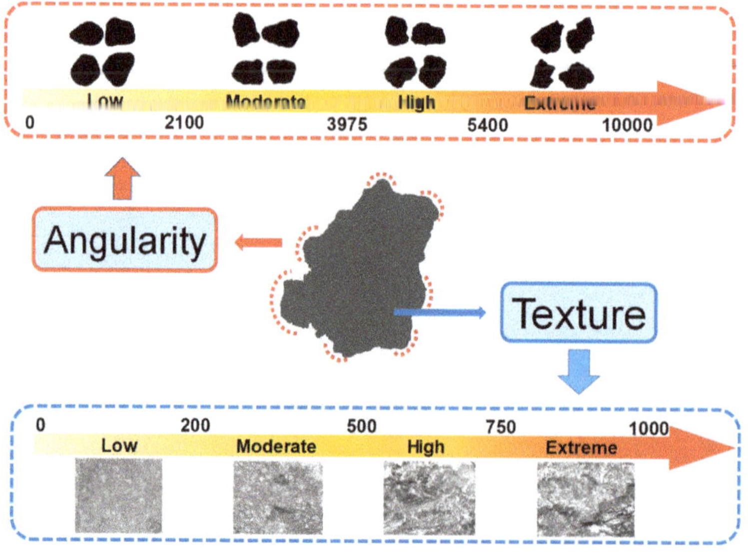

Figure 6. Schematic diagram of morphological properties.

With regard to angularity, the angularity index is calculated by the radius method [34]. Through measuring the difference between particle radius in a certain direction and that of an equivalent ellipse, the angularity index can be calculated according to the following Equation (2) [28]:

$$Angularity\ Index = \sum_{\theta=0}^{355} \frac{|R_\theta - R_{EE\theta}|}{R_{EE\theta}} \tag{2}$$

In the equation, R_θ represents the radius of the particle at an angle of θ, and $R_{EE\theta}$ represents the radius of the equivalent ellipse at an angle of θ. The equivalent ellipse has the same aspect ratio of the particle but has no angularity (smooth with no sharp corners). Normalization of the aspect ratio can minimize the effect of form on the angularity index [34]. The values of angularity index distribute from 0 to 10,000, and can also be divided into four levels: low level is from 0 to 2100 and particle is rounded, moderate level is from 2100 to 3975 and particle is sub-rounded, high level is from 3975 to 5400 and particle is sub-angular, extreme level is from 5400 to 10,000 and particle is angular. Additionally, the particle can be recognized to be more rounded if the angularity index is closer to 0.

Table 2 and Figure 7 summarize the angularity test results of different steel slags. It can be inferred that the angularity indexes of three steel slags are mainly distributed in the two levels of sub-rounded and sub-angular, and the distributions in sub-rounded account for the most. The average value and variance of the angularity of RSS are the smallest of three steel slags, while the average value and variance of LPSS are the largest. The results show that RSS owns the lowest angularity but the best control of variability. In contrast, LPSS owns the worst control of variability but the largest angularity.

Table 2. Aggregate Imaging System (AIMS) angularity index of tested steel slag particles.

Sample	Angularity Index					
	Average Value	Standard Deviation Value	Low ($\leq$2100)	Moderate (2100–3975)	High (3975–5400)	Extreme (5400–10,000)
RSS	2946.5	609.6	0%	95%	5%	0%
HBSS	3292.0	862.2	5%	75%	15%	5%
LPSS	3725.8	1226.1	0	60%	30%	10%

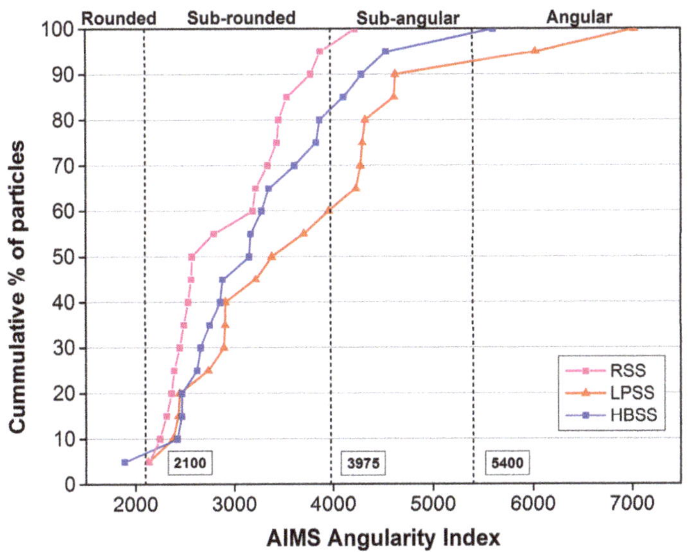

Figure 7. AIMS angularity index curve of tested steel slag particles.

Texture index is applied to characterize the texture distinction by wavelet analysis [35], the texture index at any given decomposition level is the arithmetic mean of the squared values of the detail coefficients of different level and can be calculated in the following Equation (3) [8]:

$$Texture\ Index = \frac{1}{3N} \sum_{i=1}^{3} \sum_{j=1}^{N} (D_{i,j}(x,y))^2 \tag{3}$$

where N denotes the total number of coefficients in the detailed image of the aggregate; i takes a value of 1, 2, or 3, corresponding to three detailed images of the texture; J is the index of the wavelet coefficients, D means the transformed domain and (x, y) is the position of the coefficients in the texture scan area. The texture index can be divided into four levels: 0 to 200 accounts for the low grade, 200 to 500 accounts for the medium grade, 500 to 750 accounts for the high grade, and 750 to 1000 accounts for very high grade. Higher value of texture value represents the richer surface texture of aggregate.

Table 3 and Figure 8 summarize the results of different steel slag texture indexes. Results show that the texture indexes of three steel slags are mainly distributed in the medium grade. Among the three steel slags, only the HBSS is not distributed in the low grade of texture index, only the LPSS is distributed in the high grade of texture index. As with the results of angular index, LPSS owns the highest texture index, and its average texture index is 53.24% and 19.06% higher than that of RSS and HBSS, respectively. However, the texture index value of LPSS is distributed in all four grades, its standard deviation of texture index is 94.83% and 78.58% higher than that of RSS and HBSS, respectively, showing poor texture variability.

Table 3. AIMS texture index values of tested steel slag particles.

Sample	Average Value	Standard Deviation	Texture Index			
			Low (≤200)	Moderate (200–500)	High (500–750)	Extreme (750–1000)
RSS	311.4	81.3	15%	85%	0%	0%
HBSS	400.8	88.7	0%	75%	25%	0%
LPSS	477.2	158.4	5%	60%	25%	10%

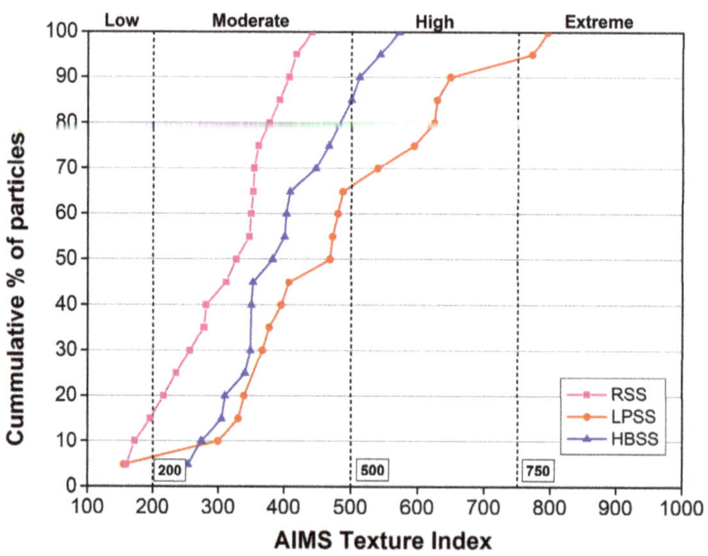

Figure 8. AIMS texture index curve of tested steel slag particles.

Both the results of angular index and texture index show that LPSS has higher geometrical characteristics, this is mainly because that the cooling and treatment of RSS and HBSS experience both aging and stable processes in the initial formation of steel slag. Most free components have been digested into calcium carbonate, the composition and material properties of the steel slag have been relatively stable. However, cooling and treatment process of LPSS cannot guarantee that the aging of free components was finished completely during the slag splashing. Therefore, during the stacking period of LPSS, the free components continuously react with the moisture and carbon dioxide in the air, and the uneven distribution of aging products lead to the rugged surface texture of the aggregate, which increases the angular index and texture index of LPSS.

3.3. Road Performance

3.3.1. High Temperature Stability

The high temperature stability of asphalt pavement represents the ability of the asphalt mixture to resist permanent deformation after being subjected to vehicle load and repeated rolling under high temperature conditions. The rutting test is currently the mainstream method for evaluating the high temperature stability. If the high temperature stability of asphalt mixture does not meet the requirements of the standard design, the road surface will be damaged due to insufficient stability and low load loading rate at high temperature. The research shows that the dynamic stability of the rutting test has a good correlation with the rutting depth of the asphalt mixture [36]. By comparing the dynamic stability, the high temperature stability of the asphalt mixture can be evaluated.

Figure 9 shows the results of high temperature stability of different BOF steel slag asphalt mixture. It shows that the dynamic stability values of three steel slag asphalt mixtures are all higher than 2800 times/mm, which meets the requirements of the specification of China properly. Excellent material characteristics of BOF steel slag make its dynamic stability far greater than the specification requirements. Among them, the RSS asphalt mixture has the highest dynamic stability, which is 25.48% and 14.44% higher than that of LPSS asphalt mixture and HBSS asphalt mixture, respectively. The high temperature stability of LPSS is the worst, representing that although the LPSS has highest geometric parameters, it also has larger geometric variability and smaller crushing value, also the residual free material in LPSS affects the high temperature stability of asphalt mixture.

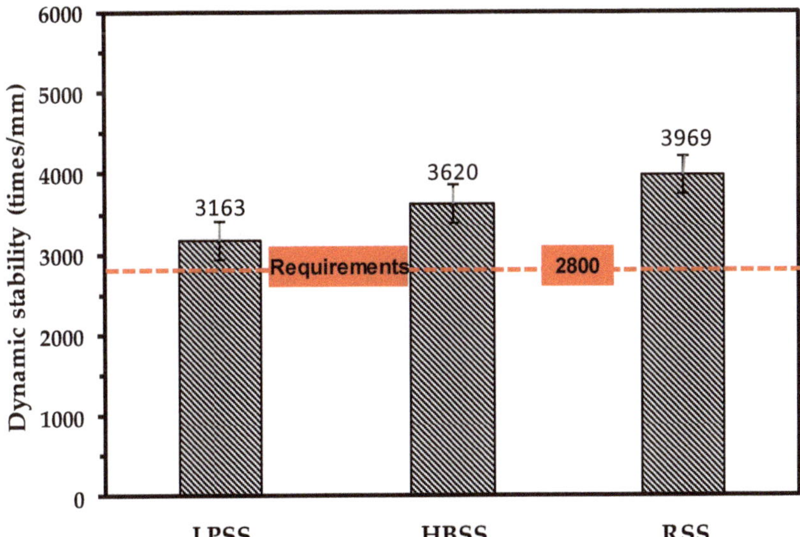

Figure 9. High-temperature stability test results of different steel slag asphalt mixtures.

3.3.2. Moisture Resistance Ability

Moisture resistance is recognized as the ability to resist the peeling of bitumen, which adhering to the aggregate surface in asphalt mixture after being corroded by water. In this paper, according to ASTM D1075 and AASHTO T283, the moisture resistance abilities of steel slag asphalt mixtures were evaluated by the water immersion Marshall test and the freeze-thaw split test. The moisture resistance test results of Superpave-13 steel slag asphalt mixtures are shown in Figures 10 and 11, it can be seen that the RMS values of three steel slag asphalt mixtures are all greater than 80%, and the TSR values of them were more than 75%, which are fully meeting the requirements of the specification. The RMS value of RSS asphalt mixture is 4.97% and 1.22% higher than that of LPSS asphalt mixture and HBSS asphalt mixture. In addition, the TSR value of RSS asphalt mixture is 5.51% and 1.68% higher than that of LPSS asphalt mixture and HBSS asphalt mixture. The moisture resistance test results of AC-13 steel slag asphalt mixtures are shown in Table 4. In consistent with the results of the Superpave-13 steel slag asphalt mixtures, the RMS values of three AC-13 steel slag asphalt mixture are all more than 80%, and the TSR values are greater than 75%. LPSS asphalt mixture still behaves the worst moisture resistance. Results show that three steel slag asphalt mixtures have good water stability, the main reason is that the steel slag shows coarse, porous surface texture that enhances the bonding performance between BOF steel slag and bitumen. In addition, the alkali metal cations such as Ca^{2+}, Mg^{2+}, Fe^{2+}, Al^{2+} and Mn^{2+} contained in the steel slag can react with the bituminous acid, and the formed materials further increases the bonding property of the steel slag asphalt mixtures. In contrast to LPSS asphalt mixture and HBSS asphalt mixture, RSS asphalt mixture behaves the best moisture resistance, and the reason is that the more advanced treatment methods make the RSS more uniform and stable than the other two steel slags during aging process, and the produced silicate compounds (like $CaCO_3$) improved the ability of asphalt mixture to resist water erosion.

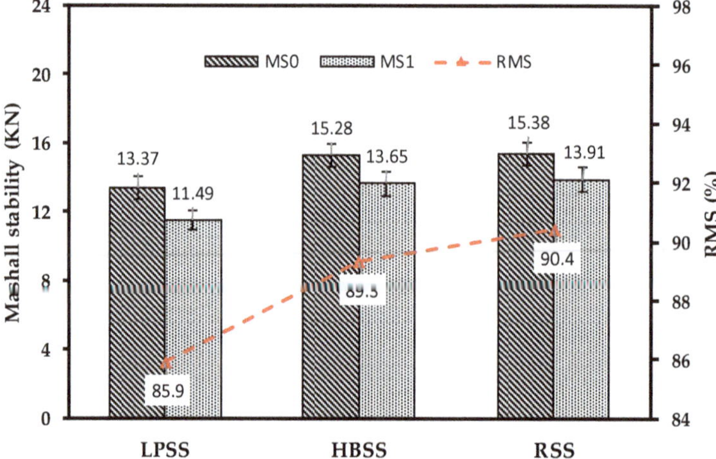

Figure 10. Water immersion Marshall test of Superpave-13 steel slag asphalt mixtures.

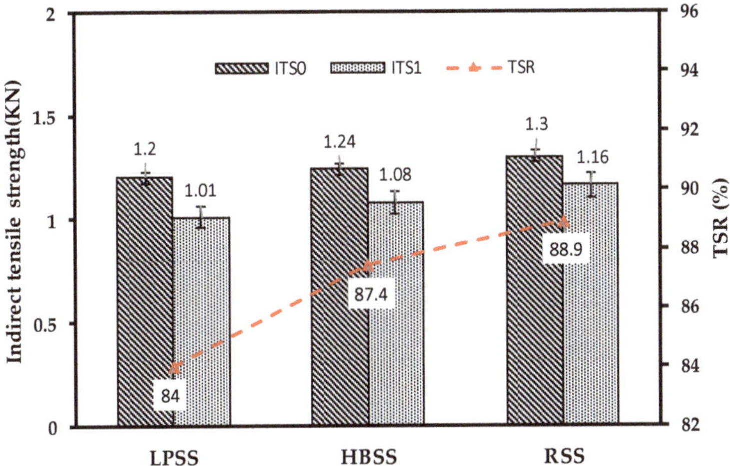

Figure 11. Freeze–thaw split test of Superpave-13 steel slag asphalt mixtures.

Table 4. Moisture resistance results of AC-13 steel slag asphalt mixtures.

AC-13 Asphalt Mixture	RMS Test			ITS Test		
	MS_0 (KN)	MS_1 (KN)	RMS (%)	ITS_0 (MPa)	ITS_1 (MPa)	TSR (%)
RSS	16.92	15.59	92.1	1.65	1.49	90.3
HBSS	15.82	14.37	90.8	1.53	1.35	88.1
LPSS	14.10	12.44	88.2	1.49	1.28	85.6

3.3.3. Low Temperature Cracking Resistance Ability

Low temperature crack resistance is the ability to characterize the asphalt mixture to resist temperature-shrinkage cracks at low temperatures. According to the measured relationship between compressive stress and strain of asphalt mixture shown in Equation (4) [14], it was found that the strain-stress relationship of asphalt mixture was quadratic parabola.

$$\sigma = A\varepsilon^2 + B\varepsilon + C \tag{4}$$

where σ is the stress, MPa: ε is the strain; A, B, and C are constants, which are related to the type of material. So, the compressive strain energy density function can be transformed into the Equation (5) [17]:

$$\frac{dW}{dV} = \int_0^{\varepsilon_0} \sigma_{ij} d\epsilon_{ij} = \frac{A}{3}\varepsilon_0^3 + B\varepsilon_0^2 + C\varepsilon_0 \tag{5}$$

where ε_0 is the strain at the peak point of compression stress–strain curve, and the stress–strain curves were obtained automatically by the data acquisition system. As shown in Figure 12, it can be clearly seen that the RSS asphalt mixture owns the largest ε_0 while LPSS asphalt mixture has the lowest ε_0. After substituting the ε_0 values into the Equation (5), the critical value dW/dV is obtained in Table 5. Observations reveal that the critical value of dW/dV is 42.7 KJ/m^3 for LPSS asphalt mixture, while RSS asphalt mixture owns the highest critical value of dW/dV and about 41.83% higher than that of LPSS asphalt mixture. Results confirm that LPSS asphalt mixture has the best ability to resist low temperature deformation.

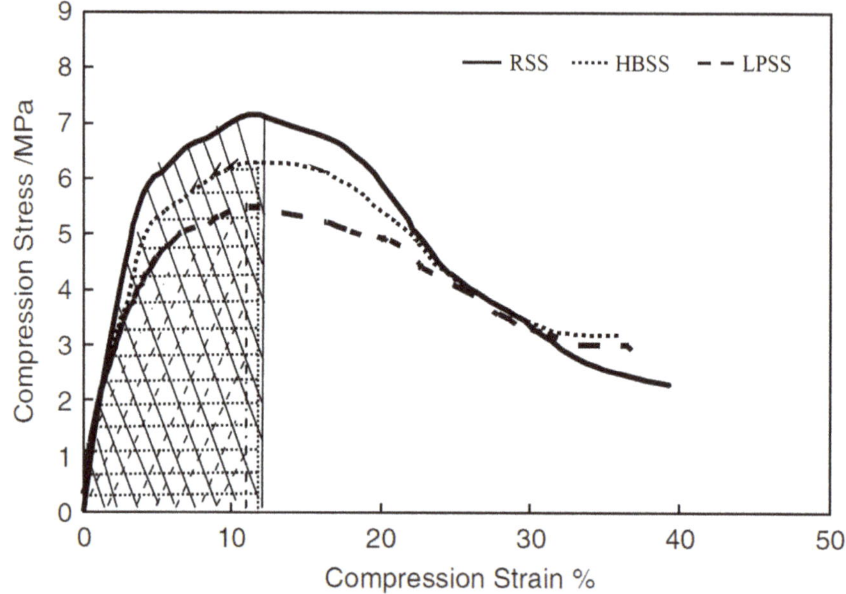

Figure 12. Compression stress–strain curves of AC-13 steel slag asphalt mixtures.

Table 5. ε_0 and critical value of dW/dV at peak poiont.

Mixture Type	Aggregate Type	dW/dV–Strain Function	ε_0	Critical Value of dW/dV (KJ/m^3)
AC-13 Asphalt mixture	LPSS	$dW/dV = -3 \times 10 - 5\varepsilon^5/5 + 0.0015\varepsilon^4/4 - 0.0882\varepsilon^3/3 + 1.128\varepsilon^2/2 + 0.5047\varepsilon$	12.0	42.7
	HBSS	$dW/dV = -3 \times 10 - 5\varepsilon^5/5 + 0.0019\varepsilon^4/4 - 0.0904\varepsilon^3/3 + 1.39\varepsilon^2/2 + 0.6116\varepsilon$	12.9	69.8
	RSS	$dW/dV = -3 \times 10 - 5\varepsilon^5/5 + 0.002\varepsilon^4/4 - 0.0959\varepsilon^3/3 + 1.44\varepsilon^2/2 + 0.8109\varepsilon$	13.3	73.4

3.4. Mechanism between Steel Slag Performance and Road Performance

The road performance of steel slag asphalt mixture is closely related to the performance of steel slag. The reasons for the difference in road performance of asphalt mixtures prepared by different steel slag mainly include two aspects: the difference of the skeleton structure after the formation of steel slag asphalt mixture, and the difference of the adhesion performance between bitumen and steel slag. According to the results of the morphological discrepancy, Figure 13 shows the structural sketch of different steel slag asphalt mixtures. In morphological discrepancy section, RSS owns the best control of variability, in contrast, LPSS owns the worst control of variability. Thus, compared with HBSS and LPSS, RSS has a lower length-to-particle ratio and a smaller number of needle-like particles. At the same time, RSS has good angularity, while the shape variability is small, and the size between the particles is relatively balanced. Therefore, after the formation of asphalt mixtures, the RSS particles have more skeleton support points and are more likely to form a tight skeleton structure. The structure of RSS asphalt mixture is advantageous for improving the ability of the asphalt mixture to resist the deformation and enhancing the stability of structure.

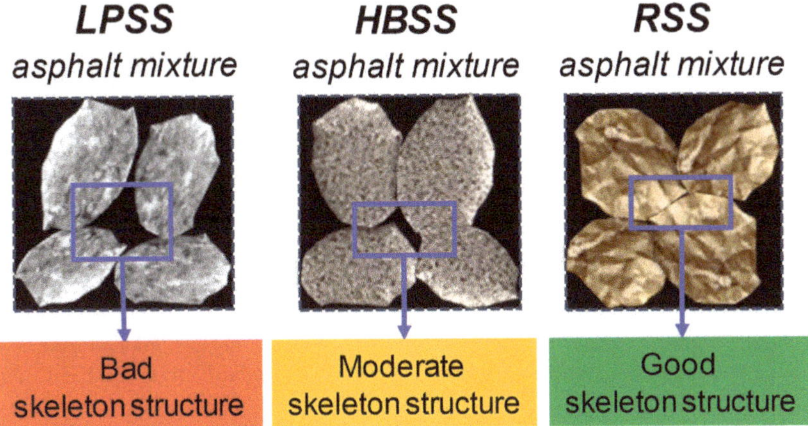

Figure 13. Structural sketch of different steel slag asphalt mixtures.

Figure 14 reveals the schematic diagram of adhesion between steel slag and bitumen. According to the XRF results, the BOF steel slag contains alkaline components like dicalcium silicate (C_2S), tricalcium silicate (C_3S), calcium hydroxide ($Ca(OH)_2$), and free calcium oxide (f-CaO). The bitumen contains some acidic groups like R-COOH, R-SO$_X$H, and R-COSH, which can react with the alkaline components in the steel slag to form an acid-base neutralization reaction, and the resulting product enhances the adhesion between the bitumen and steel slag. In addition, *M* value of RSS is 26.84% higher than that of HBSS, and 9.72% higher than that of LPSS, which caused by the higher content of CaO and lower content of SiO_2 in RSS. This make the acid-base reaction of RSS asphalt mixture most intense, which contribute to the best road performance of it [5].

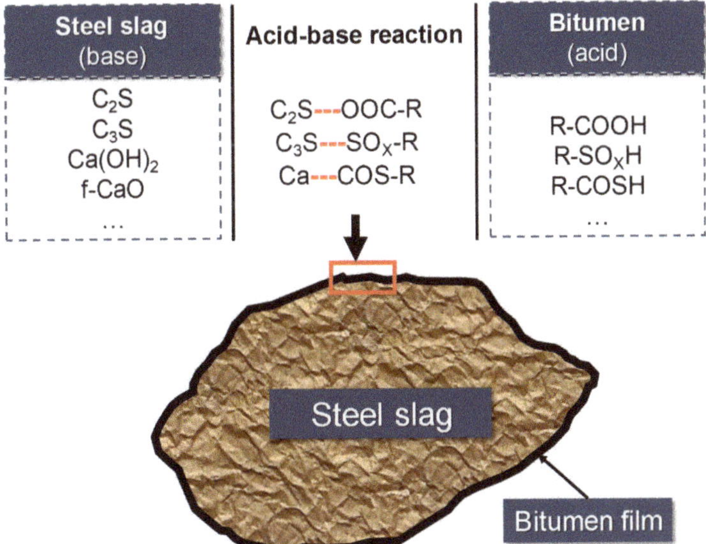

Figure 14. Schematic diagram of adhesion between steel slag and bitumen.

4. Conclusions

The primary purpose of this research was to examine the effects of different cooling and treatment processes on the morphological discrepancy of different BOF steel slag, including roller steel slag (RSS), hot braised steel slag (HBSS), and layer pouring steel slag (LPSS). Also, the road performances of corresponding asphalt mixtures were studied. On account of the results above, the following conclusions can be obtained.

RSS owns the lowest angularity but the best control of variability. In contrast, LPSS owns the worst control of variability but the highest angularity. Average texture index of LPSS is 53.24% and 19.06% higher than that of RSS and HBSS, respectively.

The RSS asphalt mixture has the highest dynamic stability, which is 25.48% and 14.44% higher than that of LPSS asphalt mixture and HBSS asphalt mixtures, respectively. The high temperature stability of LPSS asphalt mixture is the worst, representing that although the LPSS has higher geometric properties, it also has larger geometric variability and smaller crushing value, also the residual free material in LPSS affects the high temperature stability of asphalt mixture.

RSS asphalt mixture and LPSS asphalt mixture still behave the best and worst moisture resistance. Surface texture and alkali metal cations enhance the bonding performance between steel slag and bitumen. Observations reveal that the critical value of dW/dV is 42.7 KJ/m^3 for LPSS asphalt mixture, while RSS asphalt mixture owns the highest critical value of dW/dV and about 41.83% higher than that of LPSS asphalt mixture.

The structure of RSS asphalt mixture is advantageous for improving the ability of the asphalt mixture to resist the deformation and enhancing the stability of structure. Higher content of CaO and lower content of SiO$_2$ make the acid-base reaction of RSS asphalt mixture most intense, which contribute to the best road performance of it.

Author Contributions: D.K. and C.L. conceived and designed the experiments. B.S. and S.W. performed the experiments. C.L. analyzed the data. Y.Y. contributed reagents/materials/analysis tools. Y.Y. and C.L. wrote the paper.

Funding: This research was funded by National Key R&D Program of China (No. 2018YFB1600200), National Natural Science Foundation of China (71961137010), the Fundamental Research Funds for the Central Universities (No. 195201013), Fundamental Research Funds for the Central Universities (WUT:182459009) and State Key Laboratory of Silicate Materials for Architectures (Wuhan University of Technology) (No. SYSJJ2019-19).

Acknowledgments: The authors gratefully acknowledge Wuhan University of Technology and the shiyanjia lab (www.shiyanjia.com), for their materials and experimental instruments supports.

Conflicts of Interest: The authors declare no conflict of interest.

References

1. Cui, P.Q.; Wu, S.P.; Xiao, Y.; Wan, M.; Cui, P.D. Inhibiting effect of layered double hydroxides on the emissions of volatile organic compounds from bituminous materials. *J. Clean. Prod.* **2015**, *108*, 987–991. [CrossRef]
2. Skaf, M.; Manso, J.M.; Aragón, Á.; Fuente-Alonso, J.A.; Ortega-López, V. EAF slag in asphalt mixes: A brief review of its possible re-use. *Resour. Conserv. Recycl.* **2017**, *120*, 176–185. [CrossRef]
3. Chen, Z.W.; Wu, S.P.; Wen, J.; Zhao, M.L.; Yi, M.W.; Wan, J.M. Utilization of gneiss coarse aggregate and steel slag fine aggregate in asphalt mixture. *Constr. Build. Mater.* **2015**, *93*, 911–918. [CrossRef]
4. Shu, B.A.; Zhang, L.; Wu, S.P.; Dong, L.J.; Liu, Q.T.; Wang, Q. Synthesis and characterization of compartmented Ca-alginate/silica self-healing fibers containing bituminous rejuvenator. *Constr. Build. Mater.* **2018**, *190*, 623–631. [CrossRef]
5. Zongwu, C.; Yuyong, J.; Shaopeng, W.; Fubin, T. Moisture-induced damage resistance of asphalt mixture entirely composed of gneiss and steel slag. *Constr. Build. Mater.* **2018**, *177*, 332–341.
6. Yi, H.; Xu, G.; Cheng, H.; Wang, J.; Wan, Y.; Chen, H. An overview of utilization of steel slag. *Procedia Environ. Sci.* **2012**, *16*, 791–801. [CrossRef]
7. Chao, L.; Chen, Z.; Wu, S.; Bo, L.; Xie, J.; Yue, X. Effects of steel slag fillers on the rheological properties of asphalt mastic. *Constr. Build. Mater.* **2017**, *145*, 383–391.

8. Li, C.; Wu, S.L.; Chen, Z.; Tao, G.; Xiao, Y. Enhanced heat release and self-healing properties of steel slag filler based asphalt materials under microwave irradiation. *Constr. Build. Mater.* **2018**, *193*, 32–41. [CrossRef]

9. Kambole, C.; Paige-Green, P.; Kupolati, W.K.; Ndambuki, J.M.; Adeboje, A.O. Basic oxygen furnace slag for road pavements: A review of material characteristics and performance for effective utilisation in southern Africa. *Constr. Build. Mater.* **2017**, *148*, 618–631. [CrossRef]

10. Xie, J.; Wu, S.; Lin, J.; Cai, J.; Chen, Z.; Wei, W. Recycling of basic oxygen furnace slag in asphalt mixture: Material characterization & moisture damage investigation. *Constr. Build. Mater.* **2012**, *36*, 467–474.

11. Ko, M.S.; Chen, Y.L.; Jiang, J.H. Accelerated carbonation of basic oxygen furnace slag and the effects on its mechanical properties. *Constr. Build. Mater.* **2015**, *98*, 286–293. [CrossRef]

12. López-Díaz, A.; Ochoa-Díaz, R.; Grimaldo-León, G.E. Use of BOF slag and blast furnace dust in asphalt concrete: an alternative for the construction of pavements. *Dyna* **2018**, *85*, 24–30.

13. Qazizadeh, M.J.; Farhad, H.; Kavussi, A.; Sadeghi, A. Evaluating the fatigue behavior of asphalt mixtures containing electric arc furnace and basic oxygen furnace slags using surface free energy estimation. *J. Clean. Prod.* **2018**, *188*, 355–361. [CrossRef]

14. Wu, S.P.; Xue, Y.J.; Ye, Q.S.; Chen, Y.C. Utilization of steel slag as aggregates for stone mastic asphalt (SMA) mixtures. *Build. Environ.* **2007**, *42*, 2580–2585. [CrossRef]

15. Shen, D.H.; Wu, C.M.; Du, J.C. Laboratory investigation of basic oxygen furnace slag for substitution of aggregate in porous asphalt mixture. *Constr. Build. Mater.* **2009**, *23*, 453–461. [CrossRef]

16. Zhang, Z.H.; Liao, J.L.; Ju, J.T.; Dang, Y.J. Treatment process and utilization technology of steel slag in China and Abroad. *J. Iron. Steel Res.* **2013**, *25*, 1–4. [CrossRef]

17. Guo, H. Selection of steel slag processing technology. *Energy Metall. Ind.* **2011**, *25*, 14–20.

18. Juckes, L.M. The volume stability of modern steelmaking slags. *Miner. Process. Extr. Metall. Trans.* **2003**, *112*, 177–197. [CrossRef]

19. Zhou, Z.; Ming, H. Analysis of affecting factors on thermal stability of asphalt mixtures. *J. Highw. Transp. Res. Dev.* **2004**, *6*, 15–21.

20. Xue, Y.J.; Wang, C.; Hu, Z.H.; Zhou, Y.; Liu, G.; Hou, H.B. Thermal treatment on sewage sludge by electromagnetic induction heating: Methodology and drying characterization. *Waste Manag.* **2018**, *78*, 917–928. [CrossRef]

21. Li, C.; Wu, S.; Chen, Z.; Shu, B.; Li, Y.; Xiao, Y.; Liu, Q. Synthesis of Fe3O4-decorated Mg-Al layered double hydroxides magnetic nanosheets to improve anti-ultraviolet aging and microwave absorption properties used in asphalt materials. *Constr. Build. Mater.* **2019**, *220*, 320–328. [CrossRef]

22. Chen, Z.W.; Xie, J.; Xiao, Y.; Chen, J.Y.; Wu, S.P. Characteristics of bonding behavior between basic oxygen furnace slag and asphalt binder. *Constr. Build. Mater.* **2014**, *64*, 60–66. [CrossRef]

23. Ziaee, S.A.; Kavussi, A.; Jalili Qazizadeh, M.; Mohammadzadeh Moghadam, A. Avaluation of long term ageing of asphalt mixtures containing eaf and bof steel slags. *Int. J. Transp. Eng.* **2015**, *5*, 65–72.

24. Ameri, M.; Hesami, S.; Goli, H. Laboratory evaluation of warm mix asphalt mixtures containing electric arc furnace (EAF) steel slag. *Constr. Build. Mater.* **2013**, *49*, 611–617. [CrossRef]

25. Shu, B.; Wu, S.; Dong, L.; Li, C.; Kong, D.; Yang, X. Synthesis and properties of microwave and crack responsive fibers encapsulating rejuvenator for bitumen self-healing. *Mater. Res. Express* **2019**, *6*, 30–35. [CrossRef]

26. Shu, B.; Wu, S.; Dong, L.; Norambuena-Contreras, J.; Yang, X.; Li, C. Microfluidic synthesis of polymeric fibers containing rejuvenating agent for asphalt self-healing. *Constr. Build. Mater.* **2019**, *219*, 176–183. [CrossRef]

27. Prowell, B.D.; Zhang, J.; Brown, E.R. Aggregate properties and the performance of superpave-designed Hot Mix Asphalt. *Nchrp Report.* **2005**, *4*, 93–101.

28. Masad, E.A. Aggregate Imaging System (AIMS): Basics and Applications. *Blending* **2005**, *23*, 453–461.

29. Ma, T.; Wang, H.; He, L.; Zhao, Y.L.; Huang, X.M.; Chen, J. Property characterization of asphalt binders and mixtures modified by different crumb rubbers. *J. Mater. Civ. Eng.* **2017**, *29*, 50–55. [CrossRef]

30. Ma, T.; Zhang, D.Y.; Zhang, Y.; Wang, S.Q.; Huang, X.M. Simulation of wheel tracking test for asphalt mixture using discrete element modelling. *Road Mater. Pavement Des.* **2018**, *19*, 367–384. [CrossRef]

31. Pan, P.; Wu, S.P.; Xiao, Y.; Liu, G. A review on hydronic asphalt pavement for energy harvesting and snow melting. *Renew. Sustain. Energy Rev.* **2015**, *48*, 624–634. [CrossRef]

32. Mason, B. The constitution of some basic open-hearth slags. *J. Iron Steel Inst.* **1994**, *12*, 69–80.

33. Wang, Q.; Yan, P.; Han, S. The influence of steel slag on the hydration of cement during the hydration process of complex binder. *Sci. China Technol. Sci.* **2011**, *54*, 388–394. [CrossRef]

34. Masad, E.; Olcott, D.; White, T.; Tashman, L. Correlation of fine aggregate imaging shape indices with asphalt mixture performance. *Transp. Res. Rec.* **2001**, *1757*, 148–156. [CrossRef]

35. Mallat, S.G. A theory for multiresolution signal decomposition: the wavelet representation. *IEEE Trans. Pattern Anal. Mach. Intell.* **1989**, *158*, 231–242. [CrossRef]

36. Chen, Z.W.; Wu, S.P.; Xiao, Y.; Zeng, W.B.; Yi, M.W.; Wan, J.M. Effect of hydration and silicone resin on Basic Oxygen Furnace slag and its asphalt mixture. *J. Clean. Prod.* **2016**, *112*, 392–400. [CrossRef]

 *materials*

Article

Interlaminar Bonding Properties on Cement Concrete Deck and Phosphorous Slag Asphalt Pavement

Guoping Qian [1], Shunjun Li [2], Huanan Yu [1],* and Xiangbing Gong [1]

[1] National Engineering Laboratory for Highway Maintenance Technology, School of Traffic and Transportation Engineering, Changsha University of Science & Technology, Changsha 410114, China; guopingqian@sina.com (G.Q.); xbgong@csust.edu.cn (X.G.)

[2] School of Traffic and Transportation Engineering, Changsha University of Science & Technology, Changsha 410114, China; shunjun.li@stu.csust.edu.cn

* Correspondence: huanan.yu@csust.edu.cn; Tel.: +86-13714470593

Received: 11 April 2019; Accepted: 29 April 2019; Published: 1 May 2019

Abstract: The slippage damage caused by weak interlaminar bonding between cement concrete deck and asphalt surface is a serious issue for bridge pavement. In order to evaluate the interlaminar bonding of cement concrete bridge deck and phosphorous slag (PS) asphalt pavement, the shear resistance properties of the bonding layer structure were studied through direct shear tests. The impact of PS as a substitute of asphalt mixture aggregate, interface characteristics, normal pressure, waterproof and cohesive layer types, temperature and shear rate on the interlaminar bonding properties were analyzed. The test results indicated that the interlaminar bonding of bridge deck pavement is improved after asphalt mixture fine aggregate was substituted with PS and PS powder, and the result indicated that the shear strength of grooved and aggregate-exposed interfaces is significantly higher than untreated interface, the PS micro-powder or anti-stripping agent can also improve the adhesion between layers when mixed into SBS asphalt. This study provided important theoretical and practical guidance for improving the shear stability of bridge deck pavement.

Keywords: asphalt mixture; cement concrete deck; phosphorous slag; interlaminar bonding; shear strength

---

1. Introduction

The early damage, especially the slippage and upheaval of asphalt pavement over cement concrete bridge deck is a serious issue for bridge operations. When the bridge is under an extensive heavy vehicle load, large shear stress is generated inside the bridge deck pavement which caused uncertain shear failure surface; or serious shear damage and diseases are caused because the horizontal shear resistance between the pavement layer and bridge deck is weak [1]. Among the factors affecting the interlaminar bonding of bridge deck pavement structure, the mixture aggregate characteristic plays a significant role in that [2]. However, the lack of high-quality aggregates is very common in many areas, the mining of raw aggregate have cost many environmental problems, and the problem has become more and more serious with the acceleration of the infrastructure construction process.

Phosphorous slag (PS) is an industrial by-product produced in the process of phosphorus ore in a high-temperature environment. At present, the global discharge of wasted PS is about 12 to 15 million ton each year. The accumulation of a large amount of PS not only occupies the land but also seriously pollutes the environment [3]. Therefore, the effective utilization of PS in the asphalt pavement would beneficial the highway construction and the environment from all prospects [4].

Many pieces of research have been carried out on the engineering application of PS. Allahverdi, et al. [5] found that PS powder can significantly improve the quality of cement after mechanical activation and chemical catalysis. Zhao, et al. [6] demonstrated that PS with appropriate cement replacement dosage and specific area could improve the mechanical properties of sleeper concrete under steam curing. Xia, et al. [7] studied the crack resistance of PS concrete from the aspects of physical properties, hydration heat, shrinkage and creep and found that the crack resistance of PS concrete is approximate to, even to some extent better than, that of fly ash concrete. He, et al. [8] found that with the increase of granularity of PS additive, there are significant increases in the uniformity of particle sizes, slurry pH, and activity index, and the effects on cement paste are significantly mitigated. Jin [9] studied the influence of superfine PS as mineral filler on the long-term performance and durability of concrete. The results show that PS can improve the pore structure of concrete which is beneficial to the development of compressive strength and splitting tensile strength of concrete at a late age. Hamideh, et al. [10] predicted and optimized the compressive strength of PS cement at different ages (3, 7 and 28 days) based on the response surface method.

Qian, et al. [11] found that PS can be potentially used as an antistripping additive in asphalt mixture because the pH of the slag was alkaline and it was also hydrophobic and stable at high temperature. The viscoelasticity test also showed that PS filler improved the stiffness of asphalt. The mixture performance tests also indicated that PS filler significantly increases the resistance of HMA to rutting and moisture damage. Qian, et al. [12] further analyzed the influence of surface modified PS powder as a modifier on the mechanism of asphalt and asphalt mixture and found that TM-P modified PS powder can enhance the compatibility with asphalt, which improves the antiaging, rutting resistance and water damage resistance of asphalt mixture. Sheng, et al. [13] studied the effect of PS powder as mineral filler on the rheological properties of asphalt binder and the properties of asphalt mixture and found that it increased the binder viscosity resulting in enhanced mixture rutting resistance. In order to study the rutting and fatigue damage of asphalt pavement, Bazzaz, et al. [14,15] proposed a straightforward procedure to characterize the nonlinear viscoelastic response of asphalt concrete materials.

In addition, there are many pieces of research that have been carried out on asphalt pavement structure of cement concrete bridge deck which provided meaningful guidance on this research. Wang, et al. [16] have studied the interface shear characteristics between the asphalt pavement structure and the concrete bridge deck pavement and found that the shear strength of SBS modified asphalt pavement over concrete bridge deck is slightly greater than that of crumb rubber modified asphalt pavement. Li, et al. [17] studied the interlaminar failure modes and mechanisms of rubber powder modified asphalt, SBS modified asphalt and epoxy resin adhesive as waterproof bonding materials and found that the shear strength is greatly affected by the thickness of the waterproof adhesive layer. Liu, et al. [18] studied the bonding performance of waterproof bonding layer between the concrete bridge deck and asphalt mixture pavement by a lab test, field temperature monitoring and finite element method (FEM). The test results show that the safety factor (strength/stress) decreases significantly with increasing environmental temperatures. Sheng, et al. [19] established a simplified formula for calculating the extreme temperature (maximum and minimum temperature) stress of bridge deck pavement structures. The results have shown a strong linear correlation between the bridge deck pavement maximum principal stress and the elastic modulus.

He, et al. [20] proposed typical structural types of cement concrete bridge deck pavement based on waterproof cohesive layer material test and bridge deck pavement composite structure test. Jia [21] proposed using the interlayer shear test as a method to study the shear performance of the interlayer structure layer of bridge deck pavement. And have proposed design method and design standard of asphalt mixture based on that. Ren, et al. [22] analyzed the influence of chip-sprinkling interlayer treatment technology on the shear resistance of cement concrete bridge deck asphalt pavement layers, and recommended the optimal chip-sprinkling technique parameters. Xu, et al. [23] evaluated the improvement effect of aggregate-exposed interface on the stability of bridge deck pavement structure by direct shear test and pull-out test. They found that exposed-aggregate showed better shear performance than other interfacial treatment methods under various positive pressures. Liu, et al. [24] proposed the environmental simulation bubble test and used MatchID-3D structural deformation analysis system to measure bubble deformation, and studied the deformation characteristics and mechanism of bubbles in bridge deck waterproofing membrane. It found that the test temperature, initial debonding aperture, and water have great influences on the performance of bridge deck pavement. Lee, et al. [25] have studied the feasibility of reducing early temperature shrinkage crack and dry shrinkage crack of low melting point concrete with shrinkage reducing agent. And confirmed the durability can be increased without affecting other properties by adding a shrinkage reducing agent.

Although there are many pieces of research on the application of PS in HMA, research on the interlaminar shear resistance of cement concrete bridge deck PS asphalt pavement is relatively limited. In this paper, interlaminar shear strength is used as evaluation indexes, and the bonding performance of the interlayer structure is evaluated through the direct shear test with normal pressure. The impact of PS as a substitute for the asphalt mixture aggregate, interface types, normal pressure levels, waterproof bonding layer types, temperature and shear rate on interlaminar bonding shear performance are analyzed.

2. Experimental Plan

2.1. Mixture Design

In this research, "Shell" SBS modified asphalt binder was used in the design of pavement surface layer and an interlaminar bonding layer. The conventional test of asphalt and asphalt mixture was following the procedures of "standard test methods for bitumen and bituminous mixture for highway engineering" (JTG E20-2011) [26]. The conventional asphalt binder test results are shown in Table 1, as shown in the table that the asphalt binder satisfied the specification requirement. The asphalt mixture was designed following the steps of "technical specification for construction of highway asphalt pavement" (JTG F40-2017) [27], the aggregate used was limestone, the filler was limestone mineral powder, the optimum asphalt content was 5.4%, and the pavement mixture gradation was widely-used AC-16C. The gradation curve of the asphalt mixture is shown in Figure 1 and the design and volumetric parameters of asphalt mixture are shown in Table 2. In order to evaluate the influence of PS on the shear resistance of asphalt mixture, the fine limestone aggregate of 0.075 mm–4.75 mm was replaced with PS and PS powder for comparative tests.

Table 1. Technical indexes of (Styrene-Butadiene-Styrene) SBS modified asphalt binder.

	Properties	Test Value	Technical Index
	Penetration at 25 °C, 100 g, 5 s (0.1 mm)	51.5	30-60
	Ductility at 5 °C, 5 cm/min (cm)	28.5	≥20
	Softening Point (°C)	87	≥60
After RTFOT	Weight Change (%)	−0.053	≤± 1.0
	Penetration Ratio (%)	80	≥65
	Residual Ductility at 5 °C, 5 cm/min	16	≥15

Table 2. Marshall test results of SBS modified asphalt.

Asphalt Content (%)	Void Ratio (%)	VMA* (%)	VFA* (%)	Stability (kN)	Flow Value (mm)
5.4	4.7	15.2	69	16.3	31.4

* VMA = Voids in Mineral Aggregate, VFA = Voids Filled with Asphalt.

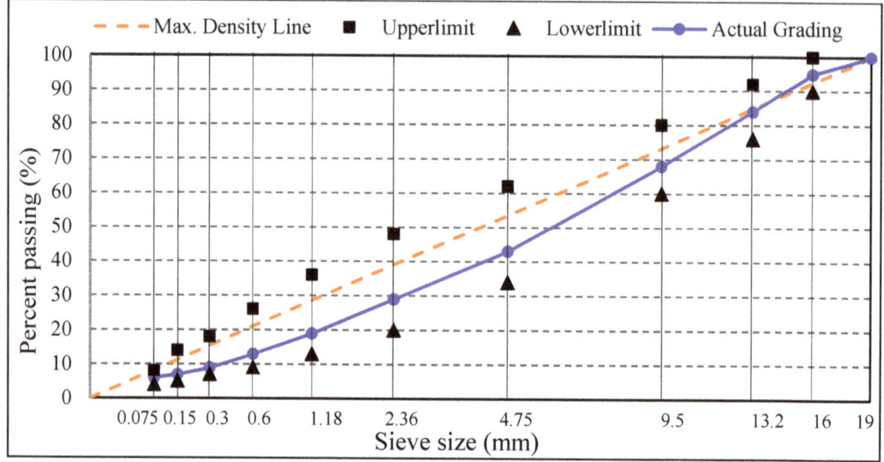

Figure 1. The gradation of dense-graded AC-16C asphalt mixtures.

2.2. Experiment Design

The interlaminar shear resistance of bridge deck pavement specimens was studied by direct shear test under normal pressure. The specimens used were prismatic specimens of 80 mm × 80 mm × 100 mm. The test design and test equipment (Material Testing Systems—MTS 810, USA) are shown in Figure 2.

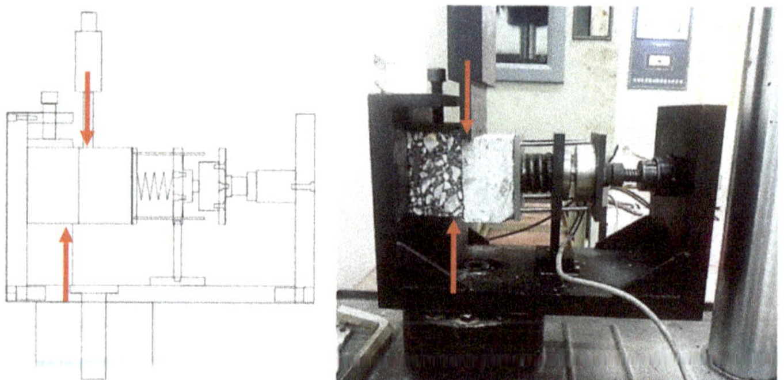

Figure 2. Direct shear test design and test device.

Four duplicate specimens were prepared for each test, and the test results were illustrated with the average of four specimens. The test specimens preparation steps were as follows:

(1) Cement concrete test specimens were formed indoor following the construction process with a size of 300 mm × 300 mm × 50 mm. Then, specimens were placed in the standard curing room for curing of 28 days, and the cement concrete panels were prepared with three types of interface: untreated (without any surface treatment), grooved and aggregate-exposed, the sample preparation process of the grooved interface and aggregate-exposed interface were shown below:

(a) *Grooved interface.* The grooves were notched according to the pavement anti-slide requirements in the technique guidelines for construction of highway cement concrete pavement [28]. The grooves were notched in a depth of 2–4 mm, in a width of 3–5 mm, and a groove spacing of about 15 mm.

(b) *Aggregate-exposed interface.* The first step prepares the aggregate-exposed concrete was to spray retarder on the surface of the cement concrete layer after paving, which delayed the setting and hydration of the surface mortar layer but did not affect the normal setting and hydration of the main body. After the main-body concrete reaches a certain strength, the surface laitance was washed out to expose part of coarse aggregate.

After curing, the cement concrete slabs of the three interface types are shown in Figure 3, which displays as the untreated interface, the grooved interface and the aggregated-exposed interface from left to right.

Figure 3. Cement concert slabs with three interface types.

(2) After the cement concrete slabs were cured, the surface of each type was coated with three different types of interlayer bonding materials separately, which included SBS modified asphalt, SBS modified asphalt mixed with PS micro-powder, and SBS modified asphalt mixed with surfactant (anti-stripping agent). The content of PS micro-powder and anti-stripping agent was 10% and 0.4% of the mass of asphalt, respectively. As the surface area of each type of interface was different, in order to make sure that all surfaces were coated well, the dosage of waterproof cohesive bonding material for untreated, grooved and aggregate-exposed interfaces was 1.0 kg/m^2, 1.2 kg/m^2 and 1.5 kg/m^2 respectively.

(3) Then, put the cement concrete specimen into a 300 mm × 300 mm × 100 mm rutting plate test mold, and poured the mixed asphalt mixture over it and applied the rutting wheel to compact it into the desired compaction level. Finally, the composite specimens were cut into 80 mm × 80 mm × 100 mm small prism specimens. The specimens with different interfacial treatment are shown in Figure 4.

Figure 4. Specimens for the direct shear test with different interfacial treatment.

For the direct shear test, the shear strength corresponding to the peak value of the load-displacement curve is the shear strength of the interface. The calculation method is shown in Equation (1):

$$\tau = P/S. \tag{1}$$

In which: τ is the interlaminar interface shear strength (MPa); P is the peak value of shear load in the direct shear test (kN); S is the interfacial area (mm^2).

As the surface roughness of cement concrete slabs with three interface types of untreated, grooved and aggregate-exposed was different, the roughness characteristics were evaluated by the texture depth (*TD*) index. The *TD* was measured by sand spreading method following the field test methods of subgrade and pavement for highway engineering (JTG E60-2008) [29] which was described below: Firstly, spread the standard sand on the cement slab into a circle, then scrape the surface of the standard sand with a scraper, measure the diameters of the two vertical directions of the circle with a ruler, then brush the standard sand on the cement board with a clean brush to weigh the quality. The *TD* of the cement board surface can be calculated by Equation (2). The determination of the *TD* of the cement slab is shown in Figure 5. The *TD* measurement results for untreated, grooved and aggregate-exposed interfaces are 0.59 mm, 1.67 mm and 4.43 mm respectively.

$$TD = 4V/\pi d^2. \tag{2}$$

In which, V is the sand volume to filling the uneven part under the measuring circle (mm^3), d is the diameter of the measuring circle (mm).

Figure 5. Method to measure the texture depth.

2.3. Experimental Schematic

In order to evaluate the impact of PS as a substitute of asphalt mixture aggregate, waterproof and cohesive layer material type, bridge deck interface treatment, normal pressure, test temperature and loading rate on the interlaminar bonding behavior of bridge deck pavement, the following tests were conducted:

(1) **Impact of PS as a substitute of asphalt mixture aggregate.** The asphalt mixture of AC-16C was used to compare the interlaminar bonding of PS as a substitute of asphalt mixture aggregate: one is named as limestone asphalt mixture in which the asphalt mixture aggregate was limestone and filler and the other was named as PS asphalt mixture in which the fine limestone aggregate of 0.075 mm to 4.75 mm and fillers were replaced by PS and PS powder in equal amounts.

(2) **Impact of waterproof and cohesive layer material type.** In order to compare the interlaminar bonding behavior of different bonding layer materials, SBS modified asphalt, SBS modified asphalt mixed with PS micro-powder and SBS modified asphalt mixed with an anti-stripping agent were selected for comparison.

(3) **Impact of bridge deck interface treatment.** In order to analyze the impact of different interface conditions on the shear resistance of interlayer, the bonding characteristics of untreated, grooved, and aggregate-exposed interfaces were evaluated.

(4) **Impact of normal pressure.** In order to study the influence of normal pressure on interlaminar shear strength, the test with normal pressures of 0 MPa, 0.3 MPa, 0.5 MPa and 0.7 MPa were conducted and compared respectively.

(5) **Impact of temperature.** The test temperature has a significant effect on the interlaminar shear strength. In order to obtain the impact of the test temperature on the interlaminar shear strength, five test temperatures of 25 °C, 40 °C, 50 °C, 60 °C and 70 °C were selected to conduct comparison tests.

(6)   **Impact of shear rate.** In order to simulate the effect of different driving speeds on the interlaminar shear performance, five different shear loading rates of 1 mm/min, 5 mm/min, 10 mm/min, 20 mm/min and 50 mm/min were selected to conduct comparison tests.

3. Results and Discussions

3.1. Impact of PS as Asphalt Mixture Aggregate Substitute

The direct shear strength between concrete deck and asphalt pavement was measured to evaluate the impact of PS as limestone substitute on the interlaminar bonding performance. The SBS modified asphalt was used in the interlaminar bonding layer, and the test temperature was 60 °C. In order to ensure the specimen had a constant test temperature, the specimens were put into the environmental chamber for 4 h before the test, the shear rate was 10 mm/min, the vertical pressure was 0.5 Mpa. The three different interface types of untreated, grooved and aggregate-exposed were selected for comparative study. The test results are shown in Table 3. The comparison figure is shown in Figure 6.

Table 3. Impact of aggregate and interlayer treatment on the interlaminar bonding.

Aggregate Type	Interface Type	Max Shear Force (kN)	Shear Strength (MPa)
Limestone Asphalt Mixture	Untreated	2.061	0.322
	Grooved	2.784	0.435
	Aggregate-exposed	3.725	0.582
PS Asphalt Mixture	Untreated	2.266	0.354
	Grooved	3.072	0.480
	Aggregate-exposed	4.089	0.639

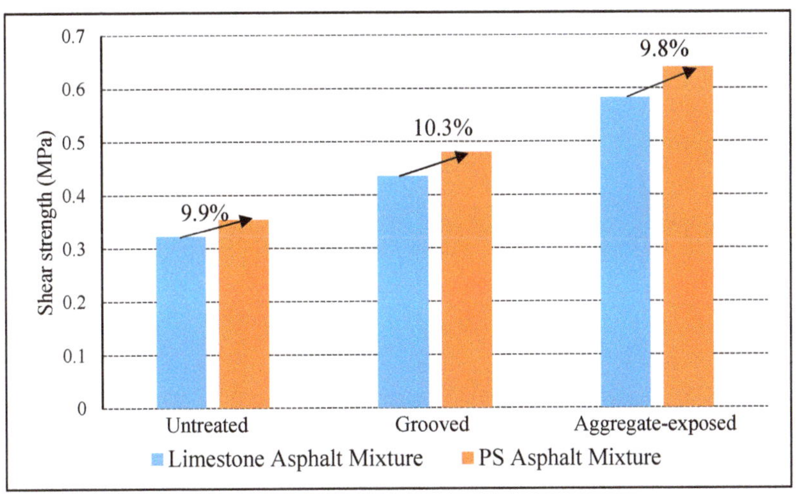

Figure 6. Impact of aggregate and interlayer treatment on the interlaminar bonding.

As shown in Table 3 and Figure 6, the interlaminar shear strength corresponding to the PS asphalt mixture is higher than that of limestone asphalt mixture for all three kinds of interface types, and the increasement corresponding to untreated, grooved and aggregate-exposed interfaces were 9.9%, 10.3% and 9.8% respectively. The main reason is that PS is alkaline and have a larger specific surface area, so it shows better adhesion to asphalt binder compared with limestone powder. As a result, the bonding effect between PS asphalt mixture and SBS modified asphalt material is improved, the shear strength and interlaminar stability of the bridge deck and asphalt pavement are improved.

3.2. Impact of Interfacial Surface on the Interlaminar Bonding

In order to study the influence of surface treatment on the shear resistance of the bridge deck, three kinds of interface types were evaluated in this paper, which including untreated interface, the grooved interface, and aggregate-exposed interface. Indoor direct shear tests were carried out on bridge deck pavement composed of two kinds of pavement materials: PS asphalt mixture and limestone asphalt mixture. The interlaminar bond coating materials were SBS modified asphalt, the test temperature was 60 °C, the shear rate was 10 mm/min, and the normal pressure was 0.5 MPa. The test results were also shown in Table 3 and Figure 6.

From Figure 6, it can be seen that both the shear strength of PS asphalt mixture and the limestone asphalt mixture shows the same relation for all three interface types: aggregate-exposed interface > grooved interface > untreated interface. When the paving layer is a limestone asphalt mixture, the shear strength of the grooved interface and the aggregate-exposed interface were increased by 35.1% and 80.7% respectively compared with the original untreated interface. In addition, for the PS asphalt mixture, the shear strength of the grooved interface and the aggregate-exposed interface were increased by 44.1% and 80.5% respectively compared with the original interface.

As mentioned earlier, the *TD* size of the three interface types is sorted as aggregate-exposed interface > grooved interface > untreated interface. The relation of *TD* on the shear strength of the composite structure is shown in Figure 7. The results showed that the shear strength closely related to the surface textural of the bridge deck, the rougher the texture of the deck surface and the greater the value of *TD*, the greater the shear strength of the corresponding deck pavement structure.

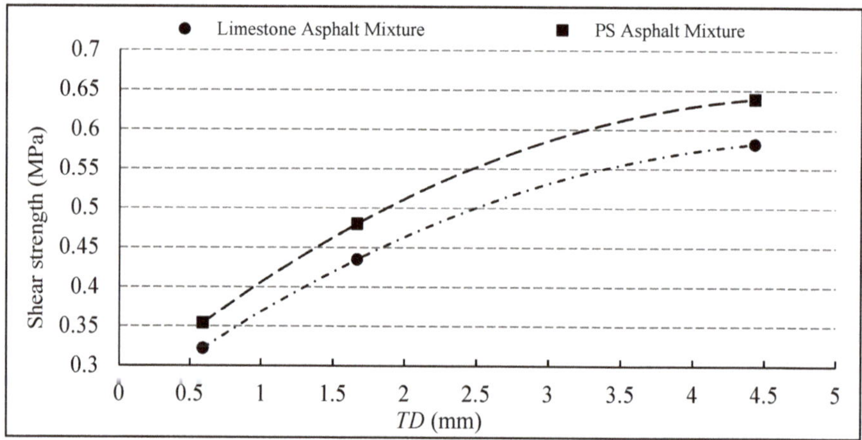

Figure 7. Impact of texture depth on the interlaminar bonding.

3.3. Impact of Waterproof and Cohesive Layer Material Type

In order to compare and study the influence of waterproof and cohesive coating material on the interlaminar shear resistance and to improve the interlaminar stability of bridge deck pavement, the interlaminar shear tests of SBS modified asphalt, SBS modified asphalt with PS micro-powder, and SBS modified asphalt with surfactant (anti-stripping agent) were carried out. The interlaminar shear test was conducted with three interface types mentioned above respectively. The PS asphalt mixture was used as a pavement layer. The test temperature was 60 °C, the shear rate was 10 mm/min, and the vertical pressure was 0.5 MPa. The test results are as shown in Table 4 and the comparative figure is shown in Figure 8.

Table 4. Interlaminar shear strength of different waterproof and cohesive materials.

Bonding Material	Interface Type	Max Shear Force (kN)	Shear Strength (MPa)
SBS asphalt	Untreated	2.266	0.354
	Grooved	3.027	0.480
	Aggregate-exposed	4.089	0.639
SBS asphalt with PS Micro-powder	Untreated	2.406	0.376
	Grooved	3.347	0.523
	Aggregate-exposed	4.224	0.660
SBS asphalt with anti-stripping agent	Untreated	2.323	0.363
	Grooved	3.226	0.504
	Aggregate-exposed	4.333	0.677

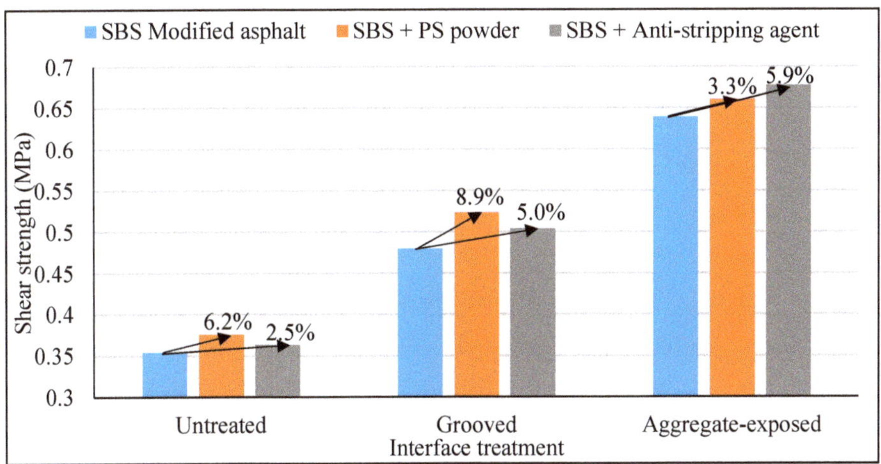

Figure 8. Interlaminar shear strength of different waterproof and cohesive materials.

As can be seen from Table 4 and Figure 8, For the original untreated interface, the shear strength increase by 6.2% and 2.5%, the grooved interface increased by 8.9% and 5.0% and the aggregate-exposed interface increased by 3.3% and 5.9%, respectively. The results show that the interlaminar shear strength of SBS asphalt with PS powder and SBS asphalt with anti-stripping agent were higher than that of SBS asphalt. The reason is that the surface of the grooves and exposed coarse aggregate increased the adhesion ability between the bridge deck and asphalt pavement, therefore, the contribution of the PS powder or anti-stripping agent on the interlayer shear stress of treated interface is greater than that of the untreated interface.

3.4. Impact of Normal Pressure

The mechanical properties of bridge deck pavement are greatly influenced by the grade of vehicle load. As the normal pressure at the interface of pavement and bridge deck is below 0.7 MPa at most cases, therefore, the tested normal pressure for the interlaminar bonding properties was in the range of 0 MPa to 0.7 MPa. In order to better study the impact of different vehicle loads on the shear stability of deck pavement, different normal pressures of 0 MPa, 0.3 MPa, 0.5 MPa and 0.7 MPa were selected respectively to conduct the direct shear test. SBS modified asphalt was used as a waterproof and cohesive layer and PS asphalt mixture was used as a bridge deck surface layer. The experimental temperature was 60 °C and the shear rate was 10 mm/min. The results of shear tests were shown in Table 5 and Figure 9.

Table 5. Test results of interlaminar shear strength under different normal pressures.

Pavement Mixture	Normal Pressure (MPa)	Max Shear Force (kN)	Shear Strength (MPa)
PS asphalt mixture	0	0.352	0.055
	0.3	2.169	0.339
	0.5	3.072	0.480
	0.7	4.166	0.651

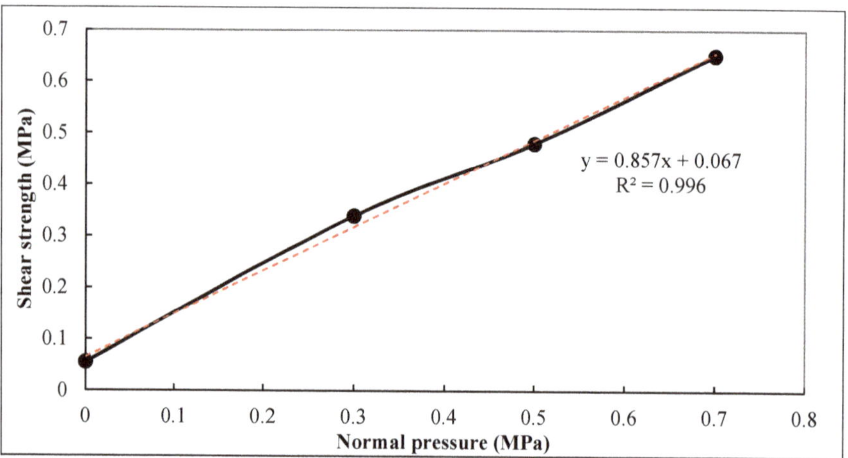

Figure 9. Relationship between normal pressure and interlaminar shear strength.

Although the factors affecting the interlaminar bonding properties are very complicated, as it can be seen from Figure 9, it is reasonable to assume it satisfied the Mohr-Coulomb strength theory when the normal pressure was below 0.7 MPa, in which the interlaminar shear strength has a linear relation with the normal pressure. In addition, based on the Mohr-Coulomb strength theory, the interlaminar shear failure will not occur if the interlaminar shear stress caused by vehicle load satisfied the following equation:

$$\tau < c + \sigma_z tg\varphi \tag{3}$$

where c is the cohesion between asphalt pavement and cement deck, σ_z is the normal compressive stress on the shear surface, and $\sigma_z tg\varphi$ is the friction between the rough surface texture structure of bridge deck and the asphalt mixture of pavement. The linear regression equation of the interlaminar shear strength and the corresponding cohesive force c and internal friction angle φ of the grooved interface under different normal pressures are shown below.

$$\tau = 0.067 + 0.857\sigma \left(R^2 = 0.996\right). \tag{4}$$

In which, the cohesive strength of $c = 0.067$ MPa, and the internal friction angle $\varphi = 40.6°$.

3.5. Impact of Temperature on Interlaminar Shear Strength

As the interlayer bonding material of bridge deck pavement is usually viscoelastic material, the bonding properties are sensitive to temperature change, therefore, the bridge deck pavement tended to occur interlaminar shear deformation due to the decrease of bond performance at summer high-temperature conditions.

In order to study the influence of temperature on interlaminar shear strength of PS asphalt mixture, the interlaminar shear tests were carried out at 25 °C, 40 °C, 50 °C, 60 °C and 70 °C respectively. In order to ensure the temperature field of the specimen was uniform, the specimens were placed in the environmental chamber for 4 h before the test. The interlaminar bonding material was SBS modified asphalt, and the interface was grooves treated. The experimental interlaminar shear rate was 10mm/min and the normal pressure was 0.5 MPa. The test results are shown in Table 6.

Table 6. Results of interlaminar shear tests at different temperatures.

Mixture Type	Temperature (°C)	Maximum Shear (kN)	Shear Strength (MPa)
	25	5.318	0.831
	40	3.757	0.587
PS asphalt mixture	50	3.347	0.523
	60	3.027	0.480
	70	2.835	0.443

As it can be seen from Figure 10, the influence of temperature on the interlaminar shear stress of PS asphalt pavement is very obvious, the interlaminar shear strength decreases gradually with the increase of temperature, and the influence slope varies in a different temperature range. It can be calculated from the diagram that the interlaminar shear strength decreased by 29.4% when the temperature increased from 25 °C to 40 °C, the shear strength decreased by 10.9% when the temperature raised from 40 °C to 50 °C, the interlaminar shear strength decreased by 8.9% when the temperature raised from 50 °C to 60 °C, and the interlaminar shear strength decreased by 7.7% when the temperature raised from 60 °C to 70 °C. From the experimental data, it can be seen that the interlaminar shear strength at 60 °C is only 57.7% of that at 25 °C, this would explain the reason why the interlaminar shear failure of bridge deck pavement occurs mostly in the high-temperature season.

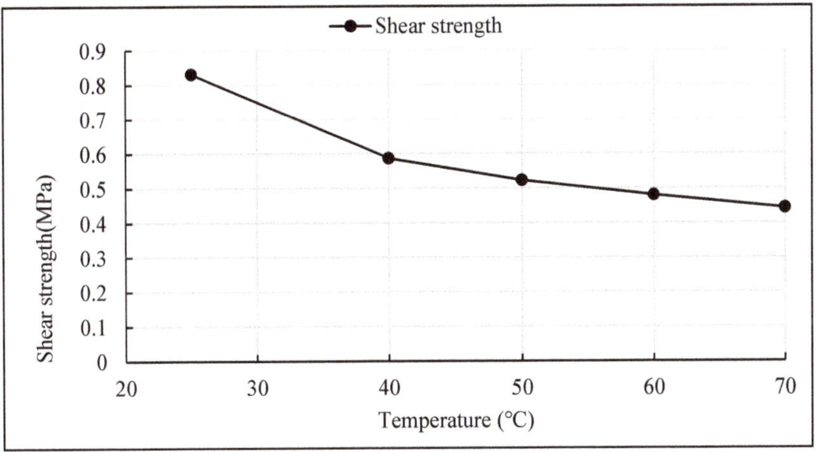

Figure 10. Relation of interlaminar shear strength with temperature.

3.6. Impact of Shear Rate on Interlaminar Shear Strength

In order to study the influence of different driving speed on the shear stress of PS asphalt pavement, various shear rates of 1 mm/min, 5 mm/min, 10 mm/min, 20 mm/min and 50 mm/min were used to simulate the different driving speeds. The grooved interface was used in these tests, the waterproof and cohesive material was SBS modified asphalt. The test temperature was 60 °C and the normal pressure was 0.5 Mpa. The test results are shown in Table 7 and Figure 10.

Table 7. Results of interlaminar shear strength tests at different shear rates.

Pavement Mixture	Shear Rate (mm/min)	Max Shear Force (kN)	Shear Strength (MPa)
	1	2.425	0.379
	5	2.765	0.432
PS asphalt mixture	10	3.072	0.480
	20	3.149	0.492
	50	2.976	0.465

As shown in Figure 11, the shear rate had a significant effect on the shear strength between layers, and the shear strength increased with the increase of shear rate when the shear rate was below 10mm/min, and when the shear rate increased from 1 mm/min to 10 mm/min, the effect of loading rate on shear strength was especially significant. As it can see from that test data that the shear strength of 10 mm/min was 26.6% higher than that of 1 mm/min. As the loading rate continues to increase, the test curve gradually tends to smooth and stable, that is, when the loading rate was large, the influence of loading rate on the interlaminar shear strength became smaller. This also shows that the vehicle's damage to the pavement of the bridge deck at high speed was less than at lower speed.

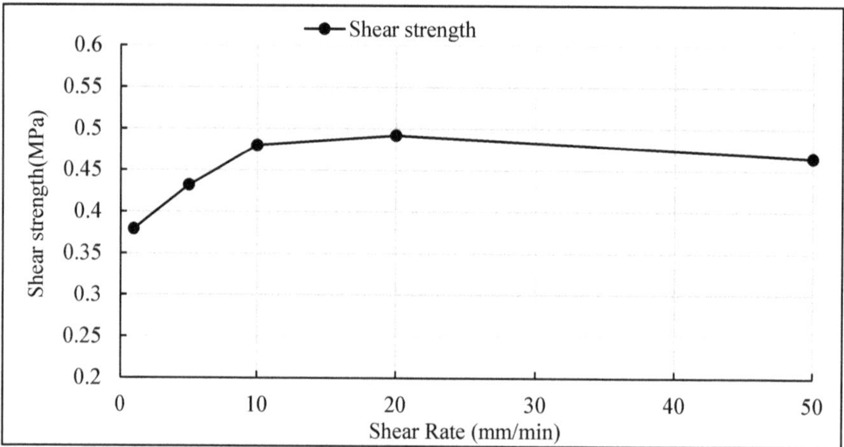

Figure 11. Relationship between shear rate and interlaminar shear strength.

4. Conclusions

This research evaluated the factors impacting the interlayer bonding and shear resistance of cement concrete bridge asphalt pavement, the comparatively tests including PS as a substitute of fine aggregate for asphalt mixture, interface characteristics, normal pressure, waterproof and cohesive layer types, temperature and shear rate. The following conclusions can be drawn:

(1) After the substitution of limestone aggregate between the size of 0.075 mm and 4.75 mm in the asphalt mixture with an equal amount of PS and PS powder, the bonding characteristics between surface asphalt mixture and bridge deck were improved, the interlaminar shear resistance and interlaminar stability were increased.

(2) The interfacial treatment of the bridge deck had a significant effect on the shear resistance of the deck. The rougher the surface texture and the greater the TD of the surface, the greater the shear resistance of the deck. The result indicated the aggregate-exposed treatment could increase the shear strength up to 80% compared with the untreated interface. It was an effective way to improve the shear resistance of bridge deck pavement by grooved or aggregate-exposed treatment.

(3)   The impact of the test conditions on the interlaminar shear strength is significant. The interlaminar shear resistance increases with the increase of normal pressure, and the shear strength decreases gradually with the increase of test temperature. The interlaminar shear strength increases with the increase of shear rate in a certain range and tends to be stable when the shear rate is getting higher than 10 mm/min.

Overall, the bonding performance of bridge deck pavement can be improved by replacing the limestone aggregate and filler of 0.075–4.75 mm in asphalt mixture with PS and PS powder respectively. The rougher the deck surface is, the better the bonding strength between layers will be. Deck surfaces treated with the grooved or aggregate-exposed method can significantly improve the interlaminar bonding performance. Both PS powder and anti-stripping agent can improve the adhesion performance of SBS modified asphalt as a bonding layer.

Author Contributions: Conceptualization, G.Q.; Data curation, S.L. and H.Y.; Formal analysis, H.Y.; Funding acquisition, G.Q.; Investigation, S.L. and X.G.; Methodology, H.Y.; Project administration, G.Q.; Resources, G.Q.; Software, X.G.; Supervision, H.Y.; Visualization, X.G.; Writing—Original Draft, S.L.; Writing—Review & Editing, H.Y.

Funding: The research was funded by the National Key R and D Program of China (SQ2018YFB160027); National Natural Science Foundation of China (51778071 and 51808058); open fund (kfj180103) of National Engineering Laboratory of Highway Maintenance Technology (Changsha University of Science and Technology).

Conflicts of Interest: The authors declare no conflict of interest.

References

1.   Zhang, Z.; Hu, C.; Wang, B. Research on the design method of asphalt concrete pavement for concrete bridge deck structure. *China J. Highw. Transp.* **2001**, *14*, 58–61. [CrossRef]

2.   European Asphalt Pavement Association. *Asphalt Pavements on Bridge Decks*; EAPA Position Paper; European Asphalt Pavement Association: Brussels, Belgium, 2013.

3.   Jiang, M.; Wang, Z.; Huang, X.; He, Y.; Zhang, L.; Ning, P.; Fu, J.; Liu, H. Secondary pollution prediction of recycling process of yellow phosphorus slag by moulding under thermal state. *J. Cent. South Univ. (Sci. Technol.)* **2016**, *47*, 1078–1084. [CrossRef]

4.   Quan, J.; Zhang, K.; Ma, B. Effect of Pore Solution pH Value and Hydration Process on Phosphorous Slag Replacing of Slag. *Bull. Chin. Ceram. Soc.* **2016**, *35*, 2513–2517, 2523. [CrossRef]

5.   Allahverdi, A.; Mahinroosta, M. Mechanical activation of chemically activated high phosphorous slag content cement. *Powder Technol.* **2013**, *245*, 182–188. [CrossRef]

6.   Zhao, G.; Gao, Y.; Li, Y.; Zheng, A.; Kang, X. Research and experiment on the mechanical property of sleeper concrete adding phosphorus slag. *Concrete* **2011**, *9*, 84–86. [CrossRef]

7.   Chen, X.; Zeng, L.; Fang, K. Anti-crack performance of phosphorus slag concrete. *Wuhan Univ. J. Sci.* **2009**, *14*, 80–86. [CrossRef]

8.   He, X.; Ye, Q.; Yang, J.; Dai, F.; Su, Y.; Wang, Y.; Bohumir, S. Physico-chemical Characteristics of Wet-milled Ultrafine-granulated Phosphorus Slag as a Supplementary Cementitious Material. *J. Wuhan Univ. Technol. Sci. Ed.* **2018**, *33*, 625–633. [CrossRef]

9.   Hu, J. Comparison between the effects of superfine steel slag and superfine phosphorus slag on the long-term performances and durability of concrete. *J. Anal. Calorim.* **2017**, *128*, 1251–1263. [CrossRef]

10.   Mehdizadeh, H.; Kani, E.N. Modeling the influence of chemical composition on compressive strength behavior of alkali-activated phosphorus slag cement using statistical design. *Can. J. Civ. Eng.* **2018**, *45*, 1073–1083. [CrossRef]

11.   Qian, G.; Bai, S.; Ju, S.; Huang, T. Laboratory Evaluation on Recycling Waste Phosphorus Slag as the Mineral Filler in Hot-Mix Asphalt. *J. Mater. Civ. Eng.* **2013**, *25*, 846–850. [CrossRef]

12.   Qian, G.; Wang, K.; Bai, X.; Xiao, T.; Jin, D.; Huang, Q. Effects of surface modified phosphate slag powder on performance of asphalt and asphalt mixture. *Constr. Mater.* **2018**, *158*, 1081–1089. [CrossRef]

13.   Sheng, Y.; Zhang, B.; Yan, Y.; Chen, H.; Xiong, R.; Geng, J. Effects of phosphorus slag powder and polyester fiber on performance characteristics of asphalt binders and resultant mixtures. *Constr. Mater.* **2017**, *141*, 289–295. [CrossRef]

14. Bazzaz, M. Experimental and Analytical Procedures to Characterize Mechanical Properties of Asphalt Concrete Materials for Airfield Pavement Applications. Ph.D. Thesis, University of Kansas, Lawrence, KS, USA, 2018.
15. Bazzaz, M.; Darabi, M.K.; Little, D.N.; Garg, N. A Straightforward Procedure to Characterize Nonlinear Viscoelastic Response of Asphalt Concrete at High Temperatures. *Transp. Res. Rec.* **2018**, *2672*, 481–492. [CrossRef]
16. Wang, W.; Liu, S.; Wang, Q.; Yuan, W.; Chen, M.; Hao, X.; Ma, S.; Liang, X. The Impact of Traffic-Induced Bridge Vibration on Rapid Repairing High-Performance Concrete for Bridge Deck Pavement Repairs. *Adv. Mater. Sci. Eng.* **2014**, *2672*, 481–492. [CrossRef]
17. Li, Y.; Li, S.; Lv, R.; Zhang, P.; Xu, Y.; Hou, G.; Cui, C. Research on failure mode and mechanism of different types of waterproof adhesive materials for bridge deck. *Int. J. Pavement Eng.* **2015**, *16*, 602–608. [CrossRef]
18. Liu, Y.; Wu, J.; Chen, J. Mechanical properties of a waterproofing adhesive layer used on concrete bridges under heavy traffic and temperature loading. *Int. J. Adhes. Adhes.* **2014**, *48*, 102–109. [CrossRef]
19. Sheng, Y.; Wang, Q.; Li, H. Theoretical Analysis of Extreme Temperatures and Thermal Stresses in Bridge Deck Pavements. *Int. J. Pavement Res. Technol.* **2011**, *4*, 238–243.
20. He, D.; Xiong, X.; Dong, Q. Typical Structure of Bridge Deck Asphalt Pavement of Cement Concrete Bridge. *J. Chongqing Jiaotong Univ. (Nat. Sci.)* **2011**, *30*, 27–30, 73.
21. Jia, J. Research on Shearing Resistance of Asphalt Pavement and Bridge Deck Pavement. Ph.D. Thesis, Chang'an University, Chang'an, China, 2011.
22. Ren, W.; Han, S.; Li, J.; Chen, D. Application of Chip-Springkling Technology to Concrete Bridge Asphalt Pavement Structure. *J. Build. Mater.* **2017**, *20*, 222–228. [CrossRef]
23. Xu, O.; Han, S.; Yu, J. Research on Improvement of Structural Stability for Concrete Bridge Deck Pavement. *J. Wuhan Univ. Technol.* **2010**, *32*, 22–24, 69.
24. Liu, H.; Li, Y.; Zhang, Q.; Hao, P. Deformation characteristic and mechanism of blisters in cement concrete bridge deck pavement. *Constr. Mater.* **2018**, *172*, 358–369. [CrossRef]
25. Lee, B.J.; Kim, Y.Y. Durability of Latex Modified Concrete Mixed with a Shrinkage Reducing Agent for Bridge Deck Pavement. *Int. J. Struct. Mater.* **2018**, *12*, 23. [CrossRef]
26. Ministry of Transport of the People's Republic of China. *Standard Test Methods of Bitumen and Bituminous Mixtures for Highway Engineering: JTG E20-2011*; Ministry of Transport of the People's Republic of China: Beijing, China, 2011.
27. Ministry of Transport of the People's Republic of China. *Technical Specification for Construction of Highway Asphalt Pavement: JTG F40-2017*; Ministry of Transport of the People's Republic of China: Beijing, China, 2017.
28. Ministry of Transport of the People's Republic of China. *Technique Guidelines for Construction of Highway Cement Concrete Pavement: JTG/T F30-2014*; Ministry of Transport of the People's Republic of China: Beijing, China, 2014.
29. Ministry of Transport of the People's Republic of China. *Field Test Methods of Subgrade and Pavement for Highway Engineering: JTG E60-2008*; Ministry of Transport of the People's Republic of China: Beijing, China, 2008.

Article

Assessment for Sustainable Use of Quarry Fines as Pavement Construction Materials: Part I—Description of Basic Quarry Fine Properties

Yinning Zhang *, Leena Katariina Korkiala-Tanttu, Henry Gustavsson and Amandine Miksic

Department of Civil Engineering, Aalto University, 00076 Aalto, Finland; leena.korkiala-tanttu@aalto.fi (L.K-T.); henry.gustavsson@aalto.fi (H.G.); amandine.miksic@aalto.fi (A.M.)

* Correspondence: yinning.zhang@aalto.fi

Received: 5 March 2019; Accepted: 10 April 2019; Published: 12 April 2019

Abstract: As a secondary material, quarry fines are a valuable material to be reused for many purposes in civil engineering projects. The aggregate source depletion, especially the lack of high quality aggregates as expected in the future, as well as the demand for a carbon-neutral society and circular economy, also promotes the high-volume utilization of secondary materials such as quarry fines. The aim of this study is to do a feasibility assessment including a series of laboratory tests and analyses to evaluate the properties of quarry fine materials to determine if this type of material could be qualified as pavement construction material in high volume. The gradation information obtained from both sieving and hydrometer tests indicates the frost susceptibility of unstabilized quarry fines, therefore frost heave tests were performed and which further suggest the necessity of stabilization to improve its properties for pavement applications, especially in structural layers such as base, subbase, or filter layers. Some other general information and properties of unbound quarry fines, especially regarding their validity for application in pavement engineering are also investigated and discussed.

Keywords: sustainable; secondary materials; quarry waste; frost susceptibility

1. Introduction

One big challenge in road engineering is the economic and ecological use of soil and aggregate materials. The design of pavement layers in Finland is mainly based on empirical methods, which are presented as tabulated model structures [1]. This has led to a situation where high quality virgin aggregates have been used in excess and the use of secondary materials has been challenging [2]. Finland, in proportion to its population, is one of biggest user of aggregates inside the European Union. Annually, 80–100 million tons of aggregates are used in Finland [3]. The conditions of cold climate and long distances caused by the relatively low density of population both increase the need to use a lot of aggregates [4]. Road and street construction consumes about 50 million tons of non-renewable aggregates and the amount of CO_2 emissions caused is 0.8 million tons annually [5]. Besides, in Europe the built environment is responsible for 40% of greenhouse gases emission, 50% of resource extraction, and 30%–45% of waste production [6]. According to Ehrukainen, the application of secondary or recycled materials was only 1% of the total use of aggregates [7]. Of the total consumption of aggregates, around 10% is used to produce concrete and 10% to produce asphalt concrete [4]. Therefore, it is of vital importance to significantly increase the utilization of "waste" materials.

Finland has a good source of natural rock for aggregate production, thus the motivation to use recycled material has been low in the past. Now, with the increase in the price of natural aggregates as well as the increased level of consciousness regarding environmental preservation and the promotion of a circular economy, strategies and targets have been proposed and are now enforced. The European Commission set four waste directives in 2018 [8]. The focus of the directives is to decrease the amount of waste, to increase recycling, and to diminish the need for landfills. A more solid aim is to increase the recycling, reuse, and other utilization of building and demolition waste up to 70% in weight by 2020 [9]. The United Nations Agenda for Sustainable Development, called Agenda 2030, also sets targets towards carbon-neutral societies by 2030 [10]. Finland has committed itself to taking action in regards to this agenda, including the aim to promote carbon-neutrality and the wise use of resources [11]. Many big cities, e.g., Helsinki, Espoo, Vantaa, Tampere, and Turku have defined their own targets for carbon neutrality and a circular economy. For example, the city of Helsinki has set a target to be carbon-neutral by 2035 [12]. These targets also require a much higher level of circular economy and promote the use of secondary materials. In summary, the reduction in natural resources and increasing consciousness regarding environment preservation have led to the established trend of utilizing secondary materials at higher levels in the near future, which is also the motivation of this study.

2. Literature Review

According to United States Environmental Protection Agency (EPA), non-hazardous secondary materials refers to any materials that are not the primary product of manufacturing or commercial processes but are the byproducts of such processes, including post-consumer material, post-industrial material, and scrap [13]. The European Commission also defines the waste that can be recycled and injected back into the economy as secondary raw materials [14]. On one hand, these materials may be considered as waste which needs to be treated and disposed of. On the other hand, many of them can have material or British Thermal Unit (BTU) value, with the possibility of being reused and managed for other industrial purposes with little or no harm to the environment.

Quarry waste is one such type of secondary material that unavoidably comes from the overburden, interburden, and extraction and processing of aggregates and includes quarry fines from processing activities [15]. As a general concept, quarry waste actually consists of different material types known as "quarry fines", "quarry dust", "stone byproducts", "recycled aggregates", "quarry powder wastes", and so forth [16]. Quarry fines have become a real problem in the quarry industry because of two main reasons. Firstly, the higher the fraction of fines contained in the material, the more easily the material can be dislocated by gravity, wind, and water, and the more difficult the material is to handle, store, and dispose of. Very fine crusher dust can be generated during crushing and screening processes and is easily dispersed into the atmosphere, contaminating the air. Secondly, the recent tax policy has promoted the possible utilization or management of large stockpiles of quarry fines to minimize the adverse social and environmental effects caused by this material, but effective high-volume applications are underdeveloped. Moreover, many places around the world will be facing problems of reduced rock resources in the future, which will result in an inadequate production of aggregates and an increase in high quality aggregate prices. Quarry fines could provide alternatives to alleviate the problem of aggregate shortages in the civil engineering industry. Consequently, methods of utilization of quarry fines are in demand, regardless of the quarry sources.

Table 1 presents a substantial amount of under-utilized quarry fines produced and accumulated at Lemminkäinen Infra Oy in Finland over the recent years. Besides Lemminkäinen Infra Oy, there are some other aggregate producers. For example, Destia Oy had a total of 194 000 tons of quarry fines around Finland in 2015 [17]. These numbers indicate a clear potential for economic and environmental benefits from the appropriate utilization and application of this material.

Table 1. The amount of quarry fines at Lemminkäinen Infra Oy [18].

Grain Size (mm)	2016 (tons)	2017 (tons)	Accumulation/Year (tons)	Accumulation since 2017 (%)
0/3	824,300	875,800	51,500	5.9
0/6	118,000	127,300	9,300	7.3
In total	942,300	1,003,100	60,800	6.1

Similarly to other granular unbound materials, one possible way of utilizing this type of quarry waste material is for landfill or pavement constructions. In Europe, the project "Alternative materials in road construction (ALT-MAT)" undertaken from 1998 to 1999 has been focused on utilizing different types of secondary/alternative materials for pavement base/subbase layers [19]. Many of the investigated secondary materials, including municipal solid waste incinerator (MSWI) bottom ashes and all types of slags, are fine-grained unbound materials that have shown positive results (both mechanically and environmentally) as pavement base/subbase materials. However, even though previous research has proven that there is a promising future for applications using many of the secondary materials in road construction, the research into quarry fines, and especially Finnish quarry fines, is still inadequate and many detailed aspects require further study. These aspects include physical and mechanical properties, durability evaluation, leaching properties, quarry fine applications in high volume, and so forth.

Since 2007, there has been an increase in research focusing on utilizing quarry waste as a construction material in the field of civil engineering. It has been implemented to fabricate artificial aggregates [20], hemp concrete [21], and mortars. In general, quarry fine materials have been found to be effective in improving the properties of conventional civil engineering materials, which has also been the most common application of this material in previous research. Quarry fines have been proven to be effective in mitigating the early decreasing of strength due to fly ash application [22]. It was also found that the partial replacement of cement by granite quarry dust is beneficial in several aspects including good durable behavior, less drying shrinkage and expansion, lower possibility of early-age cracking, while keeping strength properties comparable [23]. When used as stabilizer, quarry waste could improve the properties of soils in terms of better workability, a decrease in optimal water content and an increase in maximum dry unit weight [24,25], better swell and shrinkage properties for expansive soil [26], and satisfactory durability under freeze-thaw cycles [27]. Quarry fines were also employed as an addition to natural soils to improve properties such as grading and compaction characteristics, strength, and to reduce swelling and plasticity [24]. For flexible pavements, quarry fines have been utilized to improve physical, mechanical, and swelling properties of soil subgrade and subbase [28].

Extended research attempts have been made to utilize quarry waste in constructions for both non-bearing [21,29] and structural purposes [20,22,23,29–31]. Non-bearing blocks made of quarry waste and other additives are common for wall partitions and decorations. In addition, the most common application area of quarry fines in Finland today is as the uppermost unbound layer of yards, fields, and low-volume streets [17]. For structural purposes, there are mainly three types of effective applications utilizing quarry waste. One is to use quarry waste as a substitute for sands/filler in different percentages in concrete [22,30,32,33], and the other is to replace part of cement by the quarry waste to produce concrete [22], or even employ quarry waste as an additive to improve soil properties [23–27]. In other words, it is common to employ quarry waste as an alternative material to reduce the dosage of other materials (more expensive, more in demand, or less environmentally friendly) or as an additive, rather than using quarry waste as a major construction material. In fact, most of the previous research applies less than 50% quarry waste [19–21,23–27], and the majority applies 5% to 30% as aggregates/fillers to develop different types of concrete. This might be related to the high requirements for quality and the mechanical properties of concrete materials for adequate structural performance. Especially when applied to asphalt concrete for pavement constructions, quarry waste is only recommended by researchers for low-to-moderate volume traffic [32,34]. Another key point that can be summarized from current research is the inadequate durability evaluation of quarry-waste-based construction materials. There is only limited research conducted into durability evaluations of quarry-waste-based materials, and some of these have shown a reduction in durability with the incorporation of quarry waste [30]. However, a qualified pavement construction material should have enough durability in moist environments and also adequate frost-resistance for its application in cold regions. It should also be noted that the applications of the secondary quarry fine materials in ground water area would need permissions from the environment agencies in Finland, especially where active use of ground water is needed.

In this study, a comprehensive assessment including a series of laboratory tests and analyses was conducted to evaluate the properties of quarry fine materials, which have not been well-established in the past. Even though quarry fines have shown very promising results as an alternative or additive material in previous research, its properties as a pavement construction material are not clear. Unlike most previous studies, which apply quarry wastes only in a limited amount, the main purpose of this study was to determine if this type of byproduct could be qualified as a pavement base, subbase, or any other construction material, and explore a way to utilize it in high volumes. The benefits of utilizing this byproduct in high volumes include reducing the carbon footprint and promoting a circular economy, considering it can be easily found and obtained at most quarry sites. In addition, another inadequate aspect of previous research, the evaluation of the durability of quarry fines, was investigated here in term of a frost-susceptibility test to verify the suitability of this material for applications in cold regions. Some basic information and properties of unbound quarry fines, especially regarding their validity for application in pavement engineering were investigated and discussed here as a preliminary study and foundation to further research.

3. Material Sampling and Preparation

In general, adopting a proper sampling method using a specified apparatus helps avoid biased sampling and provides reliable test results. To obtain representative samples of the average properties of a batch at plants, the European standard SFS-EN 932-1 "Tests for general properties of aggregates. Part 1: Methods for sampling" has been followed for material sampling [35]. It should be noted that even though sampling from the stationary conveyor belt or from the stream of material is recommended, in practice, the quarry fines were only available from stockpiles at the plants when the authors collected them. However, to avoid segregation-caused biased results, sampling increments were randomly selected from as many parts of the batch as possible, and an adequate number of sampling increments was used to reduce the influence of the heterogeneity of the batch.

In total, around 100 kg of quarry fines with a 0–4 mm particle size were collected from the Koskenkylä quarry of Destia Oy, packaged into 10-kg capacity buckets, and directly transferred back to the laboratory for storage. The collected quarry fines are from high quality granite host rock according to CE marking, with major mineral compositions of plagioclase (50.4%) and quartz (39.6%). Reduction of a bulk sample by a riffle box or by the quartering method, as recommended in the standard, was conducted to obtain the amount of materials required for further research purposes. The reduced batches were heated in oven at (110 ± 5) °C to remove excess moisture and obtain a constant mass following the European standard SFS-EN 1097-5 "Tests for mechanical and physical properties of aggregates. Part 5: Determination of the water content by drying in a ventilated oven" [36]. This was important to maintain the identical status of all of the batches so that water content can be accurately controlled for any further specimen preparation and testing.

4. Description of Basic Quarry Fine Properties

4.1. Gradation of the Quarry Fines

The particle size distribution of the quarry fines with a 0–4mm grain size was obtained according to European standard SFS-EN 933-1 "Tests for geometrical properties of aggregates. Part 1: Determination of particle size distribution. Sieving method" [37]. The wet sieving method is necessary for naturally agglomerated materials or materials containing a large amount of fine fractions because fine fractions with static charges are more prone to clumping. Washing breaks down the agglomerations and also helps the separation of fines from coarse fractions [38], which improves the results of particle distribution from the successive sieving process.

4.1.1. Wet Sieving Method

A trial test using the wet sieving method was conducted to determine the fine contents of the quarry fines. It showed that the fine content of quarry fines reaches as high as 12.5%. According to the trial test results, it was decided that the wet sieving method and a hydrometer test are necessary for the quarry fines in order to get reliable results due to the large amount of fine fractions (<0.063 mm). Figure 1 presents the gradation curve of quarry fines with a 0–4 mm particle size obtained from both the wet and dry sieving methods, together with the limits for pavement base, subbase, and filter layers required by Finnish and European standards (Tables S1 and S2). The gradation curve of quarry fine materials as obtained from wet and dry sieving methods shows differences in finer grains but the difference appears to have no significant effects regarding the qualification for base/subbase/filter layer requirements. It was found that the quarry fines are not qualified for pavement base materials according to the gradation information only, whereas around half of the gradation curve stay inside the limits for subbase [39]. However, because of the gradation curve that sits on the lower limit of subbase requirements, it is hard to validate quarry fines for either base or subbase from the gradation information only, and thus a mechanical test to determine the strength of quarry fines will be planned and performed in the future. As a filter layer, the material can be qualified completely since the gradation curve sits inside the limits, stays only at one side of the c limit, and also satisfies the requirement $D_{15}/d_{85} < 5$ [39].

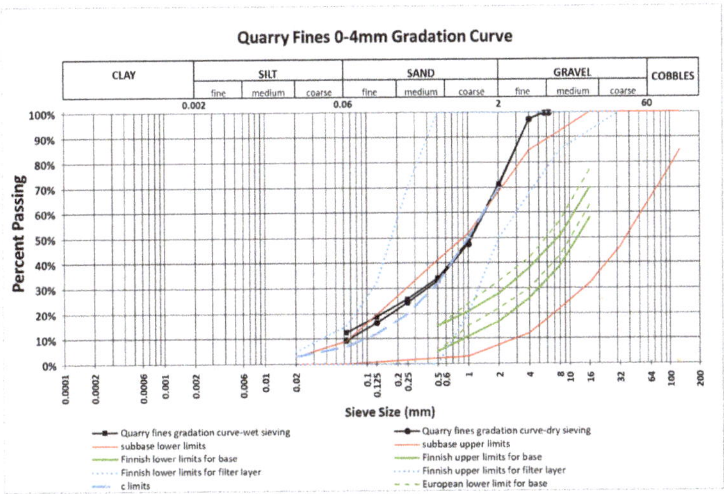

Figure 1. Gradation curves of quarry fines with a 0–4 mm particle size and the limits for pavement base, subbase, and filter layers according to Finnish and European requirements.

D_{15} is a grain size equivalent to 15 percent of the permeation of granular material (insulating layer material); d_{85} is the grain size of the finer material (bottom), having a permeation percentage of 85.

The coefficient of uniformity (C_u) and the coefficient of gradation (C_c) are calculated by Equations (1) and (2) to evaluate the grade of quarry fines. The classifications as listed in the European standard SFS-EN ISO 14688-2:2018 "Geotechnical investigation and testing" [40]. Identification and classification of soil. Part 2: Principles for a classification" were followed to categorize the grading of quarry fines used in this study.

$$C_u = \frac{D_{60}}{D_{10}} \tag{1}$$

$$C_c = \frac{(D_{30})^2}{D_{10} \times D_{60}} \tag{2}$$

For well-graded sand, $C_u > 15$ and $1 \leq C_c \leq 3$.

The coefficient of uniformity (C_u) and the coefficient of gradation (C_c) were determined to be 28.8 and 2.2, respectively. Therefore, the quarry fines can be classified as well-graded materials. The same classification is also valid according to the Unified Soil Classification System. From the 0.45 power maximum density graph shown in Figure 2, it can be observed that the quarry fines have the gradation characteristics of both fine- and dense-graded material, with the fine-graded characteristic being dominant. This characteristic is better captured when a wet sieving method is applied.

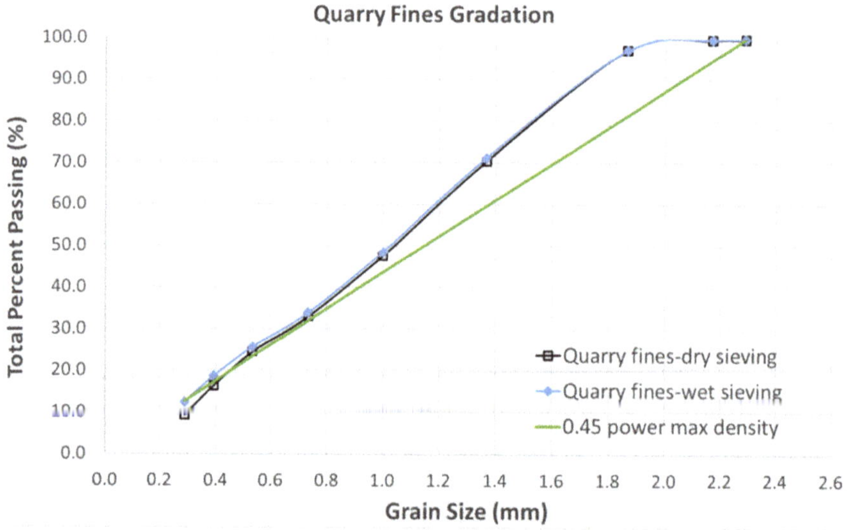

Figure 2. 0.45 power maximum density graph for quarry fines.

4.1.2. Hydrometer Method

The particle size distribution of the fine fraction significantly influences the frost susceptibility of granular materials, which can be determined by hydrometer test. A quarry fine sample containing only fine fraction (finer than 0.063 mm) was obtained by sieving the 0–4 mm quarry fine material mechanically and was tested for gradation with the hydrometer test. This method is based on the phenomenon of particles of different sizes settling by gravitation in a liquid at different rates, as well as Strokes' law, which establishes a relationship between the terminal velocity of a particle and other parameters of the fluid and material properties. The European standard SFS-EN ISO 17892-4: 2016 "Geotechnical investigation and testing" [41]. Laboratory testing of soil. Part 4: Determination of particle size distribution" was followed to perform the hydrometer test on quarry fine samples. Two samples were tested and the average was used for data analysis (Table S3).

Figure 3 shows the gradation curve of quarry fines provided by the quarry plant, the gradation curve obtained from combining the results from wet sieving and the hydrometer tests, as well as the four regions defining different potentials of frost susceptibility for different granular materials. This chart is typically used to estimate the frost susceptibility of materials according to their gradation curves, based on the criterion proposed by the ISSMGE Technical Committee on Frost, which is also the guide for the Finnish Road Administration [42]. If the gradation curve stays within region 1 only, this material can be seen as frost susceptible, but in region 1L the susceptibility is low. Gradation curves falling completely in regions 2, 3, or 4 indicate non-frost-susceptibility. However, any curves spanning one region to another in the finer part (left), similar to the gradation curve of quarry fines shown in Figure 3, can be seen as frost susceptible. The highly consistent part of the gradation curves obtained from laboratory tests and provided by the quarry plant Destia, although only available for particles coarser than 0.063mm, have shown the reliability of the gradation information. The completed gradation curve a containing finer particle distribution (<0.063mm) indicates that frost susceptibility tests are greatly necessary for quarry fine materials since untreated frost susceptible material is not appropriate to be applied in base, subbase, or filter layers in pavements. Consequently, an improvement in the quarry fine materials to validate their application in pavement constructions, such as stabilization techniques, are highly recommended.

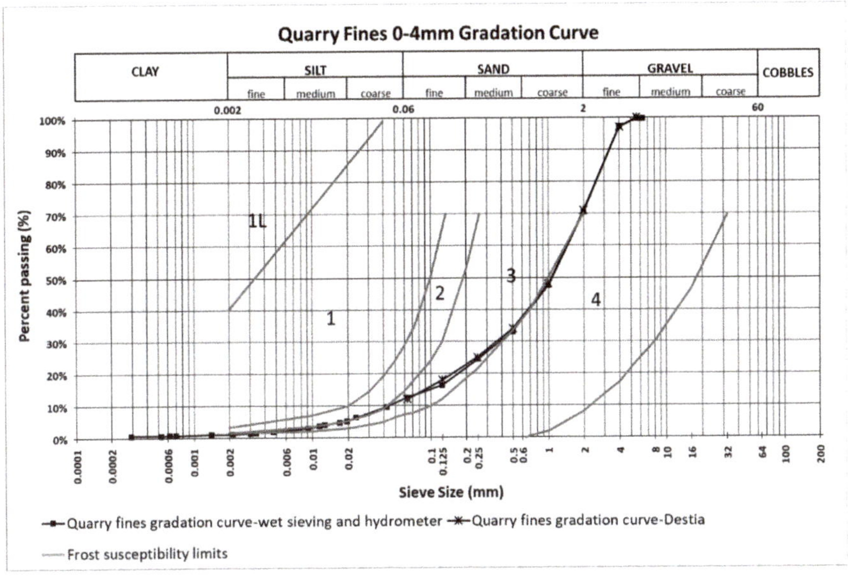

Figure 3. Estimation of frost susceptibility from the gradation curves of quarry fines according to ISSMGE [42]. Adapted with permission from [42]; 2005, Slunga, E., Saarelainen, S.

4.2. Density and Water Content

Granular material properties are closely related to moisture content. When compacted, the particles come together more closely and the dry density increases. The maximum dry density that can be achieved mainly depends on not only the effective compaction work but also the water content of the mixture. For a given degree of compaction work, there is typically an optimum water content at which the dry density reaches a maximum value for the non-free-draining materials. However, for a self-draining mixture, which is defined by the difference in water content before and after compaction being larger than 0.3%, may not have a well-defined water-density relationship. Therefore, a preliminary evaluation of the quarry fines to determine if they are a self-draining material is the first step to adoption.

The European standard EN 13286-2: 2010 "Unbound and hydraulically bound mixtures. Part 2: Test methods for laboratory reference density and water content—Proctor compaction" [43] was followed to characterize the water-density relationship of quarry fines, as shown in Figure 4. According to the difference between initial and final water content before and after compaction, quarry fines can be categorized as a self-draining material with a maximum difference in water content of 1.6%. It is also interesting to note that such variations in water content due to compaction are also related to the initial water content of the quarry fines. If more water is added to achieve a higher initial water content, the mixture is more prone to draining the excess water during compaction and thus the water content difference is larger.

(a)   (b)   (c)   (d)

Figure 4. Laboratory reference density and water content by Proctor compaction method: (**a**) Specimen with mold during compactions; (**b**) Compacted quarry fine specimen in mold; (**c**) Self-draining water during the compaction; (**d**) Compacted quarry fine specimen for water content determination.

Figure 5 and Table S4 present the relationship between the final water content and the dry density of the quarry fines. It is found that even though the quarry fines have self-draining properties, a maximum dry density still exists at the optimal water content of 9.3%. The maximum dry density of quarry fines that can be achieved from the Proctor test is 2.04 Mg/m^3, as shown in Figure 5. However, in comparison to the zero air void curve, it can be seen that water content has limited influence on the dry density of quarry fines. All of these findings have shown combined characteristics of both self-draining and fine-graded granular materials and have also indicated the complicated properties of this material.

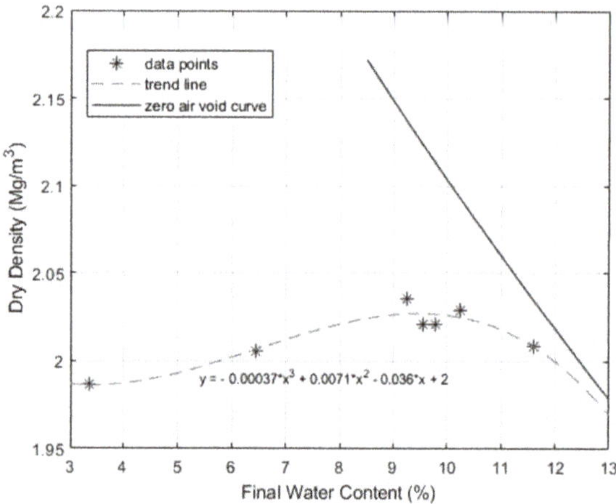

Figure 5. Variation of compacted dry density versus final water content.

4.3. Permeability

Permeability is another dominant property of pavement base and subbase materials in pavement design since excessive moisture inside the pavement structure would bring adverse effects to the performance of the system. Water-related deteriorations in highway pavements have been commonly and widely observed in the past decades, and the inadequate drainage of base and subbase layers is one of the contributors. The permeability of quarry fine material, however, has not been well established in the past. The determination of the permeability of the quarry fine specimens is very necessary and could be used as a baseline to validate the application of this material as pavement base, subbase, or even other applications.

Laboratory tests were conducted on unbound quarry fine specimens compacted at the optimum water content as determined by Proctor method. The falling head method provided by the Finnish Road and Water Engineering Board, which is suitable for fine-grained materials with intermediate-to-low permeability, was followed [44]. Whether the flow behavior of quarry fine specimens during the falling head test complies with Darcy's Law based on laminar flow is the first key question that needed to be answered, considering that Darcy's Law is adopted for permeability calculations in this method. To do so, a graph of change of head versus time was plotted to determine the linear part of the curve (if any) indicating the validity of Darcy's law (Table S5 and Figure 6).

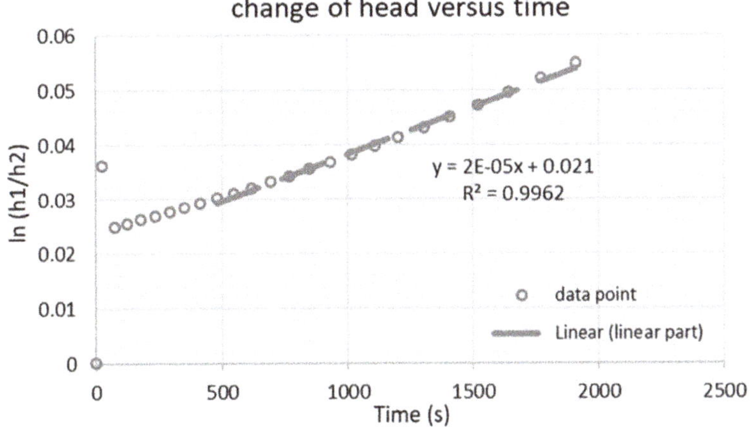

Figure 6. Change of head versus time during falling head permeability test.

As indicated in Figure 6, not all the data points obtained from the falling head permeability test are appropriate to be used for calculating permeability based on Darcy's law. The non-linear hydraulic gradient–velocity relationship at the beginning of the test shows that during this period, Darcy's law is not applicable. Therefore, only data from the successive time period (around 500 s from the beginning till the end) was adopted for data analysis. Figure 7 shows the permeability results of the unbound quarry fine specimen from a series of successive repeated tests. The coefficient of permeability of the quarry fine specimen increases slightly and then levels off in the sixth repeated test, at 5.75×10^{-5} m/s. According to Federal Highway Administration (FHWA) guidelines, this coefficient of permeability for quarry fines with a 0–4 mm particle size is still far below the requirements for a drainage aggregate base, which should have a minimum permeability of 500 ft/day (1.75×10^{-3} m/s). Instead, it is comparable to the typical coefficient of permeability of fine gravel, coarse and medium sand, and dense-graded aggregate base (4.2×10^{-5} to 1.43×10^{-4} m/s) [45] and is slightly higher than many of the reported situations from experiments or in the field nowadays [46,47]. Generally, based on the laboratory-determined permeability properties of quarry fines and the literature data, it was found that the permeability of quarry fine specimens falls within the typical permeability ranges of commonly adopted conventional pavement base and subbase materials, thus it can be validated for associated applications from the view of permeability. Since there are hardly any permeability requirements for the filter layer, the validity of quarry fines as filter layer material from permeability aspect is not possible. However, considering the major function of filter layers is to prevent fines immigration and frost actions in cold regions, lower permeability is desirable and a higher degree of compaction is consequently recommended for this purpose.

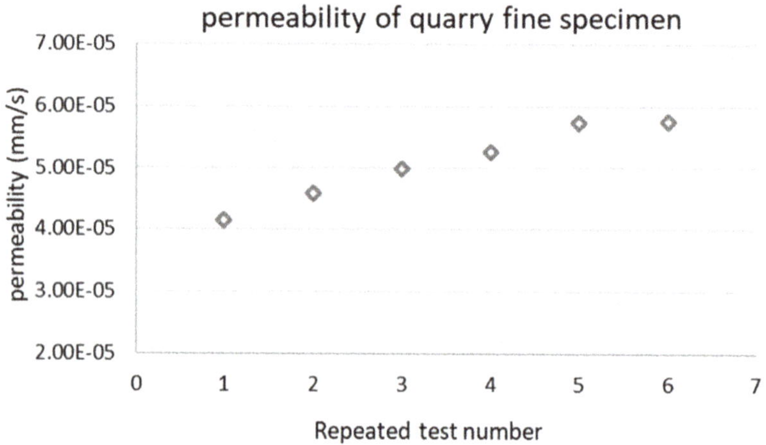

Figure 7. Permeability results of the unbound quarry fine specimen.

4.4. Capillary Rise

In capillary theory, frost heave is closely related to the rate of temperature decrease, pore sizes between particles, and the hydraulic conductivity of the unfrozen part of the material [48]. If the material can easily draw water up to the frozen zone by capillary actions and also has a high hydraulic conductivity which facilitates the delivery of a large quantity of water, large ice lenses are able to form and it is possible that the material is frost-susceptible. Therefore, to better understand the frost susceptibility of quarry fine material, capillary rise test according to Finnish test method provided by the Road and Water Engineering Board was performed [44]. It was found that the capillary rise values for virgin quarry fine samples are in range of 0.70 to 0.85 m, with an average of 0.804 m, based on four repeated laboratory tests. This capillary rise value is just in accordance with the limitation set by the Finnish railway department regarding frost susceptibility, indicating that the virgin quarry fines is possibly frost-susceptible and improvements are necessary. The capillary rise test results are consistent with the gradation information and further verified the necessity for stabilization.

4.5. Frost Heave

According to the lab-determined gradation information as well as existing information on frost susceptibility (Figure 3), the quarry fines with a 0–4 mm particle size might be frost-susceptible. It was therefore necessary to investigate frost susceptibility with a frost heave test.

The frost heave test method description TPPT-R07 by VTT communities and infrastructure [49] was followed to test unbound quarry fines. Specimens of 100 mm diameter and 100 mm height were compacted at the optimal water content using the Proctor method to reach maximum dry density. The specimen, together with its mold, is then kept frozen overnight for conditioning. The extruded frozen specimen was then placed with filter paper and a stone on the bottom of the test cell (Figure 8), covered with a rubber membrane and the top cap. The rubber membrane was fixed to the cap and the bottom of the test cell with O-rings, similar to the procedures in a triaxial test. After applying a thin layer of silicone to the surface of the membrane, the insulated split barrels were greased to the specimen and membrane. External support rings were then tightened to ensure the supporting of insulated barrels. Finally, the displacement transducers and the loading frame were attached to the cap of the test cell.

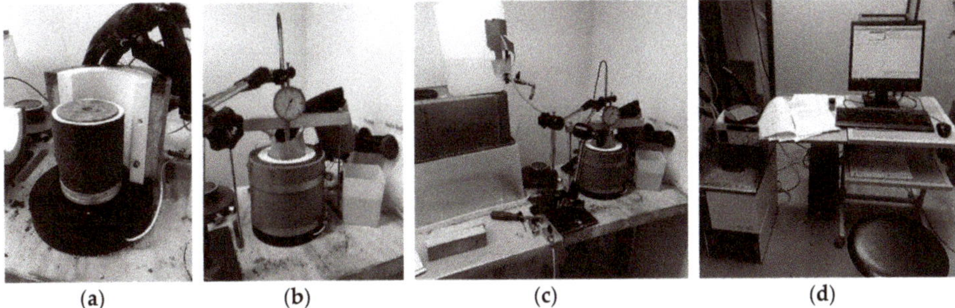

Figure 8. Frost heave test on quarry fine specimen: (**a**) Frozen quarry fine specimen mounted on the test apparatus; (**b**) Assembled specimen; (**c**) Assembled specimen with water supply; (**d**) Computer and temperature control system).

A series of frost heave tests, including three individual frost heave tests and intermediate thawing processes of different surfacing loading conditions, were performed on the specimen. The conditions adopted in the frost heave and thawing processes are listed in Table 2. Since there are six temperature sensors embedded in the insulated barrels at different depths and two on each of the top cap and the bottom of the test cell, the temperature profile throughout the specimen could be recorded during the test, and the 0 °C isotherm (assumed as the freezing front) could be calculated by interpolation. Water intake from the bottom filter stone was allowed during the entire frost heave test to simulate the moisture transfer from ground water in practice.

Table 2. Frost heave and thawing conditions adopted in the test.

Step	Loading Condition	Temperature		Time Period	Water Level of External Water Reservoir
		Top	Bottom		
Preliminary freezing	Unloaded *	−3 °C	+1 °C	24 h	N/A
Thawing and preloading	20 kPa	+3 °C	+3 °C	24 h, or until constant height **	Top level of specimen
Frost heave	unloaded	−3 °C	+1 °C	until zero net frost penetration *** (at least 24 h)	At the middle of specimen height
Intermediate thawing	20 kPa	−0.5 °C	+15 °C	until constant height	At the middle of specimen height
1st frost heave	20 kPa	−3 °C	+1 °C	until zero net frost penetration (at least 24 h)	At the middle of specimen height
Intermediate thawing	40 kPa	−0.5 °C	+15 °C	until constant height	At the middle of specimen height
2nd frost heave	40 kPa	−3 °C	+1 °C	until zero net frost penetration (at least 24 h)	At the middle of specimen height

Notes: Unloaded * means the specimen is loaded by the frame only (3.5 kPa); Constant height ** means the height of the specimen is unchanged for more than four hours; Net frost penetration *** is the frost penetration minus frost heave, for more than four hours.

Figure 8 shows the quarry fine specimen and frost heave test apparatus as applied in this study. Figure 9 presents the total heave, depth of frost front during the test, and some frost susceptibility parameters and indicators of quarry fines as obtained from the test. All these data and results are derived from the first frost heave test with a surface loading of 3.5 kPa (Table S6).

There are several parameters and frost susceptibility indicators that can be derived from the frost heave test, as shown in Figure 9:

- h is the frost heave, or the change in height of the specimen during the test, in mm;
- z is the depth of frost penetration, which is the sum of the initial height and the frost heave minus the height of the unfrozen part of the specimen, in mm;
- h/z is the frost heave ratio, which is an indicator of the relative percentage of frost heaving from the depth of frost penetration, in %;
- SP is the segregation potential, or frost heave coefficient, which is the ratio between the frost heave rate and the actual temperature gradient over the frozen part of the specimen, in mm²/Kh.

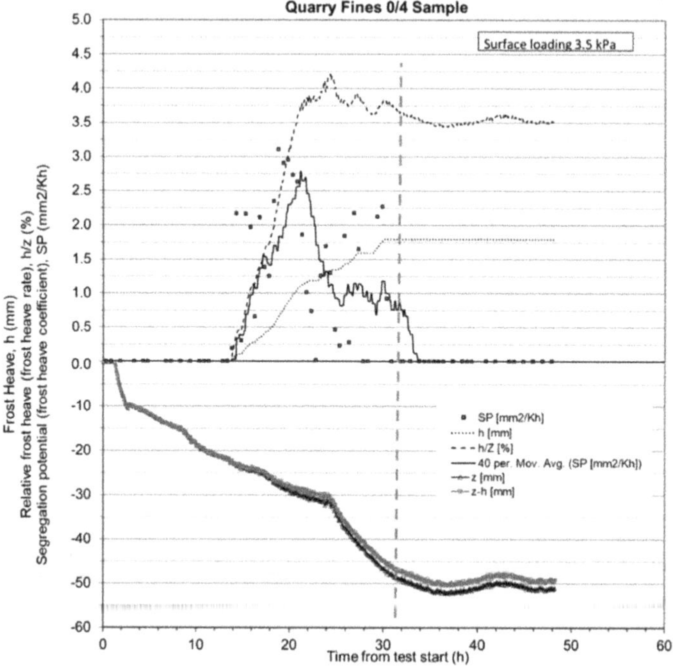

Figure 9. Frost heave parameters of a quarry fines (0–4mm) sample.

As shown in Figure 9, the total frost heave (h) of the quarry fine specimen in the unloaded frost heave period can be divided into three stages. In the very first stage, from the beginning of the test to 13.67 h, there is no obvious frost heave observed, even though the temperature inside the specimen decreased significantly. In the second stage, when the temperature at the very top of specimen dropped to around zero, total heave started to increase strongly with time. During this stage, water inside the specimen froze and expanded against the particles, which resulted in a total height increase in the specimen. Finally, in the last stage when the temperature throughout the specimen reached equilibrium status, the total frost heave leveled off at around 1.79 mm. Accordingly, the depth of frost front (height of the unfrozen part) decreased with the origin at the very bottom of the specimen, or the depth of frost part of the specimen increased as the top plate of the specimen was kept cool by the circulation of cryomat-cooled liquid. The depth of frost penetration remained stable at about 50.0mm from the time when the temperature throughout the whole specimen was stable. The frost heave ratio is also plotted as h/z in Figure 9, showing that it increased and finally leveled off at 3.5% during the unloaded frost heave test.

The net depth of frost penetration $(z–h)$, which is the depth of frost penetration minus frost heave, has the same changing tendency with time as for the depth of frost penetration. This curve is critical in finding out the segregation potential (SP) value for this quarry fine material and is used as a reference. The SP, however, was found to be a much more scattered data series than the other parameters, but the trend can still be approximated by moving average techniques. The SP value can be read from Figure 9 on the predictive moving average curve of the SP data, at the very first point where the net depth of frost penetration becomes stable. In this case, it was found that there was a transition period 30 h after the beginning of test, when the SP values suddenly decreased, becoming negligibly small. The SP value of quarry fine material in this study, was found to be just within this area. The exact point was obtained by looking for the point on the $z–h$ curve that has the same $z–h$ value as the average $z–h$ value from this point and afterwards. This point was found to be at 32.09 h from the beginning of test, and the corresponding SP value at this point is 0.75 on the moving average curve. It should be noted that this SP value is conservative because afterwards the frost heave is stable and the SP value becomes negligible.

The segregation potential of the unbound quarry fines with 0–4 mm particle sizes is then determined to be 0.75. When compared to the frost susceptibility classification in the literature, as listed in Table 3, it appears that the segregation potential values indicats a frost class of low susceptibility. The results from the laboratory capillary test further verify this low potential for frost susceptibility, showing that the capillary rise of the quarry fines with a 0–4 mm particle size is around 70 to 85 mm (Officially the limit is 1.0 meter but nowadays the Finnish agencies are more prone to use stricter requirements). Even though the unbound quarry fine materials is not very frost susceptible, it is still to some extent susceptible to frost actions and thus not safe to be used for pavements as it is, especially in cold regions. In this regard, stabilization techniques are necessary to improve the frost susceptibility of quarry fines and enable its applications in pavement base, subbase, or filter layers.

Table 3. Determination of the frost susceptibility of a soil type [42,50].

Frost Class	Segregation Potential SP (mm^2/Kh)
Negligible	<0.5
Low	0.5–1.5
Medium	1.5–3.0
Strong	>3.0

5. Discussion and Conclusions

The aggregate source depletion, as well as the demand for a carbon-neutral society and circular economy, all promote the high-volume utilization of secondary materials such as quarry fines. In this research, a feasibility assessment including a series of laboratory tests and analyses were conducted to evaluate the properties of quarry fine materials to determine if this type of material could be qualified for pavement base, subbase, or filter layers, especially in high volume applications, as well as to evaluate the durability of quarry fines in term of frost susceptibility. The main conclusions are listed as follows:

- The wet sieving method is more appropriate for determining the gradation curve of quarry fines with a 0–4 mm grain size. Based on gradation information obtained from laboratory sieving and hydrometer tests, the quarry fines used in this study can be classified as well-graded with the gradation characteristics of both fine- and dense-graded material. It also shows that virgin quarry fines can satisfy the requirements for a filter layer but not for a base or subbase, and indicates that quarry fines might be frost susceptible.

- The virgin quarry fines showed self-draining properties. Nevertheless, a maximum dry density exists at the optimal water content of 9.3%, even though only a very limited influence of water content can be observed on the dry density.

- The coefficient of permeability of unstabilized quarry fine specimen was around 5.75×10-5 m/s as determined by the falling head method. This value falls within the typical permeability ranges of commonly adopted conventional pavement base and subbase materials and can be validated for application in base, subbase, and other layers from the view of permeability. To be validated for filter layers, a higher degree of compaction is desirable to prevent fines immigration and frost actions in cold regions.

- According to the frost heave test results, it was found that the changing of total frost heave is closely related to the temperature and depth of frost penetration throughout the specimen.

- The quarry fine specimen in the unloaded frost heave test can be classified into the low frost-susceptibility class according to the segregation potential values. In addition, gradation information and capillary rise test results have all verified this finding. As a result, it is concluded that the unstabilized quarry fines should be classified as frost-susceptible to ensure sound and reliable design and good performance in the long-run. To improve the frost-susceptible properties of quarry fines, stabilization techniques are necessary to qualify their application as pavement construction materials. Further research is undergoing and will be presented in the future.

Supplementary Materials: The following are available online at http://www.mdpi.com/1996-1944/12/8/1209/s1, supplementary materials.docx. Table S1: Particle size distribution of quarry fines by sieving method, Table S2: Gradation limits for different layers, Table S3: Hydrometer test results of particle size distribution (0–0.063 mm), Table S4: Dry density versus final water content from Proctor test, Table S5: Permeability test results, Table S6: Frost heave test results.

Author Contributions: Conceptualization, L.K-T.; Methodology, Y.Z., H.G., and A.M.; Software, Y.Z.; Validation, Y.Z., H.G. and L.K-T.; Formal Analysis, Y.Z., H.G.; Investigation, Y.Z.; Resources, L.K-T.; Data Curation, Y.Z., L.K-T.; Writing—Original Draft Preparation, Y.Z.; Writing—Review and Editing, L.K-T., H.G. and A.M.; Visualization, Y.Z., L.K-T.; Supervision, L.K-T.; Project Administration, L.K-T.; Funding Acquisition, L.K-T.

Funding: This research received no external funding.

Acknowledgments: This research was supported by the School of Engineering, Aalto University, for financing the researchers, and Destia Ltd. for providing the raw materials. We would like to present our gratitude to our colleagues Hakala Veli-Antti, Peltonen Petri, Nikiforow Heli, Makowska Michalina, Eloranta Pekka, and Borén Mari from Destia Oy, who provided expertise that greatly assisted the research.

Conflicts of Interest: The authors declare no conflict of interest. The funders had no role in the design of the study, the collection, analysis, or interpretation of data, the writing of the manuscript, or in the decision to publish the results.

References

1. Finnish Road Administration. *Tierakenteen Suunnittelu*; Edita Prima Oy: Helsinki, Finland, 2004; p. 69. ISBN 951-803-403-6.

2. Kolisoja, P. Resilient Deformation Characteristics of Granular Materials. Ph.D. Thesis, Tampere University of Technology, Tampere, Finland, 1997.

3. Betoni. Kiviaines. Available online: https://betoni.com/tietoa-betonista/perustietopaketti/betoni-rakennusmateriaalina/kiviaines/ (accessed on 7 December 2018).

4. Hernesniemi, H.; Berg-Andersson, B.; Rantala, O. *Kalliosta Kullaksi—Kummusta Klusteriksi*; Elinkeinoelämän tutkimuslaitos. ETLA, Taloustieto: Helsinki, Finland, 2011; p. 241. ISBN 978-951-628-527-9.

5. Korkiala-Tanttu, L.; Tenhunen, J.; Eskola, P.; Häkkinen, T.; Hiltunen, M.-R.; Tuominen, A. *Environmental Values and Ecoindicators of the Infra Construction*; Finnra Reports; Finnish Road Administration: Helsinki, Finland, 2006; p. 53. ISBN 951-803-713-2.

6. Henrotay, C. Buildings as Materials Bank. In Proceedings of the Rakentamisen kiertotalouden ajankohtaispäivät, Helsinki, Finland, 7–8 November 2018.

7. Ehrukainen, E. Kiviainesten kestävä kierrätys ja käyttö. Available online: http://www.uusiomaarakentaminen. fi/sites/default/files/Kiviainesten%20kest%C3%A4v%C3%A4%20kierr%C3%A4tys%20ja%20k%C3% A4ytt%C3%B6%20%E2%80%93%20Eija%20Ehrukainen_0.pdf (accessed on 7 December 2018).

8. Levinen, R. Jätedirectiivin rakennusalaa koskevat muutokset, The changes for building industry caused by the waste directive. In Proceedings of the Rakentamisen kiertotalouden ajankohtaispäivät, Helsinki, Finland, 7–8 November 2018.

9. European Commission. Directive (EU) 2018/851 of the European Parliament and of the Council of 30 May 2018 Amending Directive 2008/98/EC on Waste (Text with EEA Relevance). 2018. Available online: https://eur-lex.europa.eu/eli/dir/2018/851/oj (accessed on 6 February 2019).

10. United Nations. The Sustainable Development Agenda. 2018. Available online: https://www.un.org/ sustainabledevelopment/development-agenda/ (accessed on 7 December 2018).

11. Prime Minister's Office. *Government Report on the 2030 Agenda for Sustainable Development. Sustainable Development in Finland—Long-Term, Coherent and Inclusive Action*; Prime Minister's Office: Helsinki, Finland, 2017; p. 57. ISBN 978-952-287-392-7.

12. Major of Helsinki. Hiilineutraali Helsinki 2035- Toimenpideohjelma. Available online: https://www.hel. fi/static/liitteet/kaupunkiymparisto/julkaisut/julkaisut/HNH-2035-toimenpideohjelma.pdf (accessed on 7 December 2018).

13. EPA. Identification of Non-Hazardous Secondary Materials That Are Solid Waste. Available online: https://www.epa.gov/rcra/identification-non-hazardous-secondary-materials-are-solid-waste (accessed on 8 July 2018).

14. European Commission. Raw Materials. Available online: http://ec.europa.eu/environment/green-growth/ raw-materials/index_en.htm (accessed on 8 July 2018).

15. Mitchell, C. Quarry Fines and Waste. Available online: https://www.bgs.ac.uk/downloads/start.cfm?id=1449 (accessed on 8 July 2018).

16. Galetakis, M.; Soultana, A. A review on the utilisation of quarry and ornamental stone industry fine by-products in the construction sector. *Constr. Build. Mater.* **2016**, *102*, 769–781. [CrossRef]

17. Pitkänen, I.-J. Selvitys Destia Oy:n kivituhkamääristä ja kivituhkan nykyisistä ja uusista käyttömahdollisuuksista. Bachelor's Thesis, Savonia University of Applied Science, Kuopio, Finland, 2015.

18. Melander, M. Kivituhkan soveltuminen sidotun kantavan kerroksen runkoaineeksi. Master's Thesis, Aalto University, Espoo, Finland, 2018.

19. Reid, J.M.; Evans, R.D.; Holnsteiner, R.; Wimmer, B.; Gaggl, W.; Berg, F.; Pihl, K.A.; Milvang-Jensen, O.; Hjelmar, O.; Rathmeyer, H.; et al. *ALT-MAT: Alternative Materials in Road Construction*. The European Commission under the Transport RTD Programme of the 4th Framework Programme. 2001, p. 190. Available online: https://trimis.ec.europa.eu/sites/default/files/project/documents/20040909_172706_52558_alt-mat.pdf. (accessed on 7 December 2018).

20. Thomas, J.; Harilal, B. Mechanical properties of cold bonded quarry dust aggregate concrete subjected to elevated temperature. *Constr. Build. Mater.* **2016**, *125*, 724–730. [CrossRef]

21. Dubois, V.; Wirquin, E.; Flament, C.; Sloma, P. Fresh and hardened state properties of hemp concrete made up of a large proportion of quarry fines for the production of blocks. *Constr. Build. Mater.* **2016**, *102*, 84–93. [CrossRef]

22. Rai, B.; Kumar, S.; Satish, K. Effect of Fly Ash on Mortar Mixes with Quarry Dust as Fine Aggregate. *Adv. Mater. Sci. Eng.* **2014**, *2014*, 1–7. [CrossRef]

23. Medina, G.; Sáez del Bosque, I.F.; Frías, M.; Sánchez de Rojas, M.I.; Medina, C. Durability of new recycled granite quarry dust-bearing cements. *Constr. Build. Mater.* **2018**, *187*, 414–425. [CrossRef]

24. Amadi, A.A. Enhancing durability of quarry fines modified black cotton soil subgrade with cement kiln dust stabilization. *Transp. Geotech.* **2014**, *1*, 55–61. [CrossRef]

25. Sivrikaya, O.; Kıyıldı, K.R.; Karaca, Z. Recycling waste from natural stone processing plants to stabilize clayey soil. *Environ. Earth Sci.* **2014**, *71*, 4397–4407. [CrossRef]

26. Oncu, S.; Bilsel, H. Ageing effect on swell, shrinkage and flexural strength of sand and waste marble powder stabilized expansive soil. *E3S Web Conf.* **2016**, *9*, 1–6. [CrossRef]

27. Gurbuz, A. Marble powder to stabilise clayey soils in subbases for road construction. *Road Mater. Pavement Des.* **2015**, *16*, 481–492. [CrossRef]

28. Gautam, P.K.; Kalla, P.; Jethoo, A.S.; Agrawal, R.; Singh, H. Sustainable use of waste in flexible pavement: A review. *Constr. Build. Mater.* **2018**, *180*, 239–253. [CrossRef]

29. Galetakis, M.; Piperidi, C.; Vasiliou, A.; Alevizos, G.; Steiakakis, E. Experimental investigation of the utilization of quarry dust for the production of microcement-based building elements by self-flowing molding casting. *Constr. Build. Mater.* **2016**, *107*, 247–254. [CrossRef]

30. Vijayalakshmi, M.; Sekar, A.S.S.; Prabhu, G.G. Strength and durability properties of concrete made with granite industry waste. *Constr. Build. Mater.* **2013**, *46*, 1–7. [CrossRef]

31. Galetakis, M.; Alevizos, G.; Leventakis, K. Evaluation of fine limestone quarry by-products, for the production of building elements—An experimental approach. *Constr. Build. Mater.* **2012**, *26*, 122–130. [CrossRef]

32. Karasahin, M.; Terzi, S. Evaluation of marble waste dust in the mixture of asphaltic concrete. *Constr. Build. Mater.* **2007**, *21*, 616–620. [CrossRef]

33. Karakus, A. Investigating on possible use of Diyarbakir basalt waste in Stone Mastic Asphalt. *Constr. Build. Mater.* **2011**, *25*, 3502–3507. [CrossRef]

34. Akbulut, H.; Gurer, C. Use of aggregates produced from marble quarry waste in asphalt pavements. *Build. Environ.* **2007**, *42*, 1921–1930. [CrossRef]

35. Finnish Standard Association. *SFS-EN 932-1 Tests for General Properties of Aggregates. Part 1: Methods for Sampling*; CENELEC Management Centre: Brussels, Belgium, 1997; p. 27.

36. Finnish Standard Association. *SFS-EN 1097-5 Tests for Mechanical and Physical Properties of Aggregates. Part 5: Determination of the Mater Content by Drying in a Ventilated Oven*; CENELEC Management Centre: Brussels, Belgium, 2008; p. 11.

37. Finnish Standard Association. *SFS-EN 933-1 Tests for Geometrical Properties of Aggregates. Part 1: Determination of Particle Size Distribution. Sieving Method*; CENELEC Management Centre: Brussels, Belgium, 2012; p. 18.

38. Test Sieving: Principles and Procedures. A Discussion of the Uses, Capabilities, and Limitations of Testing Sieves as Analytical Tools. Available online: http://www.advantechmfg.com/pdf/Principles%20and%20Procedures%2002_21_2013.pdf (accessed on 8 July 2018).

39. *InfraRYL 2010 Code of Building Practice, Infrastructure, Part 1: Roads and Areas*; 21300 Kantavat kerrokset; Rakennustieto Oy: Helsinki, Finland, 2010; pp. 318–320. ISBN 978-951-682-958-9.

40. Finnish Standard Association. *SFS-EN ISO 14688-2:2018 Geotechnical Investigation and Testing. Identification and Classification of Soil. Part 2: Principles for a Classification (ISO 14688-2:2017)*; CENELEC Management Centre: Brussels, Belgium, 2018; p. 18.

41. Finnish Standard Association. *SFS-EN ISO 17892-4:2016 Geotechnical Investigation and Testing. Laboratory Testing of Soil. Part 4: Determination of Particle Size Distribution*; CENELEC Management Centre: Brussels, Belgium, 2016; p. 37.

42. Slunga, E.; Saarelainen, S. Determination of frost-susceptibility of soils. In Proceedings of the 16th International Conference on Soil Mechanics and Geotechnical Engineering, Osaka, Japan, 12–16 September 2005; Millpress: Rotterdam, The Netherlands, 2005; pp. 3577–3578.

43. Finnish Standard Association. *EN 13286-2: 2010 Unbound and Hydraulically Bound Mixtures- Part 2: Test Methods for Laboratory Reference Density and Water Content- Proctor Compaction*; CENELEC Management Centre: Brussels, Belgium, 2010; p. 29.

44. Road and Water Engineering Board. *Construction Research and Design Part II Laboratory Tests*; Government of Road and Water Construction: Helsinki, Finland, 1970. (In Finnish)

45. Tangpithakkul, R. Study of Permeability of Pavement Base Materials. Master's Thesis, Ohio University, Athens, OH, USA, 1997.

46. Vermont Agency of Transportation. Permeability of Highway Base and Subbase Material. Available online: https://vtrans.vermont.gov/sites/aot/files/highway/documents/materialsandresearch/completedprojects/AOT_PermeabilityofHighwayBaseandSub-baseMaterial.pdf (accessed on 8 July 2018).

47. Khoury, N.N.; Zaman, M.M.; Ghabchi, R.; Kazmee, H. *Stability and Permeability of Proposed Aggregate Bases in Oklahoma. Report No.FHWA-OK-09-05*; Oklahoma Department of Transportation Planning and Research Division: Oklahoma City, OK, USA, 2010; p. 243.

48. Chamberlain, E.J. *Frost Susceptibility of Soil Review of Index Tests. Report No. Monograph 81-2*; Federal Aviation Administration (Systems Research and Development Service) and Federal Highway Administration (Office of Research): Washington, DC, USA, 1981; p. 121.

49. Onninen, H. *Method Description TPPT-R07 Frost Heave Test Thaw Compression Test*; VTT Communities and Infrastructure: Espoo, Finland, 1999; p. 10.
50. Slunga, E.; Saarelainen, S. Determination of Frost-Susceptibility of Soil. Available online: https://www.issmge.org/uploads/publications/1/33/1989_02_0171.pdf (accessed on 8 July 2018).

Article

Assessment for Sustainable Use of Quarry Fines as Pavement Construction Materials: Part II-Stabilization and Characterization of Quarry Fine Materials

Yinning Zhang [1,*], Leena Katariina Korkiala-Tanttu [1] and Mari Borén [2]

1   Department of Civil Engineering, Aalto University, 00076 Aalto, Finland
2   Destia Oy, Neilikkatie 17, 01301 Vantaa, Finland
*   Correspondence: yinning.zhang@aalto.fi

Received: 1 July 2019; Accepted: 24 July 2019; Published: 1 August 2019

Abstract: A secondary by-product, quarry fines, has previously been investigated for applications in high volume as pavement construction materials. Results from a series of laboratory tests suggest qualified basic properties except for the possibility of frost susceptibility for the virgin quarry fines. In Part II of the research, stabilized quarry fine specimens were prepared and investigated in view of the mechanical behavior, and the durability represented by susceptibility to freezing and thawing cycles. The unconfined compressive strength, which is also the commonly used strength indicator, was adopted to evaluate the validity of the stabilized quarry fines as pavement construction materials. The laboratory-determined parameters were then compared among specimens treated with different stabilizers and with the typical requirements for pavement base/subbase layers. The stabilized quarry fines can be qualified for applications in pavement base, subbase and filter layer depending on the types of stabilizers used and degrees of compaction achieved.

Keywords: sustainable; secondary materials; quarry fines; freezing-thawing

1. Background

Quarry fine is a secondary by-product produced from the processing of rocks or aggregates yet of great material value for many civil engineering purposes. A huge amount of quarry fine materials is produced every year as a by-product of processing aggregates. Being an under-utilized material, the majority of it is just stored. At present, the most common application area of quarry fines in Finland is to use it as the uppermost unbound layer of yards, fields and low-volume streets [1]. Some research has also found that quarry fines could be used as an alternative to Portland cement for concrete [2], a stabilizer/additive to improve soil properties [2–6], and a substitute to sands/filler in concrete [7–10]. However, compared with the under-utilized portion, the quarry fine that has found its application in practice is still far from adequate. Unfortunately, these "leftover" quarry fine materials, if not appropriately treated, will lead to environmental and economic problems in society since the fine fraction of quarry fines will be easily dispersed by gravity, water and wind. Moreover, in the light of resource depletion and the promotion of a carbon-neutral society, wise utilization of existing resources is of vital importance. Under this circumstance, an effective way to utilize the quarry fine material in large volume is needed, and research activities regarding the characterization of this material to validate the promising applications are necessary.

As verified and stated in the previous research published earlier [11], basic properties of the unbound quarry fine materials have been investigated by a series of laboratory tests and analysis. Based on the particle distribution information obtained from both sieving and hydrometer methods, as well as the frost heave test results, it was found that the unbound quarry fines of 0–4 mm are prone to be frost-susceptible. The mechanical properties of quarry fine materials alone do not satisfy most of the requirements for a base or subbase as provided by many agencies. Thus, the stabilization of quarry fines to improve its properties for pavement base/subbase applications is potentially an effective way of utilizing it in higher volumes. The main motivations of the design, improvement and application of stabilized quarry fines as base/subbase material are listed as follows.

(1) Conventional road construction materials of good quality, the majority as natural resources, are becoming scare due to the booming population, urbanization and industrial developments in the past few decades. Demanding sustainability and environmentally-friendly construction activities has become essential to all the countries encountering this problem. Numerous environmental constrains have been implemented by many countries, such as the EU Landfill Directive.

(2) An increasing amount of industrial by-products have so far been stored, leading to environmental problems and waste of resources. As this problem has become more widely recognized, there have been increasing efforts to reuse industrial by-products in infrastructure constructions. Often these by-products have been used as stabilizers for treating weak soils. Several existing researches have demonstrated very promising results from utilizing industrial by-products to stabilize and improve the properties of various soil types. Results have shown that not only material strength, but also freeze-thaw performance, and compaction can be enhanced by applications of by-product binders. [12–14].

(3) There is a need for improving the long-term performance and increasing the lifespan of current infrastructure systems to achieve sustainable development. Therefore, higher requirements for material strength, stability and durability are necessary, and the application of stabilized base/subbase to achieve better performance is welcomed. Numerous types of recycled materials have been investigated for their effectiveness in cementation and ground improvement in previous research, including but not limited to Portland cement (PC), ground granulated blast furnace slag (GGBS), fly ash, recycled gypsum, cement kiln dust (CKD), bentonite and so forth [15]. All the existing experience in different types of stabilizers has been used as instructive information for stabilizing quarry fines as a pavement construction material in this study.

The aim of this follow-up study is to compare different binder materials to widen the usability of quarry fines, to define the mechanical properties of the stabilized samples, to assess their durability in laboratory, and to access their suitability as a pavement construction material. The tested mechanical property of the stabilized quarry fines was Unconfined Compressive Strength (UCS), whereas the durability of the stabilized quarry fines was further verified by a freeze-thaw test and capillary rise test. Last but not least, the feasibility of the quarry fine materials for pavement construction in the base, subbase or filler layer has been discussed. This study concentrates mainly on the mechanical and durability properties, not on the environmental (leaching or emission) issues.

2. Virgin Quarry Fine Materials

In the previous study [11], virgin quarry fine materials of 0 to 4 mm grain size, produced at the quarry in Koskenkylä by Destia Oy, were collected and investigated through a series of laboratory tests. The host rock of quarry is very high quality rock with extremely good abrasion value [16], which is only about 1% of what Finnish host rock material fulfills. Based on the results of previous research, the unbound quarry fine materials are shown to be good-quality and well-graded aggregates with both fine- and dense-graded characteristics. It has shown self-draining properties, but still a maximum dry density can be achieved, even though the influence of water content on dry density is low. The coefficient of permeability also complies with the conventional values in pavement base and subbase layers. In general, it is a promising material to be utilized for pavement constructions with regard to gradation, compactability and permeability. However, both the gradation information and frost heave test results have shown frost-susceptibility of the virgin quarry fines.

To improve the performances of quarry fines, several types of stabilizers were used in this study. Further laboratory tests were performed to characterize the stabilized quarry fine materials and to access its feasibility in pavement constructions. Comparison analysis between cement-stabilized quarry fines (as a reference material) and quarry fines stabilized with other binders gives good evidence to show how effective different types of stabilizers are and to which part of the pavement the stabilized materials can be applied.

3. Characterization of Stabilized Quarry Fines As Construction Material for Pavements

The main issue in the stabilization is to determine what type of stabilizer would be effective for better performance, cost-effectiveness and environmental friendliness. In this regard, the cement, which is one of the most commonly used and well established stabilizers for granular materials around the world, was selected as a reference material. It provided a baseline for comparisons with other types of by-product stabilizers: Ecolan and fly-ash stabilizers. These two types of potential stabilizers were proven to be effective as stabilizers, and their detailed components as well as the adopted ratios to quarry fines are introduced in the following sections.

3.1. Specimen Preparation

The appropriate dosage or proportion of each ingredient of stabilized material is the key factor to obtain better performance. The determination of mixture proportions requires a large amount of laboratory testing including aggregate gradation tests, reference density tests to determine the relationship between water content and dry density, selection of the relevant water content, and possibly activator content for each stabilizer. Moreover, the situation is even more complicated when it comes to the stabilized materials due to potential hydraulic reactions between the stabilizer and water. In this study, the mix proportions were selected either based on earlier laboratory and field test results, empirical knowledge, or suggestions from experienced agencies. The following section describes how the mixture design was determined for laboratory experiments.

- Cement type and content: For cement-stabilized quarry fines, quick cement by Finnsementti (pikasementti, Finnsementti Oy, Parainen, Finland) [17] was chosen because it is the purest calcium carbonate of current cement types, and many older researches have been made with relatively pure cement. Cement content was selected based on Finnish guidelines of overlay structures [18]. According to the guidelines, the typical cement content for stabilized soil is 2.5% to 5.0%. Considering the general experience that lower cement content is beneficial to the long-term performance of a stabilized base [19], as well as that for weather resistance the cement content should be more than 4.5% [18], the final two levels of cement content are determined to be 2.5% and 4.5% for quarry fine stabilization.

- Ecolan stabilizer contents: Ecolan stabilizer (Kolliantal och-slag, Renotech Oy, Turku, Finland) composed of coal ash, wood biomass, lime and cement, was also selected. The contents adopted for this stabilizer were selected to be 4.5% and 6.5% with suggestions from the earlier tests [20].

- Fly ash-based stabilizer content: Fly ash-based stabilizer is composed of 100% coal combustion residuals (CCRs) with a major component of fly ash. The stabilizer content was selected based on existing experience of the typical fly ash content as suggested for a stabilized base course by FHWA (federal highway administration) [21,22]. It suggests that a content of 12% to 30% of fly ash, with an activator of lime or cement adopted together, should be added to stabilize the pavement base course. Therefore, considering economy and environmental efficiency, two levels of low fly ash-based stabilizer content, 12% and 15%, were selected. Cement was also used as an activator in the fly ash-based mixture because the further pozzolanic reactions require an alkali condition to form cementitious Calcium-Silicate-Hydrate (C-S-H) gels and to stabilize the quarry fines. A cement content of 1% was adopted as an activator in the fly ash-stabilized quarry fines according to the FHWA recommended activator content of 0.5% to 1.5% for Portland cement [21].

- Optimal water content: The optimal water content of the unbound quarry fines was determined according to European standard in previous research [11]. The as-determined optimal water content of 9.26% was used as reference to select water content for adequate compaction.

- Water to cement ratio: Water-to-cement ratio was selected for effective hydraulic reactions when the quarry fines were stabilized with different stabilizers. According to recommendations by American Concrete Institute (ACI) [21] for the fly ash concrete, two levels within the typical range of 0.4 and 0.6 were adopted for stabilized quarry fines.

Following the designed ingredient proportions, raw materials of designed dosages were mixed in an electrical mixer thoroughly to get a batch for triplicated specimens. Each stabilized quarry fine specimen was fabricated by compacting a certain mass of material into a 50-mm-diameter mold in three layers until a final height of 100 mm and a degree of compaction of 85%, was reached. The compaction energy was kept the same for the three layers of triplicated specimens by keeping the same blow number and dropping from the same height, which also ensures a homogeneous specimen as much as possible.

Since the 85% degree of compaction is the highest level that can be achieved by manual compaction, the specimens of 50 mm diameter were only used to determine the best mixture design. To achieve a higher degree of compaction, more specimens were prepared using the gyratory compactor with the best mixture design to evaluate performances of stabilized quarry fines.

3.2. Capillary Rise

Capillary actions driven by matric suction will pull water upwards to the frozen zone, whereas hydraulic conductivity will allow quantities of water to be transferred to the frozen zone. As a result, ice lenses are able to form continuously in the material and problems related with frost heave, and freezing-thawing can occur. In Part 1 [11], virgin quarry fines of grain size 0–4 mm have been proven to be frost susceptible based on laboratory-determined indicators: segregation potential. The effectiveness of applying stabilization techniques were evaluated by capillary rise test according to the European standard SFS-EN 13057: Products and systems for the protection and repair of concrete structures. Test methods. Determination of resistance of capillary absorption [23].

There are different analyzing methods to obtain the absorption coefficient, including a 30 min method, one tangent method, and two tangent method [24]. The main difference of these methods is the way of selecting effective data points for the coefficient calculation. In this study, it was found that different methods have very limited influence on the final results, and one tangent method was used to determine the indicator "absorption coefficient" by following the European standard. The absorption coefficient was defined as the gradient of the prediction line between water uptake per unit area and square root of time (Figure 1). All the groups showed flat end portions, which indicated full saturation before the end of test, thus only the initial linear part of the graph has been used for gradient determination in this study.

The results show that the capillary sorption coefficients of stabilized quarry fine specimens are comparable to but still higher than the coefficients of concrete in the literature, if the same test method (EN 13057) is adopted. The average rate of water absorption for concrete cross-beam with silane of 20-year service life is from 0.247 to 1.033 mg/cm^2 s$^{1/2}$ [25]. The reported coefficients for hardened concrete containing recycled aggregate, air-entraining agent, cement, fly ash, and superplasticizer of varied proportions were from 0.746 to 1.549 mg/cm^2 s$^{1/2}$ (5.56 to 24.0 cm^2/s) [26]. Similar ranges can be found in other studies where the coefficient is from 0.68 to 2.21 mg/cm^2 s$^{1/2}$ [27]. On the other hand, the coefficients for quick cement-stabilized quarry fine specimens were 1.24 to 3.56 mg/cm^2 s$^{1/2}$, and for quarry fine specimens stabilized with Ecolan or fly ash-based stabilizer, higher values from 5.85 to 11.36 mg/cm^2 s$^{1/2}$, and 9.27 to 14.41 mg/cm^2 s$^{1/2}$ were obtained. If compared with mixtures of river sand and natural hydraulic lime, similar water absorption coefficients between 7.9 and 19.2 mg/cm^2 s$^{1/2}$ have been published [24]. This coefficient for the building material was determined by European standard EN 1925 where a similar test method was used [28].

It can also be seen from Figure 1 that cement-stabilized quarry fines need a longer time to reach the status of full saturation compared with Ecolan-stabilized and fly ash-stabilized quarry fines, indicating that the Ecolan-stabilized and fly ash-stabilized quarry fines are more prone to absorb water and subject to freeze–thaw cycles. This finding can be further verified by the results of the freeze–thaw test presented in the following sections.

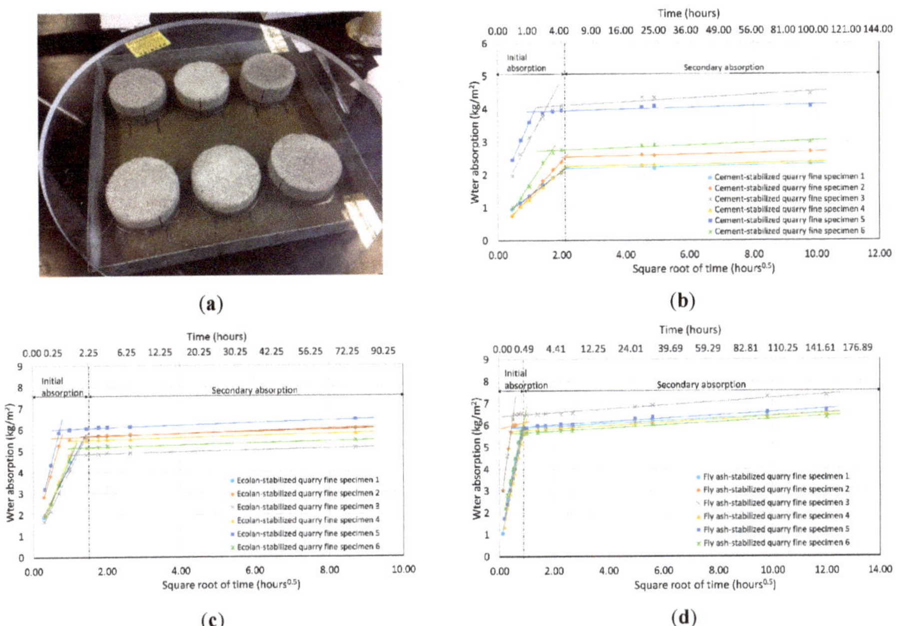

Figure 1. Capillary rise test: (**a**) specimens in one group; (**b**) water absorption of cement-stabilized quarry fines; (**c**) water absorption of Ecolan-stabilized quarry fines; (**d**) water absorption of fly ash-stabilized quarry fines.

3.3. Unconfined Compressive Strength

3.3.1. Determination of the Mix Design

Unconfined compressive test is one of the methods that is normally adopted to determine the strength of bound materials. Unconfined Compressive Strength (UCS) is a typical strength parameter used by not only researchers but also industries in civil engineering to verify if the mixture is suitable for applications with satisfactory performance. There are also empirical relationships developed to predict other material properties such as resilient modulus from the unconfined compressive strength. In pavement engineering, many research reports, specifications and standards have suggested the minimum UCS of stabilized material for base and subbase layers of pavement. In this consideration, the UCS has been selected in this study as the indicator to determine the best mix design for stabilizing quarry fines. Table 1 lists some of the requirements as provided by different agencies.

The standard test method Geotechnical investigation and testing. Laboratory testing of soil. Part 7: Unconfined compression test (ISO 17892-7:2017) was followed to determine the unconfined compressive strength. Stabilized quarry fine specimens of 50 mm diameter and 100 mm height were compressed at a strain rate of 1.0 mm per minute (between 1% and 2% of the specimen height per minute) while the load, displacement and elapsed time were recorded for further data processing. At a given mix design, triplicated quarry fine specimens were tested successively and the average was adopted in data analysis. Table 2 lists all the mixture designs of the specimens and the highest compressive strength together with the best mixture option that has been achieved.

Table 1. Unconfined compressive strength requirements for base/subbase/subgrade of pavements.

References	Minimum or Typical UCS Ranges (MPa)	Curing Period	Applications
Portland Cement Association [29]	2.068–5.516 (300–800 psi)	7 days	Medium- to high-volume roads for subbase/base
Mechanistic-Empirical Pavement Design Guide (MEPDG) [30]	1.72 (250 psi) 5.17 (750 psi)	7 days for cement and 28 days for lime-fly ash or cement fly ash	for subbase, subgrade for base
Austroads [31]	2.0	28 days curing and 4 h soaking prior to testing	bound pavement materials (stabilizer > 3%)
	1.0–2.0		subgrade, lightly-bound pavement materials (stabilizer < 3%)
Federal Highway Administration (FHWA) [32,33]	4.1	N/A	for stabilized drainable base in cold regions to resist frost deterioration
	0.7	7-day cure at 105 °F (40 °C)	subgrade lime/soil
	1.0	7-day cure at 105 °F (40 °C)	subgrade lime/fly ash/soil
	1.4	7-day cure	subgrade cement/soil, cement/fly ash/soil, or fly ash/soil
InfraRYL [34]	3.0–8.0 5.0–13.0 1.0–2.0	7-day cure 28-day cure 28-day cure	cement-stabilized cement-stabilized blast furnace slag stabilized

The maximum compressive strength of quarry fine specimens stabilized with quick cement at 85% degree of compaction reached around 2.5 MPa for a 28-day curing period in this study, compared to a lowest strength of 0.703 MPa that is only 28.1% of the maximum strength. This result indicates the profound influences of mixture design, or the stabilizer and water content, and associated compaction energy on the final strength of stabilized quarry fine materials. It is found in this study that a lower water-to-binder ratio and higher cement content added to the quarry fine specimen brings higher compressive strength of cement-stabilized quarry fine mixture. The same trends can be found for quarry fines stabilized with fly ash and an activator of cement. Nevertheless, different situations can be observed for other types of stabilizers. When the Ecolan stabilizer is applied, a higher water-to-stabilizer ratio and stabilizer content are needed to get higher compressive strength. It is also found that the fly ash itself has shown little improvement to the strength of quarry fines, but with a small amount of activator (e.g., cement) added, higher strength can be achieved. Generally, quarry fine specimens stabilized with quick cement have the highest UCS, whereas those stabilized with other binders consisting of coal ash with the same or even higher stabilizer content are less strong.

The undrained shear strength predicted from the UCS, as well as the secant modulus or modulus of deformation E_{50}, which was obtained from the data corresponding to half the maximum UCS [35,36], are also listed in Table 2. Figure 2 presents the relationship between the UCS and modulus of deformation from the laboratory test data for stabilized quarry fine specimens of different stabilizer content, water content and compaction energy. It is found that the modulus of deformation E_{50} of cement-stabilized quarry fines can be predicted as having an approximate value of 5 to 125 × UCS, even though the strength and modulus of cemented quarry fines are much higher than that of the cemented clay. For quarry fines stabilized with Ecolan and fly ash-based stabilizers (with activator), the coefficients are of 72 to 468 × UCS and of 35 to 179 × UCS respectively. The multiplier coefficients of Ecolan-stabilized quarry fines are higher than the typical coefficients of undisturbed cemented clays from 50 to 300 [37], while the fly ash-stabilized quarry fines have similar coefficients to undisturbed cemented clays.

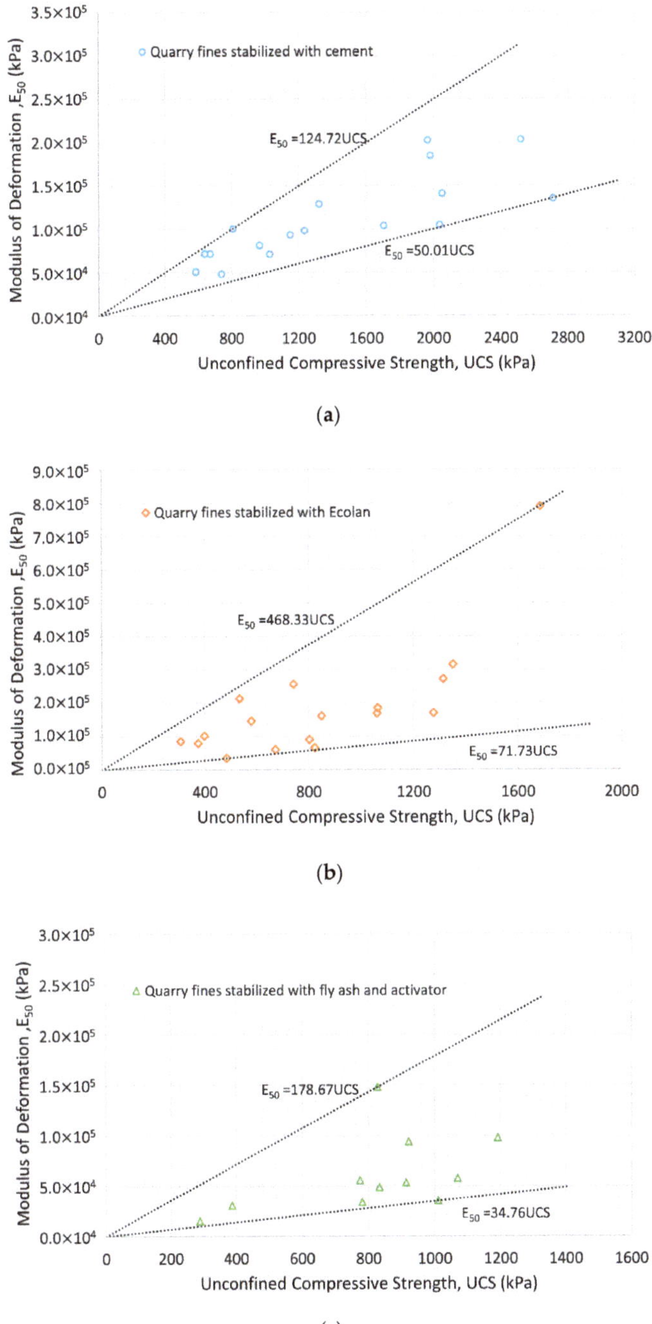

Figure 2. Modulus of deformation versus unconfined compressive strength of stabilized quarry fines: (**a**) Quarry fines stabilized with cement; (**b**) Quarry fines stabilized with Ecolan; (**c**) Quarry fines stabilized with fly ash and activator.

Table 2. Types of stabilized quarry fine specimens and their strengths.

Quick Cement Stabilized Quarry Fine Specimen #	Stabilizer Content	Water to Binder Ratio	** Unconfined Compressive Strength, kPa	Undrained Shear Strength, kPa	Secant Modulus/Deformation Modulus E50, MPa
A1–A3	4.5%	0.4	1741.2	870.59	172.08
A4–A6	4.5%	0.6	1924.4	962.22	115.10
B1–B3	2.5%	0.4	758.8	379.41	60.68
B4–B6	2.5%	0.6	703.5	351.77	82.63
C1–C3*	4.5%	0.0	2483.2	1241.59	196.64
C4–C6	2.5%	0.0	1128.0	563.98	87.92

Ecolan Stabilized Quarry Fine Specimen #	Binder Content	Water to Binder Ratio	** Unconfined Compressive Strength, kPa	Undrained Shear Strength, kPa	Secant Modulus/Deformation Modulus E50, MPa
E1–E3	4.5%	0.4	632.9	316.44	142.16
E4–E6	4.5%	0.6	713.2	356.59	95.85
F1–F3	6.5%	0.4	993.6	496.82	167.09
F4–F6*	6.5%	0.6	1316.7	658.35	225.64
G1–G3	4.5%	0.0	365.0	182.51	90.38
G4–G6	6.5%	0.0	981.7	490.83	273.83

Fly ash + Cement Stabilized Quarry Fine Specimen #	Binder Content + Activator Content	Water to Binder Ratio	** Unconfined Compressive Strength, kPa	Undrained Shear Strength, kPa	Secant Modulus/Deformation Modulus E50, MPa
H1–H3 *	12.0% + 1%	0.0	846.1	423.08	105.85
H4–H6	15.0% + 1%	0.0	571.3	285.67	42.44
H7–H9	12.0% + 1%	0.6	795.1	397.54	43.41
H10–H12	15.0% + 1%	0.6	734.8	367.40	34.56

* highest unconfined compressive strength, the best mixture; ** Curing conditions: 100% humidity, 25 °C, 28-day.

On the other hand, it can be seen from Table 2 that part of the compressive strength satisfies the requirements provided by several agencies as listed in Table 1. However, it is not convincing to qualify the best mixture for applications in pavement subbase and even base layers, considering that the quarry fines stabilized by Ecolan and fly ash-based stabilizers are around or below the lowest requirement, and much higher thresholds are required by the other agencies. Moreover, to resist frost deteriorations in cold regions, a higher compressive strength should be satisfied. Therefore, the test results as presented previously are only utilized to determine the best mixture design and the appropriate curing conditions. Considering that the degree of compaction of these specimens is 85% but most of the agencies require a higher degree (e.g., >90% in Finnish guidelines) especially for pavement base/subbase layer, higher strength is necessary and achievable for the same mix design at a higher compaction energy.

3.3.2. Influence of Degree of Compaction and Curing Days

Figure 3 shows that the degree of compaction has significant effects on the unconfined compressive strength for the cement-stabilized quarry fine specimens. The 100% degree of compaction does not imply a practical status of zero air voids, but is the highest degree of compaction that can be achieved by a gyratory compactor after 512 gyrations. It can be seen from the figure that the value of UCS has increased exponentially as the degree of compaction increased from 85% to 100% and that the UCS of stabilized quarry fine specimens of 96% degree compaction can adequately satisfy the requirements of pavement base/subbase provided by different agencies. It is therefore concluded that stabilized quarry fine materials can be qualified for pavement base/subbase applications in term of unconfined compressive strength, as long as appropriately designed and constructed. A least degree of compaction of 93% is recommended for cement-stabilized quarry fines with 0–4 mm grain size and 4.5% cement content for such purposes. For quarry fines stabilized by Ecolan, or fly ash and activator, at least a 93% or higher degree of compaction is necessary to validate their applications in base or subbase. Therefore, more stabilized quarry fine specimens with the best mixture design were prepared by gyratory compactor to achieve at least a 93% degree of compaction, and the following characterization of stabilized quarry fines is only based on these denser specimens.

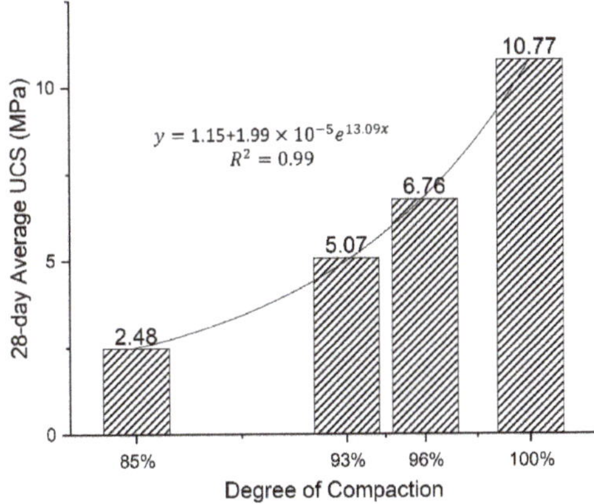

Figure 3. Effects of degree of compaction on the unconfined compressive strength.

Other than degree of compaction, the duration of curing time also influences the strength of stabilized quarry fines. In this study, the curing condition of 100% humidity was applied for all the stabilized quarry fine specimens of all the types of stabilizers. Figure 4 displays the influence of curing time on the UCS of stabilized quarry fines.

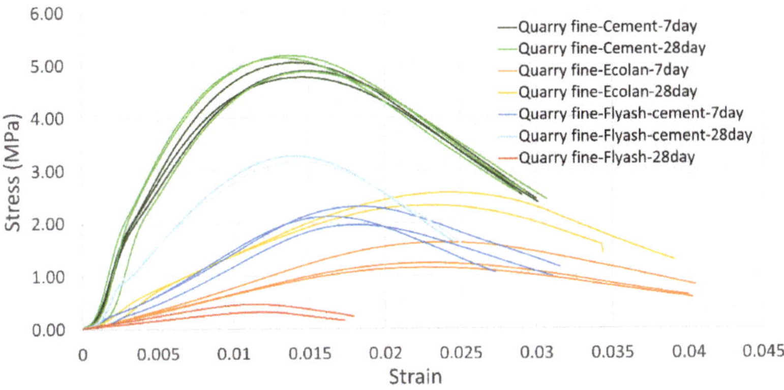

Figure 4. Effects of curing time on the unconfined compressive strength of stabilized quarry fines.

From the relationships between stress and strain in Figure 4, it can be seen that the type of stabilizer and mix design have influenced the way of deformation and failure under the same loading conditions. Quarry fines stabilized with Ecolan stabilizer have higher strains at failure whereas the cement-stabilized quarry fines demonstrate the most similar stress–strain relationship/shape of curve for different curing times. To better present the effects of curing time, an indicator based on "deformation-ductility ratio", which is originally developed for reinforced concrete, was developed and adopted. As shown in Figure 5, the strain–ductility ratio (SDR) is capable of considering the strains between the first yield point and the point where the strength is reduced to a certain percentage, serving as a good indicator for evaluating the ductility of the material. Table 3 lists the strain–ductility ratio (Equation (1)) of different stabilized quarry fines of 7-day and 28-day age, showing that cement-stabilized quarry fine is the most stable mixture regarding curing time. Ecolan-stabilized and fly ash-stabilized quarry fines all became more ductile and easier to yield, whereas the fly ash-stabilized quarry fines show considerable increase in ductility within curing time.

$$Strain-ductility\ ratio = \frac{\varepsilon_{50}}{\varepsilon_y} \tag{1}$$

where ε_{50} is the strain corresponding to 50% of the compressive strength after failure, and ε_y is the strain at first yield point.

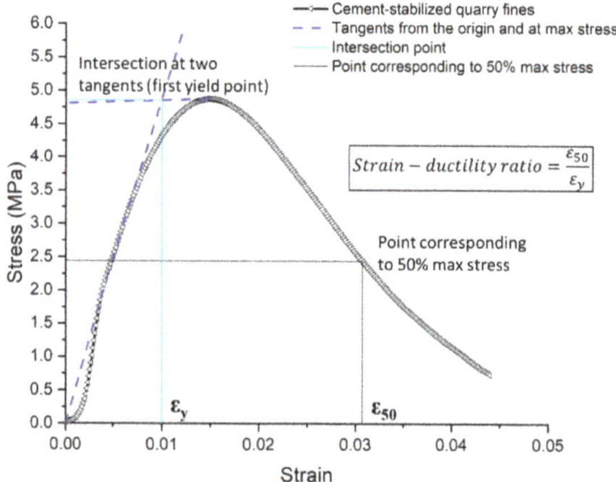

Figure 5. Strain–ductility ratio from stress–strain relationship.

Table 3. Strain–ductility ratio of stabilized quarry fines at 7-day and 28-day ages.

Type of Specimen	Cement-Stabilized Quarry Fines		Ecolan-Stabilized Quarry Fines		Fly Ash-Stabilized Quarry Fines	
Curing Days	7-day	28-day	7-day	28-day	7-day	28-day
Average SDR	3.64	3.63	2.14	2.24	1.95	2.45
* $\varepsilon_{y,ave}$	0.00817	0.00829	0.0188	0.0164	0.0153	0.0108
** $\varepsilon_{f,ave}$	0.0144	0.0139	0.0234	0.0237	0.0175	0.0148

* Average first yield strain; ** Average strain at failure.

The change in strength and ductility is linked to the continuous reaction between water and the binders during the curing period. The changes in strength due to curing time also vary among different mixtures. It is interesting to see that a longer curing time with the presence of water does not necessarily have positive effects on the development of unconfined compressive strength for stabilized quarry fine materials. In fact, the compressive strengths of cement-stabilized quarry fines are hardly changed even undergoing a longer period of curing time from 7 days to 28 days, with an average of 4.91 MPa and 5.07 MPa respectively. It is possibly related with the fast strength development of quick cement as a stabilizer for quarry fines, so that a longer period of exposure to moisture can be harmful to the strength of cement-stabilized quarry fines. However, when Ecolan or fly ash plus cement was used as a stabilizer, the strength increased by 78.6% and 42.5% from 7 days to 28 days.

3.3.3. Comparison with Requirements

To validate the stabilized quarry fine materials for pavement constructions in terms of unconfined compressive strength, unconfined compressive strength tests were conducted only on stabilized specimens of a higher degree of compaction (93% and 96%). These specimens were prepared using the gyratory compaction method, following the most promising mixture design as obtained from the previously mentioned investigation. Typical strength limits for pavement base, subbase and subgrade as listed in Table 1 were used to assess the potential applications of the stabilized quarry fine materials. Figure 6 shows the average unconfined compressive strength of each mix based on triplicated specimens.

It should be noted that only 7- and 28-day ages were used in this research so that the stabilizations mainly based on hydraulic reactions are considered, rather than the slow-setting pozzolanic and carbonation reactions.

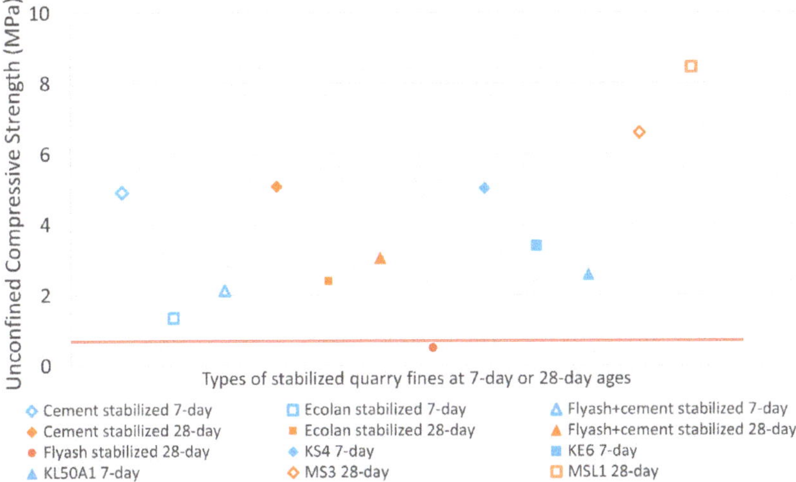

Figure 6. Average unconfined compressive strength of stabilized quarry fine specimens.

In fact, the lower limit of 0.7 MPa as plotted in Figure 6 is the minimum requirement of all the recommended lower limits by different agencies as listed in Table 1. It can be seen as the lowest limit that should be satisfied to qualify high-volume applications of stabilized quarry fine materials for pavement construction, in either base, subbase or subgrade. Fly ash itself has been proven to be not suitable to stabilize quarry fines because of its low strength. However, when a small amount of cement is used as an activator, the strength can be improved significantly and the stabilized quarry fines would have adequate strength (28-day, about 3 MPa) even for the base layer in medium to high volume roads according to Portland Cement Association. The other agency's requirements also suggest its adequate strength for applications in subbase and subgrade. However, the quarry fines stabilized with Ecolan are more suitable for subbase and subgrade according to the agencies listed in Table 1. When quick cement is used as a stabilizer, stabilized quarry fines can be satisfactorily qualified by most of the agencies as base material in medium to high volume roads, in terms of unconfined compressive strength.

Results from previous research are also plotted in Figure 6 for comparison [20,38]. The specimens KS4, KE6 and KL50A1 are quarry fines (0–3 mm) stabilized with plus cement, Ecolan stabilizer and 50% fly ash with 1% cement as activator. The 7-day unconfined compressive strength of cement-stabilized specimens is quite similar, while for the other two types of stabilizers, the previous research obtained higher strengths mainly due to higher degree of compaction (DOC, 96%). If the gradation of aggregates was improved by adding more coarse fractions such as crushed rock of 0–20mm, the strength will be improved significantly as shown for specimens MS3 (5% plus cement) and MSL1 (2% plus cement + 6% fly ash).

3.3.4. Failure Modes

The major cracking types developed in the stabilized quarry fine specimens after unconfined compressive test are shown in Figure 7, where two types of cracks caused by extension and shear have been observed. However, in this study, the combinations of tensile splitting and inclined shear cracks are the main failure modes, and the barrel-shaped specimen after testing with cracks indicates semi-plastic characteristics of the stabilized quarry fine materials. Among all 18 quarry fine specimens stabilized with quick cement, 16 of them are observed to have combined extension and shear cracks as the failure mode, while the other two are found to have only extension cracks after compression. Therefore, combined extension and shear failures can be recognized as the main failure mode of cement-stabilized quarry fines under unconfined compression conditions.

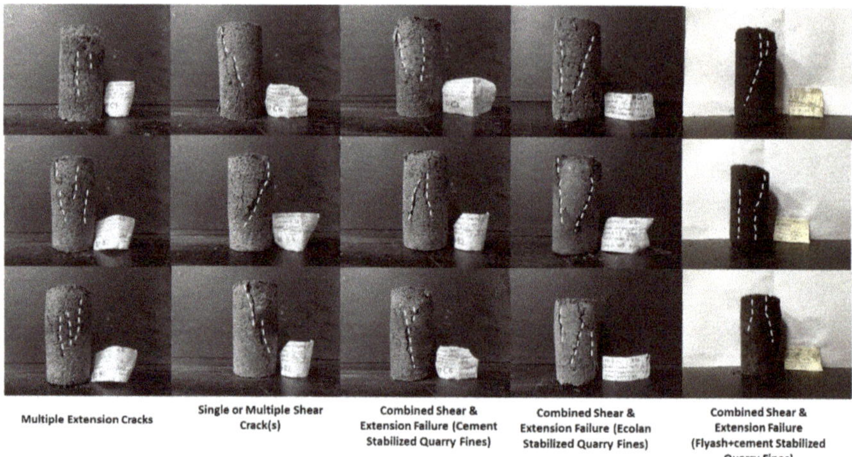

Figure 7. Cracking types (first two columns) and major failure mode (column three to five) of the stabilized quarry fine specimens in unconfined compressive test.

The same main failure mode of combined extension and shear can also be observed for the Ecolan-stabilized and fly ash plus cement-stabilized quarry fine specimens. Two out of the total 18 Ecolan-stabilized specimens were found to have multiple extension cracks while the others all have combined extension and shear failures. For quarry fines stabilized with fly ash plus cement, two out of the total 12 specimens only have failure mode of multiple extension cracks and the rest all have combined cracks. There seems to be no significant difference in failure mode for the quarry fines stabilized by the three types of stabilizers as adopted in this study.

3.4. Freeze–Thawing Properties

Ground freezing and thawing is a result of the annual cycle of temperatures, which is also a common situation encountered in Nordic countries. A recent study found that with global climate changes, the intensity of thaw weakening will be more serious [39], and the annual number of freeze–thaw cycles is increasing. As a candidate construction material, stabilized quarry fines should be capable to sustain in such severe environmental conditions. The resistance of the stabilized quarry fines to the cyclic action of freezing and thawing is critical to the performances of buildings and infrastructures in cold regions, and thus is estimated by freeze–thaw testing. The technical specification CEN/TS 13286-54 Test method for the determination of frost susceptibility. Resistance to freezing and thawing of hydraulically bound mixtures was followed to conduct the test on stabilized quarry fines [40].

In the first stage, cylindrical stabilized quarry fine specimens of 100 mm diameter and height were prepared using gyratory compactor ICT to achieve 96% degree of compaction. The specimens for freeze-thaw testing were fabricated according to the most promising mix designs to achieve the highest UCS as obtained from previous tests. For each type of stabilizer, six specimens were fabricated, kept in mould, and cured in a humidity room for 28 days. Then they were removed from the mould and divided into Set A and B with three specimens in each. As a result, the first stage was finished to allow adequate strength development for the specimens. The second stage consists of the application of freeze-thaw cycles and the strength test. Both of the two sets were placed in a humidity room to prevent the loss of moisture, and Set A was removed from the humidity room after 2 days and wrapped and placed in a low temperature cabinet for 10 freeze-thaw cycles, whereas Set B was kept in a humidity room until the conditioning of Set A was finished. Then all the specimens were used in an unconfined compressive strength test and the freeze–thaw retained strength ratio was calculated.

As shown in Figure 8, one freeze-thaw cycle takes 24 hours to finish, and the controlled temperature inside the low temperature cabinet was within the limits as suggested by the specifications. The monitored temperature was measured inside a cubic cement concrete specimen placed inside the cabinet.

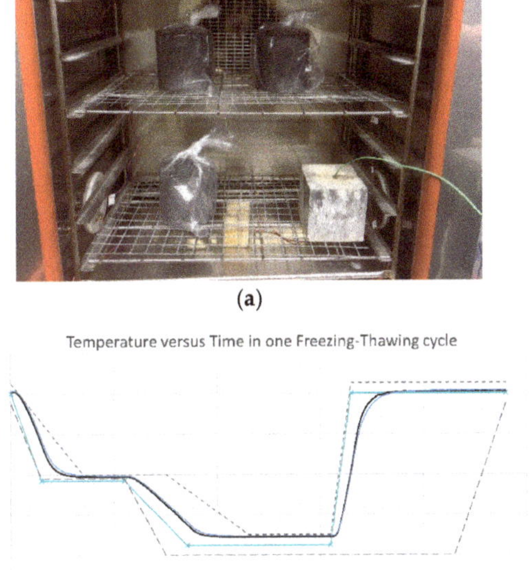

(a)

(b)

Figure 8. Freeze–thaw test: (**a**) Stabilized quarry fine specimens inside the chamber; (**b**) Temperature profile applied in freeze–thaw cycles.

Based on the unconfined compressive strength of both Set A and B quarry fine specimens subjected to none or 10 freeze-thaw cycles, no distinguished differences were found between the two sets. However, still the average UCS of regular specimens is higher than that of conditioned specimens, and the strain corresponding to the UCS at failure has increased. Based on calculations, some indicators including the previously mentioned strain-ductility ratio, the freeze-thaw retained strength ratio is presented in Table 4.

Table 4. Indicators to evaluate effects of freeze-thaw cycles.

Type of Specimen	Cement-Stabilized Quarry Fines		Ecolan-Stabilized Quarry Fines		Fly Ash-Stabilized Quarry Fines	
Freezing-Thawing Conditions	No Freezing-Thawing Cycles	10 Freezing-Thawing Cycles	No Freezing-Thawing Cycles	10 Freezing-Thawing Cycles	No Freezing-Thawing Cycles	10 Freezing-Thawing Cycles
Average SDR	1.81	1.50	1.76	1.23	1.41	1.36
$\varepsilon_{y,ave}$	0.0149	0.0201	0.0231	0.0332	0.0169	0.0190
$\varepsilon_{f,ave}$	0.0204	0.0240	0.0308	0.0379	0.0193	0.0211
FTR	0.946		0.808		0.662	

Where FTR is the freeze–thaw retained strength ratio $\frac{R_A}{R_B}$, and R_A and R_B are the mean value of strength for Set A and Set B, respectively.

Subjected to 10 freeze-thawing cycles, the specimens have been deteriorated in terms of reduced UCS and strain-ductility ratios. As shown in Table 4, given the same external loading conditions, the specimens have become less ductile after the cyclic freeze-thawing conditions, whereas the strains at first yield point and at failure all increase. This can be further interpreted as the materials subject to freeze-thawing cycles were deformed more before failure but became more brittle and easy to break after the peak stress. From the curves of cement-stabilized quarry fines, there are obvious relaxation stages after the beginning of the load for the conditioned specimens, contributing to the total strains at failure. On the other hand, based on the freeze-thaw retained strength ratio, it can be seen that the deteriorations caused by 10 freezing and thawing cycles for the cement-stabilized quarry fine specimens were not significant. This qualifies the cement-stabilized quarry fines as a pavement construction material for different layers. The deteriorations caused by freeze-thawing cycles for Ecolan-stabilized quarry fines were more serious, which resulted in a decrease in the UCS by about 20%. Yet, this amount of deterioration in the UCS is normal for stabilized granular materials such as lime-treated kaolinite [41,42] and fiber-reinforced clay [43,44]. However, improvement is still necessary to enhance the freeze-thawing resistance for quarry fines stabilized with fly ash, even though the ratio is right on the limit set by Finnish guidelines of 6.7 [18]. The fundamental mechanism behind such a difference in freeze-thawing resistance of quarry fines stabilized with different stabilizers stills need to be investigated from a finer-scale point of view. Related research and results will be introduced in the future.

4. Discussion on Potential Applications of Quarry Fine Materials

Existing research has already presented several applications of virgin quarry fine materials, including as a partial substitute to cement and as an additive to improve the properties of soils, whereas these applications are only in limited volumes. The virgin quarry fines of 0–4mm as adopted in this study have a similar particle size distribution to well-graded sand, and a coefficient of permeability around 5.75×10^{-5} m/s [11] indicates its potential material value in construction for civil engineering purposes. However, considering that the previous test results have also suggested possible frost-susceptive properties, stabilizers were applied to improve the properties of virgin quarry fines as a construction material. Unconfined compressive strengths of the stabilized quarry fine materials were investigated by laboratory tests, which further verified the potential of the stabilized quarry fines for civil engineering applications from the strength point of view.

Based on the typical strength and durability requirements for materials used in pavement structural layers, the stabilized quarry fine materials can qualify for high-volume civil engineering applications, as long as they are well designed and compacted to enough degrees. One of the most possible applications is in road construction. The stabilized quarry fines can be qualified for base, subbase, filter layers, and subgrade depending on the different stabilizers used and degrees of compaction achieved. Nevertheless, applications for pavement subbases are much more suggested. This is because higher quality material is better for base layers to ensure pavement performances in the long run, and unbound material is more favorable for the filter layers considering cost efficiency. The other possible applications include but are not limited to embankment building, landfill construction, and construction site leveling, where strength requirements are not high.

However, when the strength requirements are high, it is still possible to utilize the stabilized quarry fine materials. It is well-established that particle size distribution of the aggregate has shown important effects on the unconfined compressive strength of mixtures. This has also been seen when comparing the results of this research with the previous ones. When coarse particles are added to quarry fines to obtain a good gradation curve and compaction, the strength can be improved significantly. Therefore, in situations where higher strength is required, either modifying the gradation of quarry fines by adding more coarse fractions or increasing the DOC level (if achievable in field) would be options.

On the other hand, even though the resistance to freeze-thawing cycles of the stabilized quarry fines is acceptable based on laboratory results in this study, the detrimental effects of salt have been found somewhere else. In situations where freeze-thawing cycles are presented and deicing agent is applied, the resistance would be impaired especially for the cement-stabilized quarry fines. It is therefore suggested that for pavements where deicing agent is needed, a waterproof asphalt concrete surface layer would be helpful to reduce the salt infiltration and maintain the resistance of the underlying stabilized quarry fine layers to freeze-thawing cycles.

It should be noted that all these potential high-volume applications are based on the assumption that the quarry fine materials can satisfy all the environmental requirements set by the government. Also, there are still some uncertainties of the conclusions made in this study, mainly due to the limited number of specimens that has been tested and the possible variations of the material quality over different quarries.

5. Conclusions

- It is found that the modulus of deformation E_{50} of cement or fly ash (with activator) stabilized quarry fines can be predicted with a similar multiplier coefficient to cemented clays. For quarry fines treated with Ecolan stabilizer, the multiplier coefficients are higher than the typical coefficients of undisturbed cemented clays. The differences in multiplier coefficients among mixtures are mainly due to different virgin material properties, water content and the type of stabilizer applied.

- The UCS of stabilized quarry fine specimens of 96% degree of compaction can adequately satisfy the requirements of pavement base/subbase provided by different agencies in terms of strength, as long as appropriately designed and constructed. A least 93% degree of compaction is recommended for cement-stabilized quarry fines with the proposed mixture design, to be qualified for applications in base layers. When Ecolan or flyash with cement is used as the stabilizer, a higher degree of compaction would work better.

- Higher cement content and a lower water-to-cement ratio will result in stiffer cement-stabilized quarry fine materials. A similar situation was also observed on quarry fine specimens stabilized by the fly ash and activator. The opposite situation was found for the Ecolan stabilizer where a higher stabilizer content and water-to-stabilizer ratio will lead to higher strength.

- Combined extension and shear failures can be recognized as the main failure mode of stabilized quarry fines under unconfined compression conditions.

- The specimens subjected to freeze-thawing cycles have been deteriorated in terms of reduced bearing capacity and the increased deformation given the same external loading conditions. The freeze-thaw retained strength ratio indicates that the deteriorations caused by 10 freeze-thaw cycles were very limited for cement-stabilized quarry fines which can be qualified as a pavement construction material. For the Ecolan or fly ash-stabilized (with activator) quarry fines, higher susceptibilities, yet within the requirements, were observed, indicating a need for improvements. The results of the capillary rise test also support this finding.

- When the strength requirements are high (such as for base layer), it is still possible to utilize the stabilized quarry fine materials by modifying the gradation or increasing the DOC level (if achievable in field).

- For pavements where a deicing agent is needed, a waterproof asphalt concrete surface layer is suggested to maintain the resistance of the underlying stabilized quarry fine layers to freeze-thawing cycles.

6. Future Research

In this research, varied stress-strain relationships under the same loading conditions for the quarry fine specimens stabilized with different stabilizers have been observed. In the future research, another mechanical test, such as Repeated Load Triaxle (RLT) test, may be conducted to further investigate its mechanical properties. Fundamental mechanisms lying behind this, as well as the different UCS values caused by different binding effects of various stabilizers, still need to be investigated. A future study based on more fundamental, fine-scaled investigation with Scanning Electron Microscopy (SEM) and Energy Dispersive Spectroscopy (EDS) techniques will be conducted to compare mineral compositions of quarry fines and host rock, and to explain the binding and freeze-thawing effects on quarry fines.

Author Contributions: Ceptualization, L.K.K.-T.; Methodology, Y.Z. and L.K.K.-T.; Software, Y.Z.; Validation, Y.Z., and L.K.K.-T.; Formal Analysis, Y.Z. and L.K.K.-T.; Investigation, Y.Z., L.K.K.-T. and M.B.; Resources, L.K.K.-T. and M.B.; Data Curation, Y.Z. and L.K.K.-T.; Writing—Original Draft Preparation, Y.Z. and L.K.K.-T.; Writing—Review and Editing, L.K.K.-T. and M.B.; Visualization, Y.Z. and L.K.K.-T.; Supervision, L.K.K.-T. and M.B.; Project Administration, L.K.K.-T. and M.B.; Funding Acquisition, L.K.K.-T. and M.B.

Funding: This research received no external funding.

Acknowledgments: This research was supported by the School of Engineering, Aalto University, for financing the researchers, and Destia Ltd. for providing the raw materials. We would like to present our gratitude to our colleagues Amandine Miksic, Hakala Veli-Antti, Eloranta Pekka, and Henry Gustavsson, who provided expertise that greatly assisted the research.

Conflicts of Interest: The authors declare no conflict of interest. The funders had no role in the design of the study, the collection, analysis, or interpretation of data, the writing of the manuscript, or in the decision to publish the results.

References and Notes

1. Pitkänen, I.J. Selvitys Destia Oy: n kivituhkamääristä ja kivituhkan nykyisistä ja uusista käyttömahdollisuuksista. Bachelor's Thesis, Savonia University of Applied Science, Kuopio, Finland, November 2015.
2. Medina, G.; Sáez del Bosque, I.F.; Frías, M.; Sánchez de Rojas, M.I.; Medina, C. Durability of new recycled granite quarry dust-bearing cements. *Constr. Build. Mater.* **2018**, *187*, 414–425. [CrossRef]
3. Amadi, A.A. Enhancing durability of quarry fines modified black cotton soil subgrade with cement kiln dust stabilization. *Transp. Geotech.* **2014**, *1*, 55–61. [CrossRef]
4. Sivrikaya, O.; Kıyıldı, K.R.; Karaca, Z. Recycling waste from natural stone processing plants to stabilize clayey soil. *Environ. Earth Sci.* **2014**, *71*, 4397–4407. [CrossRef]
5. Oncu, S.; Bilsel, H. Ageing effect on swell, shrinkage and flexural strength of sand and waste marble powder stabilized expansive soil. *E3S Web Conf.* **2016**, *9*, 1–6. [CrossRef]
6. Gurbuz, A. Marble powder to stabilise clayey soils in subbases for road construction. *Road. Mater. Pavement Des.* **2015**, *16*, 481–492. [CrossRef]
7. Rai, B.; Kumar, S.; Satish, K. Eect of Fly Ash on Mortar Mixes with Quarry Dust as Fine Aggregate. *Adv. Mater. Sci. Eng.* **2014**, *2014*, 1–7. [CrossRef]
8. Vijayalakshmi, M.; Sekar, A.S.S.; Prabhu, G.G. Strength and durability properties of concrete made with granite industry waste. *Constr. Build. Mater.* **2013**, *46*, 1–7. [CrossRef]
9. Karasahin, M.; Terzi, S. Evaluation of marble waste dust in the mixture of asphaltic concrete. *Constr. Build. Mater.* **2007**, *21*, 616–620. [CrossRef]
10. Karakus, A. Investigating on possible use of Diyarbakir basalt waste in Stone Mastic Asphalt. *Constr. Build. Mater.* **2011**, *25*, 3502–3507. [CrossRef]
11. Zhang, Y.; Korkiala-Tanttu, L.; Gustavsson, H.; Miksic, A. Assessment for sustainable use of quarry fines as pavement construction materials: Part I description of basic quarry fine properties. *Materials* **2019**, *12*, 1209. [CrossRef] [PubMed]

12. Sargent, P.; Rouainia, M.; Hughes, P.N.; Glendinning, S. Alkali Activation of Industrial By-Products for use in Soil Stabilisation. In Proceedings of the ISSMGE Technical Committee TC 211 International Symposium on Ground Improvement (IS-GI BRUSSELS 2012): Recent Research, Advances & Execution Aspects of Ground Improvement Works, Proceedings of the International Symposium, Brussels, Belgium, 31 May–1 June 2012; Denies, N., Huybrechts, N., Eds.; ISSMGE TC211 and BBRI: Brussels, Belgium, 2012; Volume II, pp. 467–477.

13. Mirzaeifar, H.; Abdi, M.R. Stabilizing clays using basic oxygen steel slag (BOS). In Proceedings of the Recent Research, Advances & Execution Aspects of Ground Improvement Works, Proceedings of the International Symposium, Brussels, Belgium, 31 May–1 June 2012; Denies, N., Huybrechts, N., Eds.; ISSMGE TC211 and BBRI: Brussels, Belgium, 2012; Volume II, pp. 403–410.

14. Blanck, G.; Cuisinier, O.; Masrouri, F. A non-traditional treatment for the compaction of fine-grained soils. In Proceedings of the Recent Research, Advances & Execution Aspects of Ground Improvement Works, Proceedings of the International Symposium, Brussels, Belgium, 31 May–1 June 2012; Denies, N., Huybrechts, N., Eds.; ISSMGE TC211 and BBRI: Brussels, Belgium, 2012; Volume II, pp. 281–287.

15. Kamei, T.; Ahmed, A.; Shibi, T. Recycled basanite in conjunction with coal ash for stabilization of soft clay soil. In Proceedings of the Recent Research, Advances & Execution Aspects of Ground Improvement Works, Proceedings of the International Symposium, Brussels, Belgium, 31 May–1 June 2012; Denies, N., Huybrechts, N., Eds.; ISSMGE TC211 and BBRI: Brussels, Belgium, 2012; Volume II, pp. 373–384.

16. CE marking of quarry fines. Delivered by Destia Oy. Neilikkatie 17, 01301 Vantaa.

17. CE marking of pikacement. Delivered by Finnsementti Oy. 21600 Parainen.

18. Tiehallinto. Päällysrakenteen Stabilointi. Available online: https://julkaisut.vayla.fi/thohje/pdf/2100055-v-07paallysrakenteen_stabilointi.pdf (accessed on 24 September 2018).

19. Guthrie, W.S.; Sebesta, S.; Scullion, T. *Selecting Optimum Cement Contents for Stabilizing Aggregate Base Materials*; Report No. FHWA/TX-05/7-4920-2; Texas Department of Transportation: Austin, TX, USA, February 2002.

20. Melander, M. Suitability of by-product fines for stabilized base course. Master's Thesis, Aalto University, Helsinki, Finland, May 2018.

21. American Coal Ash Association. Fly Ash Facts for Highway Engineers. Available online: https://www.fhwa.dot.gov/pavement/recycling/fafacts.pdf (accessed on 24 September 2018).

22. Federal Highway Administration Research and Technology. User Guidelines for Waste and Byproduct Materials in Pavement Construction. Available online: https://www.fhwa.dot.gov/publications/research/infrastructure/structures/97148/cfa55.cfm (accessed on 24 September 2018).

23. Finnish Standard Association. *SFS-EN 13057 Products and Systems for the Protection and Repair of Concrete Structures. Test Methods. Determination of Resistance of Capillary Absorption*; CENELEC Management Centre: Brussels, Belgium, 1997.

24. Karagiannis, N.; Karoglou, M.; Bakolas, A.; Moropoulou, A. *Building Materials Capillary Rise Coefficient: Concepts, Determination and Parameters Involved*; Delgado, J., Ed.; New Approaches to Building Pathology and Durability. Building Pathology and Rehabilitation; Springer: Singapore, 2016; Volume 6.

25. Christodoulou, C.; Goodier, C.I.; Austin, S.A.; Webb, J.; Glass, G.K. Long-term performance of surface impregnation of reinforced concrete structures with silane. *Constr. Build. Mater.* **2013**, *48*, 708–716. [CrossRef]

26. Yeşim, T.; Remzi, Ş. Compressive Strength and Capillary Water Absorption of Concrete Containing Recycled Aggregate. *Int. Sch. Sci. Res. Innov.* **2015**, *9*, 995–999.

27. Mors, R.; Jonkers, H. Effect on Concrete Surface Water Absorption upon Addition of Lactate Derived Agent. *Constr. Build. Mater.* **2017**, *7*, 51. [CrossRef]

28. Finnish Standard Association. *SFS-EN 1925 Natural Stone Test Methods. Determination of Water Absorption Coefficient by Capillarity*; CENELEC Management Centre: Brussels, Belgium, 1999.

29. *Portland Cement Association: Soil-Cement Laboratory Handbook 1992*; Portland Cement Association: Skokie, IL, USA, 1992; p. 59.

30. American Association of State Highway and Transportation Officials. *Mechanistic Empirical Pavement Design Guide A Manual of Practice*; Interim Edition; AASHTO: Washington, DC, USA, 2008; ISBN 978-1-56051-423-7.

31. Jameson, G. *Guide to Pavement Technology Part 4D: Stabilised Materials Edition 2.1*; Austroads Ltd.: Sydney, NSW, Australia, 2019; ISBN 978-1-925854-17-6.

32. Geotechnical Aspects of Pavements - Reference Manual/Participant Workbook (Publication No. FHWA NHI-05-037). U.S. Department of Transportation. Available online: https://www.fhwa.dot.gov/engineering/geotech/pubs/05037/05037.pdf (accessed on 24 September 2018).

33. Federal Highway Administration. Standard Specifications for Construction of Roads and Bridges on Federal Highway ProjectsFP-14. Available online: https://flh.fhwa.dot.gov/resources/specs/fp-14/fp14.pdf (accessed on 24 September 2018).

34. *InfraRYL 2010 Code of Building Practice, Infrastructure, Part 1: Roads and Areas*; Rakennustieto Oy: Helsinki, Finland; ISBN 978-951-682-958-9.

35. Bieniawski, Z.T.; Bernede, M.J. Suggested Methods for Determining the Uniaxial Compressive Strength and Deformability of Rock Materials. *Int. J. Rock Mech. Min. Sci. Geomech. Abstr.* **1979**, *16*, 137–140.

36. Strozyk, J.; Tankiewicz, M. The elastic undrained modulus Eu50 for stiff consolidated clays related to the concept of stress history and normalized soil properties. *Stud. Geotechnica. Mech.* **2016**, *38*, 67–72. [CrossRef]

37. Ayeldeen, M.; Hara, Y.; Kitazume, M.; Negm, A. Unconfined Compressive Strength of Compacted Disturbed Cement-Stabilized Soft Clay. *Int. J. Geosynth. Ground. Eng.* **2016**, *2*, 10. [CrossRef]

38. Kortelainen, L. The frost resistance of stabilized base course. Master's Thesis, Aalto University, Helsinki, Finland, June 2019.

39. Saarelainen, S.; Gustavsson, H.; Korkiala-Tanttu, L. Climate change and freezing conditions in Finland. In Proceedings of the 19th International Conference on Soil Mechanics and Geotechnical Engineering, ISSMGE, Seoul, Korea, 17–21 September 2017; pp. 1411–1414.

40. Finnish Standard Association. *CEN/TS 13286-54:2014 Unbound and Hydraulically Bound Mixtures. Part 54: Test Method for the Determination of Frost Susceptibility. Resistance to Freezing and Thawing of Hydraulically Bound Mixtures*; CENELEC Management Centre: Brussels, Belgium, 2014.

41. Ismeik, M.; Shaqour, F. Effectiveness of lime in stabilising subgrade soils subjected to freeze–thaw cycles. *Road. Mater. Pavement. Design.* **2018**, 1–19. [CrossRef]

42. Firoozi, A.A.; Taha, M.R.; Firoozi, A.A.; Khan, T.A. The influence of freeze-thaw cycles on unconfined compressive strength of clay soils treated with lime. *Jurnal. Teknol. (Sci. Eng.)* **2015**, *76*, 107–113.

43. Ghazavi, M.; Roustaie, M. The influence of freeze–thaw cycles on the unconfined compressive strength offiber-reinforced clay. *Cold. Regions. Sci. Technol.* **2010**, *61*, 125–131. [CrossRef]

44. Orakoglu Firat, M.; Liu, J. Effect of freeze-thaw cycles on triaxial strength properties of fiber-reinforced clayey soil. *KSCE. J. Civil. Eng.* **2017**, *21*, 2128–2140.

Article

Effect of Short-Term Aging on Asphalt Modified Using Microwave Activation Crumb Rubber

Bo Li [1,2,*]**, Jianing Zhou** [1]**, Zhihao Zhang** [1]**, Xiaolong Yang** [1,2] **and Yu Wu** [1,3]

[1] National and Provincial Joint Engineering Laboratory of Road & Bridge Disaster Prevention and Control, Lanzhou Jiaotong University, Lanzhou 730070, China; zjn1409@163.com (J.Z.); tmgcxyyb@163.com (Z.Z.); glhyyfzx@gslq.com (X.Y.); gsgfjt@163.com (Y.W.)

[2] Research and Development Centre of Transport Industry of Technologies, Materials and Equipment of Highway Construction and Maintenance, Gansu Road and Bridge Group Co., Ltd., Lanzhou 730070, China

[3] Key Laboratory of Highway Net Monitoring in Gansu Province, Gansu Hengda Road and Bridge Group Co., Ltd., Lanzhou 730070, China

* Correspondence: libolzjtu@hotmail.com

Received: 7 March 2019; Accepted: 25 March 2019; Published: 29 March 2019

Abstract: Effective approaches are required to be developed to solve the poor compatibility and thermal stability problems of crumb rubber-modified asphalt (CRMA). This study focuses on a method called microwave activation. However, seldom researches pay attention to the properties of MACRMA after aging. The objective of this study was to prepare microwave-activated CRMA (MACRMA) and investigate the performance of asphalt after aging. The samples were subjected to thin-film oven test (TFOT) at different times and temperatures. The effect of heat aging on the properties of MACRMA was evaluated by three indicator tests: viscosity, dynamic shear rheology test (DSR), and repeat creep recovery test (RCRT). The test results indicated that the MACRMA after two aging conditions had noticeably lower performance values (e.g., penetration, ductility) compared to unaged samples, and thus, the need to control temperature and time for mixing and construction was verified to be important. In addition, the $G^*/\sin\delta$ and phase angle values were largely influenced by the TFOT aging temperature and time. The MACRMA's ability to recover was improved after aging. Compared with the aging temperature, the aging time had a more significant effect on the deformation and recovery ability of MACRMA.

Keywords: microwave activation; short-term aging; rheological property; repeated creep recovery

1. Introduction

Approximately 1.4 billion tires are produced every year for vehicles of industrialized and developing countries, which leads to more tires that need to be recycled or disposed [1–4]. Disposal of used and discarded tires have posed a significant environmental threat worldwide every year due to fire hazards and have become a habitat for insects/rodents. In recent years, many modern technologies have been invested in producing crumb tire from tire rubber while removing fibers, steel, and other contaminants [5–8]. One of the means for utilizing crumb rubber is as a modifier for producing rubberized asphalt, since crumb tires contains a variety of rubber polymers, predominantly natural rubber and styrene-butadiene rubber.

Currently, the technology of crumb rubber-modified asphalt (CRMA) has been widely used in the asphalt pavement industry [9–15]. The addition of crumb rubber is typically accomplished using the wet process, in which the crumb rubber is blended with the asphalt binder to produce a crumb rubber modified binder that is then mixed with the aggregate. In this wet process, asphalt binder modification is due to physical and compositional changes from interactions between rubber and asphalt, characterized by a swelling of the rubber particles in the asphalt binder due to absorption of some lighter fractions (aromatic oils) from asphalt. The swelling of these rubber particles and their absorption, in turn, results in a viscous gel. CRMA developed using the wet process has been proven by many research projects and field applications to be an effective method to increase the performance grade of the asphalt, improve the high temperature properties, decrease susceptibility to permanent deformation, and provide resistance against reflective cracking. In addition, using crumb rubber as a modifier instead of expensive polymer modifiers (SBS, SBR, etc.) can reduce the cost of the road asphalt [16–19].

However, CRM has poor compatibility with asphalt since the waste rubber powder is an inert material, which limits the partial performance of CRMA. The final performance of asphalt with CRM modified binder is greatly influenced by the physical and chemical properties of CRM. As crumb rubber molecules are vulcanized, the three-dimensional network is difficult to crack in the asphalt, even after blending at high temperatures for a long period of time. This leads to incompatibility between CRM and asphalt, and settlement of rubber particles at the bottom of the bulk asphalt phase, which further affects the performance of CRMA [20–24]. The literature shows that in order to minimize the problem of CRM particle sedimentation, the prepared CRM binder should be used within four hours. Many attempts have been made to improve the properties of CRM, such as regenerated desulphurization modifications and grafting to break the C–S or S–S bonds in the rubber chemical linking networks. However, these modification methods are done by adding CRM to rubber or plastic but not to asphalt. In addition, the modification process is complicated and difficult to control. In recent years, researchers have used microwaves to modify CRM, and then add the activated CRM to asphalt. Many studies have shown that microwave activation is similar to previous methods that break the internal S–S bond of the rubber powder and improve the surface activity, and the storage stability of the microwave activated CRM asphalt (MACRMA) and pavement performance is better [25–27].

Despite its significantly modified properties, MACRMA is vulnerable to aging, which is an inevitable process in its service life that can significantly deteriorate the asphalt binding properties. Asphalt has always been in adverse conditions due to excessive temperatures and long heating times in the actual production, transportation, and construction, which will inevitably cause short-term aging of the asphalt and weaken the function of the asphalt binder and mixture. Short-term aging mainly takes place during the manufacturing process of asphalt mixtures in which the aggregates are covered by a thin asphalt film at high temperatures. The thin asphalt film has large surface areas that are exposed to air, causing the oxidative aging of asphalt components and the volatilization of light fractions. It is noted that the performance of asphalt mixtures in pavement after short-term aging was determined universally [28,29].

Numerous laboratory aging methods have been proposed to simulate asphalt aging. Among these methods, the rolling thin-film oven test (RTFOT) is the most commonly applied as recommended by the Strategic Highway Research Program [30]. However, researchers have shown that the RTFOT is not suitable for CRM asphalt owing to its relatively low aging temperature of 163 °C. The aging of RTFOT relies on spreading the fluid asphalt into thin films. With a reduced asphalt film thickness, the exposure to oxygen is increased and thus the aging process is accelerated. The CRM is usually too stiff to flow at 163 °C inside the RTFOT aging bottle, and thus it cannot form thin films, resulting in insufficient aging. Many studies have indicated that the current 163 °C aging temperature should be elevated because it was established on field investigations of neat asphalt and is apparently lower than the CRM mixing temperature. As MACRMA brings notably enhanced engineering properties and higher viscosity, this results in increased mixing and compaction temperatures. Contractors in China have reported that CRM asphalt usually has a mixing temperature of 190 °C and a paving temperature of 170–185 °C, depending on the time spent in transport. RTFOT does not have a comparable aging temperature in the field and fails to form an asphalt film of a small uniform thickness for asphalt binders with different viscosities, and its application for MACRMA is questionable [31].

The objective of this study is to investigate the influence of different aging factors on the properties of MACRMA from the perspective of rheological properties, and gradually exploring the thermal oxygen aging law of MACRMA. The MACRMA was subjected to a thin film oven test (TFOT) for different times (0, 5, 15, and 20 h) and temperatures (150, 160, 170, 180, and 190 °C). A serious of lab tests including three indicator tests: viscosity, dynamic shear rheology (DSR) test, and repeated creep recovery test have been conducted.

2. Materials and Experiments

2.1. Materials

The neat asphalt binder used in this study was petroleum asphalt of penetration grade 90. Table 1 presents some detailed technical properties of neat petroleum asphalt of penetration grade 90. Each index was tested based on the Standard Test Method of Bitumen and Bituminous Mixture for Highway Engineering (JTG E20-2011) by the Chinese Ministry of Communications [32]. The 40-mesh crumb rubber powder produced by the normal temperature method was selected, and the physical and chemical properties of the test results are shown in Table 2. At the same time, considering the influence of the production method of the crumb rubber powder on the particle size, the crumb rubber powder is sieved through the 40-mesh filter prior to using.

Table 1. Technical properties of neat petroleum asphalt.

Item		Specification	Result	Test Methods
Penetration (25 °C, 100 g, 5 s) (0.1 mm)		80~100	92	T 0604-2011 [32]
Softening point (°C)		≥42	46.2	T 0606-2011 [32]
Ductility (5 cm/min, 10 °C) (cm)		≥100	>100	T 0605-2011 [32]
RTFOT (163 °C, 85 min)	Loss (%)	≤±0.8	0.07	T 0610-2011 [32]
	Penetration ratio (%)	≥54	70	T 0604-2011 [32]
	Ductility at 10 °C (cm)	≥6	9	T 0605-2011 [32]

Table 2. Physical and chemical specifications of rubber powder.

Item	Density (kg·m^{-3})	Water Content (%)	Metal Content (%)	Fiber Content (%)	Ash (%)	Acetone Extract (%)	Carbon (%)	Rubber Hydrocarbon (%)
Specification	260~460	<1	<0.03	<1	≤8	≤22	28	≥42
Result	302.5	0	0.009	0.065	7.3	7.2	30	52

2.2. Experimental Procedures

The waste tire rubber powder was placed in a constant-temperature oven at 60 °C for 30 min for drying and dehydration, and then 100 g of the rubber powder was placed in the microwave oven. In this experiment, the activated rubber powder was prepared by activating the rubber powder for 90 s at 800 W and then cooling to room temperature. The asphalt binder was heated to approximately 135 °C as measured with a thermometer and poured into a beaker. Then, the beaker was placed in a constant temperature magnetic heating stirrer to raise the temperature to 190 °C. Additionally, the activated rubber powder was slowly added at a stirring rate of 1500 r/min. After reacting with the asphalt for 60 min, the microwave rubber modified asphalt was obtained. The cigar tube test showed that the difference of softening point of the upper and lower ends of the tube was only 2.3 °C, which indicated that the modified asphalt had been uniformly mixed. Some of its performance indicators are shown in Table 3.

Table 3. Technical properties of microwave-activated crumb rubber-modified asphalt (MACRMA).

Item	Specification	Result
Penetration (25 °C, 100 g, 5 s) (0.1 mm)	30~70	40.6
Softening point (°C)	>65	68.7
Ductility (5 cm/min, 10 °C) (cm)	>5	7.7
Viscosity values (Pa·s)	2.0~5.0	2.93

To simulate the effect of short-term aging on the performance of MACRMA, the asphalt was aged by the thin film oven test (TFOT, Figure 1). This test uses a single variable method: (1), keeping the aging time at 5 h and changing the aging temperature (150, 160, 170, 180, and 190 °C); (2), keeping the aging temperature at 163 °C and changing the aging time (5, 10, 15, and 20 h).

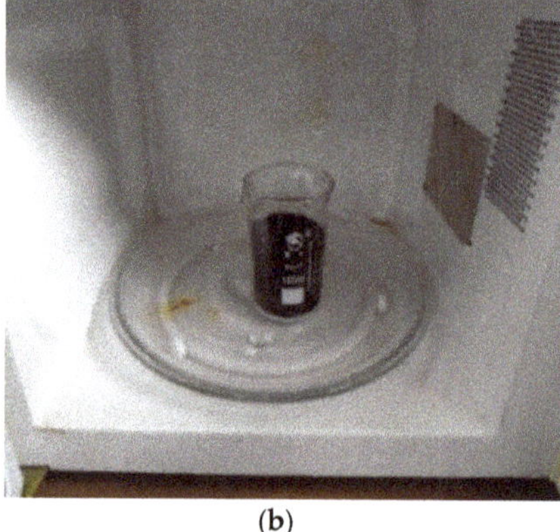

(a)   (b)

Figure 1. Rubber powder activation equipment: (**a**) microwave oven and (**b**) microwave-activated crumb rubber.

The dynamic shear rheology test (DSR) was used to test the complex shear modulus G^*, phase angle (δ), and other rheological properties with respect to AASHTO T315. The temperature sweep, frequency sweep and repeated creep recovery test of MACRMA samples were carried out by AR1500ex dynamic shear rheometer. The temperature scanning range was 58–82 °C and the frequency was 10 rad/s. The frequency sweep range was 0.1~10 rad/s with a test temperature of 70 °C and 12% strain.

The RCRT test is an abbreviation for the repeated creep recovery test specified by AASHTO MP19-1. The test temperature was set to 60 °C, and the loading stress was 100 Pa and 300 Pa, respectively. In addition, the loading parallel plate diameter was 25 mm, and the test spacing was 1 mm. One creep recovery cycle included the 1 s loading mode and the 9 s unloading mode, which gave 50 time measurements in total. This loading mode can better simulate the intermittent characteristics of actual asphalt pavement load. In other words, the characteristics of higher elastic recovery of the modified asphalt could be considered by this test.

3. Experiment Results

3.1. Penetration/Softening Point/Ductility Tests

These three indicator tests are the most widely used methods for evaluating asphalt performance in China. The penetration test is the standard method for determining the consistency of asphalt. The penetration degree indicates the consistency of the asphalt. The larger the index, the lower the viscosity of the asphalt.

The penetration values in terms of increased aging temperature and time are shown in Figure 2a,b, respectively. As shown in Figure 2a, the results indicate that the increased aging temperature results in a decrease of penetration values for all the modified asphalts. The linear fit analysis result found that the correlation coefficient between the penetration values and aging temperature was 0.905. Similar trends for the modified binders can be found in Figure 2b, where the penetration values are reduced when the aging time increases. The MACRMA with 20 h TFOT aging has the lowest penetration values. In addition, the penetration values have a better correlation with aging time.

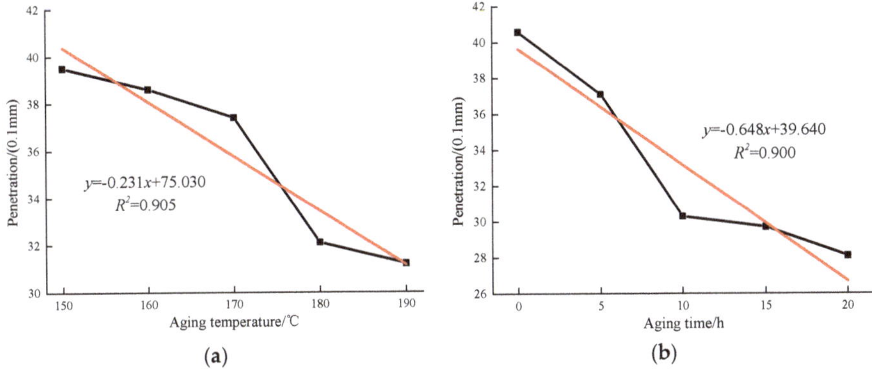

Figure 2. Penetration values of MACRMA at different aging conditions: (**a**) aging temperature and (**b**) aging time.

The ductility indicates the plastic deformation ability and the low temperature performance of the asphalt. The test results shown in Figure 3a illustrate that the ductility values of all the asphalts reduce as the aging temperature increases from 150 to 190 °C. The lowest ductility value is 7.0 cm, which correlates to the 190 °C aged asphalt. The linear fit analysis results found that the correlation coefficient between the ductility values and aging temperature is 0.939. Additionally, the results shown in Figure 3b indicate that the ductility values reduced as the aging time increased from 0 to 20 h, and the lowest ductility value is 3.6 cm when the aging time rose to 20 h. Moreover, the ductility values also have a better correlation with aging time.

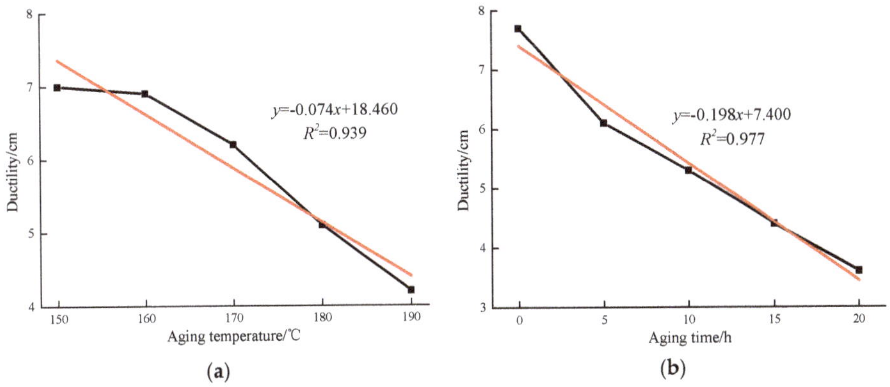

Figure 3. Ductility values of MACRMA at different aging conditions: (**a**) aging temperature and (**b**) aging time.

The softening point of the asphalt is the conditional temperature when asphalt reaching a certain viscosity, which indicates the thermal stability of the asphalt. The higher the softening point, the better the thermal stability of the asphalt. As shown in Figure 4a, all samples after aging have higher softening point values, and the highest softening point values reach 75.5 °C when the aging temperature is

190 °C. In Figure 4b, the samples exhibit similar trend results under various aging times. However, the softening point value is more than 76 °C when the aging time exceeds 10 h. In addition, the highest softening point value is 83.7 °C when the aging time is 20 h. Moreover, the linear fit analysis results found that the correlation coefficient between the softening point values and aging time was 0.990.

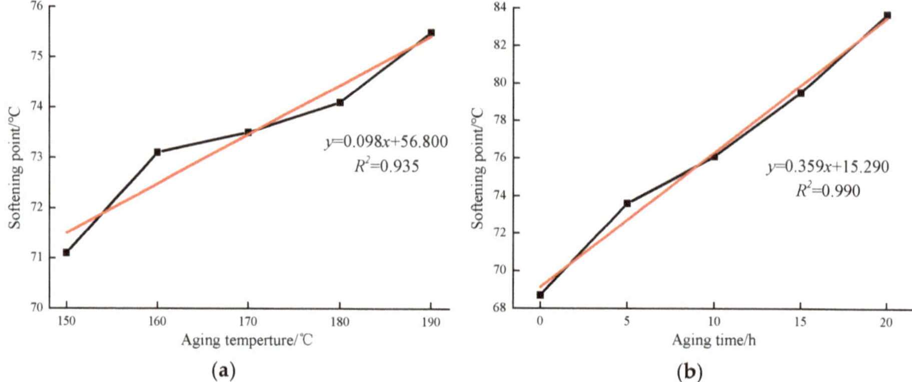

Figure 4. Softening point values of MACRMA at different aging conditions: (**a**) aging temperature and (**b**) aging time.

In general, it can be observed that the softening point of the MACRMA increased with the increase of aging temperature and time. However, the penetration and ductility gradually decreased. Consequently, it could be concluded that TFOT aging will increase the aging degree of leaching, regardless of aging time and temperature.

3.2. Viscosity

As mentioned above, the classical properties test can properly characterize the performance of crumb rubber asphalt. However, the human error of the test has large significance on its performance accuracy. Therefore, we also need to describe its performance with other indices. The viscosity is the flow characteristics of the asphalt binder to provide some assurance that it can be pumped and handled at the hot mix facility. It is also used to develop temperature–viscosity charts for estimating mixing and compaction temperature for use in mix design as a reference. Generally, a higher viscosity value results in higher mixing and compaction temperatures and may increase the energy consumption.

As shown in Figure 5a, it can be observed that the viscosity values of MACRMA increase as the aging temperature increased except at 180 °C. When the aging temperature is increased from 150 to 170 °C, the viscosity gradually increased from 3.04 Pa·s to 3.53 Pa·s. As the aging temperature rises to 180 °C, the viscosity is 3.20 Pa·s. The reason may be that the crumb rubber is degraded at high temperature, and the internal net structure is cracked into small chains, and the light component in the bitumen absorbed by the swelling action is released. However, this does not mean that the asphalt does not age. In fact, this process reflects the combined effect of the continuous swelling reaction of the rubber powder and the aging degradation on the rheology at 180 °C [33]. When the temperature continues to rise to 190 °C, the viscosity of the MACRMA increases rapidly and reaches 51 Pa·s. Therefore, it can be concluded that the production and construction temperature of microwave crumb rubber modified asphalt is recommended to be below to 190 °C.

Similarly, as shown in Figure 5b, the viscosity of MACRMA after different aging times increases rapidly from 2.93 to 4.71 Pa·s from 0 to 10 h. It might be a result of a strong swelling reaction in the modified asphalt at the initial stage of aging. It seems that the light component in the asphalt is absorbed by crumb rubbers, which cause volume expansion, and the oil content in the mixed system is reduced. Moreover, the viscosity is rapidly increased. In addition, in the 10–20 h segment, the viscosity growth tends to be stable, and reaches 5.76 Pa·s, which is the highest value of any of the modified asphalt samples. In general, the TFOT asphalts have a noticeably higher viscosity value than the unaged asphalt. This indicates that the TFOT aging causes the asphalt to stiffen at the maximum pavement service temperature, which is beneficial in improving resistance to permanent deformation.

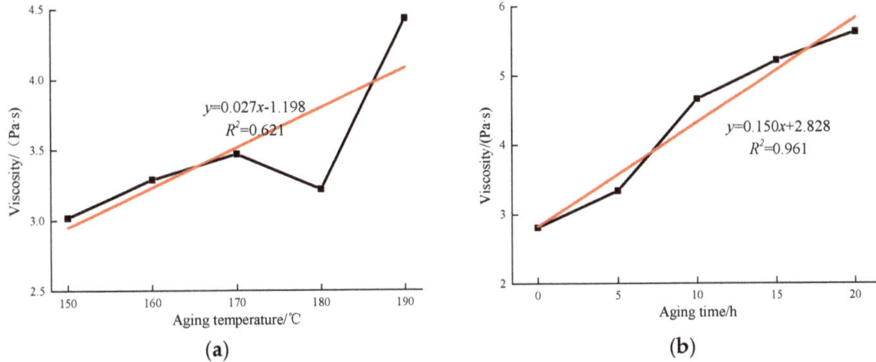

Figure 5. Viscosity values of MACRMA at different aging conditions: (**a**) aging temperature and (**b**) aging time.

3.3. $G^*/\sin \delta$

The $G^*/\sin \delta$ is often used to characterise the rutting resistance of an asphalt pavement at high temperature. A greater $G^*/\sin \delta$ value indicates a pavement with better permanent deformation resistance. It is worth mentioning that a frequency sweep is a particularly useful test as it enables the viscoelastic properties of a sample to be determined as a function of timescale. In this study, the frequency sweep test was used to determine the $G^*/\sin \delta$ values of different MACRMA samples.

The $G^*/\sin \delta$ values of the MACRMA regarding an increased frequency after the aging temperature and aging time respectively are shown in Figure 6. In Figure 6a, it can be observed that with the increase of loading frequency, the $G^*/\sin \delta$ values of MACRMA tend to increase gradually. All $G^*/\sin \delta$ values of modified asphalt at 0.1 kPa are very close, which is not significantly different between aged and unaged asphalt. In addition, the $G^*/\sin \delta$ values are higher than 1.2 kPa when the frequency increased to 1 rad/s, and some are even greater than 3.2 kPa when the aging temperature was 190 °C. At 10 rad/s, the lowest $G^*/\sin \delta$ value found was 7.29 kPa when the MACRMA was unaged. However, at 150 °C or higher to, the $G^*/\sin \delta$ values of aging asphalts are greater than 8.6 kPa. Among them, the MACRMA after 190 °C aging conditions have the greatest $G^*/\sin \delta$ values, which reach 13.37 kPa. Generally, all modified asphalts after different aging temperature have greater $G^*/\sin \delta$ values than unaged. This indicates that the aging temperature stiffens asphalts at pavement service conditions, which is beneficial to improve the resistance to permanent deformation.

Similar trends for those modified asphalts at different aging times can be found in Figure 6b. It was observed that $G^*/\sin \delta$ values of all MACRMA samples tend to increase with frequency. The reason might be that the higher loading frequencies correlate well with faster driving speeds. The load action time is so short that the load dissipated quickly. Therefore, the road surface is less prone to rutting. Compared with Figure 6a, there are more significant differences in $G^*/\sin \delta$ values among overall asphalt sources, and the trend is directly correlated with frequency. Moreover, when the aging time is long enough, the $G^*/\sin \delta$ values of MACRMA are significantly higher than the $G^*/\sin \delta$ values of MACRMA under different aging temperatures. For instance, when the aging time exceeds 10 h, the $G^*/\sin \delta$ values of MACRMA is more than 16 kPa at 10 rad/s. It was concluded that the TFOT aging results in a stiffer asphalt binder, which is beneficial in improving the rutting resistance of MACRMA, regardless of the aging temperature and time.

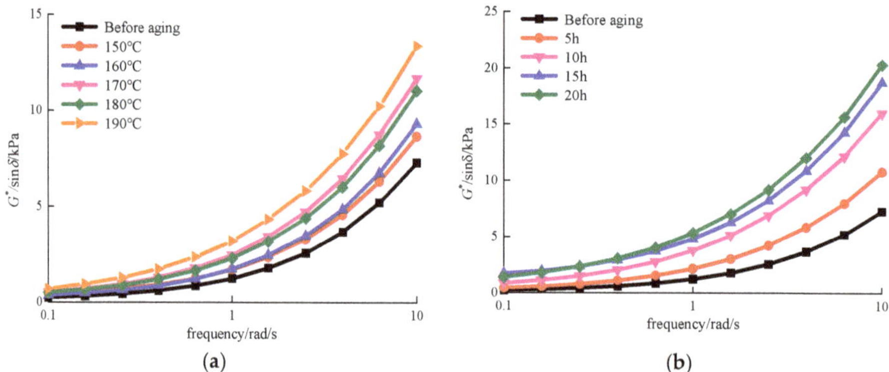

Figure 6. $G^*/\sin \delta$ values of MACRMA at different aging condition (**a**) aging temperature and (**b**) aging time.

3.4. Phase Angle

In Superpave specifications, phase angle is defined as the time lag between strain and stress under traffic loading and is highly dependent on the temperature and frequency of loading. It can be used as an indicator of viscosity and elasticity of binders. In general, for a perfectly elastic material, an applied load causes an immediate response, thus, the time lag is zero. Conversely, a viscous material has a relatively large time lag between load and response; its phase angle approaches 90°. Under normal pavement temperatures and traffic loadings, asphalt binders act with the characteristics of both viscous liquids and elastic solids. However, this property can be influenced by the nature of the polymer such as visco-elastic, visco-plastic, and elastic–plastic at a high temperature. In this study, all TFOT residues of MACRMA were tested at the high temperatures from 58 to 82 °C and thus exhibit viscoelastic properties.

As shown in Figure 7a, phase angles of various modified asphalts increase as the test temperature increased when the test temperature increased from 58 to 82 °C. In addition, at the frequency of 10 rad/s, the phase angles of the unaged sample reached its highest at 82 °C, and the peak value at this time was 2.54. The results indicated that the unaged sample has a greater phase angle than other modified binders and exhibited a more viscous behaviour. The phase angles of different aging samples reduced as the aging temperature constant increased compared with the unaged sample. However, no significantly different phase angles were found between 150 and 160 °C of modified asphalts. Therefore, for the asphalts tested, it could be concluded that the aging temperature has an impact on the viscoelastic characteristics of the MACRMA at high temperatures.

As shown in Figure 7b, all modified asphalts except unaged samples show increased phase angles when the test temperature is in the range of 58 to 82 °C, showing higher viscous characteristics. Compared with Figure 7a, when the aging time is long enough, the phase angles of MACRMA are significantly lower, and the growth trend is not obvious. It was concluded that, for the asphalts tested, the aging time has an influence on the elastic and viscous characteristics of MACRMA, which causes an increase in the elastic properties of modified asphalt. However, it also indicated that the phase angles are close to each other (5 and 10 h, 15 and 20 h). Therefore, it is necessary to do some further research to identify the visco-elastic behaviour of these polymer materials at high temperatures.

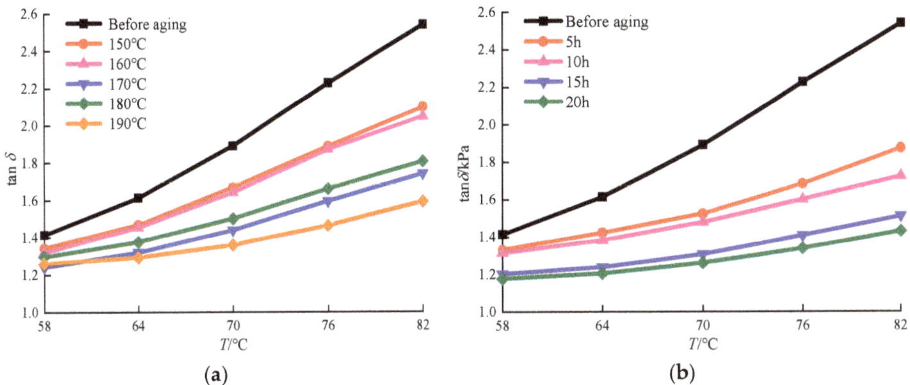

Figure 7. tan δ values of MACRMA at different aging conditions: (**a**) aging temperature and (**b**) aging time.

3.5. The Deformation Recovery Ability

The creep test is a simple test, in principle, with the application of a constant stress where the strain is measured accurately from the time of the application of the stress. Creep recovery is found when the stress is removed and the sample's recovery response is monitored with time. Generally, if the material is elastic and the elasticity has not been disturbed by the creep test, it may recoil back to its original position. Otherwise, it will stay where it is.

During the creep recovery process, the elastic portion of the asphalt is gradually recovered over time, which is due to the presence of the viscoelastic properties of the asphalt and gives a certain retardation elastic property. The residual strain in the recovery stage of asphalt is defined as εP, and the initial strain in the recovery stage of asphalt is defined as εL. The ratio of $\varepsilon P/\varepsilon L$ is used to characterise the deformation recovery ability of asphalt. The smaller the value of $\varepsilon P/\varepsilon L$, the greater the recovery ability of asphalt deformation [19,20].

Figure 8 shows the relationship between the $\varepsilon P/\varepsilon L$ and the number of loadings for MACRMA under different aging conditions at a temperature of 60 °C and pressures of 100 and 300 Pa. It can be seen from Figure 8 that all asphalts tested in this study show an increase in $\varepsilon P/\varepsilon L$ in the initial stage of loading (0~10), but with higher numbers of loads, the $\varepsilon P/\varepsilon L$ values gradually becomes stable with some slight increases. The reasons might be the asphalt becomes fatigued with an increased number of loadings, which increases the irreversible deformation and deterioration ability of the deformation recovery. As shown in Figure 8a, it can be seen that, as expected, the unaged sample has the highest $\varepsilon P/\varepsilon L$ value, which exhibits a higher viscosity than the other modified binders, whereas the asphalt under different temperatures of TFOT aging exhibits lower values with higher elastic properties. Additionally, similar trends for those modified asphalts under different aging times at 300 Pa can be seen in Figure 8b. However, there are no significant differences in $\varepsilon P/\varepsilon L$ values between any two stresses for all samples, apart from 20 h of TFOT aging time of MACRMA.

As shown in Figure 8c,d, there are similar trends for MACRMA after different aging temperatures. At a stress of 100 Pa, there is only slight differences in the $\varepsilon P/\varepsilon L$ values among 170, 180, and 190 °C of MACRMA. Additionally, no significantly different $\varepsilon P/\varepsilon L$ values were found between 100 and 300 Pa. It indicated that stress has little effect on the deformation recovery ability of asphalt. Therefore, it could be concluded that TFOT aging could make the asphalt harder, which is beneficial in improving the performance of the deformation recovery of MACRMA.

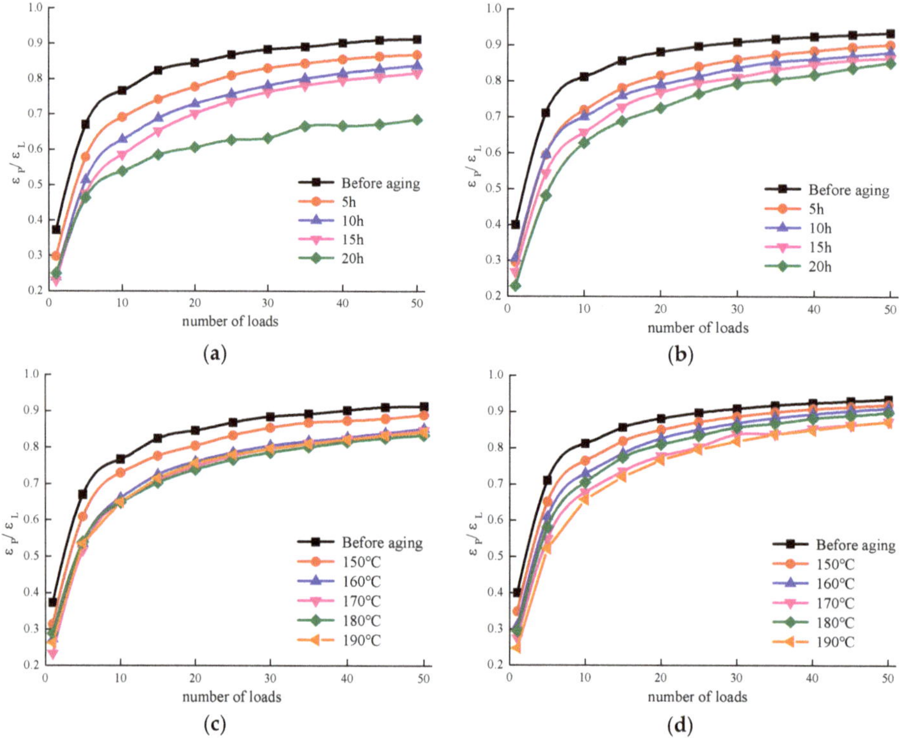

Figure 8. Relationship between $\varepsilon P/\varepsilon L$ and load times of asphalt under different aging conditions: (**a**) different aging time at 100 Pa; (**b**) different aging time at 300 Pa; (**c**) different aging temperature at 100 Pa; and(**d**) different aging temperature at 300 Pa.

3.6. Cumulative Strain

The cumulative strain reflects the total strain of the asphalt during the total process of loading and unloading. During the creep recovery process, this test is studied by monitoring the total stress processing over the decay load procedure.

Figure 9 shows the cumulative strain of the MACRMA under different stress conditions. As shown in Figure 9, the cumulative strain of asphalt increases with an increase in the number of stresses and loads, which is consistent with the fact that heavy traffic leads to rutting damage in roads. In addition, it can be noted that unaged asphalt has the highest cumulative strain regardless of the number of loads, while the cumulative strain of other aging samples is lower, although this value increases with the number of loads. In addition, the results shown in Figure 9a indicate that there are significant differences in the cumulative strain before and after aging conditions of the modified binders. The cumulative strain of aged MACRMA generally reduced as the aging time increased at the same number of loads under stress of 100 Pa. Similar trends for those modified binders can be found in Figure 9b, but the MACRMA exhibits a higher cumulative strain at a stress of 300 Pa compared to 100 Pa, which indicated that stress has a large effect on the resistance to the deformation of MACRMA. Additionally, there were no obvious differences among the 10, 15, and 20 h aging tests, but these values are noticeably different for the modified binders under stress of 100 Pa (Figure 9a).

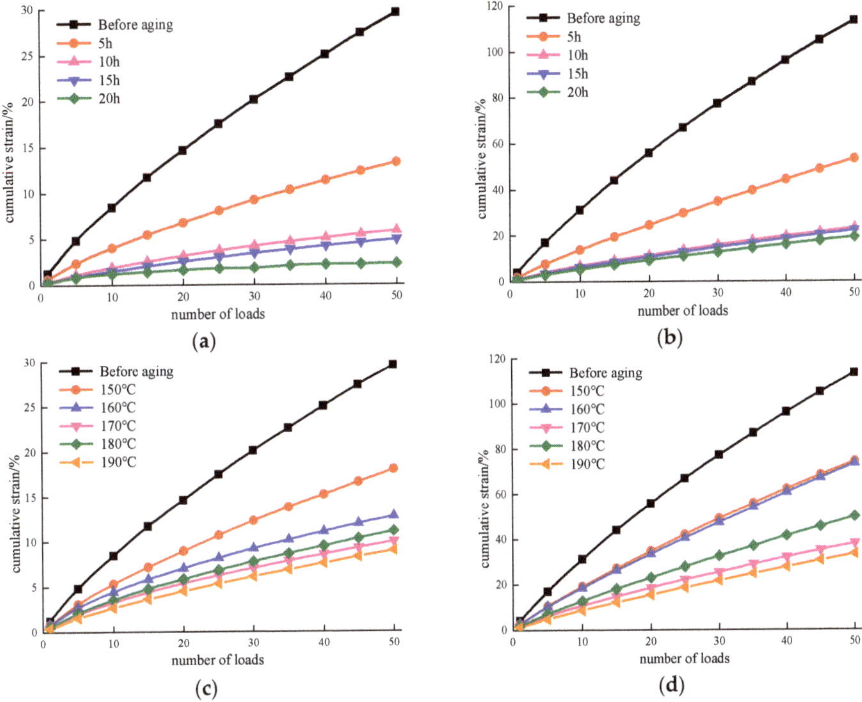

Figure 9. Relationship between cumulative strain of asphalt and number of loading times: (**a**) different aging time at 100 Pa; (**b**) different aging time at 300 Pa; (**c**) different aging temperature at 100 Pa; and (**d**) different aging temperature at 300 Pa.

Figure 9c,d show the cumulative strain of the MACRMA at aging temperatures from 150 to 190 °C at stresses of 100 Pa and 300 Pa. In Figure 9c, it was observed that the cumulative strain of aged MACRMA decreases with an increase of aging temperature at the same number of loads compared with unaged asphalt. For instance, the cumulative strain of unaged asphalt at 50 times is 29.52%, while the cumulative strain of aged asphalts is less than 20%. Thus, it seems that the anti-deformation ability of MACRMA increases. As shown in Figure 9d, the aged modified asphalts under stresses of 300 Pa show increased cumulative strain compared with 100 Pa, showing lower anti-deformation ability.

4. Conclusions

Based on the test results in this study, the following findings and conclusions can be observed of the properties of MACRMA before and after aging.

- The performance of microwave crumb rubber modified asphalt after aging is closely related to aging time and aging temperature, both of which will increase the aging degree of asphalt, so that the penetration and elongation of asphalt decrease, while the softening point and viscosity increase.
- There may be a strong swelling and degradation reaction of the rubber powder during the aging process. This effect presents the opposite viscosity characteristics of the MACRMA with the aging trend. Additionally, the MACRMA with 190 °C or 20 h TFOT aging had the highest viscosity values. A lower mixing and compaction temperature (<190 °C) could be used to produce the MACRMA mixture, and thus, reducing the energy demand and retain the performance during the mixing and compaction procedures.

- The $G^*/\sin \delta$ and phase angle values reported in this research were mainly influenced by the TFOT aging temperature and time. The aging will gradually increase the rutting factor of asphalt and reduce the phase angle, which could increase the elasticity of asphalt and improve the rutting ability at high temperatures.

- It can be observed that, in this study, the behaviours of creep and creep recovery of all modified asphalts were generally affected by TFOT aging. The aging effect reduces the cumulative deformation of the rubber modified asphalt, increases the elasticity of asphalt and improves its deformation resistance. Compared with the aging temperature, the aging time has a more significant effect on the deformation recovery ability of the MACRMA.

Author Contributions: Conceptualization, B.L. and Z.Z.; Investigation, J.Z. and Y.W.; Methodology and Formal Analysis, Z.Z. and J.Z.; Writing–Original Draft Preparation, Z.Z. and B.L.; Writing–Review and Editing, J.Z. and X.Y.; All authors of the article provided substantive comments.

Funding: This research was funded by the National Natural Science Foundation of China (51408287, 51668038,51868042), Rolls Supported by Program for Changjiang Scholars and Innovative Research Team in University (IRT-15R29), the Distinguished Young Scholars Fund of Gansu Province (1606RJDA318), the Natural Science Foundation of Gansu Province (1506RJZA064), Excellent Program of Lanzhou Jiaotong University (201606), and Foundation of A Hundred Youth Talents Training Program of Lanzhou Jiaotong University.

Acknowledgments: The authors gratefully acknowledge many important contributions from the researchers of all reports cited in our paper.

Conflicts of Interest: The authors declare no conflict of interest.

References

1. Shu, X.; Huang, B. Recycling of waste tire rubber in asphalt and Portland cement concrete: An overview. *Constr. Build. Mater.* **2014**, *67*, 217–224. [CrossRef]

2. Huang, Y.; Bird, R.N.; Heidrich, O. A review of the use of recycled solid waste materials in asphalt pavements. *Resour. Conserv. Recycl.* **2008**, *52*, 58–73. [CrossRef]

3. Wang, H.; Dang, Z.; You, Z. Effect of warm mixture asphalt (WMA) additives on high failure temperature properties for crumb rubber modified (CRM)binders. *Constr. Build. Mater.* **2012**, *35*, 281–288. [CrossRef]

4. Xu, O.; Xiao, F.; Han, S. High temperature rheological properties of crumb rubber modified asphalt binders with various modifiers. *Constr. Build. Mater.* **2016**, *112*, 49–58. [CrossRef]

5. Liu, G.; Nielsen, E.; Komacka, J. Rheological and chemical evaluation on the aging properties of SBS polymer modified bitumen: From the laboratory to the field. *Constr. Build. Mater.* **2014**, *31*, 244–248. [CrossRef]

6. Colom, X.; Faliq, A.; Formela, K. FTIR spectroscopic and thermos gravimetric characterization of ground tyre rubber devulcanized by microwave treatment. *Polym. Test.* **2016**, *52*, 200–208. [CrossRef]

7. Shen, J.N.; Amirkhanian, S.; Xiao, F.P. Influence of surface area and size of crumb rubber on high temperature properties of crumb rubber modified binders. *Constr. Build. Mater.* **2009**, *23*, 304–310. [CrossRef]

8. Singh, B.; Kumar, L.; Gupta, M. Effect of activated crumb rubber on the properties of crumb rubber-modified bitumen. *J. Appl. Polym. Sci.* **2013**, *129*, 2821–2831. [CrossRef]

9. Wang, Q.; Li, S.; Wu, X.Y. Weather aging resistance of different rubber modified asphalts. *Constr. Build. Mater.* **2016**, *106*, 443–448. [CrossRef]

10. Ibrahim, I.M.; Fathy, E.S.; Elshafie, M. Impact of incorporated gamma irradiated crumb rubber on the short-term aging resistance and rheological properties of asphalt binder. *Constr. Build. Mater.* **2015**, *81*, 42–46. [CrossRef]

11. Lee, S.J.; Chandra, K.A.; Akisetty, S.N. The effect of crumb rubber modifier(CRM) on the performance properties of rubberized binders in HMA pavements. *Constr. Build. Mater.* **2008**, *22*, 1368–1376. [CrossRef]

12. Doh, Y.S.; Amirkhanlan, S.N.; Kim, K.W. Analysis of unbalanced binder oxidation level in recycled asphalt mixture using GPC. *Constr. Build. Mater.* **2008**, *22*, 1253–1260. [CrossRef]

13. Lu, X.; Isacsson, U. Effect of ageing on bitumen chemistry and rheology. *Constr. Build. Mater.* **2002**, *16*, 15–22. [CrossRef]

14. Li, Z.; Li, C. Improvement of properties of styrene-butadiene-styrene modified bitumen by grafted eucommia ulmoides gum. *Road Mater. Pavement Des.* **2013**, *14*, 404–414. [CrossRef]

15. Heitzman, M. Design and construction of asphalt paving materials with crumb crumb rubber. Transportation Research Record No.1339. *Transp. Res. Board.* **1992**, *1339*, 1–8.

16. Maccarrone, S. Properties of polymer modified binders and relationships to mix and pavement performance. *J. Assoc. Asph. Paving Technol.* **1995**, *64*, 209–240.

17. Rasool, R.T.; Wang, S.F.; Zhang, Y. Improving the aging resistance of SBS modified asphalt with the addition of highly reclaimed rubber. *Constr. Build. Mater.* **2017**, *145*, 126–134. [CrossRef]

18. Liang, M.; Liang, P.; Fan, W.Y. Thermo-rheological behavior and compatibility of modified asphalt with various styrene-butadiene structures in SBS copolymers. *Mater. Des.* **2015**, *88*, 177–185. [CrossRef]

19. Huang, B.; Shu, X.; Li, G. Analytical modeling of three-layered HMA mixtures. *Int. J. Geomech.* **2017**, *7*, 140–148. [CrossRef]

20. Shenoy, A. Refinement of the superpave specification parameter for performance grading of asphalt. *J. Transp. Eng.* **2001**, *127*, 357–362. [CrossRef]

21. Yang, X.L.; Shen, A.Q.; Guo, Y.C. Effect of process parameters on the high temperature performance and reaction mechanism of CRMA. *Pet. Sci. Technol.* **2018**, *36*, 1537–1543. [CrossRef]

22. Liu, S.; Mo, L.T.; Wang, K. Preparation, microstructure and rheological properties of asphalt sealants for bridge expansion joints. *Constr. Build. Mater.* **2016**, *105*, 1–13. [CrossRef]

23. Wang, H.; You, Z.; Mills, B.P. Laboratory evaluation on high temperature viscosity and low temperature stiffness of asphalt binder with high percent scrap tire rubber. *Constr. Build. Mater.* **2012**, *26*, 583–590. [CrossRef]

24. Huang, B.; Shu, X.; Cao, J. A two-staged surface activation to improve properties of rubber modified cement composites. *Constr. Build. Mater.* **2013**, *40*, 270–274. [CrossRef]

25. Wang, Y.H.; Wen, Y.; Zhao, K. Connections between the Rheological and Chemical Properties of Long-Term Aged Asphalt Binders. *J. Mater. Civ. Eng.* **2015**, *27*, 04014248. [CrossRef]

26. Kim, H.; Lee, S.J.; Amirkhanian, S.N. Quantification of oxidative aging of polymer-modified asphalt mixes made. *J. Mater. Civ. Eng.* **2013**, *25*, 1–8. [CrossRef]

27. Zhang, Z.; Li, B.; Wei, Y. Effect of short-term aging on high-temperature properties of crumb rubber modified asphalt. *Mod. Chem. Ind.* **2018**, *38*, 105–109, 111.

28. Zhang, F.; Yu, Y. The research for high-performance SBR compound modified asphalt. *Const. Build. Mater.* **2010**, *24*, 410–418. [CrossRef]

29. Zhang, Z.; Li, B.; Li, P.; Yang, J. Influence of surface area of microwave activated crumb rubber on rheological properties of rubber asphalt. *Pet. Process. Petrochem.* **2018**, *49*, 79–83.

30. LI, B.; Zhang, Z.; Liu, X. Adhesion in SBS modified asphalt containing warm mix additive and aggregate system based on surface free theory. *Mater. Rev.* **2017**, *31*, 115–120.

31. Xiao, F.; Amirkhanian, S.; Wang, H. Rheological property investigations for polymer and polyphosphoric acid modified asphalt binders at high temperatures. *Const. Build. Mater.* **2014**, *64*, 316–323. [CrossRef]

32. *JTG E20-2011 Standard Test Methods of Asphalt and Asphalt Mixtures for Highway Engineering: JTG E20-2011*; Ministry of Communications of the People's Republic of China: Beijing, China, 2011.

33. Li, P.; Ding, Z.; Chen, C. Analysis of swelling and degradation characteristics of crumb rubber in hot asphalt. *J. South China Univ. Technol. Nat. Sci. Ed.* **2016**, *44*, 97–103.

Article

Determination of Construction Temperatures of Crumb Rubber Modified Bitumen Mixture Based on CRMB Mastic

Yanan Li [1], Yuchao Lyu [1], Meng Xu [1], Liang Fan [2,*] and Yuzhen Zhang [1,*]

[1] College of Chemical Engineering, China University of Petroleum, Qingdao 266580, Shandong, China; liyanan2992@163.com (Y.L.); B15030098@s.upc.edu.cn (Y.L.); xumeng04@126.com (M.X.)

[2] Shandong Transportation Research Institute, Jinan 250000, Shandong, China

* Correspondence: fanliang218@sina.com (L.F.); zhangyuzhen1959@163.com (Y.Z.); Tel.: +86-137-9101-2426 (L.F.); +86-138-0101-7308 (Y.Z.)

Received: 27 September 2019; Accepted: 19 November 2019; Published: 22 November 2019

Abstract: Crumb rubber modified bitumen (CRMB) has been widely used in pavement construction and provides an effective way to recycle waste tires and helps alleviate the "black pollution" problem. There are no current specifications regarding the appropriate mixing and compaction temperatures of the CRMB mixture. There is a direct relationship between the mixing and the compaction temperatures of the CRMB mixture and the viscosity of the CRMB mastic. In this study, we first prepared CRMB using crumb rubber powder and penetration grade 70 neat bitumen, then prepared the CRMB mastic using CRMB and fillers (limestone mineral powder and cement). Finally, we used the CRMB mastic and aggregate to make mixture specimens. The best air void of the specimens was subsequently used to demarcate the viscosity of the CRMB mastic, and the construction temperatures (including the mixing temperature and the compaction temperature) were calculated based on the viscosity of the CRMB mastic from the viscosity–temperature curves. Test results indicated that the best viscosity of the CRMB mastic was 2.7 ± 0.2 Pa·s and 3.9 ± 0.3 Pa·s that corresponded to the mixing and compaction temperatures, respectively.

Keywords: crumb rubber modified bitumen mastic; Brookfield viscosity; mixing and compaction temperatures

---

1. Introduction

Increasing attention has been paid to modified bitumen and modified bitumen mixture over time. Pavement materials require a higher quantity of bitumen mastic because of the higher volume of traffic, load, and dynamic water pressure they are exposed to. Rubber powder made from old tires is added to neat bitumen to obtain modified bitumen, which is termed crumb rubber modified bitumen (CRMB) or rubber bitumen. The CRMB mixture has a favorable ageing resistance, fatigue restraint ability, high-temperature stability, low-temperature anti-cracking performance and water resisting properties. It also contributes to alleviating the "black pollution" problem. Stone mastic asphalt (SMA) displays good pavement performance and has been widely applied in pavement construction. Bitumen mastic plays a key role in SMA; however, there is increasing concern about the use of it. Improved adhesive attraction between the bitumen and aggregate occurs only after the formation of the bitumen mastic by bitumen absorbing onto the surface of the filler. Bitumen mastic is one of the three space grid structure dispersed phases in modern mastic theory, being the first and most important phase. The filler is the dispersed phase and the bitumen is the medium. The structure, composition and performance of the bitumen mastic have a direct bearing on the performance of the pavement composed of the bitumen mixture [1]. Therefore, research only on the bitumen binder is inadequate and it is more important to research the bitumen mastic. However, most road workers ignore the dispersed effect of the filler and the interface chemistry effect between the filler and the bitumen and just treat the filler as an inert filler. In addition, although the bitumen binder itself is good, the actual pavement performance is less than satisfactory. Therefore, it is necessary to study the material made from the filler and the bitumen, i.e., the bitumen mastic.

Puzinauskas, V. found that the filler has two functions in the bitumen mixture when he researched cement–filler bitumen mixtures [2]. He found that the filler was not just a pure filler, but was combined with the bitumen to obtain the bitumen mastic. Tunnicliff, D.G. found that mineral powder below 0.075 mm had a great effect on the water stability and the anti-aging performance of bitumen mixtures [3]. Huang Baoshan et al. chose three filler types and four different filler–bitumen ratios to study bitumen mastics, and the results showed that the filler–bitumen ratio must be controlled by the overall performance of the bitumen mixture [4]. Wu Yuhui used a dynamic shear rheometer (DSR) and a bending beam rheometer (BBR) to study the effect of mineral powder content on the high- and low-temperature performance of penetration grade 90 neat bitumen mastic [5]. The results showed that a filler–bitumen ratio between 0.8 and 1.2 could provide a balance between high- and low-temperature performance.

Previous studies showed that the construction temperatures of the CRMB mixture are significantly higher than that of the road bitumen mixture due to the higher viscosity of the CRMB, which leads to higher levels of energy consumption and air pollution [6–10]. There are no specifications for determining the construction temperatures of the CRMB mixture. Several researchers have concentrated on studying the construction temperatures of modified bitumen in recent decades. Yildirim, Y. et al. used high shear rate viscosity–temperature curves to determine the construction temperatures of modified bitumen mixture [11]. In 2001, Bahia, H.U. et al. used the change in zero shear viscosity to determine the construction temperatures by changing the shear rate [12]. In 2003, Reinke, G. obtained the construction temperatures by the shear stability viscosity of DSR [13]. Akisetty, C.K. et al. used the phase angle and the angular frequency of DSR to determine the construction temperatures [14]. West, R.C. et al. [15] and Cai Jun et al. [16] used the air void to determine the construction temperatures of warm mix rubber bitumen mixtures. Ozturk, H. I. et al. found that the compaction temperature could be lowered by 15 °C and warm mix asphalt additive can be used as a compaction aid to lower the compaction effort [17].

The research regarding bitumen mastic to date has been concentrated on mostly non-modified bitumen mastic. There is limited research on the construction temperatures of the CRMB mixture, if any, based on the air void of the CRMB mixture. Hence, studying the determination of the construction temperatures of the CRMB mixture based on the viscosity of the CRMB mastic is required. In the present study, the viscosity of the CRMB mastic was used to determine the construction temperatures of the CRMB mixture, and the temperatures were verified by the air void of the CRMB mixture.

2. Materials and Methods

2.1. Raw Materials

Penetration grade 70 neat bitumen produced in SsangYong, Incheon, Korea, was used in the present study. Table 1 summarizes the properties of this bitumen binder based on the Chinese standard JTG E20-2011 [18] and the test method by the Strategic Highway Research Program A [19].

Table 1. Properties of the penetration grade 70 neat bitumen used in the present study.

Properties		Requirements	Test Results
Penetration (25 °C; 0.1 mm)		60–80	68.9
Ductility (15 °C; cm)		≥40	68
Softening point (°C)		≥46	48.3
Flash point (°C)		≥230	263
Wax content (%)		≤3	1.84
Density (g/cm^3)		report	1.032
Solubility (trichloroethylene; %)		≥99.5	99.7
	Mass change (%)	≤0.8	0.3
RTFOT (163 °C, 75 min)	Retained penetration (%)	≥58	78
	Ductility (15 °C; cm)	≥15	44
PG grade		64–22	

The rubber powder (20 mesh size) was made from truck tires in Changzhou, China. Tables 2 and 3 summarize the physical properties and the chemical properties of the rubber powder, respectively.

Table 2. Physical properties of the rubber powder.

Testing Items	Relative Densities	Water Ratio (%)	Metal Content (%)	Fiber Content (%)	Screenings (%)
Test results	1.16	0.6	0.01	0.5	8.3
Requirements	1.10–1.30	<1	<0.05	<1	<10

Table 3. Chemical properties of the rubber powder.

Testing Items	Ash Content (%)	Acetone Extract (%)	Carbon Black Content (%)	Rubber Hydrocarbon Content (%)
Test results	7.2	4.9	31.3	55.2
Requirements	≤8	≤22	≥28	≥42

The fillers were limestone mineral powder and cement (strength degree of 42.5). Table 4 summarizes their various properties.

Table 4. Properties of the mineral powder and the cement.

Kinds of Filling	Apparent Density (g/cm^3)	Hydrophilic Coefficient	<0.075 mm Content (%)
Mineral powder	2.71	0.89 (<1)	90.4
Cement	2.80	0.78 (<1)	92.3

The diameter of the mineral powder and cement used in the present study was smaller than 0.075 mm. Table 5 summarizes the screening results of the mineral powder.

Table 5. Sieve analysis of the mineral powder.

Screen Size (mm)	Passing percent (%)
0.075	83.4
0.15	91.2
0.3	97.3
0.6	100

The aggregate used in the CRMB was basalt. Tables 6 and 7 summarize the technology parameters of the basalt.

Table 6. Test results of the aggregates' properties.

Aggregate Properties	#1 (9.5–16 mm)	#2 (4.75–9.5 mm)	#3 (2.36–4.75 mm)	#4 (0–2.36 mm)
Apparent density (g/cm^3)	2.868	2.857	2.839	2.865
Bulk density (g/cm^3)	2.797	2.766	2.757	2.749
Water absorption (%)	0.89	1.15	1.05	1.47

Table 7. Analysis results of different mineral aggregates.

Screen Size (mm)		16	13.2	9.5	4.75	2.36	1.18	0.6	0.3	0.15	0.075
	#1	100	83.2	10.4	0.3	0.3	0.3	0.3	0.3	0.3	0.3
Passing-percent (%)	#2	100	100	98.5	12.1	0.6	0.6	0.6	0.6	0.6	0.6
	#3	100	100	100	85.3	25.7	14.0	8.1	4.0	2.9	1.9
	#4	100	100	100	100	86.3	64.1	39.2	18.5	10	7.6

2.2. Preparation of Bitumen Mastic

The present study used CRMB binder and CRMB mastic to compare the accuracy of determination based on viscosity. The CRMB mastic was made by adding the filler to the CRMB binder. Hence, the preparation process parameters did not influence the comparison results and the parameters have been commonly used.

The CRMB binder was made by adding rubber powder (20 mesh size) to penetration grade 70 neat bitumen. The mass ratio of rubber powder to penetration grade 70 neat bitumen was defined as the generalized filler–bitumen ratio. The ratios used were 0.16, 0.18, 0.20, and 0.22. The preparation parameters of the generalized base bitumen mastic were as follows: the mixing temperature was 180 °C, the mixing time was 45 min, and the shear rate was 1000 r/min.

The CRMB mastic was made by adding fillers to the CRMB binder. The ratio of filler to CRMB binder was defined as the filler–bitumen ratio. The ratio of filler was between the weight of the mineral powder and the rubber bitumen. The ratios used were 0.10, 0.25, 0.40, 0.50, 0.60, and 0.80. We used a small mixer in the laboratory to mix the samples by mechanical agitation. The preparation parameters of the rubber bitumen were as follows: the mixing temperature was 180 °C, the mixing time was 45 min, the shear rate was 1000 r/min and they were the same for different ratios of samples. The preparation parameters of the CRMB mastic were as follows: the mixing temperature was 200 °C, the mixing time was 20 min when the ratio was less than 0.6 and was 30 min when the ratio was less than 0.8, and the shear rate was 1000 r/min. The preparation processes for the CRMB binder and the CRMB mastic are shown in Figure 1.

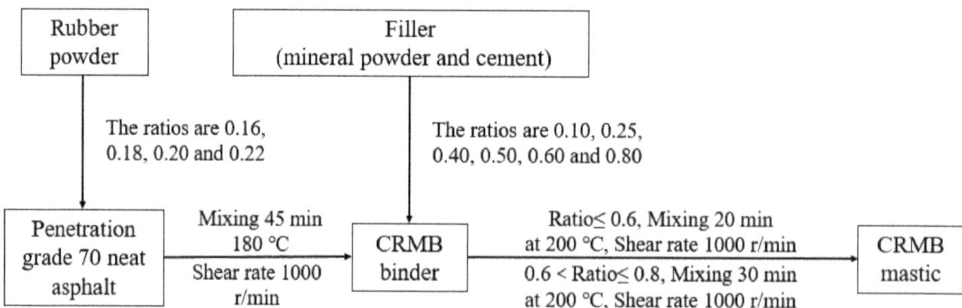

Figure 1. Preparation processes of the crumb rubber modified bitumen (CRMB) binder and the CRMB mastic.

2.3. Research Methodology

There are three commonly used methods for determining the viscosity of bitumen. The present study used the Brookfield viscosity method, which tested the viscosity of the bitumen binder and the bitumen mastic at different temperatures. Then, the construction temperatures were calculated by the relationship between the viscosity and temperature. Table 8 summarizes the test parameters for the CRMB that is used mainly in America [20].

Table 8. Test parameters of the CRMB binder and the CRMB mastic.

Test Parameters	Rotor	Revolving Speed (r/min)
CRMB binder	27#	20
CRMB mastic	27#	30

3. Results and Discussion

3.1. Determination Based on the CRMB Binder

Four groups of the CRMB binder were tested based on the different ratios described previously and the viscosity–temperature curves were calculated with the Saal formula (Equation (1)) from the ASTM D2493 standard:

$$\text{Lglg}\,(\eta \times 1000) = n - m \times \lg(\,T + 273.15)\tag{1}$$

where, η is the viscosity (Pa·s) and T is the temperature (°C).

The viscosity–temperature curves calculated by Equation (1) using the viscosity of the CRMB binder at 165 °C and 177 °C are shown in Figure 2.

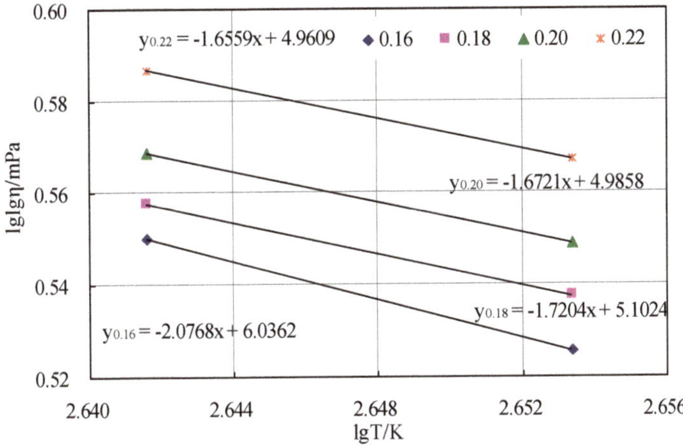

Figure 2. Viscosity–temperature curves of the CRMB binder.

The viscosity ranges of the mixing and the compaction temperatures were 0.17 ± 0.02 Pa·s and 0.28 ± 0.03 Pa·s, respectively, based on the Chinese standard JTG E20-2011 [18]. The construction temperatures were obtained from Figure 2 and are shown in Table 9.

Table 9. Construction temperatures of the CRMB mixture.

Generalized Filler–Bitumen Ratio	0.16	0.18	0.20	0.22
Mixing temperature (°C)	269–281	299–315	313–329	329–346
Compaction temperature (°C)	246–256	270–283	282–295	298–311

Table 9 shows that the construction temperatures were too high and against real-life engineering. Thus, these construction temperatures that were determined based on the viscosity of the CRMB binder are inaccurate.

3.2. Determination Based on the CRMB Mastic

3.2.1. The Relationship between Viscosity and Temperature of the CRMB Mastic

Six groups of the CRMB mastic were tested based on the different ratios described previously and the viscosity–temperature curves were calculated with Equation (1) from the ASTM D2493 standard. Figure 3 shows the viscosity–temperature curves made with the Brookfield viscosity method at 165 °C and 177 °C.

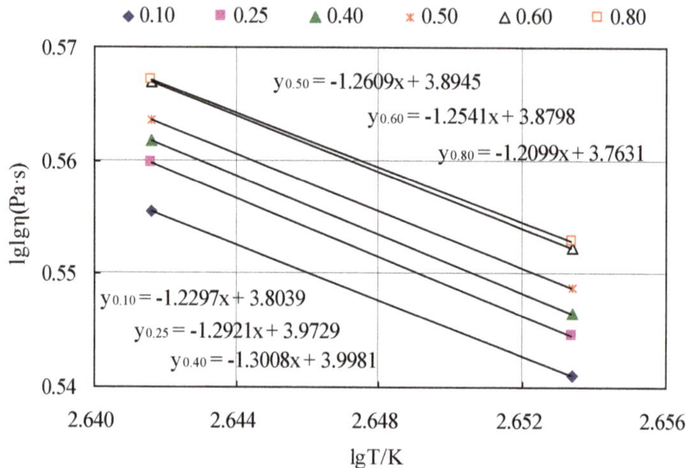

Figure 3. Viscosity–temperature curves of the CRMB mastic.

3.2.2. Demarcating the Viscosity Ranges of the CRMB Mastic

The air void is a very important index of the bitumen mixture. The type of CRMB mixture was AR-AC-13. The generalized filler–bitumen ratio was 0.18. The shaping method used was the superpave gyratory compactor (SGC). The best air voids of the CRMB mixture were used to calculate the viscosity ranges of the construction temperatures. The other construction temperatures of the mixture at the different generalized filler–bitumen ratios were calculated using the viscosity ranges.

Table 10 shows three types of gradation (A, B and C) of the CRMB mixture. The bitumen aggregate ratio was 8.2%.

Table 10. Three grading results of the CRMB mixture.

Types of Gradation	Screen Size (mm)									
#1:#2:#3:#4	16.0	13.2	9.5	4.75	2.36	1.18	0.6	0.3	0.15	0.075
A (35:38:7:20)	100.0	94.1	68.1	30.7	19.4	14.1	8.7	4.3	2.5	2.0
B (37:38:7:18)	100.0	93.8	66.3	28.7	17.7	12.9	8.0	3.9	2.3	1.8
C (39:38:7:16)	100.0	93.4	64.5	26.7	16.0	11.6	7.2	3.6	2.1	1.7

The air voids of three types of gradation are shown in Table 11.

Table 11. Volume analysis of preliminary grading.

Types of Gradation	Bitumen Aggregate Ratio (%)	Air Void (%)
Gradation A	8.2	3.1
Gradation B	8.2	4.8
Gradation C	8.2	5.2
Requirements	-	4.5–6.5

Table 11 shows that the air voids of gradation B and gradation C were satisfactory. The present study chose gradation C as the experimental gradation next, as shown in Figure 4, and Table 12 summarizes the volume index of the CRMB mixture.

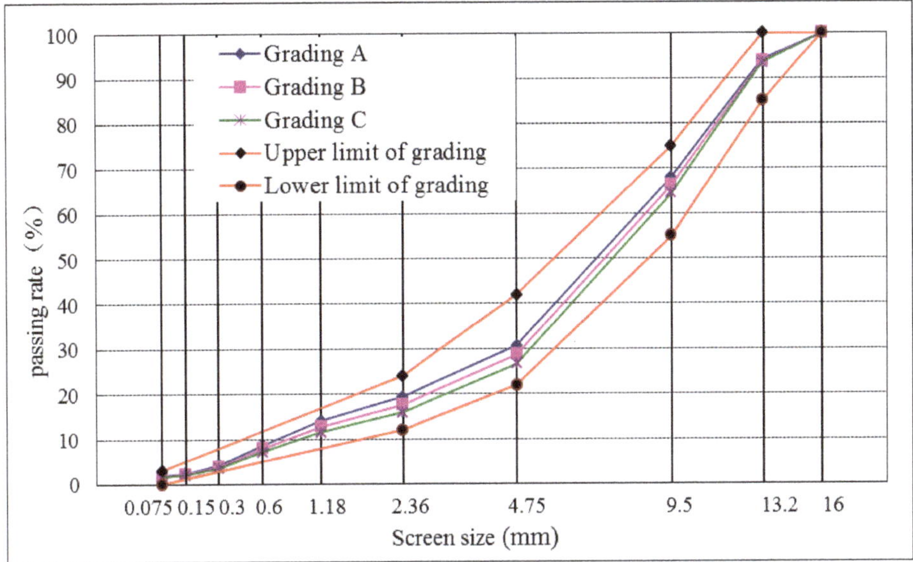

Figure 4. Gradation curves of the CRMB mixture.

Table 12. Volume indexes of gradation B.

Bitumen Aggregate Ratio (%)	Air Void (%)	VMA (%)	VFA (%)
8.2	5.2	21.59	77.77
Requirements	5.5 ± 1	≥19	—

The mixing temperature was approximately 170–180 °C for the actual construction. Therefore, the mixing temperature was approximately 170–180 °C in the test and the compaction temperatures were 180 °C, 170 °C, 160 °C, and 150 °C using the SGC. The results are shown in Figure 5.

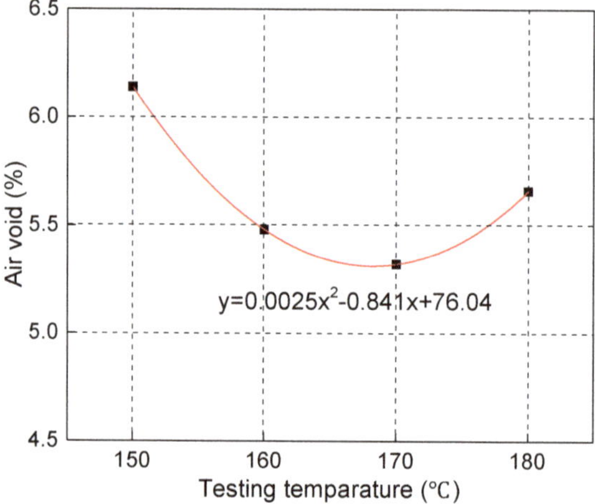

Figure 5. Relationship between the air voids and the compaction temperatures.

Figure 5 shows that the relationship between the air voids and the compaction temperatures was a hyperbolic curve. It can be seen that with increasing the compaction temperature, the air voids first decreased, reaching a minimum level, and then began to increase gradually. Furthermore, it can be calculated from the above figure that the CRMB mixture has the lowest air void when the compaction temperature is 168.2 °C. Therefore, a compaction temperature range of 164–172 °C was determined. Generally, the mixing temperature is higher than the compaction temperature by approximately 10 °C [21–24]. Therefore, a mixing temperature range of 174–182 °C was determined. Corresponding to the viscosity–temperature curve y0.18 = −1.7204x + 5.1024 in Figure 2, the viscosity ranges of the mixing and compaction temperatures were selected as 2.4–3.1 Pa·s and 3.3–4.5 Pa·s, respectively. In order to avoid the range of construction temperatures being too wide, the final viscosities for mixing and compaction were selected as 2.7 ± 0.2 Pa·s and 3.9 ± 0.3 Pa·s, respectively.

3.2.3. Determination of Construction Temperatures Using Viscosity of the CRMB Mastic

Corresponding to the viscosity–temperature curves in Figure 3, the mixing and compaction temperatures were calculated by the viscosity of the CRMB mastic and the results are shown in Table 13.

Table 10. Construction temperatures of the CRMB mixture.

Filler–Bitumen Ratio	0.10	0.25	0.40	0.50	0.60	0.80
Mixing temperature (°C)	178–185	181–188	183–189	185–192	188–195	189–196
Compaction temperature (°C)	162–169	165–172	167–174	169–180	171–178	171–179

Table 13 shows that the difference between the mixing and the compaction temperatures was about 12–17 °C, which agrees with engineering practice overall. However, the difference between the mixing and the compaction temperatures was not approximately 10 °C, with the reason for this requiring further study.

4. Conclusions

The viscosity ranges of the mixture binder at the mixing and compaction temperatures were 0.17 ± 0.02 Pa·s and 0.28 ± 0.03 Pa·s based on the Chinese standard JTG E20-2011, respectively. However, it can be seen from the data in this paper that the construction temperatures calculated based on the viscosity of the CRMB binder were too high and were inconsistent with the actual construction parameters, that is, the construction temperatures of the CRMB mixture determined by this method may be inaccurate. The results of this paper show that the construction temperatures of the CRMB mixture can be determined based on the viscosity of the CRMB mastic instead of the CRMB binder and the construction temperatures determined according to this recommendation are consistent and feasible with engineering practice overall. In detail, it is recommended that the viscosity ranges of the CRMB mastic in the mixing and compaction process should be 2.7 ± 0.2 Pa·s and 3.9 ± 0.3 Pa·s, respectively.

Author Contributions: Investigation, Y.L. (Yanan Li) and Y.L. (Yuchao Lyu); Methodology, Y.L. (Yanan Li) and Y.Z.; Validation, L.F. and Y.Z.; Writing—original draft, Y.L. (Yanan Li); Writing—reviewing and editing, M.X.

Funding: This research was funded by the Applied Basic Research Project of the Ministry of Transport (Project No. 2013319781140).

Conflicts of Interest: The authors declare no conflict of interest.

References

1. Szydlo, A.; Mackiewicz, P. Asphalt mixes deformation sensitivity to change in rheological parameters. *J. Mater. Civ. Eng.* **2005**, *17*, 1–9. [CrossRef]
2. Puzinauskas, V. Filler in asphalt mixtures. In *The Asphalt Institute Research Report*; Transportation Research Board: Washington, DC, USA, 1983.
3. Tunnicliff, D.G. Binding effects of mineral filler. In Proceedings of the Association of Asphalt Paving Technologists, Washington, DC, USA; 1967, 2, p. 15.
4. Huang, B.S.; Shu, X.; Chen, X.W. Effects of mineral fillers on hot-mix asphalt laboratory-measured properties. *Int. J. Pavement Eng.* **2007**, *8*, 1–9. [CrossRef]
5. Wu, Y.H. Effect of mineral powder content on the performance of asphalt mastic. *J. Highw. Transp. Res. Dev.* **2008**, *9*, 35–38.
6. Huang, B.S.; Mohammad, L.N.; Graves, P.S.; Abadie, C. Louisiana experience with crumb rubber-modified hot-mix asphalt pavement. In *Bituminous Paving Mixtures 2002: Materials and Construction*; Transportation Research Board: Washington, DC, USA, 2002; pp. 1–13. [CrossRef]
7. Pouranian, M.R.; Shishehbor, M. Sustainability Assessment of Green Asphalt Mixtures: A Review. *Environments* **2019**, *6*, 73. [CrossRef]
8. Chen, T.; Ma, T.; Huang, X.M.; Guan, Y.S.; Zhang, Z.X.; Tang, F.L. The performance of hot-recycling asphalt binder containing crumb rubber modified asphalt based on physiochemical and rheological measurements. *Constr. Build. Mater.* **2019**, *226*, 83–93. [CrossRef]
9. Khosravani, M.R.; Weinberg, K. A review on split Hopkinson bar experiments on the dynamic characterisation of concrete. *Constr. Build. Mater.* **2018**, *190*, 1264–1283. [CrossRef]
10. Yang, X.; You, Z.; Hasan, M.R.M.; Diab, A.; Shao, H.; Chen, S.; Ge, D. Environmental and mechanical performance of crumb rubber modified warm mix asphalt using Evotherm. *J. Clean. Prod.* **2017**, *159*, 346–358. [CrossRef]
11. Yildirim, Y.; Solaimanian, M.; Kennedy, T.W. *Mixing and Compaction Temperatures for Hot Mix Asphalt Concrete*; University of Texas at Austin. Center for Transportation Research: Austin, TX, USA, 2000.
12. Bahia, H.U.; Hanson, D.; Zeng, M.; Zhai, H.; Khatri, M.; Anderson, R. *Characterization of Modified Asphalt Binders in Superpave Mix Design*; Transportation Reaserch Board: Washington, DC, USA, 2001.
13. Reinke, G. *Determination of Mixing and Compaction Temperature of PG Binders Using a Steady Shear Flow Test*; Superpave Binder Expert Task Group, AASHTO: Washington, DC, USA, 2003.
14. Akisetty, C.K.; Lee, S.-J.; Amirkhanian, S.N. Effects of Compaction Temperature on Volumetric Properties of Rubberized Mixes Containing Warm-Mix Additives. *J. Mater. Civ. Eng.* **2009**, *21*, 409–415. [CrossRef]
15. West, R.C.; Watson, D.E.; Turner, P.A.; Casola, J.R. *NCHRP Report 648: Mixing and Compaction Temperatures of Asphalt Binders in Hot-Mix Asphalt*; Transportation Research Board of the National Academies: Washington, DC, USA, 2010.
16. Cai, J.; Ge, Z.; Yuan, J. Determining on construction temperature of warm rubberized asphalt mixture. *Sci. Technol. Eng.* **2011**, *11*, 1386–1388.
17. Ozturk, H.I.; Kamran, F. Laboratory evaluation of dry process crumb rubber modified mixtures containing Warm Mix Asphalt Additives. *Constr. Build. Mater.* **2019**, *229*, 116940. [CrossRef]

18. Research Institue of Highway Ministry of Transport. *M.O.T. Standard Test Methods of Bitumen and Bituminous Mixture for Highway Engineering*; China Communications Press: Beijing, China, 2011.
19. Kennedy, T.W.; Huber, G.A.; Harrigan, E.T.; Cominsky, R.J.; Hughes, C.S.; Von Quintus, H.; Moulthrop, J.S. *Superior Performing Asphalt Pavements (Superpave): The Product of the SHRP Asphalt Research Program*; AASHTO: Washington, DC, USA, 1994.
20. ASTM D4402. *Standard Test Method for Viscosity Determination of Asphalt at Elevated Temperatures using a Rotational Viscometer*; ASTM International: West Conshohocken, PA, USA, 2002.
21. Yang, G.L. Experimental study on compaction temperature of organic viscosity of CRMA. *Chin. Foreign Highw.* **2013**, *32*, 224–227.
22. Zhang, Z.Q.; Li, N.; Chen, H.X. Determining method of mixing and compaction temperatures for modified asphalt mixture. *J. Traffic Transp. Eng.* **2007**, *7*, 36–40.
23. Wang, P.Z. Investigation of the mixing construction temperature of SBS modified asphalt. *Shanxi Traffic Sci. Technol.* **2008**, *9*, 12–13.
24. Wang, Y.; Sun, W.H. Research on the determination method of the construction temperature of SBS modified asphalt. *Highw. Automot. Appl.* **2013**, *11*, 92–96.

 *materials*

Article

Mesostructural Modeling of Dynamic Modulus and Phase Angle Master Curves of Rubber Modified Asphalt Mixture

Linhao Gu [1,2], Luchuan Chen [3], Weiguang Zhang [1], Haixia Ma [4] and Tao Ma [1,]*

[1] School of Transportation, Southeast University, Nanjing, Jiangsu 210096, China; gulinhao@seu.edu.cn (L.G.); wgzhang@seu.edu.cn (W.Z.)

[2] State Engineering Laboratory of Highway Maintenance Technology, Changsha University of Science and Technology, Changsha 410114, China

[3] Qilu Transportation Development Group, 1 Longaoxi, Jinan, Shandong 200101, China; chenluchuan01@163.com

[4] Shandong Guilu Expressway Construction Co. LTD., 23 Changrun, Liaocheng, Shandong 252000, China; 230129425@seu.edu.cn

* Correspondence: matao@seu.edu.cn; Tel.: +86-158-0516-0021

Received: 1 April 2019; Accepted: 20 May 2019; Published: 22 May 2019

Abstract: The main objective of this paper was to develop a mesostructure-based finite element model of rubber modified asphalt mixture to predict both the dynamic modulus master curve and phase angle master curve under a large frequency range. The asphalt mixture is considered as a three-phase material consisting of aggregate, asphalt mortar, and air void. The mesostructure of the asphalt mixture was digitized by a computed tomography (CT) scan and implemented into finite element software. The 2S2P1D model was used to obtain the viscoelastic information of an asphalt mortar under a large range of frequencies and temperatures. The continuous spectrum of the 2S2P1D model was converted to a discrete spectrum and characterized by the generalized Maxwell model for numerical simulation. The Prony series parameters of the generalized Maxwell model and the elastic modulus of the aggregates were inputted into the finite element analysis as material properties. The dynamic modulus tests of a rubber modified asphalt mortar and asphalt mixture were conducted under different temperatures and loading frequencies. The dynamic modulus master curve and phase angle master curve of both asphalt mortar and asphalt mixture were constructed. The frequency of the finite element simulations of the dynamic modulus tests ranged from 10^{-6} to 10^4. The dynamic modulus and phase angle of the asphalt mixture was calculated and the master curves were compared with the master curves obtained from the experimental data. Furthermore, the effect of the elastic modulus of aggregates on the master curves was analyzed. Acceptable agreement between dynamic modulus master curves obtained from experimental data and simulation results was achieved. However, large errors between phase angle master curves appeared at low frequencies. A method was proposed to improve the prediction of the phase angle master curve by adjusting the equilibrium modulus of the asphalt mortar.

Keywords: rubber modified asphalt mortar; asphalt mixture; continuous and discrete spectrum; finite element model; dynamic modulus; phase angle; master curves

---

1. Introduction

The asphalt mixture exhibits linear viscoelasticity (LVE) under small strain conditions (150 $\mu\varepsilon$). The dynamic complex modulus test was introduced by NCHRP 9–19 to characterize the viscoelastic mechanical behavior of asphalt mixture. The dynamic modulus and phase angle master curves can be constructed by employing the time-temperature superposition principle (TTSP) [1–3]. However, laboratory tests are time-consuming and require expensive advanced testing equipment. In the past few decades, scholars have been working on virtual test methods to obtain mechanical properties more efficiently [4–8]. Meanwhile, with the development of CT technology and digital image processing, virtual test simulations based on the mesostructure or microstructure model were conducted using a numerical algorithm like the finite element model and discrete element model [9–14]. You [15] employed DIC (digital image correlation) to reconstruct a mechanical model of the asphalt mixture. A two-dimensional (2D) clustered DEM (discrete element method) model was built to simulate the dynamic modulus test of sand mastic and asphalt mixture. You and Dai [16,17] used a micromechanical-based FEM (finite element method) model to simulate the dynamic modulus of the hot mix asphalt mixture. A two-phase model consisting of aggregates and sand mastic was modeled with finite element method. The results had good agreement with that of the DEM models. Dai [18] employed the same procedure to predict both the dynamic modulus and phase angle of the stone-based composites of the asphalt mixture. Ying et al. [19] reconstructed a 3D heterogeneous FEM model by X-ray tomography images. Dynamic modulus test simulations were conducted. It indicated that the mastic had more influence on the dynamic modulus than the aggregates. Liu and You [20] analyzed the creep stiffness of the three-dimensional heterogeneous structure of the asphalt mixture. It was found that the geometric characteristics of aggregate orientation and sphericity have a certain comprehensive influence on the creep stiffness of the mixture. Chen proposed a random packing algorithm to generate a 3D virtual asphalt mixture sample called RAGS (random aggregate structures). The dynamic modulus and phase angle were simulated using both the 2D and 3D finite element model. It was shown that the 2D model underestimates the dynamic modulus while overestimating the phase angle. Cao [6] generated 2D FEM models of asphalt mixture by random packing. A large number of simulations were conducted and a theoretical statistic formula for describing the size effort of a dynamic modulus was proposed.

The studies mentioned above mainly focused on the prediction of a dynamic modulus at one temperature, which means in a small frequency range. In order to obtain the full viscoelastic information of an asphalt mixture, the prediction of a dynamic modulus and phase angle master curves under a wide range of frequency needs to be studied.

Many attempts were made to obtain master curves of an asphalt mixture from mechanical properties of an asphalt mortar or asphalt binder. Olard [21] proposed a relationship between the binder and the mixture complex modulus. The dynamic modulus of the asphalt mixture can be calculated using the glassy modulus and equilibrium modulus of both the asphalt mixture and asphalt binder. Kuna [22] proposed a method to construct stress dependent master curves of foamed bitumen treated mixes (FBMs). The laboratory data were shifted both horizontally and vertically according to the TTSP to create a single master curve. Blasi [23] employed the Christensen and Anderson model (CAM) and the Olard-Di Benedetto model (2S2P1D) to characterize the master curves of the bitumen samples extracted from stone mastic asphalt mixtures. Different objective functions were used to fit the master curves, but the master curves of the asphalt mixture were not predicted. Nobakht [24] proposed a kinetic-based aging model to predict the dynamic modulus master curve of asphalt mixtures. The influence of aging on the dynamic modulus and phase angle master curves was analyzed. Yin [25] conducted dynamic modulus tests of four asphalt binders and corresponding asphalt mixtures. The dynamic modulus master curves were constructed using both the Williams–Landel–Ferry (WLF) equation and the Arrhenius equation. It can be seen that the relationship between the master curves of the asphalt mixture and asphalt mortar was not fully studied. In these studies, the influence of the mesostructure of an asphalt mixture on the dynamic modulus and phase angle was not considered.

In this study, dynamic modulus test simulations using a mesostructure-based finite element model were conducted under a large frequency range to calculate the dynamic modulus and phase angle of asphalt mixture. To be able to accurately predict the master curves of the asphalt mixture, the mechanical behavior of the asphalt mixture and asphalt mortar under a wide range of frequency needs to be accurately characterized by viscoelastic constitutive models.

In the past decades, many viscoelastic constitutive models have been proposed. The generalized Maxwell model (GMM) was widely used to characterize the viscoelastic mechanical behavior of asphalt concretes. It consists of a combination of springs and linear dashpots and is known as the discrete spectrum model. The relaxation models in the time domain can be expressed by the Prony series, which can be analytically converted into the frequency domain and conveniently implemented into the hereditary integrals of numerical simulation [1,3,26]. However, the determination of model parameters is a big problem for the generalized model, especially when employed to characterize the dynamic modulus and phase angle master curves of the asphalt mixture. The number of relaxation times in the discrete spectrum needs to be large enough to cover a wide range in the frequency domain to ensure accuracy. Since the fitting process needs to solve a set of ill-posed nonlinear equations, a large number of parameters will cause severe difficulties in the numerical calculation [27–31].

Other than discrete spectrum models, the continuous spectrum models can characterize the dynamic modulus and phase angle master curves in a wide frequency range with few parameters. The continuous spectrum model is usually expressed by a complex function and can represent the dynamic modulus and phase angle simultaneously. A lot of continuous spectrum models have been proposed. Superpave-A-357 proposed a creep equation in the form of a power law function to characterize the viscoelastic mechanical behavior of asphalt concretes. Zeng [32], based on the CAM (Christensen–Anderson–Marasteanu) constitutive model of an asphalt binder [33], proposes a model for characterizing the master curve and phase angle master curve of the dynamic shear modulus of asphalt concrete. Olard [21] proposed a continuous spectrum model for asphalt concrete based on the Huet model and Huet–Sayegh (HS) model. The 2S2P1D model can accurately describe the mechanical behavior of the asphalt concrete over the entire frequency range and temperature range. Havriliak and Negami [34] proposed the HN model based on the Cole–Cole model and Davison–Cole model. However, the numerical implementation is difficult for the continuous spectrum model since the analytical expression of the relaxation modulus in the time domain does not exist.

The main objective of this study is to propose a method to predict the dynamic modulus and phase angle master curves of rubber modified asphalt mixture. The continuous spectrum model is used to capture the viscoelastic information of an asphalt mortar and then converted into a discrete spectrum model for finite element implementation. Simulations with a large frequency range are conducted and the dynamic modulus and phase angle master curves are calculated. The effect of the aggregate modulus on the dynamic modulus and phase angle of the asphalt mixture is studied. The prediction of the phase angle is improved by adjusting the equilibrium modulus of the asphalt mortar.

2. Theoretical Background

2.1. Continuous Spectrum Model for an Asphalt Mortar and Asphalt Mixture

The rheological models consisting of a combination of linear springs and linear dashpot can characterize the viscoelastic behavior of an asphalt binder, asphalt mortar, and asphalt mixture. The relaxation modulus and creep compliance are commonly represented by the generalized Maxwell model and generalized Kevin–Voigt model, which can be expressed by the Prony series and Dirichlet series, respectively.

The constitutive relation for linear viscoelastic materials can be written as a convolution integral:

$$\sigma(t) = \int_0^t E(t-\xi)\frac{d\varepsilon}{d\xi}d\xi \tag{1}$$

where σ is stress; E is the relaxation modulus; ε is the strain; t is time; and ξ is the integral variable.

The relaxation modulus of the generalized Maxwell model is written as:

$$E(t) = E_e + \sum_{i=1}^{n} E_i e^{-t/\tau_i} \tag{2}$$

where E_e is called the equilibrium modulus, which is the relaxation modulus when the time approaches infinity; E_i and τ_i are the relaxation modulus and the relaxation time of the ith Maxwell model; and n is the number of the sub-models.

The set of E_i and τ_i is called the discrete spectrum, which represents the distribution of relaxation modulus over relaxation time. When the number of the sub-models increases to infinity and the interval between relaxation time decreases to zero, the Prony series becomes an integral and the discrete spectrum becomes the continuous spectrum. The relaxation modulus function expressed by a continuous spectrum model is:

$$E(t) = E_e + \int_0^\infty \frac{H(\xi)}{\xi} e^{-t/\xi} d\xi \tag{3}$$

where $H(\xi)$ is the continuous spectrum function.

The relaxation modulus represents the viscoelastic behavior in the time domain. It can be converted into the complex modulus to represent the viscoelastic behavior in the frequency domain by the Laplace transform. The complex modulus, storage modulus, and loss modulus can be expressed as:

$$E^*(i\omega) = L\{E(s)\}|_{s=i\omega} = E'(i\omega) + iE''(i\omega) \tag{4}$$

$$E'(i\omega) = E_e + \int_0^\infty \frac{H(\xi)}{\xi} \frac{\omega^2 \xi^2}{1 + \omega^2 \xi^2} d\xi \tag{5}$$

$$E''(i\omega) = \int_0^\infty \frac{H(\xi)}{\xi} \frac{\omega \xi}{1 + \omega^2 \xi^2} d\xi \tag{6}$$

where E^* is the complex modulus; L is the Laplace transform; s is the Laplace transform variable; E' is the storage modulus; E'' is the loss modulus; and $\omega = 2\pi f$ is the angular frequency, and f is the frequency.

The 2S2P1D model was proposed to characterize the viscoelastic properties of both the asphalt binder and asphalt mixture. It is expressed by a complex function and can characterize the dynamic modulus and phase angle simultaneously. It consists of two springs, two parabolic elements, and one dashpot. The phenomenological structure of the 2S2P1D model is shown in Figure 1.

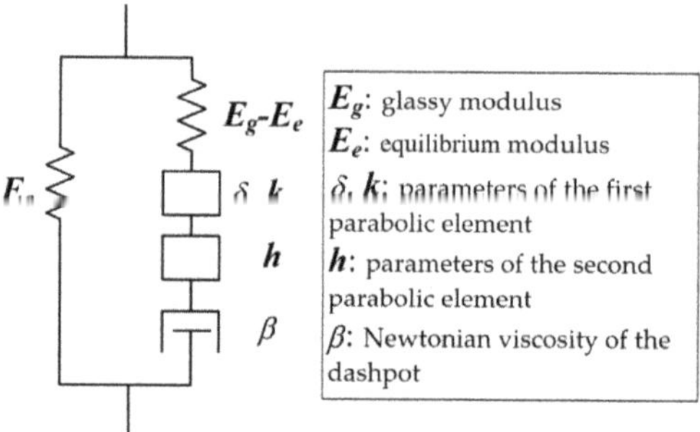

Figure 1. Phenomenological structure of the 2S2P1D model.

The parabolic element, also called the Abel model, can be seen as a rheological element between the linear spring and the linear dashpot. Its constitutive relation in the time domain can be expressed by a fractional order derivation, while the constitutive relations of the spring and the dashpot are zero order and first order respectively.

The parabolic element is defined as:

$$\text{Creep compliance}: \ J(t) = \delta \left(\frac{t}{\tau_{ref}} \right)^k \tag{7}$$

$$\text{Complex modulus}: \ E^*(i\omega) = \frac{\left(i\omega\tau_{ref}\right)^k}{\delta\,\Gamma(k+1)} \tag{8}$$

where δ and k are material constants; ω is the angular frequency; Γ is the Gamma function; and τ_{ref} is the characteristic time that accounts for the time-temperature superposition principle and determined by the WLF (Williams–Landel–Ferry) equation:

$$\tau(T) = \alpha_T * \tau_{ref} \tag{9}$$

where α_T is the shift factor at temperature T; and τ_{ref} is the characteristic time at the reference temperature.

α_T is determined by the WLF equation:

$$\log(\alpha_T) = \frac{-C_1\left(T - T_{ref}\right)}{C_2 + \left(T - T_{ref}\right)} \tag{10}$$

where C_1 and C_2 are constants of the WLF equation; and T_{ref} is the reference temperature.

According to the phenomenological structure, the complex modulus function of the 2S2P1D model can be easily derived in the Laplace domain:

$$E^*(i\omega) = E_e + \frac{E_g - E_e}{1 + \delta\left(i\omega\tau_{ref}\right)^{-k} + \left(i\omega\tau_{ref}\right)^{-h} + \left(i\omega\tau_{ref}\beta\right)^{-1}} \tag{11}$$

where E_e is the equilibrium modulus; E_g is the glassy modulus or instantaneous modulus, which is the relaxation modulus when time equals zero; ω is the angular frequency; δ, k, and h are parameters of the parabolic elements and set as $0 < k < h < 1$; β is the Newtonian viscosity of the dashpot; and τ_{ref} is the characteristic time.

2.2. Determination of the Discrete Spectrum

As mentioned above, the continuous spectrum models can characterize the viscoelastic behavior of the asphalt concrete in the frequency domain more accurately and effectively. The continuous spectrum models can capture the viscoelastic information under a large range of temperatures and frequencies with fewer model parameters than that of the generalized Maxwell model. The accuracy of the generalized Maxwell model can be improved by increasing the number of sub-models. However, when parameters of the generalized Maxwell model are directly obtained from the dynamic modulus test, the fitting process needs to solve a set of ill-posed nonlinear equations. As the number of parameters increases, the difficulty of fitting calculation increases.

The continuous spectrum models have disadvantages in numerical calculation and finite element analysis. Some continuous spectrum models, like the HS model and the 2S2P1D model, do not have an analytical expression of the relaxation modulus in the time domain. Other continuous spectrum models, like the generalized fractional Maxwell model, can be analytically converted into a relaxation modulus using the Mittag–Leffler function [35]. In this case, the incremental strain history and incremental time history need to be stored during the numerical calculation, which requires a large storage space and decreases computational efficiency. The generalized Maxwell model expressed by the Prony series, however, can be conveniently written as an incremental form in the finite element algorithm. Only the incremental strain and incremental time of the last incrementation need to be stored, so the generalized Maxwell model is more suitable for finite element analysis.

In this paper, the 2S2P1D model is used to capture the viscoelastic information of the asphalt mixture from dynamic modulus test data and then the continuous spectrum is converted into a discrete spectrum in the form of Prony series for finite element implementation.

The continuous spectrum function of a continuous spectrum model can be derived by the following formula [3]:

$$H(\tau) = \pm \frac{1}{\pi} \mathrm{Im} E^* \left(\frac{1}{\tau} e^{\pm i\pi} \right) \tag{12}$$

where Im denotes the imaginary part of the complex number; E^* is the complex modulus function; and τ is relaxation time.

The continuous spectrum function of the 2S2P1D model is:

$$H(\tau) = \frac{E_g}{\pi \sqrt{A^2 + B^2}} \sin \varphi \tag{13}$$

where

$A = 1 + \delta \left(\frac{\tau_{ref}}{\tau} \right)^{-k} \cos k\pi + \left(\frac{\tau_{ref}}{\tau} \right)^{-h} \cos h\pi - \left(\frac{\tau_{ref}}{\tau} \beta \right)^{-1}$;

$B = \delta \left(\frac{\tau_{ref}}{\tau} \right)^{-k} \sin k\pi + \left(\frac{\tau_{ref}}{\tau} \right)^{-h} \sin h\pi$;

$\varphi = \arctan \frac{B}{A}$.

By interpreting the integral in Equation (3) as discrete approximations, the relaxation modulus function can be written as:

$$\begin{aligned} E(t) &= E_e + \int_0^\infty \frac{H(\xi)}{\xi} e^{-t/\xi} d\xi \\ &= E_e + \sum_{i=1}^{n} \left[\frac{H(\tau_i)}{\tau_i} \cdot \Delta\tau_i \right] e^{-t/\tau_i} \end{aligned} \tag{14}$$

where the set of τ_i is discrete relaxation times as mentioned in Equation (2); and $\Delta\tau_i$ is the interval between relaxation times.

Comparing Equation (14) with Equation (2), the discrete spectrum can be determined as follow:

$$E_i = \frac{H(\tau_i)}{\tau_i} \cdot \Delta\tau_i \tag{15}$$

where E_i is the discrete relaxation modulus as mentioned in Equation (2).

The set of relaxation time τ_i needs to be artificially selected. When the relaxation times are evenly distributed on the time axis, the convergence of the discrete spectrum to the continuous spectrum is very slow [36]. The convergence can be greatly improved by evenly distributing the relaxation times on the logarithmic time axis. In this case, Equation (3) and Equation (15) need to be rewritten as:

$$E(t) = E_e + \int_{-\infty}^{\infty} H(\xi) e^{-t/\xi} d\ln \xi \tag{16}$$

$$E_i = H(\tau_i) \cdot \Delta \ln \tau_i \tag{17}$$

The details of the parameter calibration of the 2S2P1D model and the determination of the discrete spectrum will be elaborated in the next section.

3. Experiment and Parameter Acquisition

3.1. Dynamic Modulus Tests of the Asphalt Mortar and Asphalt Mixture

In this study, dynamic modulus tests of both the asphalt mortar and asphalt mixture were conducted. The test data of the asphalt mortar were used to obtain the parameters of the 2S2P1D model and the generalized Maxwell model. The test data of the asphalt mixture were used to verify the finite element simulation.

The asphalt mixture was dense graded with a 13.2 mm nominal maximum aggregate size. The gradation is shown in Figure 2.

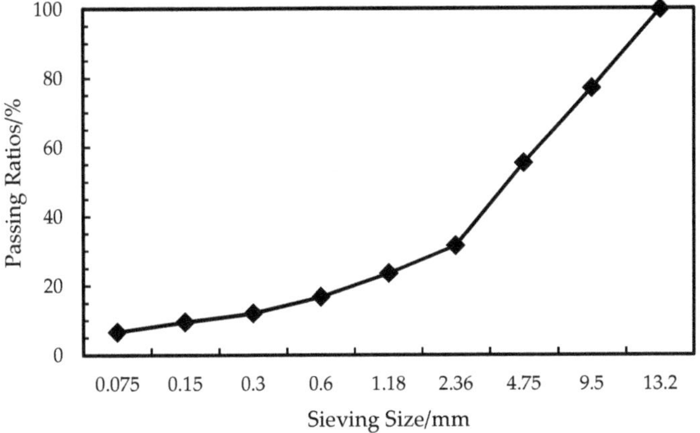

Figure 2. Aggregate gradation of the asphalt mixture.

The asphalt binder used was a rubber modified asphalt. The asphalt content was 4.5% by weight and the air void is 3.1%.

The gradation of the asphalt mortar was the same as that of asphalt mixture finer than 2.36 mm. The asphalt content of the asphalt mortar was calculated based on the specific surface area of the aggregates. It was assumed that the asphalt was uniformly dispersed and attached to the surface of the aggregates. The amount of asphalt coated on the aggregates was proportional to the specific surface area of the aggregate. Table 1 shows the specific surface area of aggregates with different size. Table 2 shows the calculation of the asphalt content of asphalt mortar.

Table 1. Specific surface area of aggregates.

Sieving Size (mm)	≥4.75	2.36	1.18	0.6	0.3	0.15	0.075
Specific surface area (m^2/kg)	0.41	0.82	1.64	2.87	6.14	12.29	32.77

Table 2. Calculation of the asphalt content of the asphalt mortar.

Properties	Coarse Aggregates				Fine Aggregates					
Sieving size (mm)	16	13.2	9.5	4.75	2.36	1.18	0.6	0.3	0.15	0.075
Passing Ratio (%)	100.0	99.8	77.0	55.3	31.4	23.5	16.7	12.0	9.5	6.6
Specific surface area (m^2/kg)	0.41	0.41	0.41	0.41	0.82	1.64	2.87	6.14	12.29	32.77
Surface area (m^2)	0.00	9.34	8.88	9.81	6.49	11.10	13.43	15.47	35.40	217.59
Aggregate weight (kg)	68.36				31.41					
Surface area (m^2)	28.03				299.49					
Proportion of surface area	0.09				0.91					
Coated asphalt (kg)	0.43				4.57					
Asphalt content (%)				12.71						

The superpave gyratory compactor (SGC) was utilized to compact specimens with 150 mm diameter and 170 mm height. The asphalt mixture samples were cut into a cylinder with 100 mm diameter and 100 mm height. The asphalt mortar samples were cut into a smaller cylinder with 50 mm diameter and 50 mm height to reduce the effect of potential plastic deformation at high temperatures.

The dynamic modulus tests of the asphalt mixture were conducted under four temperatures and six loading frequencies. The test temperatures were −10 °C, 10 °C, 30 °C, and 50 °C and the loading frequencies were 0.1 Hz, 0.5 Hz, 1 Hz, 5 Hz, 10 Hz, and 20 Hz.

The dynamic modulus tests of the asphalt mortar were conducted under five temperatures and nine loading frequencies. The test temperatures were $-10\,°C$, $0\,°C$, $10\,°C$, $20\,°C$, and $30\,°C$ and the loading frequencies were 0.1 Hz, 0.2 Hz, 0.5 Hz, 1 Hz, 2 Hz, 5 Hz, 10 Hz, 20 Hz, and 25 Hz.

All the dynamic modulus tests were performed under stress-control conditions using a UTM-25 testing machine (IPC Global, Melbourne, VIC, Australia). Teflon film was used to reduce the friction between the sample and the indenter.

The dynamic modulus is the absolute value of the complex modulus:

$$|E^*| = \sqrt{(E')^2 + (E'')^2} \tag{18}$$

It can be determined by calculating the ratio of stress amplitude and strain amplitude at a steady state of the dynamic modulus test, which is the average value of the last five cycles according to the Chinese specification JTG E20-2011 [37].

$$|E^*| = \frac{\sigma_0}{\varepsilon_0} \tag{19}$$

where σ_0 is the stress amplitude; and ε_0 is the strain amplitude.

Phase angle is the argument of the complex modulus and determined by the average time delay of the last five cycles:

$$\phi = \left(\frac{T_i}{T_p}\right) \times 360 \tag{20}$$

where T_i is the time delay; and T_p is the sinusoidal load cycle.

The number of parallel samples is required larger than three by the Chinese specifications and the number of samples tested in this study is four. The dynamic modulus and phase angle are calculated using the following equation based on the T-distribution.

$$|E^*| = \overline{|E^*|} - t \times \frac{S_E}{\sqrt{n}} \tag{21}$$

$$\phi = \overline{\phi} - t \times \frac{S_\phi}{\sqrt{n}} \tag{22}$$

where $\overline{|E^*|}$ and $\overline{\phi}$ are the average dynamic modulus and phase angle of parallel samples; S_E and S_ϕ are the standard deviation of dynamic modulus and phase angle; n is the number of parallel samples; and t is the parameter of the T-distribution and determined according to the number of samples and the confidence rate. When the number of samples is four and the confidence rate is 95%, t equals 2.354.

3.2. Parameter Acquisition

Seven parameters of the 2S2P1D model and two parameters of the WLF equation need to be determined. Equations (9), (10), and (11) were utilized to fit the dynamic modulus test data in the frequency domain. A nonlinear minimization algorithm using differential evolution method was performed on the target error function F as follow:

$$\min F\left(E_e, E_g, \delta, k, h, \beta, \tau_{ref}, C_1, C_2\right) = \frac{1}{N}\left(\sqrt{\sum_{i=1}^{N}\left(1 - \frac{|E^*_{c,i}|}{|E^*_{t,i}|}\right)^2} + \sqrt{\sum_{i=1}^{N}\left(1 - \frac{\phi_{c,i}}{\phi_{t,i}}\right)^2}\right) \tag{23}$$

where $|E^*_{c,i}|$ and $|E^*_{t,i}|$ are the calculated dynamic modulus and test dynamic modulus, respectively; $\phi_{c,i}$ and $\phi_{t,i}$ are the calculated phase angle and test phase angle, respectively; and N is the number of the test data points at all frequencies.

The fitting procedure is conducted using the Mathematica software package. Table 3 lists the calculated model parameters of the asphalt mortar at a reference temperature of 0 °C.

Table 3. Model parameters of the asphalt mixture and the asphalt mortar.

Materials	E_e (MPa)	E_g (MPa)	δ	k	h	β	τ_{ref}	C_1	C_2	T_{ref} (°C)	Relative Error
Asphalt mixture	127.33	53,993.61	2.62	0.08	0.45	57,370.2	0.07	82.04	804.57	0	4.96%
Asphalt mortar	0.00	40,166.80	5.45	0.20	0.52	1247.25	0.13	83.58	712.76	0	2.02%

The fitting result is shown in Figures 3 and 4. It shows that the 2S2P1D model can characterize the dynamic modulus and the phase angle master curves of both the asphalt mixture and asphalt mortar accurately.

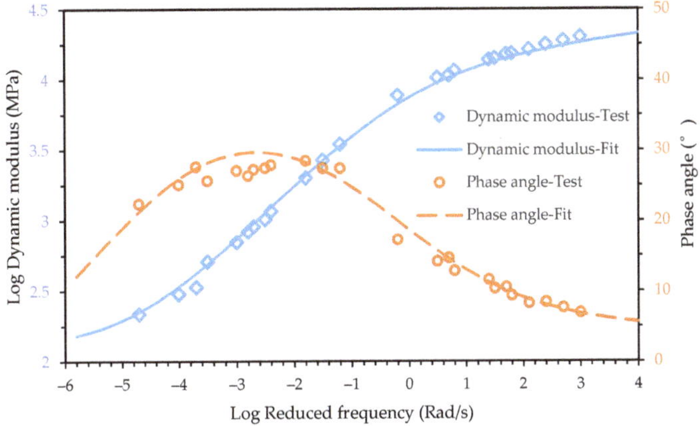

Figure 3. Test data and fitting curve of the asphalt mixture for the dynamic modulus and phase angle.

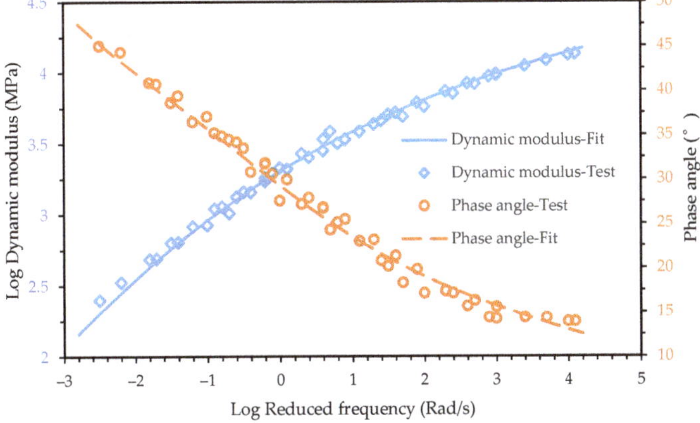

Figure 4. Test data and fitting curve of the asphalt mortar for the dynamic modulus and phase angle.

Comparing the phase angle main curve of the asphalt mixture and asphalt mortar in Figures 3 and 4, it can be seen that the phase angle of the asphalt mixture and asphalt mortar was significantly different at low frequencies. The phase angle of the asphalt mixture decreased when the reduced frequency approached zero and infinity and peaked at 10^{-3} Rad/s while the phase angle of the asphalt mortar decreased as the reduced frequency decreased. The maximum phase angle of the asphalt mixture is 30°, which indicates that the storage modulus accounts for a larger proportion of dynamic modulus. Due to the inconvenience of conducting the dynamic modulus test for an asphalt mortar at high temperatures, the minimum reduced frequency of test data in Figure 4 only reached 10^{-3} Rad/s but the corresponding phase angle already reached 45°. With the decrease of the reduced frequency, the phase angle of the asphalt mortar would keep increasing. The loss modulus would become larger than the storage modulus.

As mentioned above, to convert the continuous spectrum into the discrete spectrum efficiently, the relation times are uniformly distributed on the logarithmic time axis, so Equation (17) can be rewritten as:

$$E_i = H(\tau_i) \cdot \Delta \ln 10^{1/n} \tag{24}$$

where n is the number of relaxation time τ_i per decade.

According to Equations (13) and (17), the continuous spectrum function and the discrete spectrum can be calculated. Figure 5 shows the continuous spectrum and discrete spectrum with different values of n. It can be seen that as n increased, the difference between the discrete spectrum and the continuous spectrum curve decreased. When $n = \ln(10)$ and $\Delta \ln \tau_i = 1$, the discrete spectrum curve coincides with the continuous spectrum curve, so $n = \ln(10)$ was used to calculate the discrete spectrum.

Since $\Delta \tau_i$ was obtained, the set of τ_i could be calculated once the range of τ_i on the relaxation time axis is chosen. As shown in Figure 5, the range was from 10^{-20} to 10^{10}. It covered the reduced frequency range of the dynamic modulus tests shown in Figure 4. Additionally, the continuous spectrum $H(\tau)$ was less than 1 MPa outside the range and was considered small enough to be ignored. The details of the Prony series is listed in Appendix A.

To verify the effectiveness of the conversion from the continuous spectrum to the discrete spectrum, the dynamic modulus and phase angle master curves of the calculated generalized Maxwell model were compared with the experimental data, as shown in Figure 6. It can be seen that the master curves had good agreement with the experimental data, which means the conversion was effective. The relative error between the master curves and the test data in Figure 6 was 2.53%, which was also calculated using Equation (23). It was slightly larger than the relative error of 2S2P1D but acceptable.

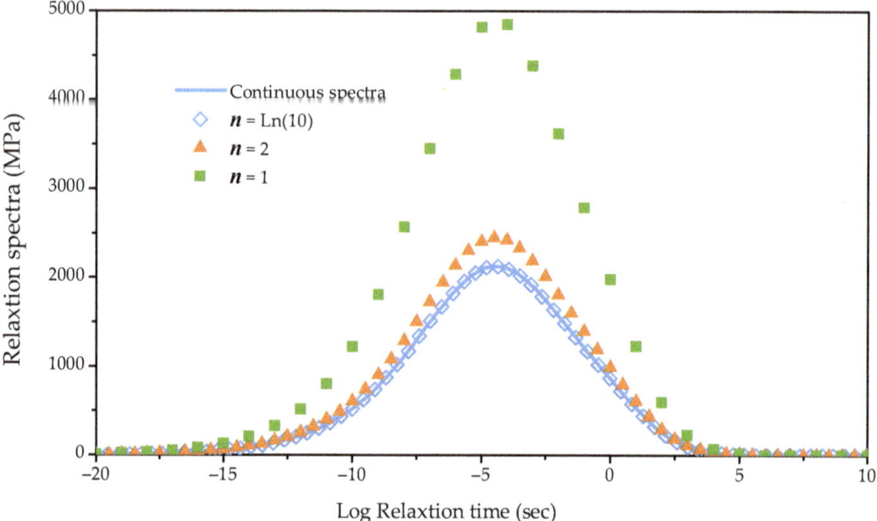

Figure 5. Continuous spectrum and discrete spectrum with different values of n.

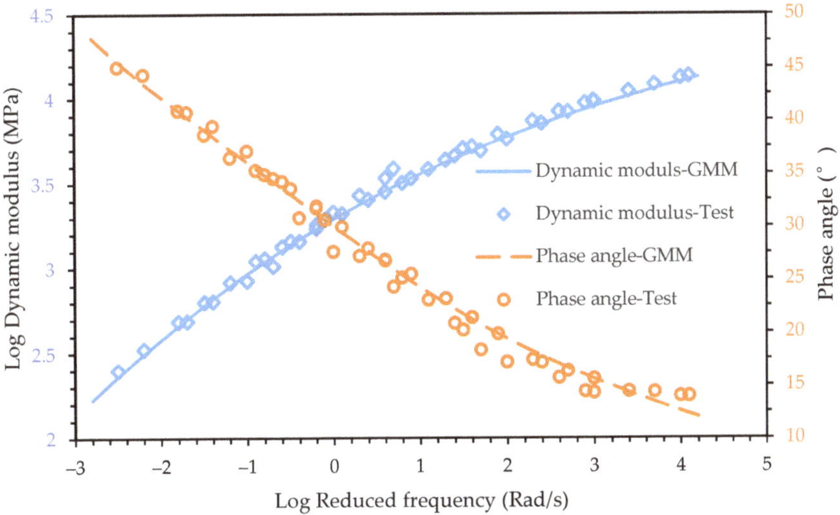

Figure 6. Dynamic modulus and phase angle master curve of the calculated generalized Maxwell model.

4. Construction of the Finite Element Model

4.1. Mesostructure of the Asphalt Mixture

The mixture sample with test data closest to the average value of four parallel samples was chosen to be scanned by an X-Ray CT. A set of two-dimensional scan images of the horizontal sections perpendicular to the cylinder axis was first captured. These two-dimensional images were manipulated by image processing software and reconstructed into a three-dimensional mesostructured as shown in Figure 7a.

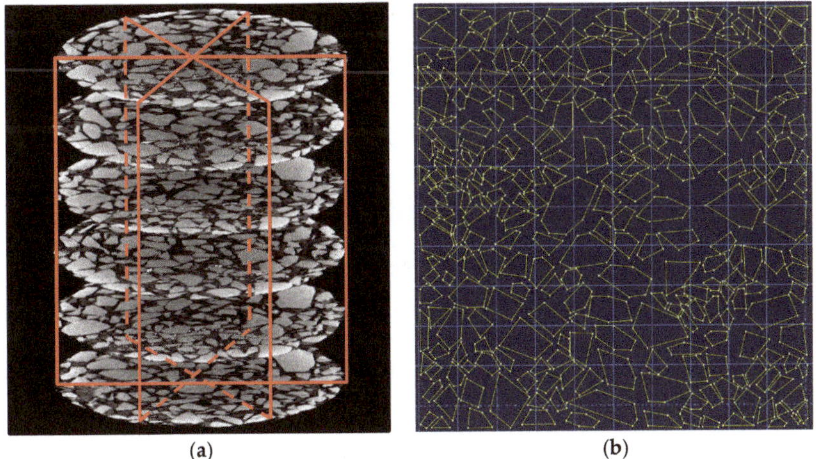

Figure 7. Mesostructure of the asphalt mixture: (**a**) Vertical section of the mixture sample; (**b**) aggregate skeleton sketch in ABAQUS.

Many studies were devoted to analyzing the differences between 3D models and 2D models in simulating the dynamic modulus. In general, the 3D models have higher accuracy while the 2D models tend to underestimate the dynamic modulus and overestimate phase angles [5]. The differences between 3D models and 2D models are related to both the contrast between the inclusions and matrix complex modulus and the contrast between the surface filling rate in 2D models and the volumetric filling rate in 3D models. Firstly, as the modulus contrast between the inclusions and matrix reduces, the difference between models reduces. The modulus contrast between the inclusions and matrix of mastic, mortar, and mixture is studied [38]. It indicates the asphalt mixture has the smallest modulus contrast. Secondly, the filling rate of 2D models and 3D models of the asphalt mixture is very close. Additionally, the element number of 2D models in the following simulation is larger than 20,000. The use of 3D models will dramatically increase the element number and reduce the efficiency of numerical simulation. Based on the above reasons, only 2D models were used in this study.

Three vertical sections of the mixture sample were obtained from the 3D mesostructure, as shown in Figure 7a. The image processing techniques were employed to identify the aggregate skeleton of the 2D image. The coordinate information of aggregate polygon was saved as input files(*.sat), which can be implemented into the finite element software. The aggregate skeleton sketch in ABAQUS is shown in Figure 7b.

4.2. Finite Element Model

The size of the 2D finite element model was the same as that of the asphalt mixture test sample, with 100 mm width and 100 mm height, as shown in Figure 8. The element type was CPS8R, an eight-node biquadratic plane stress quadrilateral, reduced integration. The element number was 23,185.

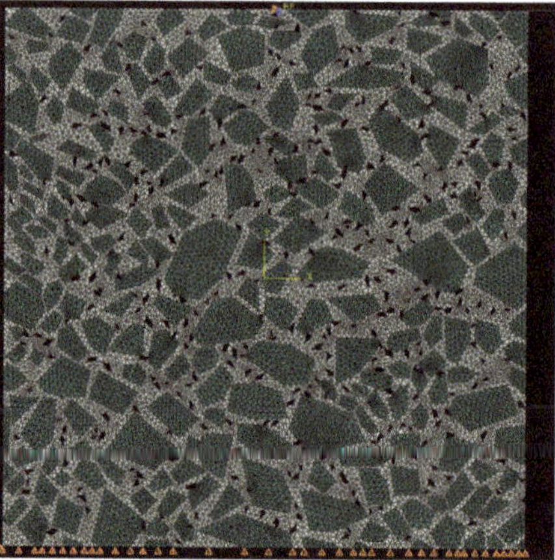

Figure 8. Finite element model of the asphalt mixture.

The definition of the constraints was on the basis of the real test conditions. The vertical displacement of the bottom edge was restraint. In order to avoid a large lateral displacement of the model, the nodes in the lower left corner were fixed. A discrete rigid wire was used to represent the indenter of the testing machine to simulate a more realistic stress-control loading condition. The horizontal displacement and the rotation of the indenter was restraint and only the vertical displacement was allowed. The normal behavior of the contact between the sample and the indenter was pressure-overclosure relationships. The tangential behavior was considered frictionless since Teflon film was used in the laboratory test to largely reduce friction between the sample and the indenter. Considering the lateral deformation of the sample under the compressive load, the rigid wire was longer than the top edge of the sample to avoid numerical problems in the contact analysis.

A haversine concentrate force was applied to the reference point of the rigid wire to simulate the stress-control loading condition. The magnitude of the concentrated force was calculated by multiplying the loading stress of the tests by the area of the sample.

The material properties of the asphalt mortar are given in Appendix A. The elastic modulus of aggregates, 58,413 MPa, was obtained from the uniaxial compression test.

Visco Step with the implicit algorithm was adopted. The number of load cycles at each frequency is required in Chinese specification JTG E20-2011. The displacement of the rigid wire was read from the simulation results and then converted into the strain of the mixture sample. The average strain amplitude of the last five cycles was used to calculate the dynamic modulus.

The increment size affects the accuracy of the simulation results especially the phase angle. The error of the phase angle can be written as:

$$\text{Error of the phase angle} = \pm \frac{360°}{2N_{\text{inc}}} \tag{25}$$

where N_{inc} is the number of incrementations in one cycle.

Automatic incrementation size was set with a max size of 1/500 cycle, which ensured that N_{inc} was greater than 500 and the error of the phase angle was less than 0.36°.

All three 2D mesostructures of the asphalt mixture sample were used in the following simulations. The average value of the predicted dynamic modulus and phase angle from different mesostructures were presented as the simulation results.

5. Simulation Results and Analysis

5.1. Simulations of Dynamic Modulus and Phase Angle Master Curves

The loading frequencies of the dynamic modulus test simulations ranged from 10^{-6} to 10^{4} with an interval of 0.5 on the logarithmic frequency axis. The predicted dynamic modulus and phase angle is plotted in Figure 9, together with the experimental data. It is shown that the predicted dynamic modulus master curve was consistent with experimental data over the entire frequency range, while the magnitude of dynamic modulus was slightly less than the test results. As discussed in the previous section, the 2D finite element models tend to underestimate the dynamic modulus due to neglect the geometric characteristics of aggregates in the third dimension.

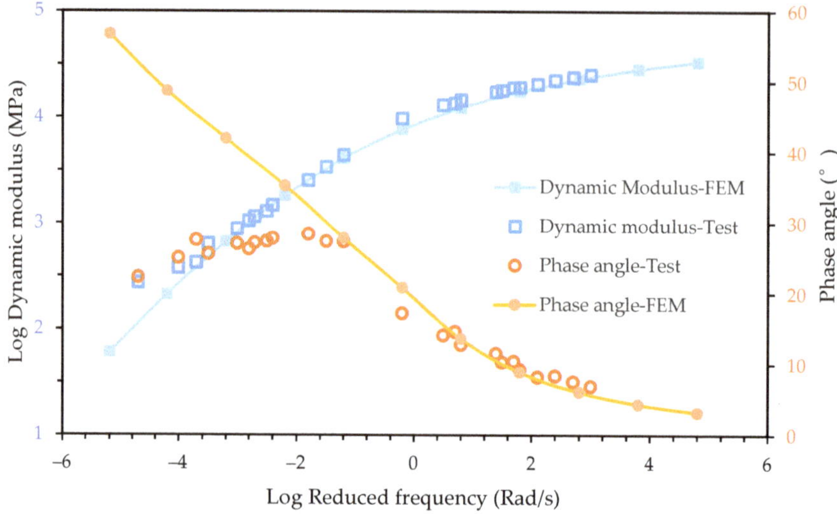

Figure 9. Comparison of simulation results and test data.

However, the predicted phase angle master curve was not consistent with experimental data over the entire frequency range. Good agreement was achieved between the predicted phase angle and the test data when frequency was larger than 0.1 Rad/s. A significant difference appeared as the frequency decreased. In previous studies, the frequency range considered in simulations was usually from 0.1 Hz to 25 Hz as the frequency range of the laboratory tests. It can be seen in Figure 9 that the simulation results and test data of the phase angle in this frequency range were in a good agreement. The significant difference at lower frequency range was ignored.

In the frequency domain, the phase angle master curve of a viscoelastic liquid and a viscoelastic solid was different. When the frequency approached infinity, both the viscoelastic liquid and viscoelastic solid exhibited pure elastic mechanical behavior, which means the phase angle approached zero. When the frequency approached zero, the viscoelastic liquid showed pure viscous mechanical behavior while the viscoelastic solid still showed pure elastic mechanical behavior [3]. In general, the phase angle of a viscoelastic liquid monotonically decreased as the frequency increased while the phase angle master curve of a viscoelastic solid exhibited a bell shape. It can be seen in Figure 4 that the phase angle of the asphalt mortar showed the mechanical behavior of a viscoelastic liquid.

In the finite element simulation, the viscoelastic mechanical behavior was represented only by asphalt mortar since the aggregates were considered pure elastic. Therefore, as shown in Figure 9, the viscoelastic mechanical response of the mesostructure was consistent with the asphalt mortar and the difference of rheology properties between the asphalt mixture and asphalt mortar caused a significant error in predicting phase angle master curves.

5.2. Influence of Elastic Modulus of Aggregates on Master Curves

The elastic modulus of the aggregates used in the previous simulations was obtained in the uniaxial compressive test, but in many studies, the elastic modulus is assumed to range from 50 GPa to 70 GPa and is artificially preselected. The influence of the elastic modulus of aggregate on the dynamic modulus and phase angle master curve needs to be analyzed.

Three different elastic moduli of aggregates were considered. The loading frequencies were the same as simulations in Section 5.1. The results are plotted in Figure 10.

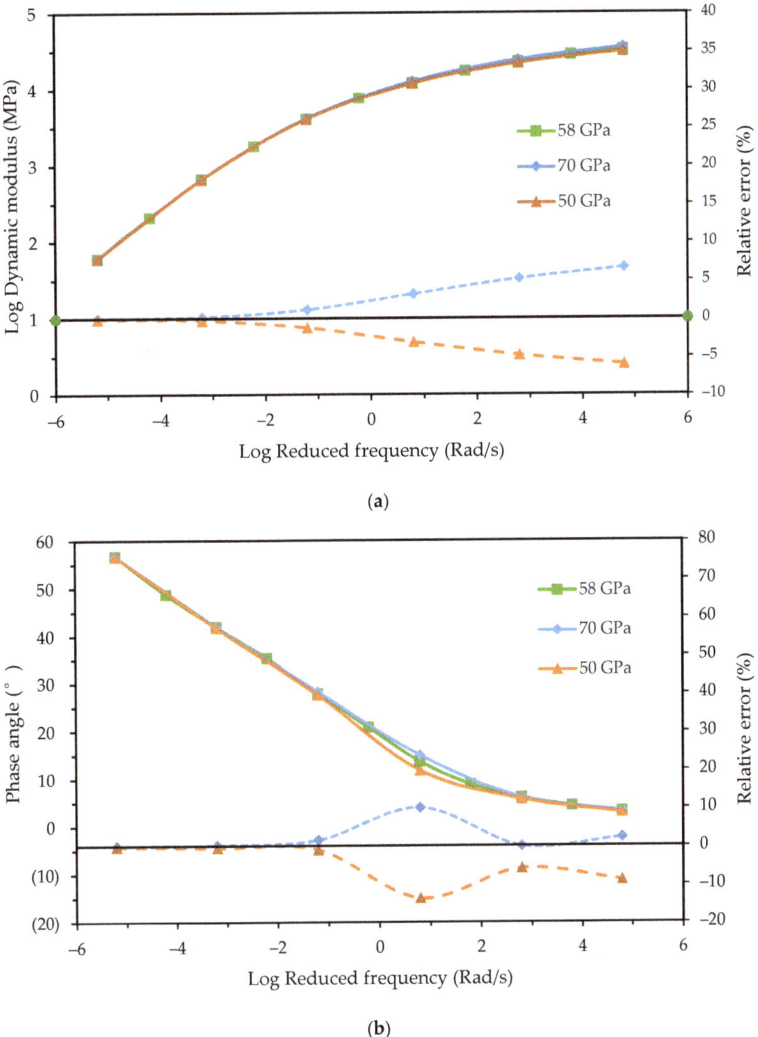

Figure 10. Influence of the elastic modulus of aggregate: (**a**) Dynamic modulus; and (**b**) phase angle.

It shows that the elastic modulus of aggregate had an effect on both the dynamic modulus and phase angle. As the frequency increased, the effect of the modulus on the dynamic modulus monotonically increased while the relative error of the phase angles had fluctuated and reached a peak at 1 Rad/s. The relative error of the dynamic modulus test was less than that of the phase angle. It indicates that the elastic modulus of aggregate had a greater effect on the phase angle.

5.3. Improved Prediction of the Phase Angle Master Curve

As discussed in Section 5.1, the finite element simulation cannot accurately predict the phase angle at low frequencies. To solve this problem, a new attempt was made in this study by modifying the equilibrium modulus of the asphalt mortar.

The equilibrium modulus E_e of the asphalt mortar in the 2S2P1D model was very close to zero. The equilibrium modulus of the asphalt mixture, however, could be calibrated using a similar procedure given in Section 3.2 and was 127 Mpa.

The effect of the equilibrium modulus on the master curves is shown in Figure 11.

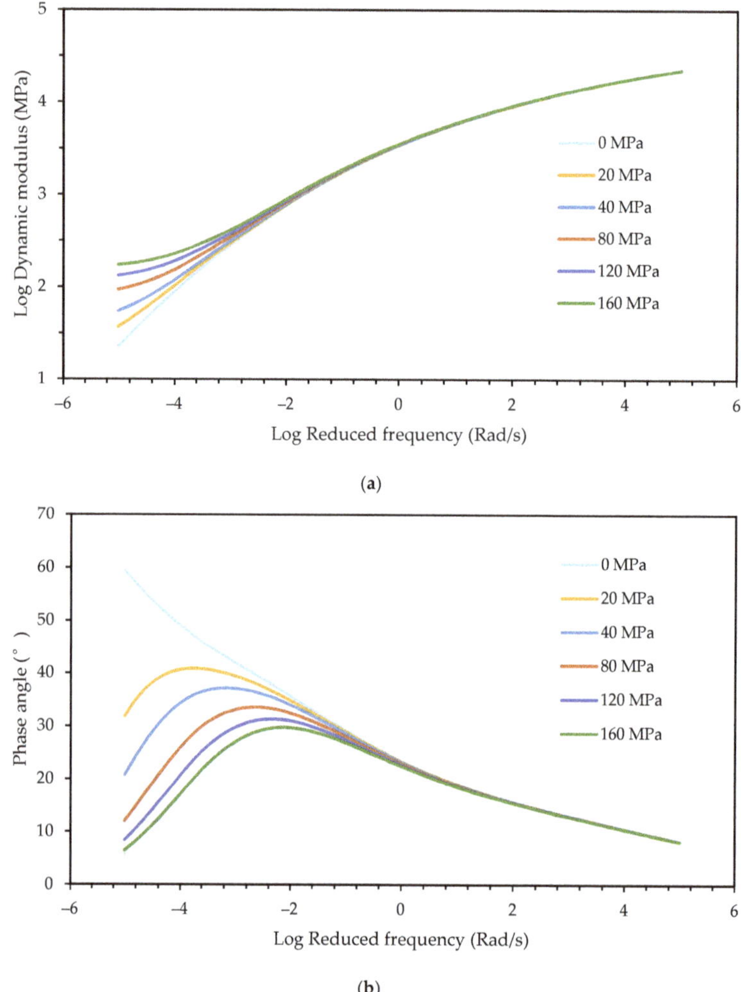

(a)

(b)

Figure 11. Effect of the equilibrium modulus: (**a**) Dynamic modulus; and (**b**) phase angle.

It is seen that the equilibrium modulus had effects on both the dynamic modulus and phase angle at low frequencies. As the equilibrium modulus increased, the dynamic modulus increased and the phase angle decreased. The frequency at which the phase angle reached the peak also increased with the equilibrium modulus.

In order to improve the accuracy of finite element simulation for the phase angle, the equilibrium modulus of the asphalt mixture was input into the finite element model as the equilibrium modulus of the asphalt mortar. Simulates with the same loading conditions as above were conducted and the results are shown in Figure 12.

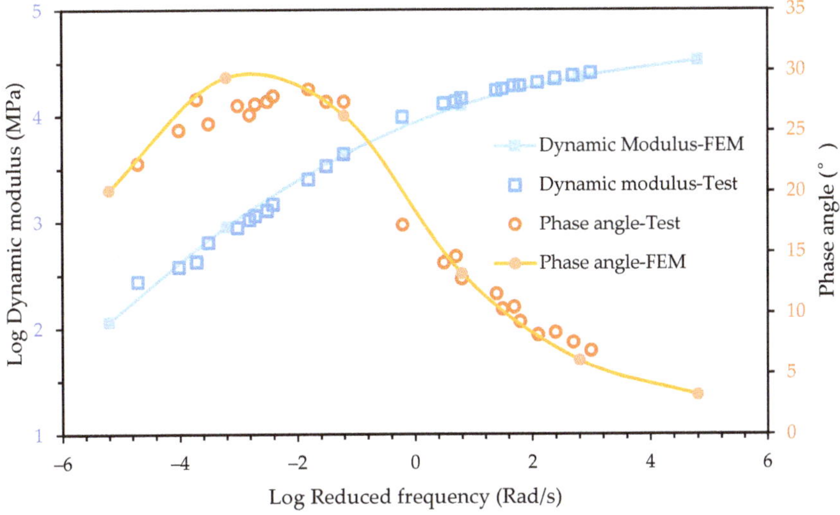

Figure 12. Improved prediction of the dynamic modulus and phase angle master curves.

It can be seen that the predicted phase angle was consistent with experimental data at all the frequency range. The increase of the dynamic modulus caused by the change of the equilibrium modulus at low frequency was not obvious. The simulated dynamic modulus was still in good agreement with the experimental data. It may be because that that the dynamic modulus of the mixture was mainly affected by the elastic modulus of aggregate at low frequencies. The result shows that the proposed method was effective to improve the prediction of the phase angle master curve. The finite element model could accurately predict the dynamic modulus and phase angle master curves of the asphalt mixture.

6. Conclusions

This study proposed a mesostructure-based finite element model of an asphalt mixture to predict both the dynamic modulus and phase angle master curves under a large frequency range. The difference between test data and simulation results of phase angle was observed. A new attempt was made to improve the accuracy of finite element simulation for the phase angle by modifying the equilibrium modulus of the asphalt mortar. Several conclusions can be drawn as follows:

(1) The 2S2P1D model originally proposed for the asphalt binder and asphalt mixture was used to characterize the viscoelastic mechanical response of the rubber modified asphalt mortar. The fitting results proved that the 2S2P1D model could accurately describe the dynamic modulus and phase angle master curves of rubber modified asphalt mortar. Furthermore, the continuous spectrum function of the 2S2P1D model was efficiently converted into a discrete spectrum for finite element implementation. The simulation results of the mesostructured-based finite element model showed that the discrete spectrum model could precisely predict the dynamic modulus of asphalt mixture under a large frequency range.

(2) The elastic modulus of aggregates demonstrated a stronger correlation with the phase angle of an asphalt mixture than the dynamic modulus. As the frequency increased, the effect on the dynamic modulus monotonically increased while the effect of the phase angle had fluctuated and reached a peak at 1 Rad/s.

(3) The test data of the phase angle indicated that the rubber modified asphalt mortar exhibited mechanical behavior of a viscoelastic liquid while the asphalt mixture was known as a viscoelastic

solid. Then the mesostructure-based finite element simulation results proved that the prediction of the phase angle master curve strongly correlated with the mechanical behavior of the asphalt mortar. Therefore, a significant difference between test data and simulation results was observed at low frequency when directly using model parameters of the asphalt mortar. This difference was ignored by the existing studies in which simulations were conducted only under a small frequency range.

(4)   The equilibrium modulus was found to have a great influence on the phase angle master curve through parameter sensitivity analysis. Further, simulation results proved that by replacing the equilibrium modulus of asphalt mortar with that of the asphalt mixture, the prediction of the phase angle could be significantly improved while the prediction of dynamic modulus was still acceptable.

In future studies, simulations for more types of asphalt mixtures with different asphalt binders and gradations need to be conducted to further verify the proposed method. The error fluctuations observed in Figure 10 need to be further analyzed.

Author Contributions: L.G.: Conceptualization, Methodology, Writing-Original Draft; L.C.: Investigation, Project administration; W.Z.: Validation, Writing-Review & Editing; H.M.: Resources, Visualization; T.M.: Conceptualization, Validation, Supervision.

Funding: The study is financially supported by the National Natural Science Foundation of China (No. 51878164) and Natural Science Foundation of Jiangsu (BK20161421).

Acknowledgments: The authors would like to appreciate anonymous reviewers for their constructive suggestions and comments to improve the quality of the paper.

Appendix A

Table A1. The details of the Prony series.

$\mathrm{Log}\tau_i$ (s)	E_i (MPa)	$\mathrm{Log}\tau_i$ (s)	E_i (MPa)	$\mathrm{Log}\tau_i$ (s)	E_i (MPa)	$\mathrm{Log}\tau_i$ (s)	E_i (MPa)
−20.00	14.05	−13.05	354.98	−6.10	4570.31	0.85	1451.90
−19.57	17.86	−12.62	431.15	−5.67	4894.69	1.28	1115.67
−19.13	21.83	−12.18	522.85	−5.23	5139.13	1.71	816.32
−18.70	26.69	−11.75	632.90	−4.80	5284.67	2.15	568.91
−18.26	32.62	−11.31	764.38	−4.37	5320.12	2.58	379.64
−17.83	39.87	−10.88	920.67	−3.93	5244.20	3.02	244.40
−17.39	48.73	−10.45	1105.24	−3.50	5065.62	3.45	152.75
−16.96	59.53	−10.01	1321.51	−3.06	4801.28	3.89	92.83
−16.53	72.73	−9.58	1572.51	−2.63	4472.97	4.32	54.41
−16.09	88.82	−9.14	1860.37	−2.19	4103.59	4.75	29.95
−15.66	108.44	−8.71	2185.74	−1.76	3713.66	5.19	14.67
−15.22	132.34	−8.27	2546.96	−1.33	3318.95	5.62	5.97
−14.79	161.44	−7.84	2939.19	−0.89	2929.14	6.06	1.96
−14.35	196.82	−7.41	3353.53	−0.46	2547.93	6.49	0.54
−13.92	239.80	−6.97	3776.39	−0.02	2174.79	6.93	0.13
−13.49	291.91	−6.54	4189.42	0.41	1808.48	7.36	0.03

References

1.   Schapery, R.A. A simple *collocation* method for fitting viscoelastic models to experimental data. *Calif. Inst. Technol.* **1962**. Available online: http://resolver.caltech.edu/CaltechAUTHORS:20141114-115330896 (accessed on 6 December 2018).

2.   Ferry, J.D. *Viscoelastic Properties of Polymers*; John Wiley Sons: Hoboken, NJ, USA, 1980.

3.   Tschoegl, N.W. *The Phenomenological Theory of Linear Viscoelastic Behavior: An Introduction*; Springer Science Business Media: Berlin, Germany, 1989.

4. Ma, T.; Zhang, D.; Zhang, Y.; Zhao, Y.; Huang, X. Effect of air voids on the high-temperature creep behavior of asphalt mixture based on three-dimensional discrete element modeling. *Mater. Des.* **2016**, *89*, 304–313. [CrossRef]

5. Ma, T.; Zhang, D.; Zhang, Y.; Wang, S.; Huang, X. Simulation of wheel tracking test for asphalt mixture using discrete element modelling. *Road Mater. Pavement Des.* **2018**, *19*, 367–384. [CrossRef]

6. Peng, C.; Feng, J.; Zhou, C.; Decheng, F. Investigation on statistical characteristics of asphalt concrete dynamic moduli with random aggregate distribution model. *Constr. Build. Mater.* **2017**, *148*, 723–733.

7. Ma, T.; Wang, H.; Zhang, D.; Zhang, Y. Heterogeneity effect of mechanical property on creep behavior of asphalt mixture based on micromechanical modeling and virtual creep test. *Mater. Des.* **2017**, *104*, 49–59.

8. Ma, T.; Wang, H.; He, L.; Zhao, Y.; Huang, X.; Chen, J. Property characterization of asphalt binders and mixtures modified by different crumb rubbers. *J. Mater. Civ. Eng.* **2017**, *29*, 04017036. [CrossRef]

9. Ding, X.; Ma, T.; Huang, X. Discrete-element contour-filling modeling method for micromechanical and macromechanical analysis of aggregate skeleton of asphalt mixture. *J. Transp. Eng. Part B Pavements* **2018**, *145*, 04018056. [CrossRef]

10. Ding, X.; Chen, L.; Ma, T.; Ma, H.; Gu, L.; Chen, T.; Ma, Y. Laboratory investigation of the recycled asphalt concrete with stable crumb rubber asphalt binder. *Constr. Build. Mater.* **2019**, *203*, 552–557. [CrossRef]

11. Zhang, Y.; Luo, X.; Onifade, I.; Huang, X.; Lytton, R.L.; Birgisson, B. Mechanical evaluation of aggregate gradation to characterize load carrying capacity and rutting resistance of asphalt mixtures. *Constr. Build. Mater.* **2019**, *205*, 499–510. [CrossRef]

12. Tang, F.; Zhu, S.; Xu, G.; Ma, T.; Kong, L.; Kong, L. Influence by chemical constitution of aggregates on demulsification speed of emulsified asphalt based on UV-spectral analysis. *Constr. Build. Mater.* **2019**, *212*, 102–108. [CrossRef]

13. Jiang, J.; Ni, F.; Dong, Q.; Yao, L.; Ma, X. Investigation of the internal structure change of two-layer asphalt mixtures during the wheel tracking test based on 2D image analysis. *Constr. Build. Mater.* **2019**, *209*, 66–76. [CrossRef]

14. Cai, X.; Zhang, H.; Zhang, J.; Chen, X.; Yang, J.; Hong, J. Investigation on reinforcing mechanisms of semi-flexible pavement material through micromechanical model. *Constr. Build. Mater.* **2019**, *198*, 732–741. [CrossRef]

15. You, Z.; Buttlar, W.G. Discrete Element Modeling to Predict the Modulus of Asphalt Concrete Mixtures. *J. Mater. Civ. Eng.* **2004**, *16*, 140–146. [CrossRef]

16. Dai, Q.; You, Z. Micromechanical finite element framework for predicting viscoelastic properties of asphalt mixtures. *Mater. Struct.* **2008**, *41*, 1025–1037. [CrossRef]

17. Dai, Q.; Sadd, M.H.; Parameswaran, V.; Shukla, A. Prediction of damage behaviors in asphalt materials using a micromechanical finite-element model and image analysis. *J. Eng. Mech.* **2005**, *131*, 668–677. [CrossRef]

18. Dai, Q. Prediction of dynamic modulus and phase angle of stone-based composites using a micromechanical finite-element approach. *J. Mater. Civ. Eng.* **2010**, *22*, 618–627. [CrossRef]

19. Ying, H.; Elseifi, M.A.; Mohammad, L.N.; Hassan, M.M. Heterogeneous finite-element modeling of the dynamic complex modulus test of asphalt mixture using x-ray computed tomography. *J. Mater. Civ. Eng.* **2013**, *26*, 04014052. [CrossRef]

20. Liu, Y.; You, Z. Discrete-element modeling: Impacts of aggregate sphericity, orientation, and angularity on creep stiffness of idealized asphalt mixtures. *J. Eng. Mech.-Asce* **2011**, *137*, 294–303. [CrossRef]

21. Olard, F.; Di Benedetto, H. General "2S2P1D" model and relation between the linear viscoelastic behaviours of bituminous binders and mixes. *Road Mater. Pavement* **2003**, *4*, 185–224.

22. Kuna, K.; Gottumukkala, B. Viscoelastic characterization of cold recycled bituminous mixtures. *Constr. Build. Mater.* **2019**, *199*, 298–306. [CrossRef]

23. Blasl, A.; Khalili, M.; Falla, G.C.; Oeser, M.; Liu, P.; Wellner, F. Rheological characterisation and modelling of bitumen containing reclaimed components. *Int. J. Pavement Eng.* **2019**, *20*, 638–648. [CrossRef]

24. Nobakht, M.; Sakhaeifar, M.S. Dynamic modulus and phase angle prediction of laboratory aged asphalt mixtures. *Constr. Build. Mater.* **2018**, *190*, 740–751. [CrossRef]

25. Yin, Y.; Huang, W.; Lv, J.; Ma, X.; Yan, J. Unified construction of dynamic rheological master curve of asphalts and asphalt mixtures. *Int. J. Civ. Eng.* **2018**, *16*, 1057–1067. [CrossRef]

26. De Araújo, P.C.; Soares, J.B.; Holanda, Á.S.; Parente, E.; Evangelista, F. Dynamic viscoelastic analysis of asphalt pavements using a finite element formulation. *Road Mater. Pavement Des.* **2010**, *11*, 409–433. [CrossRef]

27. Honerkamp, J.; Weese, J. Determination of the relaxation spectrum by a regularization method. *Macromolecules* **1989**, *22*, 4372–4377. [CrossRef]

28. Kaschta, J. On the calculation of discrete retardation and relaxation spectra. In *Theoretical and Applied Rheology*; Elsevier: Amsterdam, The Netherlands, 1992; p. 155.

29. Emri, I.; Tschoegl, N.W. Generating line spectra from experimental responses. *Rheol. Acta* **1997**, *36*, 303–306. [CrossRef]

30. Ramkumar, D.; Caruthers, J.M.; Mavridis, H.; Shroff, R. Computation of the linear viscoelastic relaxation spectrum from experimental data. *J. Appl. Polym. Sci.* **1997**, *64*, 2177–2189. [CrossRef]

31. Park, S.W.; Kim, Y.R. Fitting Prony-series viscoelastic models with power-law presmoothing. *J. Mater. Civ. Eng.* **2001**, *13*, 26–32. [CrossRef]

32. Zeng, M.; Bahia, H.U.; Zhai, H.; Anderson, M.R.; Turner, P. Rheological modeling of modified asphalt binders and mixtures. *Asph. Paving Technol. Assoc. Asph. Paving Technol.-Proc. Tech. Sess.* **2001**, *70*, 403–441.

33. Christensen, D.W., Jr.; Anderson, D.A. Interpretation of dynamic mechanical test data for paving grade asphalt. *Asph. Paving Technol. Assoc. Asph. Paving Technol.-Proc. Tech. Sess.* **1992**, *61*, 67–116.

34. Havriliak, S.; Negami, S. A complex plane analysis of α—Dispersions in some polymer systems. *J. Polym. Sci. Part C Polym. Symp.* **1966**, *14*, 99–117. [CrossRef]

35. Katicha, S.W.; Flintsch, G.W. Fractional viscoelastic models: Master curve construction, interconversion, and numerical approximation. *Rheol. Acta* **2012**, *51*, 675–689. [CrossRef]

36. Adolfsson, K.; Enelund, M.; Olsson, P. On the fractional order model of viscoelasticity. *Mech. Time-Depend. Mater.* **2005**, *9*, 15–34. [CrossRef]

37. JTG E20-2011. *Standard Test Methods of Bitumen and Bitumenous Mixtures for Highway Engineering*; Ministry of Transport of the People's Republic of China: Beijing, China, 2011. (In Chinese)

38. Tehrani, F.F.; Absi, J.; Allou, F.; Petit, C. Micromechanical modelling of bituminous materials' complex modulus at different length scales. *Int. J. Pavement Eng.* **2018**, *19*, 685–696. [CrossRef]

 *materials*

Article

Aging Characteristics of Bitumen from Different Bituminous Pavement Structures in Service

Xiaofeng Wang [1,2], Haoyan Guo [3], Bo Yang [1,2], Xingwen Chang [1,2], Chenguang Wan [1,2] and Zhenjun Wang [3,4,*]

[1] Research and Development Center of Transport Industry of Technologies, Materials and Equipments of Highway Construction and Maintenance, Zhengzhou 450052, China; wangxf0351@sina.com (X.W.); yangbohnrbi@foxmail.com (B.Y.); changxw1963@aliyun.com (X.C.); wancg1989@aliyun.com (C.W.)
[2] Henan Engineering Research Center of Intelligent Highway Big Data, Zhengzhou 450052, China
[3] School of Materials Science and Engineering, Chang'an University, Xi'an 710061, China; hyguo@chd.edu.cn
[4] Engineering Research Center of Pavement Materials, Ministry of Education of P.R. China, Chang'an University, Xi'an 710064, China
* Correspondence: zjwang@chd.edu.cn

Received: 20 January 2019; Accepted: 31 January 2019; Published: 10 February 2019

Abstract: The aging of bitumen seriously affects the service life of bituminous pavements. At present, there are many related researches on bitumen aging, but most of them focus on aging endured in indoor surroundings and conditions. Therefore, the conclusions obtained cannot reflect the actual aging changes of bitumen in bituminous pavements in service. In order to study the comprehensive aging process and mechanism of bitumen under the influence of service, we studied bridge deck, traffic lane, and ramp with bituminous pavement structures in service. The bitumen samples obtained from the core samples in different bituminous pavement structures were characterized by gel permeation chromatography (GPC), Fourier transform infrared spectroscopy (FTIR), dynamic shear rheometer (DSR), and fluorescence microscope (FM). The aging degree of different bitumen was analyzed, and conclusions were drawn on changes to bitumen aging from different pavement structures. The results showed that the aging degree of bitumen from the upper layer was the most serious, the aging degree of bitumen at the middle layer was weaker than that of bitumen from the upper layer, and the aging degree of bitumen from the bottom layer was the weakest for the different bituminous pavement structures. The aging of bitumen mainly occurred due to oxygen absorption. After aging, viscoelastic components of bitumen changed, and bitumen became harder. The macromolecule of bitumen could be divided into small molecules, and the small molecular weight of bitumen became large. The styrene-butadiene-styrene (SBS) modifier in the modified bitumen became granular after aging, and it appeared as a single phase in bitumen. The aging changes characterized by different analytical methods showed that the aging degree of bitumen from different layers of bituminous pavement structures in service was different. Effective measures should therefore be taken in time to decrease further aging of bitumen from the upper layer of bituminous pavements due to its inevitable early aging in service.

Keywords: aging characteristics; bitumen; bituminous pavement structures; in service; microscopic characterizations

1. Introduction

Bituminous pavements face comprehensive aging effect from vehicle load and the natural environment (such as temperature, oxidation, light, rainwater), which can result in the aging of bitumen in the bituminous mixture. Bitumen can become harder and brittle after a series of physical and chemical changes during the aging process [1]. In addition, the aging of bitumen can directly affect the service life of pavements. Macroscopic changes during bitumen aging include a decrease in penetration and ductility and an increase in the softening point. This is accompanied by a reduction in other properties of bituminous pavements, such as anti-rutting, water damage resistance, and crack resistance [2–4]. Therefore, the aging of bitumen has become one of the main factors affecting the service life of bituminous pavement structures.

In recent years, many researchers have studied the aging process and mechanism of bitumen [5–7]. Wang et al. [8] studied the aging process of polymer–bitumen composite systems by chemical functional group, molecular size, and rheological characteristics. Lana et al. [9] evaluated the effects of aging on micro mechanical and chemical properties of foam warm mix bitumen using gel permeation chromatography (GPC) and atomic force microscope (AFM). Bowers et al. [10] studied the fusion degree of aged bitumen and new bitumen in reclaimed bituminous pavements using GPC and Fourier transform infrared spectroscopy (FTIR). Menapace et al. [11] processed bitumen using an accelerated aging tester and analyzed the chemical composition of aged bitumen by FTIR and X-ray photoelectron spectroscopy. Lee et al. [12] tested the variations in large molecular particle size (LMS) of rubber-modified bitumen before and after aging using GPC. Wang et al. [13] studied the surface morphology of five kinds of bitumen in different aging situations using AFM. Chen et al. [14] used AFM to study the aging mechanism of bitumen and found that the light component content of bitumen was greatly reduced during the aging process. Dai et al. [15] processed styrene-butadiene-styrene (SBS)-modified bitumen using rotating film oven aging and pressurized aging vessel (PAV) and then obtained nanoscale morphology and rheological properties of SBS-modified bitumen after aging using AFM and the rheometer test. Zhang et al. [16] studied the impact of various aging methods on the physical properties and chemical composition of SBS-modified bitumen using the film furnace test, PAV, and ultraviolet (UV) radiation.

In addition, there have been some innovations in the study of bitumen aging, which are reflected in both devices and testing methods. Some scholars have developed new devices to study the aging mechanism of bitumen under different aging conditions, while others have applied traditional testing methods or instruments in their studies. Ye et al. [17], for example, developed a strong UV aging box with reference to the actual UV light intensity of field bituminous pavements. Yu et al. [18] used AFM technology to quantitatively analyze the morphology, adhesion, and modulus of traditional rotating film oven aging + PAV-aged bitumen and all weather-aging bitumen. Yan et al. [19] found that the rolling film oven test was not suitable for polymer-modified bitumen because the high viscosity polymer-modified bitumen did not roll in the glass bottle during the test. Hou et al. [20] used the spectrophotometric method to determine the aging characteristics of bitumen.

Researchers are also increasingly using simulation techniques to analyze the aging mechanism of bitumen [21–23]. Zhang et al. [24], for example, used the finite element model to study the aging process of bitumen mortar and studied the effect of aging on fatigue resistance. Chen [25] analyzed changes in the quality of bitumen components before and after aging using thermogravimetric analysis and established a dynamic model between quality and aging time. Xu et al. [26] used molecular dynamics to simulate the oxidative aging reaction of bitumen. Ding et al. [27] established a corresponding molecular model of components of original and aged bitumen by studying the functional group distribution of unaged and aged bitumen using liquid chromatography transformation with GPC and FTIR.

Some scholars have considered the aging conditions of indoor tests to be unsuitable for simulating the true aging mechanism of roads in service and have therefore proposed an aging test of bitumen under the coupling of various conditions. Ying et al. [28] put up a concept of bitumen coupling aging and discussed the reduction in bitumen performance. Liu et al. [29] analyzed the changes in viscosity and microscopic composition of bitumen after indoor, accelerated aging progress and after natural aging using dynamic shear rheometer (DSR) and GPC and concluded that there is still some difficulty in simulating the true aging mechanism of bitumen in the laboratory.

In summary, the study of bitumen aging at present is mostly limited to indoor, simulated aging, which is obviously different from the actual aging of bituminous pavements under complex service conditions. The differences in the aging of different structural layers of bituminous pavements and aging characteristics are rarely distinguished. Therefore, in order to study the actual aging characteristics and aging degree of bitumen from different structural layers in bituminous pavements, three kinds of bituminous pavement structures—bridge deck, traffic lane, and ramp—under different service condition were selected in this study. The bitumen samples obtained from the core samples in different bituminous pavement structures were characterized by GPC, FTIR, DSR, and fluorescence microscope (FM). The aging degree of different bitumen was analyzed, and conclusions were drawn on changes to bitumen aging from different pavement structures. The conclusions obtained in this work about the physical and chemical changes in the aging process of bituminous pavements with different structures are expected to provide theoretical and technical support for the improvement of the antiaging ability and durability of bituminous pavement engineering in service.

2. Materials and Methods

2.1. Materials

2.1.1. Bituminous Pavement Structures in Service

Bridge deck, traffic lane, and ramp with bituminous pavement structures were used in this work. The thickness of the bridge deck pavement structure was 5 cm AC-13 I fine-grained bitumen mixture + 5 cm AC-16 I medium-grain bitumen mixture, and the total thickness of the bituminous pavement structural layer in the bridge deck was 10 cm. The thickness of the bituminous pavement structural layer in the traffic lane was 4 cm AK-16A medium-grain asphalt mixture + 7 cm AC-20 I medium-grain asphalt mixture + 7 cm AC-25 I coarse-grain asphalt mixture, and the total thickness of the bituminous pavement layer at the traffic lane was 18 cm. The thickness of the bituminous pavement structural layer in the ramp was 3 cm AK-16A medium-grain asphalt mixture + 5 cm AC-20 I medium-grain asphalt mixture + 5 cm AC-25 I coarse-grain asphalt mixture, and the total thickness of the bituminous pavement layer in the ramp was 13 cm.

2.1.2. Core Sample Selection of Bituminous Mixture

In accordance with the Field Test Methods of Subgrade and Pavement for Highway Engineering (JTG E60-2008), the core samples of the bituminous mixture from different pavement structures were obtained with a core drilling rig machine, and the samples were cylindrical specimens with a diameter of 150 mm. The descriptions of different bituminous pavement structures and the core samples are shown in Table 1.

Table 1. Descriptions of different bituminous pavement structure and core samples.

Nos.	Samples		Pavement Thickness	Service Years	Pavement Diseases	Core Sample Description	Pictures of Pavements and Core Samples
	Bituminous Pavement Structures	Sample Labels					
1	Bridge deck	Bridge deck, upper layer of bituminous pavement (BD-UL)	5	15	Longitudinal cracks	The core sample is complete; there are tiny cracks on the surface, and there is no crack inside.	
2		Bridge deck, bottom layer of bituminous pavement (BD-BL)	5				
3	Traffic lane	Traffic lane, upper layer of bituminous pavement (TL-UL)	4	15	Longitudinal cracks	The core sample is intact, and the interlayer adhesion is good.	
4		Traffic lane, middle layer of bituminous pavement (TL-ML)	7				
5		Traffic lane, bottom layer of bituminous pavement (TL-BL)	7				
6	Ramp	Ramp, upper layer of bituminous pavement (R-UL)	3	15	Mesh cracks	The core sample is intact; there are surface cracks, and the adhesion between the two layers is poor.	
7		Ramp, middle layer of bituminous pavement (R-ML)	5				
8		Ramp, bottom layer of bituminous pavement (R-BL)	5				

2.2. Methods

2.2.1. Bitumen Extraction

In this work, the centrifugal separation method (T 0722-1993) in the Standard Test Methods of Bitumen and Bituminous Mixtures for Highway Engineering in China (JTG E20-2011) was used to extract the bitumen solution from the core sample. The steps were as follows: (1) place the core sample in an oven and heat it to a loose state; (2) pour the sample and trichloroethylene into the centrifugal separator, soak for 30 min, and wait for bitumen to dissolve completely; (3) start the centrifuge, gradually increasing the rotational speed to 3000 r/min, and flow bitumen solution into the recycled bottle through the discharge port; (4) add clean trichloroethylene to the centrifuge and repeat step (3) until bitumen solution turns pale yellow.

Next, the Abson method (T 0726-2011) in the Standard Test Methods of Bitumen and Bituminous Mixtures for Highway Engineering in China (JTG E20-2011) was used to recover the bitumen. The steps were as follows: (1) inject the extract in the recycled bottle into the centrifuge tube and remove the mineral powder in the extract by high-speed centrifugation; (2) inject the extract into the distillation flask and place the distillation flask in the distillation unit; (3) distill the trichloroethylene out of the extract under the conditions of oil bath and CO_2 atmosphere and control the temperature in the distillation flask to 160–166 °C; (4) supply CO_2 gas for 5 minutes continuously after the trichloroethylene solvent stops dropping to avoid secondary aging of bitumen; (5) obtain and analyze the recycled bitumen.

2.2.2. GPC Test

GPC is an accurate and efficient method to classify polymers by molecular weight. This method is convenient and intuitive and is widely used in the study of molecular weight distribution of polymers. The test steps used in this work were as follows: (1) wash 10 mL bottles with tetrahydrofuran (THF); (2) weigh 20 mg of extracted bitumen sample into the 10 mL bottle and mark them; (3) drop 10 mL THF to the bottle using a dropper, cover the lid, shake well and stand, and wait for the bitumen sample to dissolve in THF completely; (4) filter with 0.45 um sieve to eliminate influences of other particles on bitumen; (5) absorb 0.4–0.5 mL filtrate with tube, inject the filtrate into the GPC instrument, and carry out the test; (6) analyze the obtained results with software. In this test, the temperature was set at 35 °C, the flow rate was 1.0 mL/min, and the concentration of the sample solution was 2.0 mg/mL.

A GPC diagram of bitumen is shown in Figure 1 [30]. According to Reference [30], the peak time is labeled as the retention time (R_t), LMS time (L_t) can be determined by subtracting a fixed time interval from R_t, and the fixed time interval is given as 0.2 of total time (T_t). The total time can be calculated by subtracting the starting time (S_t) from the ending time (E_t). The location of L_t is $S_t + 12/30T_t$. The area value of LMS below the curve can be calculated by a special software after the L_t value is determined.

The molecular weight of bitumen can change during the aging process. Therefore, the aging degree of bitumen can be analyzed by the change in the molecular weight of bitumen. The average molecular weight of polymer was calculated using Equations (1)–(4) [31].

$$\overline{M_w}(polymer) = \sum w_i M_i \tag{1}$$

$$\overline{M_n}(polymer) = \sum n_i M_i \tag{2}$$

where,

$$w_i = \frac{W_i}{\sum\limits_i W_i} \tag{3}$$

$$n_i = \frac{N_i}{\sum\limits_i N_i} \tag{4}$$

where $\overline{M_w}(polymer)$ is the weight-average molecular weight, g/mol; $\overline{M_n}(polymer)$ is the number-average molecular weight, g/mol; w_i and n_i are the weight fraction and the number fraction, respectively, which can be calculated by Equations (3) and (4); M_i represents the molecular weight of each molecular fraction, g/mol; W_i is the weight of the fraction with molecular weight of M_i, g; and N_i represents the number (molar number) of each molecular fraction, mol.

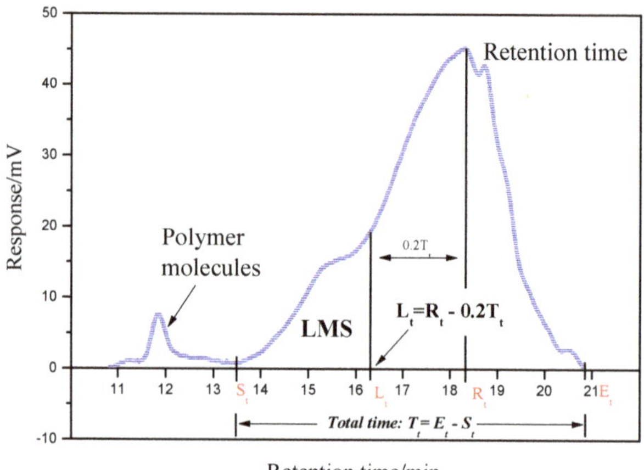

Figure 1. Gel permeation chromatography (GPC) analysis diagram of bitumen.

2.2.3. FTIR Test

FTIR can be used to analyze the variety and content of various chemical bonds in bitumen. The test steps in this work were as follows: (1) clean the sample bench with THF and alcohol; (2) place about 1 g bitumen sample on the sample bench, facing the diamond bit; (3) press the knob to make the diamond bit compact the bitumen sample and start the test; (4) obtain the test result. The temperature for the test was 20 °C. The spectra were collected in the wave number range from 4000 cm^{-1} to 500 cm^{-1} at the resolution of 4 cm^{-1}.

The content of certain functional groups in SBS-modified bitumen usually changes to some extent during the aging process, especially for some oxygen-containing groups, such as carbonyl and sulfoxide functions, which appear as a change in characteristic peaks in the infrared spectrum. Of course, there are also groups that are not affected by aging in the aging process, and their content and peaks of characteristic peaks in the infrared spectrum do not change, such as C–H. Therefore, according to References [8,32], the aging degree of bitumen can be quantitatively analyzed by Equations (5)–(7).

$$L_{(-CH=CH-)} = \frac{Area\left(966 \text{ cm}^{-1}\right)}{Area\left(1376 \text{ cm}^{-1}\right)} \tag{5}$$

$$L_{(S=O)} = \frac{Area\left(1030 \text{ cm}^{-1}\right)}{Area\left(1376 \text{ cm}^{-1}\right)} \tag{6}$$

$$L_{(C=O)} = \frac{Area\left(1700 \text{ cm}^{-1}\right)}{Area\left(1376 \text{ cm}^{-1}\right)} \tag{7}$$

2.2.4. DSR Test

DSR (GEMINI 2, Malvern Instruments Co., Ltd., Malvin, UK) was used to analyze the viscoelastic behavior of bitumen. The test steps in this work were as follows: (1) heat bitumen and then pour it into mold; (2) demold after cooling and wait for testing; (3) place the sample on a plate with diameter of 25 mm and wait for the sample temperature to rise to 64 °C for at least 10 min; (4) lower rotation axis of DSR to adjust gap to 1 mm and carefully trim out bitumen beyond the edge of the plate; (5) start the test and obtain the testing results with complex shear modulus (G*) and phase angle of bitumen samples. The method used in this test was frequency sweep with frequency ranges of 0.1–10.0 Hz. Each sample was tested three times, and the average value was adopted as the testing result.

2.2.5. FM Test

FM is used to analyze microstructures of bitumen in different aging stages. FM (LW300LFT, Xi'an Cewei Photoelectric Technology Co., Xi'an, China) is an instrument that uses ultraviolet light as a light source to irradiate the detected object and make it fluorescent. Then, the shape and location of the sample can be observed under the microscope. The experimental steps used in this work were as follows: (1) heat bitumen samples; (2) coat thin bitumen samples on microscope slide; (3) observe the coated slide under FM and obtain pictures of the tested samples. The test was carried out at a temperature of 20 °C.

3. Results and Discussion

3.1. Aging Degree Analyses of Bitumen

3.1.1. Changes in Molecular Weight

The number-average molecular weight (M_n) and weight-average molecular weight (M_w) of the polymer from the aged bitumen samples are shown in Figure 2. M_w and M_n can reflect the molecular weight of the tested samples. Polydispersity index (PDI) represents the molecular weight distribution width, which can be calculated by Equation (8). The PDI of the following were calculated: bitumen at bridge deck, upper layer of bituminous pavement (BD-UL); bridge deck, bottom layer of bituminous pavement (BD-BL); traffic lane, upper layer of bituminous pavement (TL-UL); traffic lane, middle layer of bituminous pavement (TL-ML); traffic lane, bottom layer of bituminous pavement (TL-BL); ramp, upper layer of bituminous pavement (R-UL), ramp, middle layer of bituminous pavement (R-ML); and ramp, bottom layer of bituminous pavement (R-BL). The results were 1.1654, 1.0264, 1.1263, 1.0169, 1.0456 and 1.057, 1.0151, 1.0138, respectively.

$$PDI = \frac{M_W}{M_N} \tag{8}$$

For the bridge deck, in contrast to the SBS molecular weight before aging, the molecular weight of bitumen from BD-UL and BD-BL ranged from 10,000 to 14,000 and obviously decreased. This indicated that the upper and bottom layers were both aged, but the aging degree was different. The M_n and M_w of bitumen from the upper layer were lower than those of bitumen from the bottom layer. This meant that, in contrast to bitumen from the bottom layer, a large part of SBS molecules of bitumen from the upper layer were degraded into small molecules. The higher PDI for bitumen from the upper layer indicated a broader molecular weight distribution. This showed that the aging degree of bitumen from the upper layer was more serious than that of bitumen from the bottom layer under the effect of load and the environment.

For the traffic lane, results showed that the aging degree of bitumen from the three layers was slightly different. The minimal M_n and M_w values and the maximum PDI value suggested that the aging degree of bitumen from the upper layer was significantly more serious than that of bitumen from the other two layers. However, as can be seen in Figure 2, M_n and M_w of bitumen from the middle layer was slightly higher than that of bitumen from the bottom layer. The PDI of bitumen from the middle layer was slightly less than that of bitumen from the bottom layer, which indicated that the aging degree of the bottom layer was more serious. This could be due to cracks caused by the aging of the upper layer, where moisture and air can flow into the pavement structure layer. Oxygen and moisture accumulate in the bottom layer and react with bitumen, which can result in breaking of the bitumen molecular chain in the bottom layer. Therefore, the aging degree of bitumen from the bottom layer is more serious than that of bitumen from the middle layer.

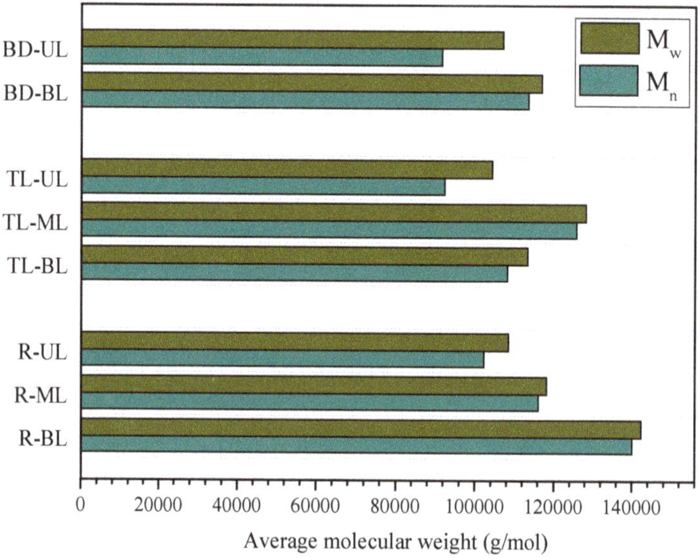

Figure 2. The M_w and M_n of bitumen after aging.

For the ramp, the three indexes of the different layers were ranked as M_n: the upper layer < the middle layer < the bottom layer; M_w: the upper layer < the middle layer < the bottom layer; and PDI: the upper layer > the middle layer > the bottom layer. This meant the aging degree of bitumen from the upper layer was the most serious, followed by the middle layer, and finally the bottom layer.

3.1.2. GPC Curves of Aged Bitumen

Figure 3 shows the GPC test results for the upper and bottom layers of the bridge deck. It can be seen that the trend of the two curves was basically the same, except that the size of the peak area was different at some specific positions. The bitumen used for the bridge deck was SBS-modified bitumen. As can be seen in Figure 3, regardless of the GPC curves of bitumen from BD-UL or that of bitumen from BD-BL, the SBS characteristic peak that should appear at about 12 min almost disappeared, indicating that the SBS modifier degradation of both was serious. At this time, judging the aging degree by characteristic peaks would not be reliable, so the LMS values of bitumen were used to judge the degree of aging. According to the calculation method (Figure 1), the LMS value of bitumen from BD-UL was calculated as 25.32, and the LMS value of bitumen from BD-BL was 20.82. This dramatic conclusion could be due to the fact that naphthenic and polar aromatics form more asphaltenes during aging, while the size of the asphaltene structure also increases [33].

Figure 4 shows the GPC test results of different layers of the traffic lane. It can be seen that the general trend of the three curves was consistent. At 12 min, the TL-ML layer showed a sharp peak and the peak of TL-BL was less intense than TL-ML, while the peak of TL-UL disappeared completely. This showed that TL-UL had the most serious aging, and the macromolecular material was broken completely. The peak in TL-ML indicated that the macromolecules were much less broken than TL-UL and TL-BL. At 18 min, the intensity of the TL-UL peak was significantly less than that of TL-ML and TL-BL, probably because the degradation of bitumen and SBS modifier from TL-UL was severe, and the macromolecule ruptured to form small molecules. However, the loss of small molecules in the upper layer was more serious under the influence of vehicle load and the environment. Therefore, the intensity of the peak here was weaker than that of bitumen from TL-ML and TL-BL. Overall, the aging degree of bitumen in the traffic lane was TL-UL > TL-BL > TL-ML.

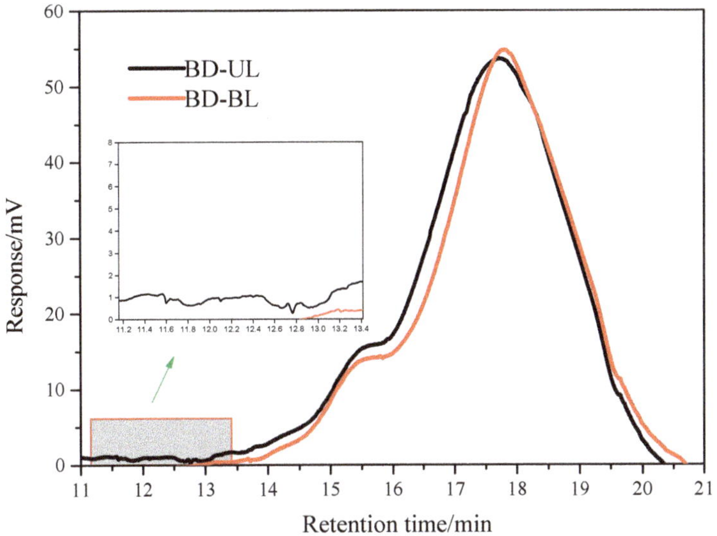

Figure 3. GPC curves of bitumen from BD-UL and BD-BL.

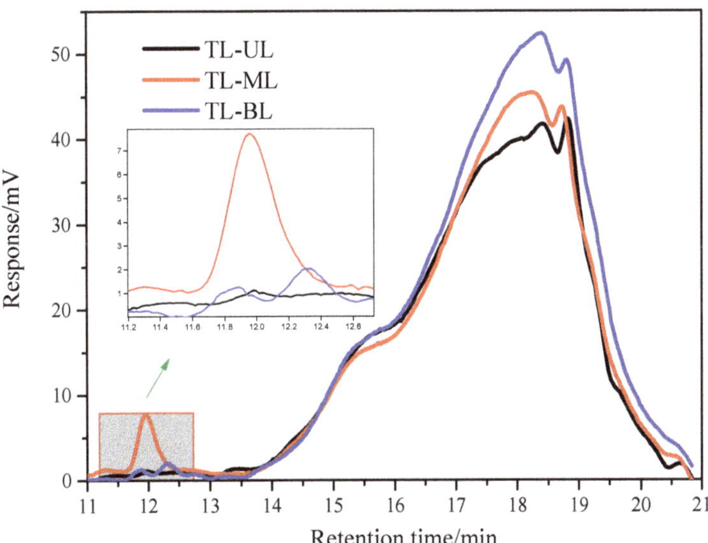

Figure 4. GPC curves of bitumen from TL-UL, TL-ML, and TL-BL.

Figure 5 shows the GPC curve results for different layers of the ramp. In combination with the GPC curves, it can be seen that the general trend of the three curves was consistent. At 12 min, the peak that should have appeared here completely disappeared, indicating that the aging degree of bitumen from R-UL was the most serious. R-ML and R-BL showed sharp peaks here, and the peak of R-BL was sharper, indicating that the degradation degree of SBS modifier of the bitumen sample from R-BL was weaker, and the aging degree of bitumen was lower. As can be seen from Figure 5, the intensity of the

R-BL peak was slightly higher than that of bitumen from R-UL and R-ML at 18 min, which could be due to the migration of a part of the small molecule material generated by the aging of the upper R-ML to the R-BL under the action of water flow. In summary, the aging degree of bitumen in the ramp was R-UL > R-ML > R-BL.

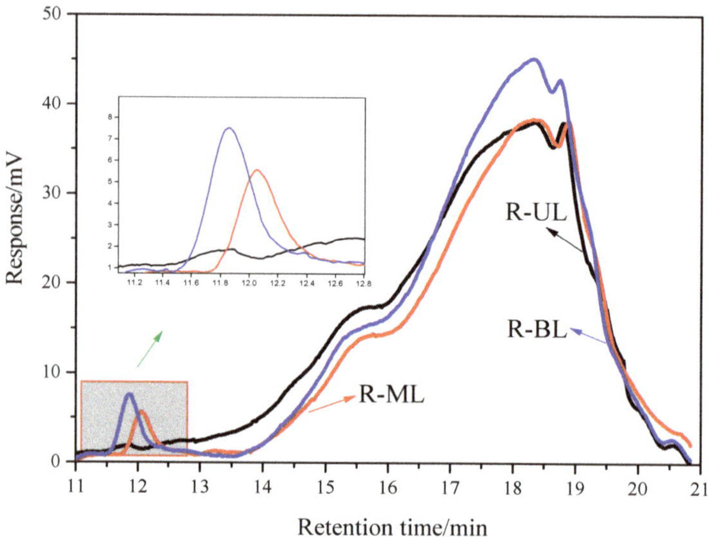

Figure 5. GPC curves of bitumen from R-UL, R-ML, and R-BL.

3.2. Composition Change Analyses of Bitumen

3.2.1. Bitumen of Bridge Deck

Figure 6 shows the infrared spectrums and quantitative analysis diagrams of bitumen from different structural layers of the bridge deck. It can be seen from Figure 6 that the absorption peaks of samples from BD-UL and BD-BL appeared near 966 cm^{-1} and 698 cm^{-1}, which were the two characteristic peaks of SBS modifier. The peak at 966 cm^{-1} and 698 cm^{-1} was the characteristic peak of C–C bond in butadiene and styrene, respectively. The absorption peak near 966 cm^{-1} of sample from BD-UL was weaker, which could be due to the direct contact of the SBS modifier in the upper layer with the environment, so the degradation of SBS was more serious. The absorption peak at 910 cm^{-1} was the absorption peak of benzene ring C–H vibration, reflecting that other components were substituted on the benzene ring of bitumen. The intensity of the absorption peak at 910 cm^{-1} of sample from BD-UL was slightly higher than that of the sample from BD-BL. The results showed that the substitutions of benzene ring in bitumen occurred more frequently in the sample from BD-UL. The absorption peaks of the sulfoxide group (S=O) of the samples from BD-UL and from BD-BL were near 1030 cm^{-1}, indicating that both of them had aged, and the peak strength of sample from BD-UL was higher than that of the sample from BD-BL. The peak of 1600–1700 cm^{-1} was caused by the oxygen absorption of unsaturated carbon chain. The peak of the sample from BD-UL produced more peaks here, and the intensity of the peak was larger, indicating that there were more oxygen absorption reactions and more serious aging of sample from BD-UL. The comprehensive analyses showed that the aging degree of bitumen was BD-UL > BD-BL.

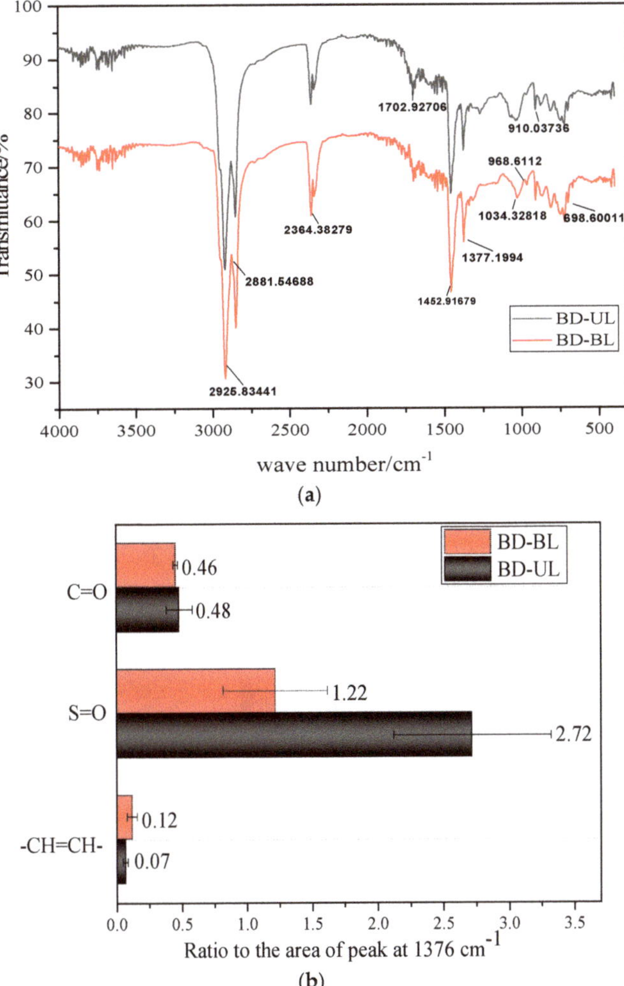

Figure 6. Composition changes of bitumen from BD-UL and BD-BL. (**a**) Infrared spectra; (**b**) quantitative analysis diagram.

3.2.2. Bitumen of Traffic Lane

According to the spectra shown in Figure 7, the intensity of the three absorption peaks at $1030\ \text{cm}^{-1}$ and $1700\ \text{cm}^{-1}$ of the upper layer was significantly higher than those of the middle layer and the bottom layer, indicating that the aging degree of bitumen from the upper layer was much greater than that of bitumen from the middle layer and the bottom layer. In contrast to the aging degree of bitumen from the middle layer and the bottom layer, the absorption peak intensity of the C–O bond formed by the oxygen absorption of unsaturated carbon chain at $1700\ \text{cm}^{-1}$ of bitumen from the middle layer was slightly higher than that of bitumen from the bottom layer. The sulfoxide group absorption peak at $1030\ \text{cm}^{-1}$ was also higher than that of bitumen from the bottom layer. Therefore, the aging degree of bitumen from the middle layer was slightly higher than that of bitumen from the bottom layer. According to the analyses of the infrared spectra, the aging degree of bitumen from the traffic lane was upper layer > middle layer > bottom layer.

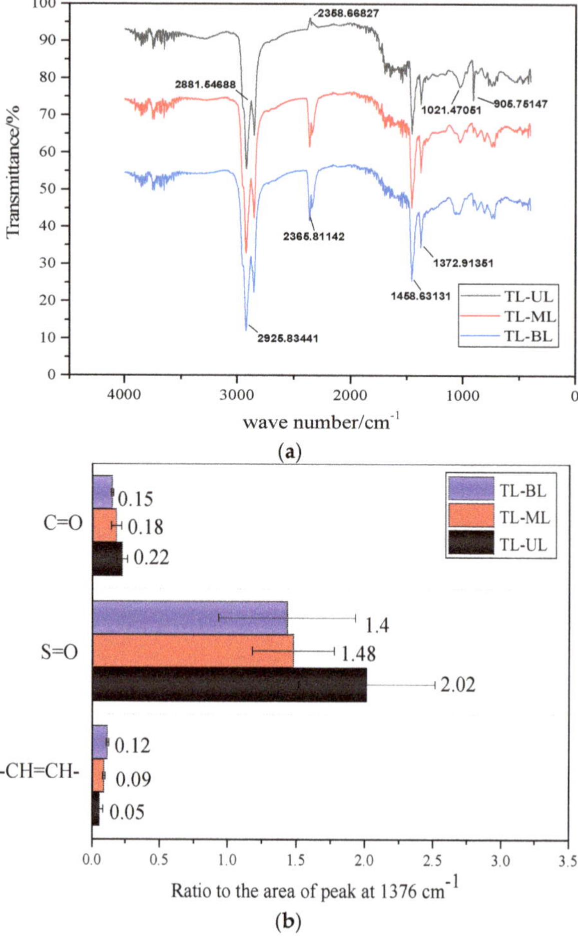

Figure 7. Composition changes of bitumen from TL-UL, TL-ML and TL-BL. (**a**) Infrared spectra; (**b**) quantitative analysis diagram.

3.2.3. Bitumen of Ramp

As shown in Figure 8, the infrared spectra of bitumen from the upper, middle, and bottom layers in the ramp were analyzed and compared. The absorption peaks of 966 cm^{-1} and 698 cm^{-1} indicated that SBS-modified bitumen was used in the ramp. The absorption peak at 1700 cm^{-1} indicated that bitumen began to absorb oxygen gradually, and the curves of samples from R-ML and R-BL decreased less before the absorption peak of 1700 cm^{-1}. However, the curve of the sample from R-UL decreased sharply when it appeared at the peak. The sulfoxide group bond (S=O) (1030 cm^{-1}) was present in the three samples. In contrast to the area of the peaks of the three samples, it was found that R-UL had the largest peak at 1030 cm^{-1}, R-ML was the next, and the area of the peak of R-BL at 1030 cm^{-1} was the smallest. The results showed that there were more oxygen absorption aging reactions and more serious aging degree of bitumen from the upper layer. Therefore, the aging degree of bitumen from the ramp was R-UL > R-ML > R-BL, that is, the upper layer > the middle layer > the bottom layer.

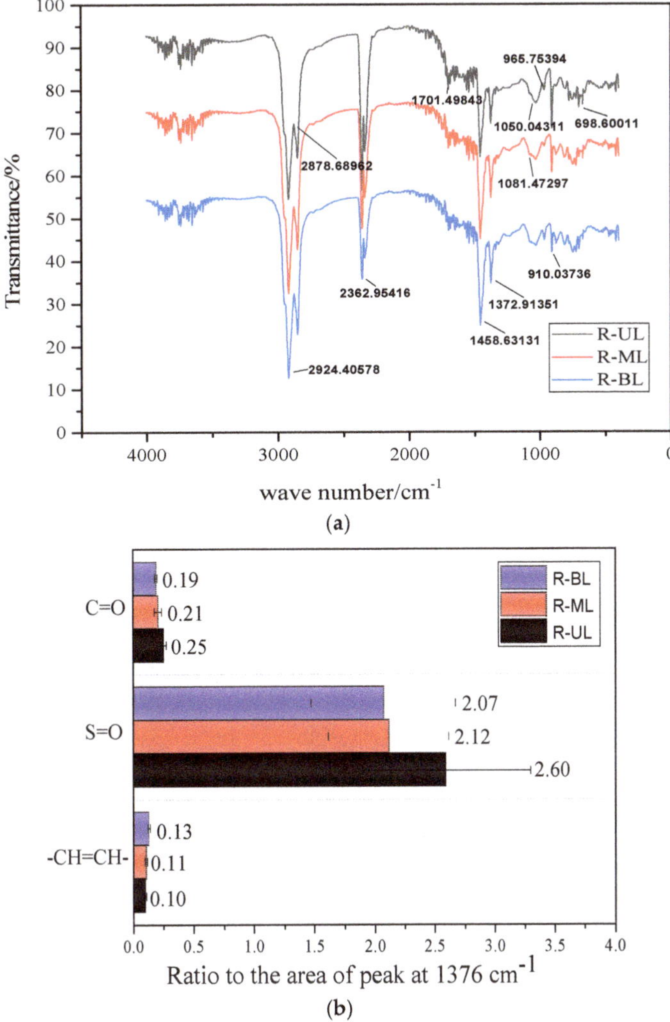

Figure 8. Composition changes of bitumen from R-UL, R-ML and R-BL. (**a**) Infrared spectra; (**b**) quantitative analysis diagram.

According to the analyses of the aging degree of bitumen from the upper, middle, and bottom layers of the three road sections, the aging of bitumen from the upper layer was more serious than that of bitumen from the middle and lower layers because the upper layer was directly affected by the vehicle load and the environment. The oxidation of the unsaturated carbon chain and the sulfur element in bitumen was caused by the oxygen absorption reaction during the aging process, which resulted in the corresponding characteristic peaks.

3.3. Viscoelastic Behavior Analyses of Bitumen

3.3.1. Bitumen of Bridge Deck

Complex shear modulus can be used to evaluate the stiffness of bitumen. For the bridge deck, it can be seen from Figure 9 that the complex shear modulus of bitumen from BD-UL and BD-BL showed a slight downward trend in the low frequency state, while there was an increasing trend in the high frequency state. The sample conformed to the general trend of low to high growth of the complex shear modulus from low to high frequency. In both low and high frequency states, the complex shear modulus of bitumen from BD-UL was always higher than that of bitumen from BD-BL. This indicated that bitumen from BD-UL was harder than that from BD-BL, proving that the aging of bitumen from BD-UL was more serious than that of bitumen from BD-BL.

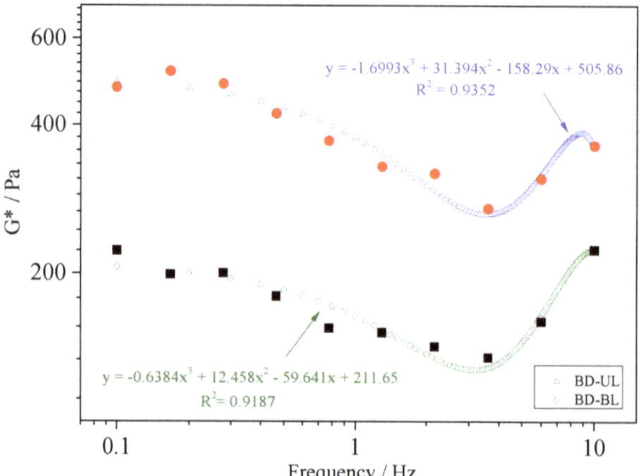

Figure 9. Complex shear modulus of bitumen from BD-UL and BD-BL.

3.3.2. Bitumen of Traffic Lane

For the traffic lane, it can be seen from Figure 10 that complex shear modulus curves of bitumen from TL-UL, TL-ML, and TL-BL showed an upward trend with the increase in frequency. The order of the complex shear modulus in both low and high frequency states was TL-UL > TL-ML > TL-BL, indicating that bitumen from the traffic lane conformed to the aging degree, that is, the upper layer > the middle layer > the lower layer.

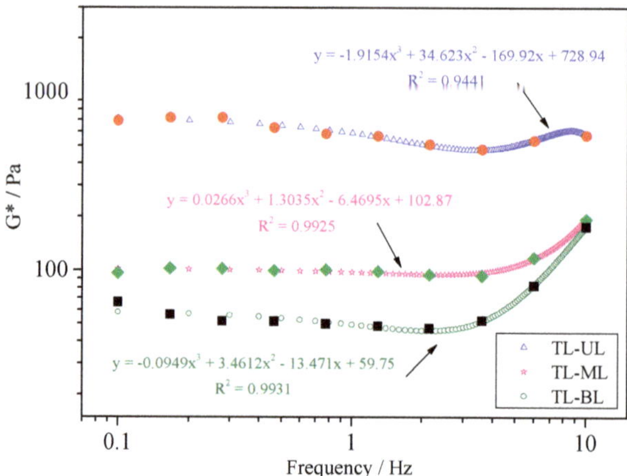

Figure 10. Complex shear modulus of bitumen from TL-UL, TL-ML, and TL-BL.

3.3.3. Bitumen of Ramp

For the ramp, it can be seen from Figure 11 that the order of the complex shear modulus was R-ML > R-UL > R-BL in the low frequency ranges and R-UL > R-ML > R-BL in the high frequency ranges. The complex shear modulus of bitumen from the middle layer in the low frequency region was higher than that of bitumen from the upper layer, indicating that bitumen from the middle layer was harder than that from the upper layer in the high temperature environment. The complex shear modulus of bitumen from the middle layer in the high frequency ranges was lower than that of bitumen from the upper layer, indicating that bitumen from the middle layer was more flexible than that from the upper layer in the low temperature surroundings. This indicated that the performance of bitumen from the middle layer was better than that of bitumen from the upper layer in both low and high temperature conditions, that is, the aging degree of bitumen in the ramp was upper layer > middle layer > lower layer.

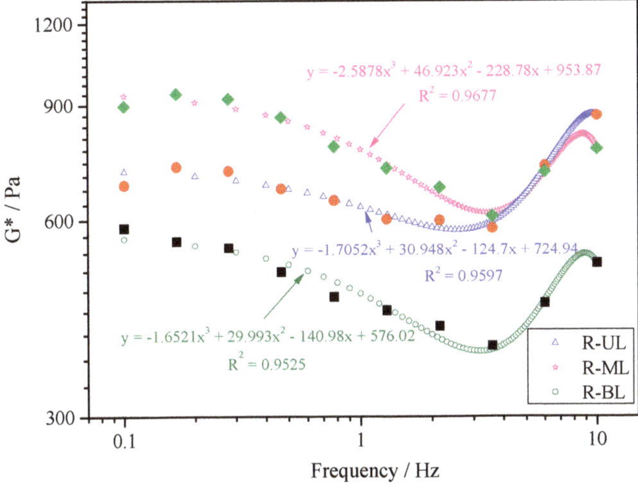

Figure 11. Complex shear modulus of bitumen from R-UL, R-M, and R-BL.

3.3.4. Phase Angle of Different Bitumen

It can be seen from Figure 12 that the phase angles of the samples were U-shaped. The phase angle is an indicator of the viscoelasticity of the materials. At medium–high temperature, the bitumen was in the region transitioning from a high elastic state to a viscous flow state. When the bitumen sample was approaching the viscous flow state, the phase angle could increase and approach 90 °C. At the same time, when the temperature was high and the load frequency was low, the phase angle decreased with the increase in temperature due to the influence of the mineral skeleton. When the load frequency was lower, the phase angle started to rise with the decrease in frequency due to the more complicated viscoelastic behavior of the bitumen in the phase transition stage. At lower temperatures and higher load frequencies, the phase angle decreased rapidly, and bitumen changed from a viscous state to a high-elastic state.

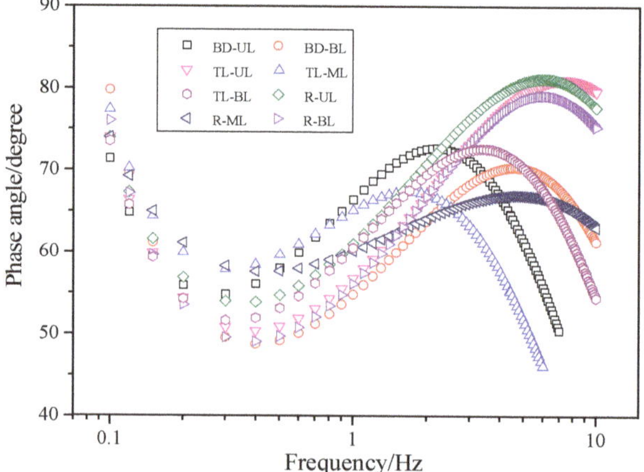

Figure 12. Phase angles of bitumen in different bituminous pavement structures.

3.4. Microscopic Morphology Analyses of Bitumen

3.4.1. Bitumen of Bridge Deck

Figure 13 shows the fluorescence pictures of bitumen extracted from the bridge deck. The original bitumen was black in the fluorescence microscopy image because it was not excited by fluorescence, while the SBS modifier was bright yellow. It can be seen from Figure 13 that the SBS modifiers from BD-UL and BD-BL were in the dispersed phase of the particles, indicating that bitumen had aged due to the influence of load and the environment. However, the difference between the two pictures was very small, meaning the aging degree of bitumen from the upper and middle layers in the bridge deck could not be evidently distinguished from the fluorescence pictures.

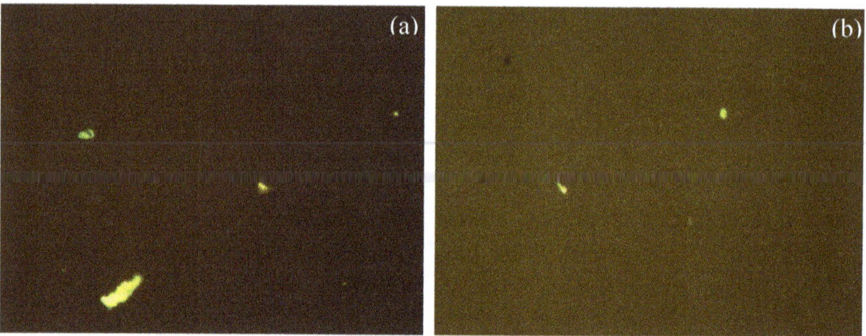

Figure 13. Fluorescence microscope (FM) pictures of bitumen from BD-UL and BD-BL (100×): (**a**) BD-UL and (**b**) BD-BL.

3.4.2. Bitumen of Traffic Lane

It can be seen from Figure 14 that the SBS modifier particles of bitumen from TL-UL and TL-ML were sparse, which indicated that the aging degree of bitumen from the upper and lower layers of the lane was relatively serious. However, the aging degree of the two bitumen could not be evidently distinguished in the fluorescence pictures due to the very close number of particles and dispersion state. The bitumen from TL-BL showed a continuous phase, and the picture was doped with SBS modifier, proving that bitumen from TL-BL possessed the lowest aging degree.

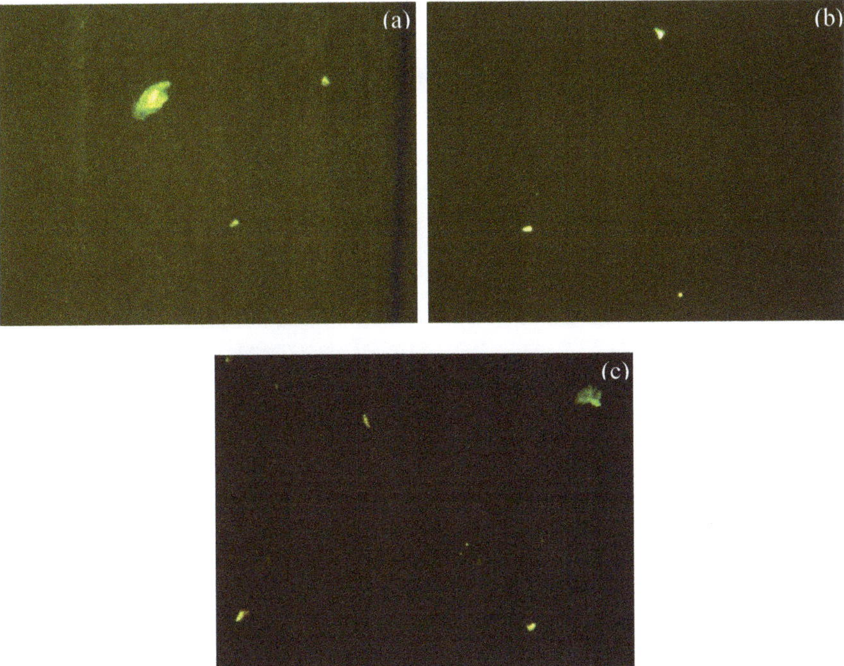

Figure 14. FM pictures of bitumen from TL-UL, TL-ML, and TL-BL (100×): (**a**) TL-UL; (**b**) TL-ML; and (**c**) TL-BL.

3.4.3. Bitumen of Ramp

The SBS modifier can absorb the light components in the original bitumen to form a continuous phase shortly after the SBS modifier is added to bitumen. It can be seen from Figure 15 that, after long-term aging, the continuous phase formed by the SBS modifier had varying degrees of damage. In bitumen from R-UL, the SBS modifier was in the form of granular-dispersed phase, and SBS-modified bitumen became a single-phase continuous structure. The quantity of SBS modifier particles in bitumen from R-ML increased and existed in the state of network structure compared with bitumen from R-UL. The number of bitumen from R-BL reached the maximum, indicating that the SBS modifier from bitumen from R-BL had the lowest degradation degree. Therefore, the aging degree of bitumen in the ramp was R-UL > R-ML > R-BL, that is, the upper layer > the middle layer > the bottom layer.

3.5. Comparison of Results with Different Bituminous Pavement Structure

The aging rules characterized by different analytical methods showed that, while the aging degree of SBS-modified bitumen from different structural layers in service was different, they basically met the same order of aging, that is, upper layer > middle layer > bottom layer. The results showed that using different analytical methods to analyze the aging rules of the same sample could give slightly different conclusions. The conclusions of the aging changes obtained by different characterizations are detailed in Table 2.

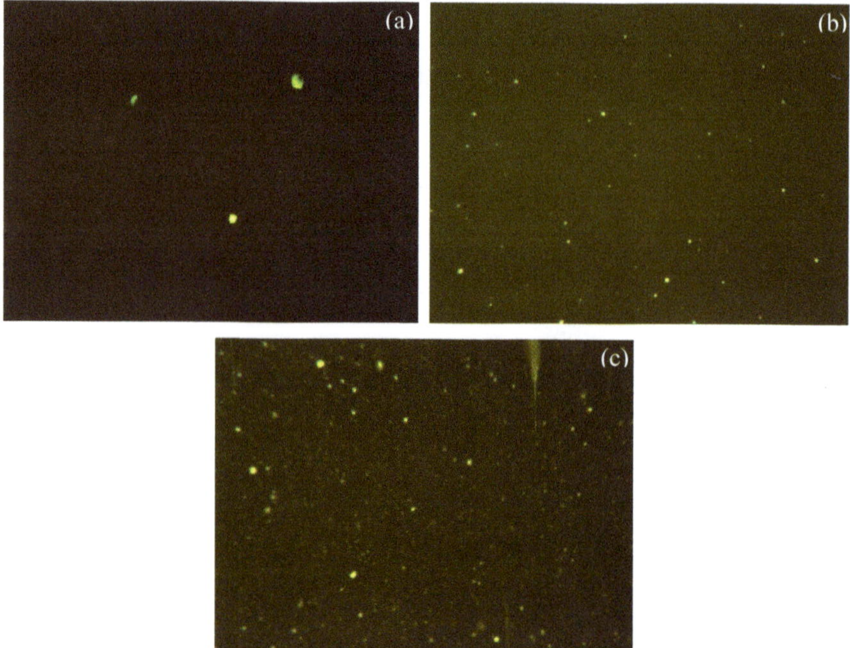

Figure 15. FM pictures of bitumen from R-UL, R-ML, and R-BL (100×): (**a**) R-UL; (**b**) R-ML; and (**c**) R-BL.

Table 2. Aging degree sequences of bitumen with different analytical methods.

Bituminous Pavement	Analytical Methods			
	GPC	FTIR	DSR	FM
Bridge deck	upper > middle	upper > middle	middle > upper	middle ≈ upper
Traffic lane	upper > bottom > middle	upper > middle > bottom	upper > middle > bottom	upper ≈ middle > bottom
Ramp	upper > middle > bottom	upper > middle > bottom	upper > middle > bottom	upper > middle > bottom

4. Conclusions and Recommendations

In this work, the aging characteristics of bitumen from different structural layers in bituminous pavement were characterized and analyzed, and differences in the aging degree of bitumen in different structural layers were analyzed from their chemical compositions and microstructures. The following conclusions were drawn:

(1) Bitumen from the upper and middle layers of the bridge deck was aged. The aging changes conformed to the rule that the bitumen macromolecular chain breaks into small molecules, the small molecule content increases, and the molecular weight width therefore increases. SBS modifiers existed in the particle dispersed phase after aging. The aging type was oxygen absorption aging, which was mainly at peaks of 1600–1700 cm^{-1}. According to the analysis results of GPC and FTIR, the aging degree of bitumen from the upper layer of the bridge deck was more serious than that of bitumen from the middle layer.

(2) The aging degree of bitumen from the three surface layers of the traffic lane was slightly different. According to the results of GPC, bitumen from the upper layer had the smallest molecular weight and the most severe aging, bitumen from the bottom layer had less aging, and bitumen from the

middle layer had the lowest degree of aging. However, according to FTIR, DSR, and FM, bitumen from the bottom layer of the lane had the lowest degree of aging as indicated by oxygen aging, the lowest complex shear modulus at different frequencies, and the continuous phase structure. This could be due to the migration of small molecular chains formed by the aging of bitumen from the upper layer to the middle layer, which had not yet migrated to the bottom layer.

(3) For bitumen from the upper, middle, and bottom layers of the ramp, different testing methods could obtain the same changes in aging degree in the different surface layers of the ramp, and they were upper layer > middle layer > bottom layer. This was because the traffic volume at the ramp would be small, and the aging would gradually expand from the upper layer to the bottom layer of the bituminous pavement structure in service.

(4) The aging changes characterized by different analytical methods showed that the aging degree of bitumen in different layers of the bituminous pavement structure in service was different, but they met the same order of aging degree, that is, upper layer > middle layer > bottom layer. After aging, the microstructures and the compositions of bitumen changed, which could affect properties of pavement structures. When diseases such as longitudinal joints occur, maintenance measures should be taken in time to avoid further deepening of aging. Due to the inevitable early aging of the upper layer, further research and optimization is recommended on the structure and material aspects of the upper layer of bituminous pavements.

Author Contributions: Conceptualization, X.W., X.C., and Z.W.; Data curation, B.Y.; Formal analysis, H.G., X.C., and Z.W.; Investigation, H.G., B.Y., and C.W.; Methodology, X.W. and Z.W.; Writing—original draft, X.W., H.G., C.W., and Z.W.

Funding: This research received no external funding.

Acknowledgments: This work is supported by the State Key Laboratory of High Performance Civil Engineering Materials (No. 2018CEM010), the Science and Technology Project of Henan Transportation Department (No. 2017J1), and Fundamental Research Funds for the Central Universities of China (No. 300102318402). The authors also thank the reviewers for their valuable comments and suggestions concerning our manuscript.

Conflicts of Interest: There are no conflicts of interest regarding the publication of this paper.

References

1. Lesueur, D. The colloidal structure of bitumen: Consequences on the rheology and on the mechanisms of bitumen modification. *Adv. Colloid Interface Sci.* **2009**, *145*, 42–82. [CrossRef] [PubMed]

2. Zhao, X.; Wang, S.; Wang, Q.; Yao, H. Rheological and structural evolution of SBS modified asphalts under natural weathering. *Fuel* **2016**, *184*, 242–247. [CrossRef]

3. Xu, S.; Yu, J.; Zhang, C.; Sun, Y. Effect of ultraviolet aging on rheological properties of organic intercalated layered double hydroxides modified asphalt. *Constr. Build. Mater.* **2015**, *75*, 421–428. [CrossRef]

4. Gao, J.; Guo, H.; Wang, X.; Wang, P.; Wei, Y.; Wang, Z.; Huang, Y.; Yang, B. Microwave deicing for asphalt mixture containing steel wool fibers. *J. Clean. Prod.* **2019**, *206*, 1110–1122. [CrossRef]

5. Li, P.; Ding, Z.; Zhang, Z.; Zhang, D. Comparative evaluation of laboratory and field ageing of asphalt binder using a non-linear differential model. *Road Mater. Pavement Des.* **2016**, *17*, 434–445. [CrossRef]

6. Cong, P.; Hao, H.; Zhang, Y.; Luo, W.; Yao, D. Investigation of diffusion of rejuvenator in aged asphalt. *Int. J. Pavement Res. Technol.* **2016**, *9*, 280–288. [CrossRef]

7. Wu, S.; Zhao, Z.; Xiao, Y.; Yi, M.; Chen, Z.; Li, M. Evaluation of mechanical properties and aging index of 10-year field aged asphalt materials. *Constr. Build. Mater.* **2017**, *155*, 1158–1167. [CrossRef]

8. Wang, J.; Yuan, J.; Kim, K.W.; Xiao, F. Chemical, thermal and rheological characteristics of composite polymerized asphalts. *Fuel* **2018**, *227*, 289–299. [CrossRef]

9. Abu Qtaish, L.; Nazzal, M.D.; Abbas, A.; Kaya, S.; Akinbowale, S.; Arefin, M.S.; Kim, S.S. Micromechanical and chemical characterization of foamed warm-mix asphalt aging. *J. Mater. Civ. Eng.* **2018**, *30*, 04018213. [CrossRef]

10. Bowers, B.F.; Huang, B.; Shu, X.; Miller, B.C. Investigation of reclaimed asphalt pavement blending efficiency through GPC and FTIR. *Constr. Build. Mater.* **2014**, *50*, 517–523. [CrossRef]

11. Menapace, I.; Yiming, W.; Masad, E. Chemical analysis of surface and bulk of asphalt binders aged with accelerated weathering tester and standard aging methods. *Fuel* **2017**, *202*, 366–379. [CrossRef]

12. Lee, S.J.; Hu, J.; Kim, H.; Amirkhanian, S.N.; Jeong, K.D. Aging analysis of rubberized asphalt binders and mixes using gel permeation chromatography. *Constr. Build. Mater.* **2011**, *25*, 1485–1490. [CrossRef]

13. Wang, P.Y.; Zhao, K.; Glover, C.; Chen, L.; Wen, Y.; Chong, D.; Hu, C. Effects of aging on the properties of asphalt at the nanoscale. *Constr. Build. Mater.* **2015**, *80*, 244–254. [CrossRef]

14. Chen, A.; Liu, G.; Zhao, Y.; Li, J.; Pan, Y.; Zhou, J. Research on the aging and rejuvenation mechanisms of asphalt using atomic force microscopy. *Constr. Build. Mater.* **2018**, *167*, 177–184. [CrossRef]

15. Dai, Z.; Shen, J.; Shi, P. Influence of SBS modification on the asphalt aging based on nano-sized topography and rheological properties. *Acta Petrol. Sin.* **2017**, *33*, 578–587.

16. Zhang, D.; Zhang, H.; Shi, C. Investigation of aging performance of SBS modified asphalt with various aging methods. *Constr. Build. Mater.* **2017**, *145*, 445–451. [CrossRef]

17. Ye, F.; Sun, D.; Huang, P.; Zhu, J.; Zhou, Z. Analysis of strong UV photoaging performance of asphalt. *China J. Highw.* **2006**, *19*, 35–38+44.

18. Yu, J.; Dai, Z.; Shen, J.; Zhu, H.; Shi, P. Aging of asphalt binders from weathered asphalt mixtures compared with a SHRP process. *Constr. Build. Mater.* **2018**, *160*, 475–486. [CrossRef]

19. Yan, C.; Huang, W.; Tang, N. Evaluation of the temperature effect on Rolling Thin Film Oven aging for polymer modified asphalt. *Constr. Build. Mater.* **2017**, *137*, 485–493. [CrossRef]

20. Hou, X.; Xiao, F.; Wang, J.; Amirkhanian, S. Identification of asphalt aging characterization by spectrophotometry technique. *Fuel* **2018**, *226*, 230–239. [CrossRef]

21. Hou, Y.; Wang, L.; Wang, D.; Guo, M.; Liu, P.; Yu, J. Characterization of bitumen micro-mechanical behaviors using AFM, phase dynamics theory and MD simulation. *Materials* **2017**, *10*, 208. [CrossRef]

22. Qu, X.; Wang, D.; Hou, Y.; Oeser, M.; Wang, L. Influence of Paraffin on the microproperties of asphalt binder using MD simulation. *J. Mater. Civ. Eng.* **2018**, *30*, 04018191. [CrossRef]

23. Bhasin, A.; Motamed, A. Analytical models to characterise crack growth in asphaltic materials and healing in asphalt binders. *Int. J. Pavement Eng.* **2011**, *12*, 371–383. [CrossRef]

24. Zhang, Y.; Leng, Z. Quantification of bituminous mortar ageing and its application in ravelling evaluation of porous asphalt wearing courses. *Mater. Des.* **2017**, *119*, 1–11. [CrossRef]

25. Chen, F. Study on Aging Behavior and Mechanism of Original Asphalt and SBS Modified Asphalt. Master's Thesis, Chang'an University, Xi'an, China, 2012.

26. Xu, X.; Yu, J.; Xue, L.; He, B.; Du, W.; Zhang, H.; Li, Y. Effect of reactive rejuvenating system on physical properties and rheological characteristics of aged SBS modified bitumen. *Constr. Build. Mater.* **2018**, *176*, 35–42. [CrossRef]

27. Ding, Y.; Huang, B.; Shu, X. Investigation of functional group distribution of asphalt using liquid chromatography transform and prediction of molecular model. *Fuel* **2018**, *227*, 300–306. [CrossRef]

28. Ying, R.; Zheng, J.; Chen, X. Experimental study on aging effect of asphalt. *Highw. Transp. Technol.* **2007**, *24*, 20–23.

29. Liu, G.; Nielsen, E.; Komacka, J.; Greet, L.; van de Ven, M. Rheological and chemical evaluation on the ageing properties of SBS polymer modified bitumen: From the laboratory to the field. *Constr. Build. Mater.* **2014**, *51*, 244–248. [CrossRef]

30. Kim, S.; Lee, S.H.; Kwon, O.; Han, J.Y.; Kim, Y.S.; Kim, K.W. Estimation of service-life reduction of asphalt pavement due to short-term ageing measured by GPC from asphalt mixture. *Road Mater. Pavement Des.* **2016**, *17*, 153–167. [CrossRef]

31. Li, H.; Zhang, C. *Polymer Physics*; Chemical Industry Press: Beijing, China, 2007.

32. Dondi, G.; Mazzotta, F.; Simone, A.; Vignali, V.; Sangiorgi, C.; Lantieri, C. Evaluation of different short term aging procedures with neat, warm and modified binders. *Constr. Build. Mater.* **2016**, *106*, 282–289. [CrossRef]

33. Hao, G.; Huang, W.; Yuan, J.; Tang, N.; Xiao, F. Effect of aging on chemical and rheological properties of SBS modified asphalt with different compositions. *Constr. Build. Mater.* **2017**, *156*, 902–910. [CrossRef]

Article

Evaluating the Effects of High RAP Content and Rejuvenating Agents on Fatigue Performance of Fine Aggregate Matrix through DMA Flexural Bending Test

Chenchen Zhang [1,2], **Qi Ren** [3], **Zhendong Qian** [1,*] **and Xudong Wang** [2]

[1] Intelligent Transport System Research Center, Southeast University, Nanjing 210096, China; 230129418@seu.edu.cn
[2] Research Institute of Highway Ministry of Transport, Beijing 100088, China; xd.wang@rioh.cn
[3] Jiangsu Transportation Research Institute, Nanjing 210017, China; qrendavis@gmail.com
*   Correspondence: qianzd@seu.edu.cn

Received: 1 April 2019; Accepted: 3 May 2019; Published: 9 May 2019

Abstract: High percentage reclaimed asphalt pavement (RAP) is prevailing in pavement engineering for its advantages in sustainability and environmental friendliness, however, its fatigue resistance remains a major concern. Fine aggregate matrix (FAM) is a crucial part in the fatigue resistance of asphalt mixtures with high RAP content. Hence, the linear amplitude sweep (LAS) test of FAM has been developed to study the fatigue resistance of asphalt mixtures. However, the torsional loading mode of the LAS test with a dynamic shear rheometer (DSR) is a limitation to simulate traffic load. In this paper, an alternative LAS test for FAM with high RAP content was proposed. Beam FAM specimens were tested using a dual-cantilever flexural loading fixture in a dynamic mechanical analyzer (DMA). To investigate the influence of RAP content and the rejuvenating agent (RA), four kinds of FAM mixes were tested with this method to study their fatigue resistance. The test results suggested that the repeatability of this alternative approach was reliable. A fatigue failure criterion based on maximum $C \times N$ was defined. Then, fatigue life prediction models based on viscoelastic continuum damage (VECD) analysis were established according to the LAS test results and validated by a strain-controlled time sweep (TS) test. It turned out that as RAP content increased, the modulus of FAM would be significantly raised, accompanied with a drop in the phase angle. The fatigue life of FAM would be greatly shortened when the RAP binder replacement rate reached 50%. Adding RA could considerably improve the dynamic properties of FAM mixes with high RAP content, resulting in a decrease in modulus, increase in phase angle and elongating fatigue life, but could not recover to the level of virgin binder.

Keywords: reclaimed asphalt pavement; fatigue; linear amplitude sweep; fine aggregate matrix; flexural bending; viscoelastic continuum damage; rejuvenating agent

---

1. Introduction

Reclaimed asphalt pavement (RAP) technology is an ideal and widely used sustainable technology in pavement engineering. RAP showed significant advantages against traditional hot mix asphalt (HMA) pavement, including non-renewable natural resources preservation, relief of landfill pressure, reducing energy consumption and greenhouse gas emissions [1]. As RAP content increases from 15% to 40% in HMA, pavement cost will be cut by $3.40 to $6.80 per ton [2]. Based on life cycle assessment (LCA), when RAP content rises from 30% to 50%, the energy consumption can be reduced by 16% to 25% [3]. Massive RAP is generated during pavement maintenance and rehabilitation all cross the world. According to the National Asphalt Pavement Association (NAPA), 76.2 million tons of RAP were reclaimed in the United States in 2017 [4]; in all European Union member countries, a total amount of 50 million tons of RAP were reclaimed in 2015 [5]; in China, this number is over 60 million tons in recent years [6]. Researchers are working on further raising RAP content, from 20–30% to 40–50%, even up to 100% [7], in order to achieve better sustainability in pavement construction. Studies have shown that HMA will become stiffer and more brittle with higher RAP content because aged binder exists in the mixture. Aged binder can improve the permanent deformation resistance, but also has a negative effect on the fatigue resistance of pavement [8]. To mitigate stiffness and improve fatigue cracking resistance, rejuvenating agent (RA) is commonly used in HMA with high RAP content [9].

In order to study the fatigue performance of asphalt pavement with RAP, several classical laboratory fatigue testing methods have been developed. Since the aged binder in RAP is considered as the main contributor to the fatigue life, the majority of researchers choose to study its fatigue properties through a method known as extraction-recovery [10]. However, this approach is not preferred by researchers for many reasons. The chemical reactions between bitumen and solvent can make the binder stiffer; studies have also proved that even a small amount of residual solvent can significantly alter the rheologic properties of the binder [11]. Meanwhile, the aged and virgin binders will be totally blended, which can camouflage the actual blending level in HMA [12]. Besides, asphalt solvent is also considered to be harmful to operators and may cause hazardous waste disposal issues.

The classical four-point beam test is another common approach to study the fatigue resistance of HMA. This test has been widely used by researchers and proved to be one of the most effective approaches in fatigue study. The only flaw is that four-point beam fatigue test is quite expensive and time-consuming. In addition, the repeatability of fatigue tests on full-graded HMA is not good due to the complexity and heterogeneity of materials [13]. A solution to solve the repeatability problem is to test the fine aggregated matrix (FAM) instead of full-graded HMA [14]. FAM is a mixture of binder, fillers and fine aggregates, which is supposed to have better homogeneity than coarse aggregate particles. Micro-cracking in HMA is usually considered to initiate and propagate in the FAM phase, so studying the fatigue resistance of FAM can help better understand the mechanism of fatigue cracking [15]. The linear amplitude sweep (LAS) test of cylindrical specimens was developed to study the fatigue performance of FAM mixes. A solid torsion bar fixture is used to fix the specimen in dynamic shear rheometer (DSR) [16]. Then, a strained-controlled oscillatory shear is applied to the FAM cylinder. Since the amplitude of strain is increasing linearly to accelerate internal damage, this method is called the linear amplitude sweep test. According to LAS test results, a mathematical model to predict the fatigue lives of FAMs can be established based on viscoelastic continuum damage (VECD) analysis, which can effectively distinguish the fatigue performance of different FAM mixes [17]. Compared with the traditional four-point beam fatigue test, the LAS test exhibits obvious advantages, such as time efficiency, repeatability and simplicity, which made it a prevalent test method for fatigue characterization [18]. However, complex viscoelasticity and anisotropic behaviors can be observed in asphalt materials, which means its mechanical behavior would be different under different loading modes. The LAS test with DSR applies torsional load on FAM specimens, while the fatigue of asphalt pavement is generally subjected to flexural bending caused by traffic load, which are two totally different loading modes. To address this issue, a LAS test of FAM mixes under flexural bending should be developed. In the field of polymer study, a solid dual-cantilever flexural loading fixture has been successfully applied to characterize the dynamic behavior of materials [19]. Since the viscoelasticity of FAM mixes is similar to polymer materials, the same fixture is employed to study the fatigue resistance of FAM with high RAP content.

The major objective of this study is to study the influence of RAP content on FAM mixes. A flexural bending LAS test was employed to study the fatigue performance. In detail, the following tasks need to be accomplished:

1. Assess the LAS test method for beam FAM specimens under flexural bending mode which serves as an alternative approach for the torsional LAS test with DSR.
2. Define a reasonable failure criterion of FAM mixes for LAS test under flexural bending mode.
3. Establish fatigue life prediction models based on VECD analysis according to LAS tests, and validate the models with measured data from time sweep (TS) tests.
4. Study the influence of high RAP content (over 25%) on the fatigue lives of FAMs and evaluate the effectiveness of rejuvenating agent.

2. Materials and Test Procedures

2.1. Materials

In this study, a typical virgin bituminous binder AH-70 with 60/80 penetration grade was selected, which has been widely used in heavy traffic freeways in China. Table 1 lists the technical index of the virgin binder. Limestone was used as the virgin aggregate. RAP passing through a 2.36 mm sieve was supplied by a local asphalt mixing plant in Beijing. A rejuvenating agent (RA) based on petroleum technology was used in this study. According to the recommendations of the producer, the RA content was 10% of the total weight of asphalt.

Table 1. Technical index of virgin asphalt binder.

Technical Index	Unit	Value
Penetration	0.1 mm (25 °C)	68
Softening point	°C	49
Ductility	cm (5 cm/min, 5 °C)	27.8
Viscosity	Pa•s (60 °C)	0.51
Flash point	°C	271
Wax content	%	1.1
Density	g/cm^3 (15 °C)	1.027

The maximum aggregate size of the FAM was 2.36 mm in this study. The target gradation and binder content of FAM mixes were determined by solvent extraction of the HMA fine portion (<2.36 mm), which was suggested by Yuan et al. [20]. This procedure aimed to ensure that the FAM mixes could represent the fine portion of HMA. The binder content of FAM and RAP were 9.0% and 7.3%, respectively. Figure 1 shows the gradations of HMA, FAM and RAP.

Four FAM mixes were tested as summarized in Table 2. In the first group, FAM mixes were batched with virgin binder, which served as the control group. In the second group, a 25% RAP binder replacement was applied, which is commonly used in maintenance and rehabilitation. In the third group, this rate was raised to 50% to examine the influence of RAP content at higher levels. In the last group, the RAP binder replacement rate was still 50%, which was consistent with the third group, but RA was added into the FAM to assess its effect. For all four FAM mixes, the gradation was kept the same with the target FAM gradation, no matter what the RAP and RA contents were. The virgin binder and aggregate contents were adjusted according to RAP binder replacement rates in order to match the target FAM gradation.

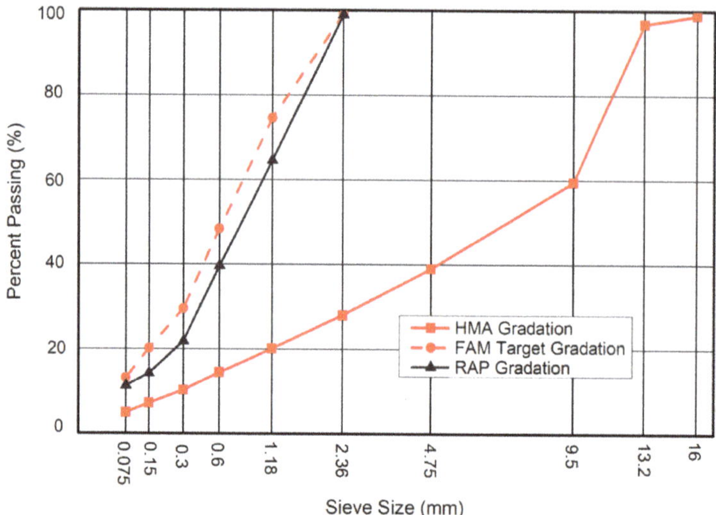

Figure 1. Gradations of hot mix asphalt (HMA), fine aggregate matrix (FAM) and reclaimed asphalt pavement (RAP).

Table 2. Summary of FAM mixes. RA = rejuvenating agent.

Mix	Target Binder Content (%)	Binder Replacement Rate (%)	Binder Replacement Content (%)	Virgin Binder Content (%)	RAP Content (%)	RA Content * (%)
0% RAP		0	0.0	9.0	0.0	-
25% RAP	9.0	25	2.2	6.8	29.7	-
50% RAP		50	4.5	4.5	61.6	-
50% RAP + RA		50	4.5	4.5	61.6	10

* By weight of target binder content.

2.2. FAM Specimen Preparation

FAM cylindrical specimens with 150 mm diameter and 50 mm height were fabricated using a Superpave Gyratory Compactor. The target air void content of cylindrical specimens was 6%. The FAM cylinders were cut by a Presi MECATOME precision cutting instrument (Figure 2a) to 60 mm × 45 mm × 15 mm rectangles, then sliced to 60 mm × 15 mm × 3.5 mm beam specimens. Since the passing percentage of the 2.36 mm sieve is 100% for FAM specimens, a thickness of 3.5 mm was considered enough to avoid the scale effect for most of the aggregates smaller than 1.18 mm. Figure 2b shows the compacted cylindrical specimen, rectangular specimen and beam specimens.

2.3. Test Setup and Procedures

A TA Instruments dynamic mechanical analyzer (DMA) Q800 apparatus (TA Instruments, New Castle, DE, USA) was used in this study along with a dual-cantilever flexural bending fixture. The FAM specimen was gripped by two clamps at its ends, then a movable head could apply load in the center of the beam, as illustrated in Figure 3. In order to examine the repeatability of the tests, three replicate tests were conducted for each group of FAM mixes.

Figure 2. FAM specimen preparation: (**a**) precision cutting machinery; (**b**) FAM specimens.

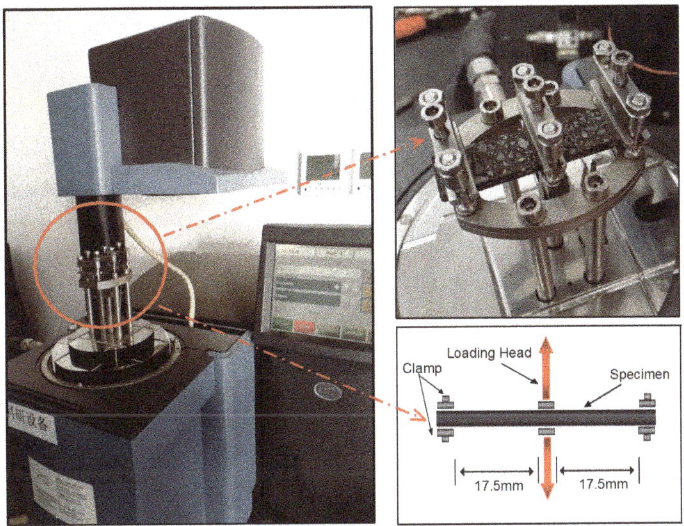

Figure 3. Dynamic mechanical analyzer (DMA) dual-cantilever bending fixture.

2.3.1. Frequency Sweep Test

To evaluate the undamaged dynamic properties of FAMs within the linear viscoelastic (LVE) region, a frequency sweep test was conducted to test the dynamic modulus and phase angle. The test temperature was 20 °C and loading frequencies ranged from 0.1 Hz to 25 Hz. A viscoelastic parameter m is indispensable to characterize damage property in VECD analysis, which is defined as the slope of the master curve (dynamic modulus versus loading frequency on log–log diagram) in the LVE region [21]. Recent studies have investigated the LVE region of similar FAM mixes and suggested that a strain level of 0.002% would be small enough to ensure that FAMs remained undamaged during the test [17,20]. Therefore, the constant amplitude strain of frequency sweep test was set to 0.002%.

2.3.2. Linear Amplitude Sweep Test

In the LAS test, the testing temperature was 20 °C and loading frequency was 10 Hz. The applied strain was linearly increased from 0.001% to 0.5% in 2000 s, as shown in Figure 4.

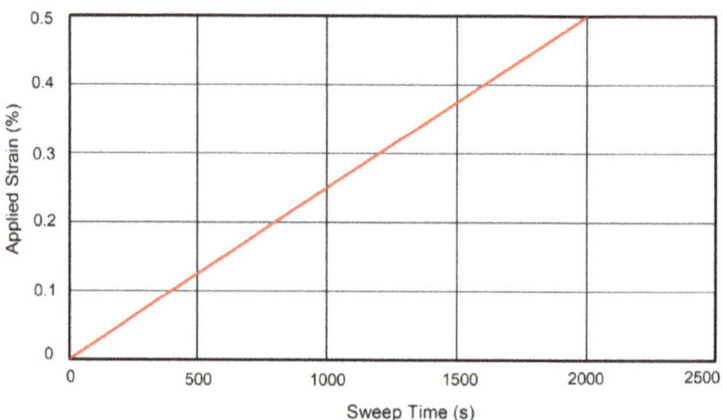

Figure 4. Strain applied in the linear amplitude sweep (LAS) test.

The LAS test can be treated as an approach to accelerate fatigue damage. Combined with VECD analysis, its results can be used to efficiently determine the regression parameters for the traditional fatigue equation, which is commonly used in strained-controlled fatigue tests. The relationship between fatigue life N_f and applied strain amplitude ε_p can be expressed as:

$$N_f = A\left(\varepsilon_p\right)^{-B} \tag{1}$$

According to Shapery's work potential theory [22], the damage intensity S of viscoelastic materials can be expressed in terms of the work performed W^R:

$$\frac{dS}{dt} = \left(-\frac{\partial W^R}{\partial S}\right)^{\alpha} \tag{2}$$

where: t is the time, s; α is equal to $1 + 1/m$ [23].

According to elastic–viscoelastic correspondence principle [24], the flexural pseudo strain ε^R can be defined as follows:

$$\varepsilon^R = \frac{1}{E_R}\int_0^t E(t-\xi)\frac{\partial \varepsilon}{\partial \xi}d\xi \tag{3}$$

where: E_R is a constant reference modulus, usually selected as 1; $E(t)$ is LVE relaxation modulus, MPa; ε is the flexural strain, percentage; ξ is an integral variable. The flexural LVE stress σ_{LVE} in MPa can be expressed as:

$$\sigma_{LVE} = \int_0^t E(t-\xi)\frac{\partial \varepsilon}{\partial \xi}d\xi \tag{4}$$

Combining Equations (3) and (4), the flexural LVE stress σ_{LVE} can be rewritten as:

$$\sigma_{LVE} = E_R\varepsilon^R \tag{5}$$

In order to quantify material integrity, pseudo stiffness $C(S)$ can be defined as a damage variable, as follows:

$$C(S) = \frac{\sigma}{\sigma_{LVE}} = \frac{\sigma}{\varepsilon^R} \tag{6}$$

where: σ is the flexural stress, MPa. For viscoelastic materials under periodical loading, the pseudo flexural strain amplitude ε_{Pi}^{R} in percentage and pseudo stiffness $C(S)$ in cycle i can be separately rewritten as:

$$\varepsilon_{Pi}^{R} = \frac{1}{E_R}\varepsilon_{Pi}\left|E_{LVE}^{*}\right| \tag{7}$$

$$C(S) = \frac{\sigma_{Pi}}{\varepsilon_{Pi}^{R}} \tag{8}$$

where: ε_{Pi} is the flexural strain amplitude in cycle i, percentage; $\left|E_{LVE}^{*}\right|$ is the flexural dynamic modulus in the LVE region, MPa; σ_{Pi} is the flexural stress amplitude in cycle i, MPa.

The work performed W^R in Equation (2) can be quantified using pseudo strain energy density [25], as follows:

$$W^R = \frac{1}{2}C(S)\left(\varepsilon_{Pi}^{R}\right)^{2} \tag{9}$$

Combining Equations (2), (7), (8) and (9) and integrating with a numerical approach, the damage intensity can be written as a function of time t:

$$S(t) \cong \sum_{i=1}^{n}\left[\frac{1}{2}(C_{i-1}-C_i)\left(\varepsilon_{Pi}^{R}\right)^2\right]^{\frac{\alpha}{1+\alpha}}(t_i - t_{i-1})^{\frac{1}{1+\alpha}} \tag{10}$$

where: n is the number of loading cycles. The curve of pseudo stiffness $C(S)$ versus damage intensity S typically follows a power model as follows [15]:

$$C(S) = 1 - C_1(S)^{C_2} \tag{11}$$

where: C_1, C_2 are regression coefficients.

Finally, by combining Equations (2), (7), (9) and (11) and integrating, the fatigue life prediction model can be described as [15]:

$$N_f = \frac{f\left(S_f\right)^{1+\alpha(1-C_2)}}{[1+\alpha(1-C_2)](0.5C_1C_2)^{\alpha}\left(\left|E_{LVE}^{*}\right|\right)^{2\alpha}}\left(\varepsilon_p\right)^{-2\alpha} \tag{12}$$

where: f is the loading frequency, Hz; S_f is the damage intensity S at failure point. The fatigue model coefficients A and B can be written as:

$$A = \frac{f\left(S_f\right)^{1+\alpha(1-C_2)}}{[1+\alpha(1-C_2)](0.5C_1C_2)^{\alpha}\left(\left|E_{LVE}^{*}\right|\right)^{2\alpha}} \tag{13}$$

$$B = 2\alpha \tag{14}$$

2.3.3. Time Sweep Test

To validate the fatigue life prediction model above, strain-controlled TS tests were employed. Four strain levels (0.07%, 0.08%, 0.09% and 0.1%) were selected with no rest period. The testing temperature and loading frequency of the TS test was consistent with the LAS test (20 °C, 10 Hz).

3. Results and Discussion

3.1. Frequency Sweep Test

The frequency sweep test results of four FAM mixes are illustrated in Figure 5. The dynamic modulus and phase angle exhibited significant relativities with loading frequency. When RA was not present in FAM mixes, the dynamic modulus would increase with the RAP content, especially at lower loading frequencies. At 0.1 Hz, when the RAP binder replacement rate was 25% and 50%, the dynamic modulus of FAM mixes was increased by approximately 1.5 and 2.5 times compared with virgin binder, respectively. However, once RA was added, FAM mixes were softened which resulted in a drastic drop of modulus. Phase angle displayed a reverse trend of dynamic modulus, which also implied the effect of RAP and RA. Furthermore, in the linear viscoelastic region, the dynamic modulus showed a linear relationship with loading frequency on a log–log scale as expected. It can be inferred from Figure 5a that the slope *m* was smaller for FAMs with a higher modulus.

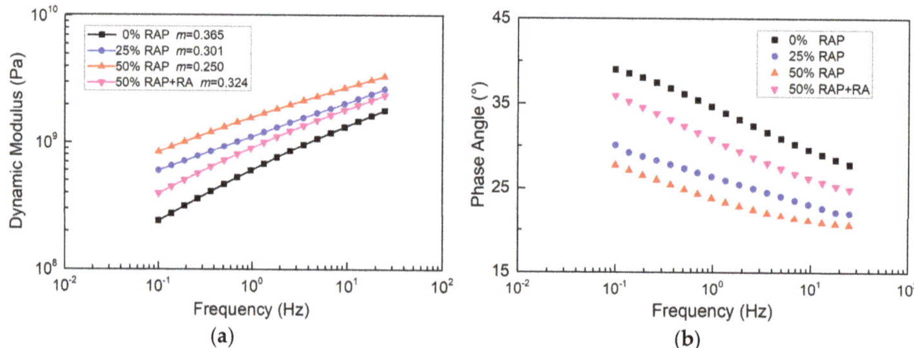

Figure 5. Frequency sweep results of FAM mixes: (**a**) Dynamic modulus; (**b**) Phase angle.

3.2. Failure Criterion Definition of LAS and TS Test

The fatigue process of materials starts with the presence of invisible internal microcracks, and then those microcracks gradually grow up and propagate to macro fractures under repeated loading. Failure criterion defines the critical point from stable to unstable damage growth stage. A reasonable criterion is a necessary part of fatigue life prediction models.

As an illustration, Figure 6 exhibits the replicate LAS test results of FAM mixes with virgin binder. Well-repeatable behaviors were observed for both stress–strain and phase angle–strain curves, with a mean absolute error (MAE) of 9.2% and 7.3%, respectively. The LAS test results of all four groups of FAM mixes are plotted in Figure 7. Stress and phase angle kept increasing with strain until a critical peak point was reached. The corresponding strains for peak stress and peak phase angle were close, but phase angle reached its peak slightly slower than stress. The maximum stress could be regarded as a yielding point, beyond which the specimen could not stand more loading. After the specimen yielded, phase angle started to drop, which means the fatigue damage reached a limit [26]. This phenomenon followed the yield-failure pattern of materials, and the final failure of FAM specimens was represented by the drop of phase angle. Thus, the peak phase angle could be regarded as the fatigue failure criterion in LAS test, which is also a typical failure indicator extensively used in the strain-controlled TS test [27]. It can be inferred from Figure 7 that FAM mixes with higher RAP content showed higher peak stress, lower peak phase angle and lower failure strain. Once RA was added into the FAM mix with high RAP content, a significant decrease of peak stress, increase of peak phase angle and higher failure strain could be clearly observed.

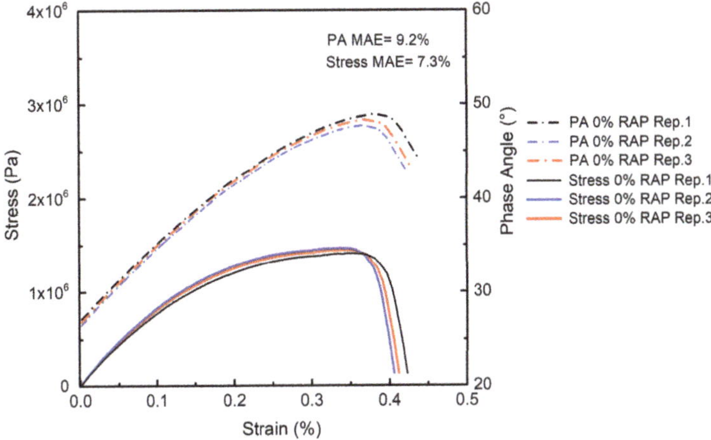

Figure 6. Typical replicate results of LAS test (0% RAP). MAE = mean absolute error.

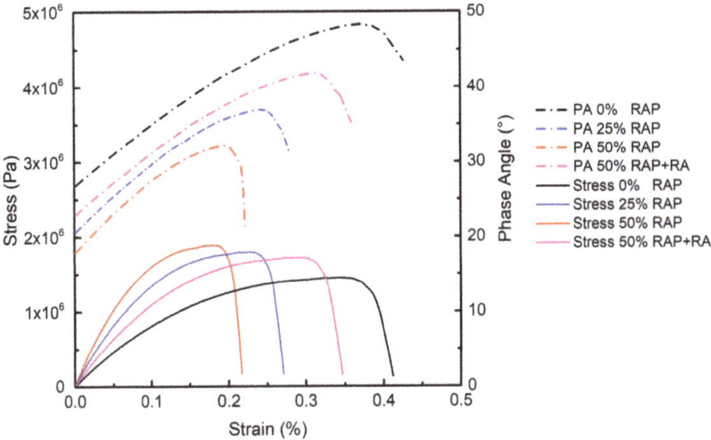

Figure 7. LAS test results of FAM mixes.

Another common phenomenological definition to assess fatigue failure is the maximum $C \times N$. C refers to the pseudo stiffness $C(S)$ in Equation (6), indicating material integrity. N is the number of loading cycle. Maximum $C \times N$ is considered as a reasonable failure indicator in both LAS and TS test [28]. The numbers of loading cycles at peak phase angle and at maximum $C \times N$ are shown in Figure 8. Obviously, the maximum $C \times N$ and the peak phase angle appeared simultaneously in both LAS tests and strain-controlled TS tests. According to American Association of State Highway and Transportation Officials (AASHTO) T321-17, maximum $S \times N$ (S refers to stiffness) is used to define fatigue failure for traditional four-point beam fatigue test of HMA, which is a similar parameter to maximum $C \times N$ [26]. Therefore, maximum $C \times N$ was selected as the failure criterion of FAM mixes in this study to keep consistency with AASHTO standards (T321-17).

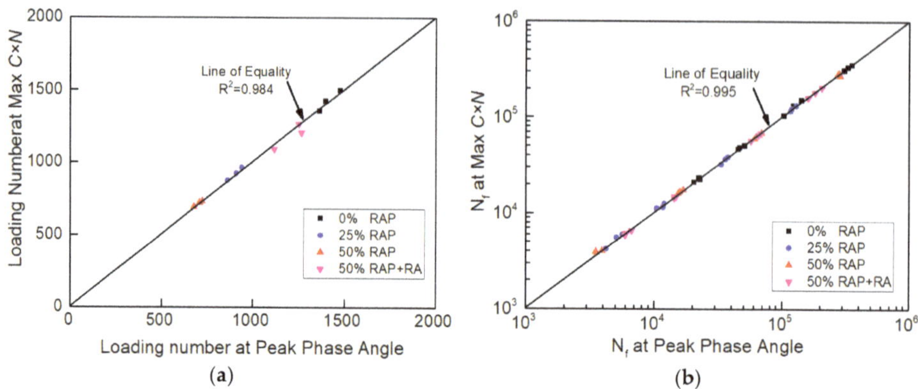

Figure 8. Comparison of failure criterion between peak phase angle and max $C \times N$: (**a**) LAS test; (**b**) time sweep (TS) test.

3.3. Fatigue Prediction Model Based on VECD Analysis

Figure 9 shows the damage characteristic curves ($C - S$) from LAS tests. As damage intensity S increased, material integrity C gradually reduced from 1 to 0.3–0.4, which was the defined failure point at maximum $C \times N$. A higher RAP binder replacement rate would lead to faster reduction of C compared with virgin binder. Also, the addition of RA in FAM mixes with high RAP content would substantially improve damage resistance but could not fully recover it to the level of virgin binder.

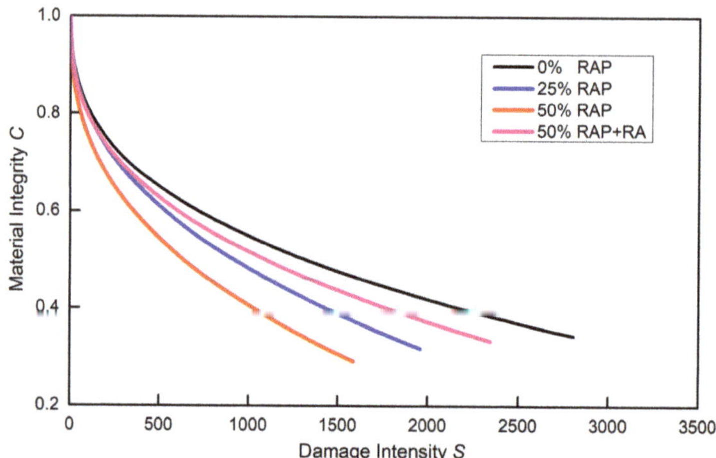

Figure 9. Damage characterization curves of FAM mixes from LAS tests.

According to the $C - S$ curve, the fatigue life prediction model could be established as Equation (12) based on VECD analysis. The regression parameters are listed in Table 3.

Table 3. Parameters of fatigue life prediction models. VECD = viscoelastic continuum damage.

Mix	VECD-Based Fatigue Model Parameters							
	$\lvert E^{*}_{LVE}\rvert$ (MPa)	m	α	S_f	C_1	C_2	A	B
0% RAP	1361	0.365	3.743	2735	3.95×10^{-2}	0.358	5.66×10^{-4}	7.487
25% RAP	2060	0.301	4.322	1940	1.51×10^{-2}	0.365	1.42×10^{-5}	8.647
50% RAP	2705	0.250	5.008	1550	6.11×10^{-2}	0.413	1.30×10^{-7}	10.016
50% RAP+RA	1812	0.324	4.085	2225	3.37×10^{-2}	0.318	5.64×10^{-5}	8.169

Note: $\lvert E^{*}_{LVE}\rvert$ is the flexural dynamic modulus in the linear viscoelastic region; m is the slope of dynamic modulus versus loading frequency. α equals to $1/(1+m)$. S_f is damage intensity at failure point. C_1 and C_2 are regression coefficients. A and B are fatigue model parameters described in Equations (13) and (14).

The predicted fatigue lives for FAM mixes at four different strain levels are displayed in Figure 10. As expected, a higher RAP content would negatively influence the fatigue performance, resulting in a much shorter fatigue life. Compared with virgin binder, the fatigue lives of FAMs with 50% RAP binder replacement rate were reduced by 80.8%, 86.3%, 89.8% and 92.2% at strain levels of 0.07%, 0.08%, 0.09% and 0.10%, respectively. However, when RA was added into the FAM mixes, their fatigue lives were greatly extended by 3.2, 4.1, 5.1 and 6.2 times, which were recovered to 61.1%, 55.8%, 51.5% and 47.9% of virgin binder FAM mixes, respectively. Obviously, the existence of RA in FAM mixes with high RAP contents would significantly mitigate stiffness and improve cracking resistance, especially at higher strain levels.

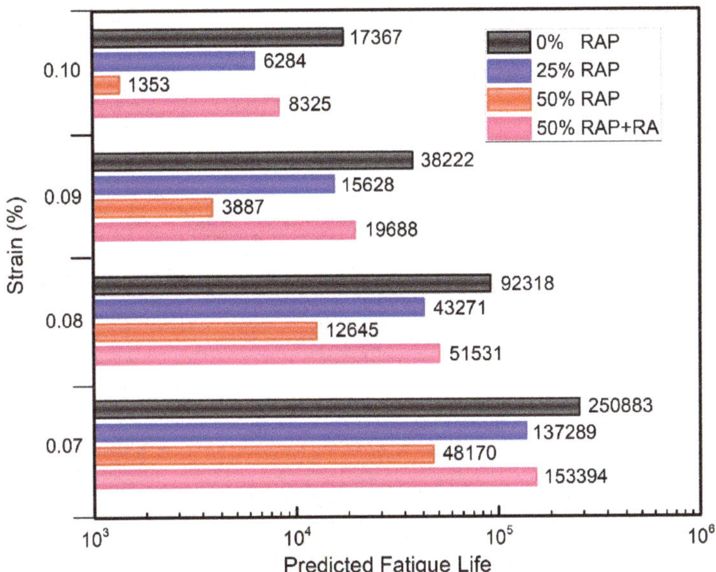

Figure 10. Predicted fatigue lives for FAM mixes at four strain amplitudes.

3.4. Validation of Fatigue Prediction Model from TS Test

The results of strain-controlled TS tests are listed in Table 4. The coefficients of variation ranged from 2.3% to 15.9%, which was smaller compared with fatigue tests of full-graded HMA. The TS test results suggested that the fatigue performance ranking of four groups of FAM mixes was virgin binder, 50% RAP + RA, 25% RAP and 50% RAP from best to worst, which was consistent with model predictions.

Table 4. Results of time sweep fatigue test.

Mix	Strain level (%)	Fatigue Life N_f	Standard Deviation	Coefficient of Variation (%)	Fatigue Performance Ranking	
					Measured	Predicted
0% RAP	0.10	22,432	1664	7.4	1	1
	0.09	46,310	3801	8.2		
	0.08	122,571	19,473	15.9		
	0.07	329,224	21,940	6.7		
25% RAP	0.10	5062	740	14.6	3	3
	0.09	11,932	1626	13.6		
	0.08	35,858	2142	6.0		
	0.07	116,548	14,384	12.3		
50% RAP	0.10	3524	440	12.5	4	4
	0.09	15,425	1630	10.6		
	0.08	65,568	4608	7.0		
	0.07	282,052	6405	2.3		
50% RAP + RA	0.10	6049	575	9.5	2	2
	0.09	14,782	736	5.0		
	0.08	64,868	7005	10.8		
	0.07	182,665	22,795	12.5		

The predicted fatigue lives are plotted along with the measured results in Figure 11. Fairly good consistency with the line of equality could be observed with a correlation coefficient R^2 of 0.975 and MAE of 17.6%. The fatigue life prediction models based on LAS test results are considered reasonable.

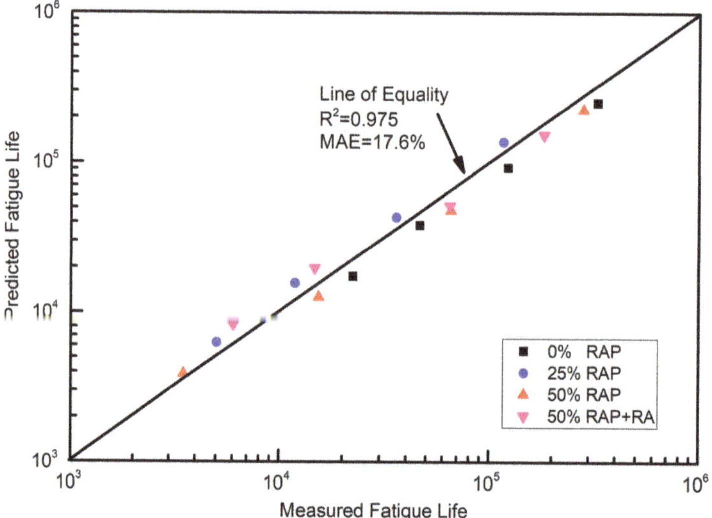

Figure 11. Measured fatigue lives versus predicted fatigue lives. R^2 = correlation coefficient.

Table 5 shows the estimated efficiency of the LAS test and strain-controlled TS test. In general, to acquire the curve of fatigue life for a given FAM mix, the TS test requires 12 individual tests and will take approximate 27 h, while the LAS test only needs three tests and 3 h. Based on the estimates, the LAS test is considered more efficient than the TS test.

Table 5. Efficiency comparison between TS and LAS tests.

Test Method	Number of FAM Specimens Required for Each Mix	Average Total Testing Time for Each FAM Mix (h)
Time sweep	12	27
Linear amplitude sweep	3	3

4. Conclusions

In this study, a flexural bending test method using DMA for FAM mixes with high RAP content was proposed. Four groups of FAM mixes were tested with this method to investigate the impact of RAP content on the fatigue properties and effectiveness of RA. The following conclusions are made:

1. As an alternative test method for torsion bar test with a DSR, the LAS test of FAM mixes under flexural bending mode can provide acceptable data with good repeatability.
2. The phase angle peak and the maximum appeared simultaneously in both LAS tests and strain-controlled TS tests. In this study, the maximum was selected as a reasonable parameter for defining fatigue failure criterions.
3. Based on the maximum failure criterion and VECD analysis, fatigue life prediction models can effectively capture the fatigue resistance of different FAMs. The predicted fatigue lives were well-consistent with the measured results of TS tests.
4. Higher RAP content will considerably increase the stiffness of FAM mixes, resulting in a decrease in phase angle and fatigue resistance. The presence of petroleum-based rejuvenating agents will soften FAMs, resulting in a significant recovery of the lost fatigue resistance.

To conclude, the LAS test under flexural bending mode is considered as a novel method to test the dynamic properties and fatigue behaviors of FAM mixes. Its effectiveness and reliability were demonstrated by multiple tests. The fatigue resistance of FAM mixes could also be greatly influenced by other factors, such as binder grades, asphalt film thickness, RAP sources, aggregate gradation superposition and RA type, which should be taken into consideration in further studies.

Author Contributions: Conceptualization, C.Z., X.W., and Z.Q.; Data curation, C.Z.; Formal analysis, C.Z., Q.R. and Z.Q.; Funding acquisition, X.W. and Z.Q.; Investigation, C.Z., Q.R., and Z.Q.; Methodology, C.Z. and Q.R.; Writing—original draft, C.Z., Q.R. and Z.Q.

Funding: This research was supported by National Key R&D Program of China (2017YFC0840200).

Acknowledgments: The authors would like to appreciate the support of RIOHTrack Research Center of the Research Institute of Highway Ministry of Transport (RIOH).

Conflicts of Interest: The authors declare no conflict of interest.

References

1. Copeland, A. *Reclaimed Asphalt Pavement in Asphalt Mixtures: State of the Practice*; FHWA-HRT-11-021; Turner-Fairbank Highway Research Center: McLean, VA, USA, 2011; pp. 1–5.
2. McDaniel, R.S.; Shah, A.; Huber, G.A.; Copeland, A. Effects of reclaimed asphalt pavement content and virgin binder grade on properties of plant produced mixtures. *Road Mater. Pavement* **2012**, *13*, 161–182. [CrossRef]
3. Yang, R.; Kang, S.; Ozer, H.; Al-Qadi, I.L. Environmental and economic analyses of recycled asphalt concrete mixtures based on material production and potential performance. *Resour. Conserv. Recycl.* **2015**, *104*, 141–151. [CrossRef]
4. Williams, B.A.; Copeland, A.; Ross, T.C. *Asphalt Pavement Industry Survey on Recycled Materials and Warm-Mix Asphalt Usage: 2017*, 8th ed.; National Asphalt Pavement Association (NAPA): Lanham, MD, USA, 2018; pp. 1–11.
5. EAPA. *EAPA's Position Statement on the Use of Secondary Materials, by-Products and Waste in Asphalt Mixtures*; European Asphalt Pavement Association: Brussels, Belgium, 2017; pp. 4–9.

6. Lin, J.; Hong, J.; Xiao, Y. Dynamic characteristics of 100% cold recycled asphalt mixture using asphalt emulsion and cement. *J. Clean. Prod.* **2017**, *156*, 337–344. [CrossRef]

7. Zaumanis, M.; Mallick, R.B. Review of very high-content reclaimed asphalt use in plant produced-pavements: State of the art. *Int. J. Pavement Eng.* **2015**, *16*, 39–55. [CrossRef]

8. Tang, S.; Williams, R.C.; Cascione, A.A. Reconsideration of the fatigue tests for asphalt mixtures and binders containing high percentage RAP. *Int. J. Pavement Eng.* **2017**, *18*, 443–449. [CrossRef]

9. Moghaddam, T.B.; Baaj, H. The use of rejuvenating agents in production of recycled hot mix asphalt: A systematic review. *Constr. Build. Mater.* **2016**, *114*, 805–816. [CrossRef]

10. You, Z.; Mills-Beale, J.; Fini, E.; Goh, S.W.; Colbert, B. Evaluation of low-temperature binder properties of warm-mix asphalt, extracted and recovered RAP and RAS, and Bioasphalt. *J. Mater. Civil Eng.* **2011**, *23*, 1569–1574. [CrossRef]

11. Ma, T.; Mahmoud, E.; Bahia, U.H. Development of testing procedure for the estimation of rap binder low temperature properties without extraction. *Transport. Res. Rec.* **2010**, *2179*, 58–65. [CrossRef]

12. Peterson, R.; Soleymani, H.; Anderson, R.; McDaniel, R. Recovery and testing of RAP binders from recycled asphalt pavements. *J. Assoc. Asph. Paving Technol.* **2000**, *69*, 72–91.

13. Kim, Y.; Lee, H.J.; Little, D.N.; Kim, Y.R. A simple testing method to evaluate fatigue fracture and damage performance of asphalt mixtures. *J. Assoc. Asph. Paving Technol.* **2006**, *75*, 755–788.

14. Suresha, S.N.; Ningappa, A. Recent trends and laboratory performance studies on FAM mixtures: A state of the art review. *Constr. Build. Mater.* **2018**, *174*, 496–506. [CrossRef]

15. Kim, Y.R.; Little, D.N.; Lytton, R.L. Use of dynamic mechanical analysis (DMA) to evaluate the fatigue and healing potential of asphalt binders in sand asphalt mixtures. *J. Assoc. Asph. Paving Technol.* **2002**, *71*, 176–206.

16. Lee, H.; Kim, Y.R.; Lee, S. Prediction of asphalt mix fatigue life with viscoelastic material properties. *Transport. Res. Rec.* **2003**, *1832*, 139–147. [CrossRef]

17. Zhu, J.; Alavi, M.Z.; Harvey, J.; Sun, L.; He, Y. Evaluating fatigue performance of fine aggregate matrix of asphalt mix containing recycled asphalt shingles. *Constr. Build. Mater.* **2017**, *139*, 203–211. [CrossRef]

18. Gudipudi, P.P.; Underwood, B.S. Use of fine aggregate matrix experimental data in improving reliability of fatigue life prediction of asphalt concrete: Sensitivity of this approach to variation in input parameters. *Transport. Res. Rec.* **2017**, *2631*, 65–73. [CrossRef]

19. Shanmugasundaram, N.; Rajendran, I.; Ramkumar, T. Static, dynamic mechanical and thermal properties of untreated and alkali treated mulberry fiber reinforced polyester composites. *Polym. Compos.* **2018**, *39*, 1908–1919. [CrossRef]

20. He, Y.; Alavi, M.Z.; Jones, D.; Harvey, J. Proposing a solvent-free approach to evaluate the properties of blended binders in asphalt mixes containing high quantities of reclaimed asphalt pavement and recycled asphalt shingles. *Constr. Build. Mater.* **2016**, *114*, 172–180. [CrossRef]

21. Kutay, M.E.; Gibson, N.; Youtcheff, J. Conventional and viscoelastic continuum damage (VECD)-based fatigue analysis of polymer modified asphalt pavements. *J. Assoc. Asph. Paving Technol.* **2008**, *77*, 395–434.

22. Schapery, R.A. Correspondence principles and a generalized J integral for large deformation and fracture analysis of viscoelastic media. *Int. J. Fract.* **1984**, *25*, 195–223. [CrossRef]

23. Masad, E.; Castelo Branco, V.T.F.; Little, D.N.; Lytton, R. A unified method for the analysis of controlled-strain and controlled-stress fatigue testing. *Int. J. Pavement Eng.* **2008**, *9*, 233–246. [CrossRef]

24. Kim, Y.R.; Lee, Y.C.; Lee, H.J. Correspondence principle for characterization of asphalt concrete. *J. Mater. Civ. Eng.* **1995**, *7*, 59–68. [CrossRef]

25. Park, S.W.; Richard Kim, Y.; Schapery, R.A. A viscoelastic continuum damage model and its application to uniaxial behavior of asphalt concrete. *Mech. Mater.* **1996**, *24*, 241–255. [CrossRef]

26. Wang, C.; Zhang, J.X. Characterization on fatigue failure definition and failure criterion of asphalt binder. *J. B. Univ. Technol.* **2015**, *41*, 1574–1583.

27. Zhang, J.; Sabouri, M.; Guddati, M.N.; Kim, Y.R. Development of a failure criterion for asphalt mixtures under fatigue loading. *J. Assoc. Asph. Paving Technol.* **2013**, *82*, 1–22. [CrossRef]
28. Safaei, F.; Lee, J.; Nascimento, L.A.H.D.; Hintz, C.; Kim, Y.R. Implications of warm-mix asphalt on long-term oxidative ageing and fatigue performance of asphalt binders and mixtures. *Road Mater. Pavement* **2014**, *15*, 45–61. [CrossRef]

Article

Synthesis and Effect of Encapsulating Rejuvenator Fiber on the Performance of Asphalt Mixture

Benan Shu [1], Shiwen Bao [1], Shaopeng Wu [1,*], Lijie Dong [2], Chao Li [1], Xu Yang [3], José Norambuena-Contreras [4], Quantao Liu [1,*] and Qing Wang [2]

[1] State Key Laboratory of Silicate Materials for Architectures, Wuhan University of Technology, Wuhan 430070, China; shuba@whut.edu.cn (B.S.); baosw@whut.edu.cn (S.B.); lic@whut.edu.cn (C.L.)
[2] State Key Laboratory of Advanced Technology for Materials Synthesis and Processing, and School of Materials Science and Engineering, Wuhan University of Technology, Wuhan 430070, China; lijie@whut.edu.cn (L.D.); 254546@whut.edu.cn (Q.W.)
[3] Department of Civil Engineering, Monash University, Clayton, VIC 3800, Australia; xu.yang@monash.edu
[4] LabMAT, Department of Civil and Environmental Engineering, University of Bío-Bío, Concepción 4030000, Chile; jnorambuena@ubiobio.cl
* Correspondence: wusp@whut.edu.cn (S.W.); liuqt@whut.edu.cn (Q.L.)

Received: 23 March 2019; Accepted: 16 April 2019; Published: 17 April 2019

Abstract: The idea of prolonging the service life of asphalt mixture by improving the self-healing ability of asphalt has received extensive attention in recent years. In view of this, this work synthesized three kinds of encapsulating rejuvenator fibers to improve self-healing properties of asphalt mixtures. A series of characterizations were performed to study the morphology, chemical structure and thermal stability of the three kinds of fibers. Subsequently, the road performance of asphalt mixture containing the fiber were investigated, which included high and low temperature, water sensitivity and fatigue performances. Finally, the self-healing performance of asphalt mixture containing the fiber was investigated by 3PB test. The results revealed that the three kinds of encapsulating rejuvenator fibers were successfully synthesized. The fibers had excellent thermal stability, which met temperature requirements in the mixing and compaction process of asphalt mixtures. Road performance of asphalt mixture containing the fiber met the requirements. Self-healing ability of asphalt mixture containing the fiber was improved. Synergistic action of temperature and rejuvenator could further significantly improve the self-healing ability of the asphalt mixture.

Keywords: asphalt mixture; encapsulated rejuvenator; road performance; self-healing

---

1. Introduction

Asphalt mixture, which is one of the most widely used pavement materials worldwide, contains bitumen, filler, fine aggregate, and coarse aggregate. After several years of service, cracking, as one of the most common diseases, will be generated in the interior of asphalt pavements due to the hardening and brittleness of aged asphalt, vehicle loading and so on [1–4]. Although asphalt binder has inherent self-healing ability, the ability was limited under the action of continuous traffic loading, moisture ingress and other factors. Thus, continuous development of cracks will lead to bitumen pavement failure [5–9]. It is urgent to improve the self-healing of asphalt to extend the service life and reduce maintenance cost of asphalt pavement.

At present, there are several ways to improve the self-healing capability of asphalt, which include nanoparticles [10,11], electromagnetic induction [12,13] and microwave heating [14,15]. The application of these technologies could effectively improve the self-healing of asphalt. In recent years, capsule method have attracted wide attention because of its crack response ability. Cracks can induce capsules ruptured, which we called it "crack response ability". Under the action of capillary, the healing agent that was encapsulated in capsules thus fills up cracks and heal the cracks [16].

The mechanism by which capsules improve the self-healing ability of asphalt is that: when the cracks inside an asphalt mixture develop to the surface of capsules, capsules were ruptured under the action of stress concentration. Because of capillary siphon, the encapsulated rejuvenator flows out and fills up the micro-cracks. Rejuvenator can supplement the light components that was lost in the service of asphalt binder [17,18]. In addition, the infiltration and diffusion of rejuvenator can significantly reduce the viscosity of asphalt binder around the cracks, and then the flow ability of asphalt is improved [19]. Therefore, the addition of encapsulating rejuvenator capsules could improve the self-healing properties of asphalt mixture.

A lot of studies have been done to synthesize various types of encapsulations containing rejuvenator to improve self-healing capability of asphalt [20–22]. For example, Sun et al. [23,24] synthesized melamine-urea-formaldehyde microcapsules by an in situ polymerization method. The optimal capsules were obtained by adjusting the parameters, which include the ratio of core to shell materials, reaction temperature, type and content of emulsifier. The optimal capsule met the temperature requirement in mixing and compaction process of asphalt mixture. In addition, the four-point bending fatigue test of asphalt mixture revealed that the fatigue life of asphalt mixture containing 3% microcapsules was double than that of asphalt mixture without microcapsules. Li et al. [25] successfully synthesized urea–formaldehyde microcapsules encapsulating asphalt rejuvenator. The parameters that affect structure and size of microcapsules were considered, for example, the stirring speed and reaction time. In addition, the healing efficiency of asphalt with the content of 0.3% microcapsules increased by 38.67%. Zhang et al. [26] synthesized urea formaldehyde self-healing microcapsules, and then the rheological properties of asphalt binder with the microcapsules was deeply studied.

Norambuena-Contreras [27] fabricated multinuclear calcium alginate capsules encapsulating sunflower oil by ionotropic gelation. The oil content in capsules could be easily controlled. Furthermore, the factors that can affect the self-healing efficiency of asphalt mixture were investigated, for example the aging effect, addition order of capsules and temperature. The result showed that the self-healing ability of asphalt mixtures with 0.5% capsules is several times higher than that of asphalt mixtures without capsules. The author's group synthesized calcium alginate microcapsules and compartmented calcium alginate/silica fiber encapsulating rejuvenator by using a microfluidic device [28,29]. The encapsulations had excellent thermal and mechanical properties. Furthermore, experimental results confirmed from the view of the macroscopic and microscopic scales that the addition of encapsulations could improve the self-healing ability of asphalt.

Up to now, most studies have focused on the synthesis of encapsulations and the self-healing performance of asphalt mixtures containing encapsulations, and few studies concerned the road performance of asphalt mixtures containing encapsulations. Before studying the self-healing performances, the road performance of asphalt mixture with encapsulations should meet the technology requirements. For example, the high and low temperature performances, water sensitivity and fatigue ability. In view of this, the three kinds of fibers encapsulating asphalt rejuvenator were synthesized in this study. Road performances of asphalt mixture with different kinds of fiber were studied. Subsequently, three-point bending test was performed to evaluate the self-healing capability of asphalt mixture.

2. Materials and Test Methods

2.1. Gradation Design of Asphalt Mixture

AC-13 asphalt mixture was designed in this work. Aggregate gradation is shown in Table 1. 70# bitumen was used in this work, and its penetration (20 °C, 0.1 mm) and softening point were 68.7 and 48.5 °C respectively. Its ductility (15 °C) was larger than 100 cm. Basalt was used as aggregate. The optimum bitumen content and the content of the fiber were 4.7% and 0.235%, respectively, based on Marshall design method. air void (VV), voids in mineral aggregate (VMA) and voids filled with asphalt (VFA) were 3.9%, 14.5% and 73.1%, respectively.

Table 1. Aggregate gradation of AC-13 asphalt mixture.

Sieve Size/mm	Designed Gradation/%
16	100
13.2	96.2
9.5	75.2
4.75	47.4
2.36	30.8
1.18	23.9
0.6	16.6
0.3	12.3
0.15	9.1
0.075	6.9

2.2. Synthesis of the Three Kinds of Fibers

Alginate, dehydrated calcium chloride, silica (SiO_2) nanoparticles, graphene oxide (GO) and asphalt rejuvenator were purchased from SINOPHARM GROUP CO. LTD. Asphalt rejuvenator was mainly composed of 67.4% aromatics and 21.07% saturates.

Alginate solution (2 weight% alginate) was prepared to synthesize calcium alginate fiber encapsulating asphalt rejuvenator. Alginate/SiO_2 (alginate: SiO_2 = 1:1) solution was prepared to synthesize calcium alginate/SiO_2 composite fiber encapsulating rejuvenator. Alginate/GO (alginate: GO = 1:10) solution was prepared to synthesize calcium alginate/GO composite fiber encapsulating rejuvenator. The purpose of synthesizing the three kinds of fibers is that: the synthesis of calcium alginate/SiO_2 composite fiber is going to improve thermal and mechanical properties, and decrease the leakage of calcium alginate fiber; the synthesis of calcium alginate/GO composite fiber is going to combine the advantages of the induction heating method and the capsule method. The fiber can be heated by microwave to increase the temperature of asphalt, so as to achieve the purpose of quickly healing cracks inside asphalt. Further, the asphalt can be rejuvenated by the encapsulated rejuvenator. Therefore, calcium alginate/GO composite fiber can make the cracks inside the asphalt double healed.

The three kinds of fibers were synthesized by using a self-assembled microfluidic device. The synthetic process was shown in a previous study [23].

2.3. Characterization of the Three Kinds of Fibers

Morphology of the three kinds of fibers was characterized by scanning electron microscope (SEM) test. SEM test was conducted on an S4800 machine. Before SEM test, the specimens were sprayed with gold for 50 s.

Chemical structure of the three kinds of fibers was evaluated by Raman spectra test. An InVia instrument with a wavelength ranging from 200 to 2000 cm^{-1} was used to complete Raman spectra test.

Thermal properties of the three kinds of fibers were tested by thermogravimetric analyzer (TGA) experiment. The testing temperature raised from 30 to 600 °C with a heating rate of 10 °C/min.

2.4. Road Performances of Asphalt Mixture with the Fibers

Rutting test, water stability test, freeze-thaw splitting test and low temperature bending test of asphalt mixture were carried out according to the specification "Standard Test Methods of Bitumen and Bituminous Mixtures for Highway Engineering JTG E20-2011".

Rutting test was conducted on a YLDCZ-6S machine with the testing temperature of 60 °C. The wheel speed was 21 times/min, and loading was 0.7 MPa. The length, width and height of specimen were 300 mm, 300 mm and 50 mm, respectively. Before the testing, the sample was kept 5 h at 60 °C. The dynamic stability (DS) was defined as follows:

$$DS = 15 \times N / (d_{60} - d_{45})$$

(1)

where $N = 42$ cycles/min, which means the wheel tracking speed; d_{60} and d_{45} are the rutting depth at the wheel tracking time of 45 min and 60 min, respectively.

The low temperature crack resistance of asphalt mixture was evaluated by flexural strength, flexural strain and flexural modulus. The length, width and height of specimen were 250 mm, 30 mm and 35 mm, respectively. The test temperature was −10 °C and the loading rate was 50 mm/min.

The water stability of asphalt mixture with the fibers was evaluated by freeze-thaw split strength ratio (TSR). For that, the specimen was firstly freezing at −18 °C for 16h, and then suffered a water bath for 24 h at 60 °C.

$$TSR = \frac{TS_2}{TS_1} \times 100\%$$

(2)

where TS_2 was splitting tensile strength of the specimen after freeze-thaw experiment; TS_1 was splitting tensile strength of the specimen before freeze-thaw experiment.

Four-point bending fatigue test was carried out to study the fatigue performance of asphalt mixture containing the fiber. Beams with length, width and height of 400 mm, 50 mm and 50 mm respectively were used. The test temperature and working frequency were 20 °C and 30 Hz respectively.

2.5. Self-Healing Capability of Asphalt Mixture with the Fibers

Three-point bending test on the asphalt mixture beams with length, width and height of 95 mm, 40 mm and 50 mm respectively, was performed by using a universal testing machine (UTM-25). Before the test, the specimens were kept at −10 °C for 4 hours. Then the specimens were tested at −10 °C with a loading rate of 0.5 mm/min. After that, the two parts of a tested specimen were rejoined by three rubber bands. Healing temperature and healing time were 30 °C and 3 days respectively. A microwave oven with the power of 800 W was used to complete microwave heating test. The self-healing ability of each asphalt mixture beam was quantified by the Healing Index (HI) according to Equation (3).

$$HI = \left(\frac{F_r}{F_i}\right) \times 100$$

(3)

where, F_r is recovery of peak strength of an asphalt beam after healing rest, kN; F_i is the initial peak strength of the same asphalt beam, kN.

Types symbology of fibers and asphalt mixtures is shown in Table 2. A1 was the asphalt mixture without fiber. Three sets of replicate samples were tested in each experiment.

Table 2. Types symbology of fibers and asphalt mixtures.

Types of Fiber	Symbology	Types of Asphalt Mixture	Symbology
Ca–alginate fiber	AF	Asphalt mixture containing Ca–alginate fiber	A2
Ca–alginate/SiO$_2$ fiber	ASF	Asphalt mixture containing Ca–alginate/SiO$_2$ fiber	A3
Ca–alginate/GO fiber	MRF	Asphalt mixture containing Ca–alginate/GO fiber	A4

3. Results and Discussion

3.1. Characterization of Three Kinds of Self-Healing Fibers

3.1.1. Morphology

Figure 1 shows the synthesis process and synthesis mechanism of the three kinds of self-healing fibers. Alginate is composed of a series of G units and M units. Sodium ions on G units can be rapidly replaced by calcium ions to form calcium alginate gels. The whole reaction process is completed in an instant, so, nano SiO_2 particles and graphene oxide particles can be completely incorporated into calcium alginate gel. The three kind of self-healing fibers, which include Ca–alginate fiber, Ca–alginate/SiO_2 composite fiber and Ca–alginate/GO composite fiber were thus fabricated.

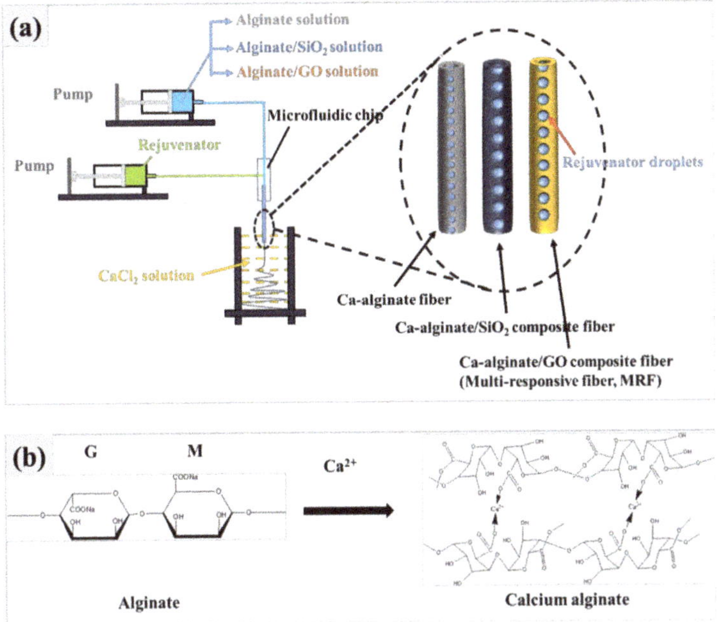

Figure 1. Synthesis process (**a**) and synthesis mechanism (**b**) of the three kinds of fibers.

The morphology of three kinds of self-healing fibers is shown in Figure 2. It can be seen from Figure 2a–c that AF and ASF had a light-yellow optical appearance, while MRF had a black appearance. From Figure 2(a1–c1), it can be observed that the wall of three kinds of fibers was intact and there were no visible holes and cracks, which indicated that rejuvenator could be completely encapsulated inside the fibers, and without leakage. The diameter of the fibers was about 500 μm. Figure 2(a2–c2) reveal that the surface of AF was the smoothest. ASF fibers had an uneven surface with many small bulges, which was attributed to the incorporation of partially agglomerated SiO_2 particles. A large number of regular folds appeared on the surface of MRF, which were peculiar to the specific structure of graphene. Over all, the three kinds of self-healing fibers encapsulating rejuvenator were successfully synthesized from the analysis of morphology.

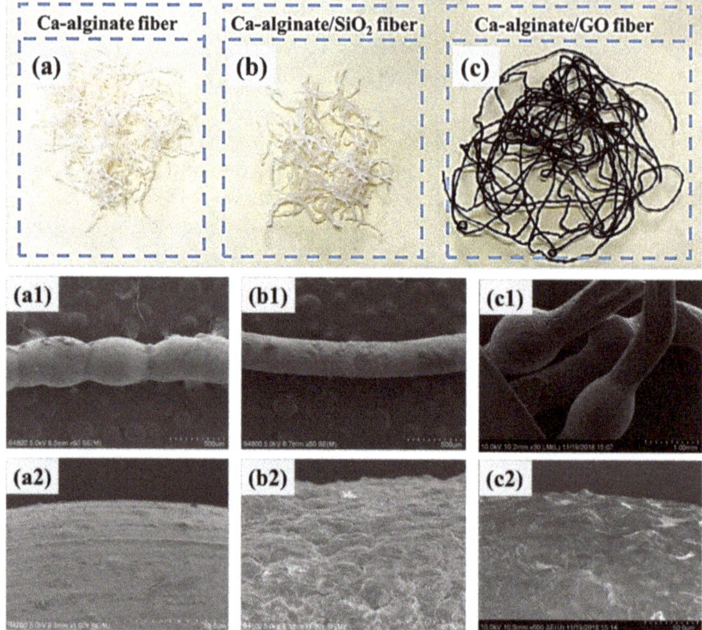

Figure 2. Morphology of the three kinds of self-healing fiber: Optical morphology of Ca-alginate fiber (**a**), Ca-alginate/SiO$_2$ fiber (**b**) and Ca-alginate/GO fiber (**c**); Micromorphology of Ca-alginate fiber (**a1,a2**), Ca-alginate/SiO$_2$ fiber (**b1,b2**) and Ca-alginate/GO fiber (**c1,c2**).

3.1.2. Chemical Structure

Raman test was conducted to confirm the successful synthesis of the three kinds of self-healing fibers. Figure 3a revealed that GO had two characteristic peaks, namely peak D and peak G. The two peaks didn't appear in the Raman spectra of calcium alginate fiber, while appeared in Ca–alginate/GO fiber Raman spectra. A characteristic peak at 1070 cm^{-1} appeared in both Ca–alginate fiber and Ca–alginate/GO fiber. The three kinds of characteristic peaks appeared in the Raman spectra of Ca–alginate/GO fiber, which indicated that Ca–alginate/GO fiber was successfully synthesized. Similarly, SiO$_2$ had a series of characteristic peaks in the Raman shift from 200 cm^{-1} to 800 cm^{-1} in Figure 3b. These peaks also appeared at the same wavelengths in Raman spectrum of Ca–alginate/SiO$_2$ fiber, which revealed that SiO$_2$ was successfully incorporated inside the Ca–alginate structure, and Ca–alginate/SiO$_2$ composite fiber was successfully synthesized.

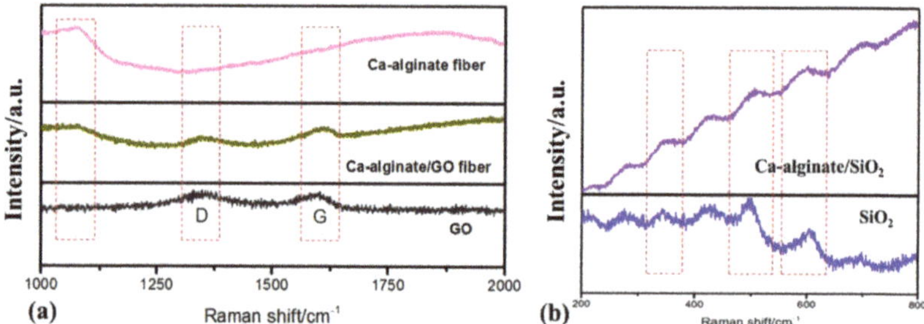

Figure 3. Raman test of the three kinds of fibers: (**a**) Ca–alginate fiber and Ca–alginate/GO fiber; (**b**) Ca–alginate/SiO$_2$ fiber.

3.1.3. Thermal Stability

The thermal stability of three kinds of self-healing fibers was studied by TGA test and the results is shown in Figure 4. It can be seen that the initial decomposition temperature of rejuvenator was about 250 °C, which means that the rejuvenator meet the temperature requirement in the mixing and compaction process of asphalt mixture. Between 100 °C and 180 °C, the loss mass of Ca–alginate self-healing fiber was attributed to degradation of crystalline water and other components of calcium alginate wall. The decomposition mass and decomposition rate of ASF and MRF were both significantly lower than those of AF, which indicated that the addition of nano-SiO$_2$ and GO particles could improve the thermal stability of Ca–alginate fiber. The loss mass of fiber between 200 and 400 °C was mainly attributed to the decomposition of rejuvenator and calcium alginate materials. Based on the residual mass at 600 °C, the relative mass of encapsulated rejuvenator could be calculated out. The relative mass of rejuvenator that was encapsulated in AF, ASF and MRF were 56.65%, 47.25% and 32.94% respectively. Based on the above analysis, the three kinds of self-healing fiber meet the temperature requirement in the mixing and compaction process of asphalt mixture.

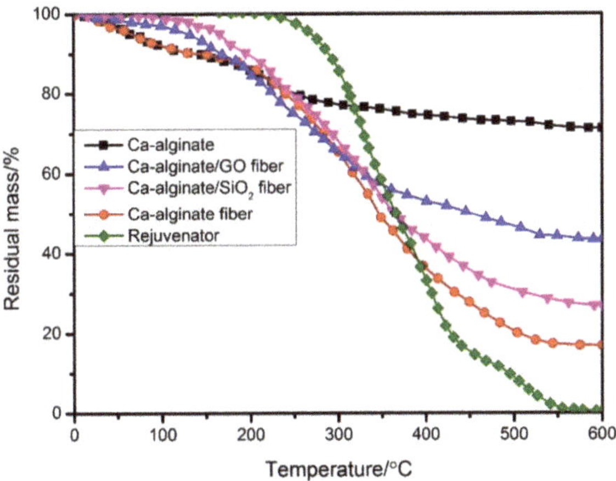

Figure 4. Thermal stability of the three kinds of fiber.

3.2. Road Performance of the Asphalt Mixture with Fibers

3.2.1. High-Temperature Performance

The high temperature performance of asphalt mixture containing the different kinds of fiber was investigated by wheel tracking test and the result is shown in Figure 5. Dynamic stability (DS) and rutting depth (RD) were tested to evaluate high temperature anti-rutting ability of asphalt mixture. The larger DS and smaller rutting depth, the better anti-rutting ability asphalt mixture has. It can be seen that the addition of fibers led to an improved DS and decreased rutting depth. For example, the DS of A2, A3 and A4 increased from 2450 times/mm to 3014 times/mm, 3223 times/mm and 3127 times/mm respectively. The rutting depth of A2, A3 and A4 decreased from 2.415 mm to 1.831 mm, 1.672 mm and 1.698 mm respectively. The improved anti-rutting ability may be attributed to that the entanglement of fibers in asphalt mixture reduced the content of free asphalt.

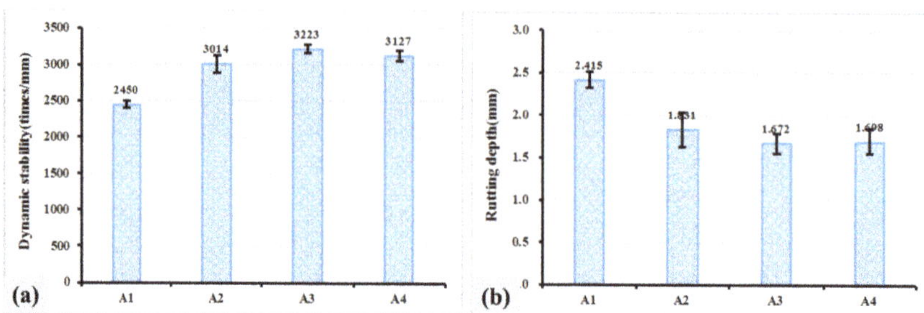

Figure 5. DS and rutting depth of asphalt mixture with the fiber: DS (**a**) and RD (**b**) of asphalt mixture.

3.2.2. Low Temperature Performance

Low temperature bending test was conducted, and flexural strength, flexural strain and flexural modulus were obtained to evaluate low temperature anti-crack ability of asphalt mixture containing the fiber. Figure 6a revealed that the addition of the fiber increased flexural strength of asphalt mixture. The flexural strength of A2, A3 and A4 increased from 9.28 MPa to 10.23 MPa, 10.68 MPa and 10.01 MPa, respectively. It can be seen from Figure 6b that the flexural strain of asphalt mixture containing the fiber was slightly smaller than that of asphalt mixture without fibers. Nevertheless, the flexural strains of all asphalt mixtures containing the fiber were larger than 2800$\mu\varepsilon$, which indicates that the flexural strain of asphalt mixture containing the fiber meets the requirements. In addition, the addition of the fiber could improve the flexural modulus of asphalt mixture. The flexural modulus of A2, A3 and A4 increased from 2968 MPa to 3325 MPa, 3518 MPa and 3447MPa, respectively.

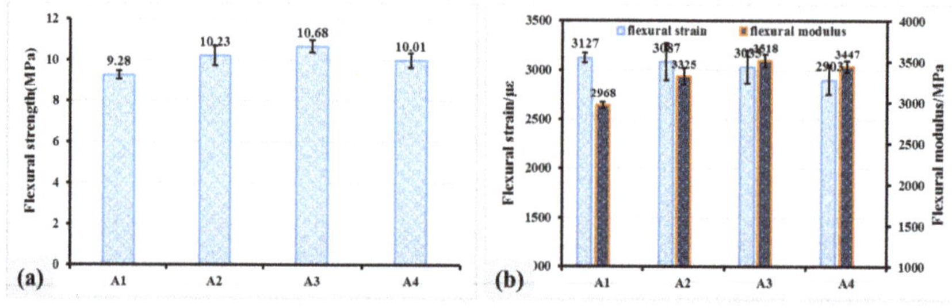

Figure 6. Flexural strength (**a**), Flexural strain and flexural modulus (**b**) of asphalt mixture containing the fiber.

3.2.3. Water Sensitivity

Freeze-thaw splitting test was performed to evaluate the water sensitivity of asphalt mixture containing the fibers, and the result is shown in Figure 7. It can be observed that the TSR of asphalt mixture slightly decreased when the fiber was added. In details, the TSR of A2, A3 and A4 decreased from 89.7% to 82.5%, 84.3% and 83.6%, respectively. The slightly decreased TSR may be attributed to the relatively strong hydrophilicity of the fiber. Nevertheless, the TSR of asphalt mixture with the fiber was still larger than 75%. Thus, moisture stability of asphalt mixture with the fiber met the requirement (according to the specification "Technical Specification for Construction of Highway Asphalt Pavements JTG F40-2004").

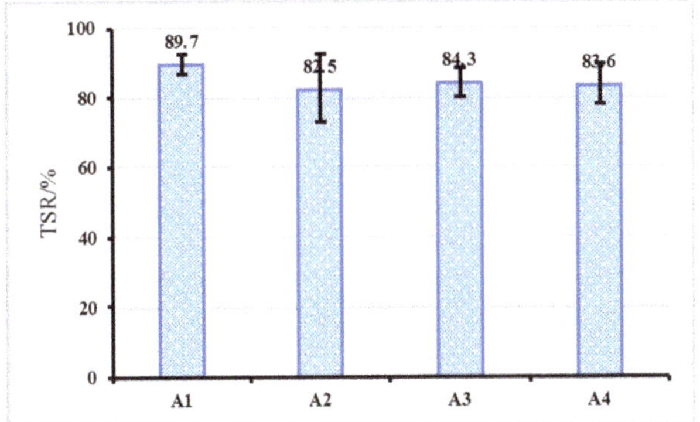

Figure 7. TSR of asphalt mixture with the fiber.

3.2.4. Fatigue Performance

Four-point bending fatigue test was conducted to evaluate fatigue performance of asphalt mixture with the fiber. It can be seen from Figure 8 that there was no obvious difference in initial stiffness for asphalt mixture with, and without the fiber, which meant that the fibers could remain intact in asphalt mixture after the mixing and compaction process. The stiffness of asphalt mixture with the fiber was smaller than that of asphalt mixture without the fiber from 10,000 numbers of load cycles to 60,000 numbers of load cycles, which indicated that the fiber was ruptured, rejuvenator that was encapsulated in the fiber flowered out, and then rejuvenated and softened bitumen. Fatigue failure of asphalt mixture without the fiber appeared when the number of load cycles was larger than 60,000. While due to the regeneration of rejuvenator, asphalt mixture with the fiber could continue to withstand fatigue loading. Thus the addition of the fibers could prolong the fatigue life of asphalt mixture.

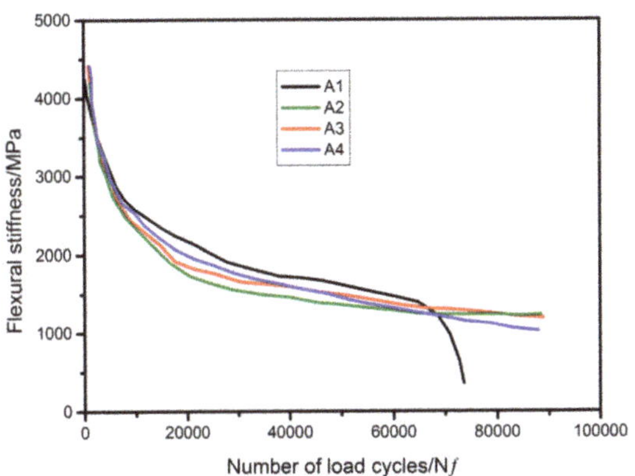

Figure 8. Fatigue performance of asphalt mixture with the fibers.

3.3. Self-Healing Capacity of the Asphalt Mixture with Fibers

The self-healing capacity of asphalt mixture with the fiber was tested by 3PB experiment and the result is shown in Figure 9. As shown in Figure 9, the HI of all fiber containing asphalt mixtures increased to varying degrees. Under the rejuvenation effect of rejuvenator, the HI of A2, A3 and A4 increased from 50.7% to 68.8%, 65.3% and 61.4%, respectively. Under the synergistic action of microwave heating and rejuvenator, the HI of A2, A3 and A4 increased from 53.2% to 74.5%, 72.1% and 89.2%, respectively. The result revealed that the addition of three kinds of fiber could improve the self-healing ability of asphalt mixture under the action of rejuvenator that was encapsulated in fiber. Furthermore, because of the incorporation of microwave absorbent (GO) into self-healing fiber, under the synergistic action of microwave heating and rejuvenator, the self-healing ability of asphalt mixture containing MRF could be significantly improved. GO is an excellent microwave heating material that generates a large amount of heat under the action of microwave heating. As the temperature of the asphalt increases rapidly, the fluidity of the asphalt is greatly improved. Under the action of the rejuvenator and temperature, the crack inside the asphalt mixture will be quickly filled by the flowing asphalt, so the self-healing properties of the asphalt mixture containing Ca–alginate/GO fiber are significantly improved after microwave heating (as shown in Figure 9 A4).

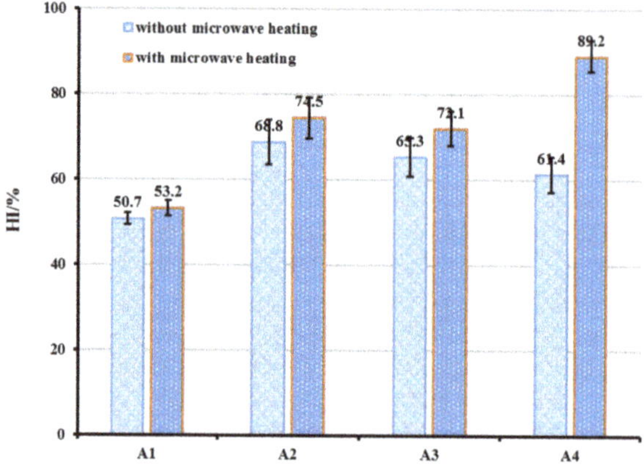

Figure 9. Self-healing capacity of asphalt mixture containing the fiber.

4. Conclusions

In this study, three kinds of self-healing fiber were synthesized. A series of characterizations were performed to study the morphology, chemical structure and thermal stability of the fibers. Then, road performance and self-healing capacity of asphalt mixture containing the fiber were investigated. Based on the tests and results, the follow conclusions can be drawn:

(a) Ca–alginate fiber, Ca–alginate/SiO$_2$ composite fiber and Ca–alginate/GO composite fiber were successfully synthesized by microfluidic device. Nano SiO$_2$ and GO particles were incorporated into the fiber wall. Rejuvenator was encapsulated inside those fibers in the form of droplets. The three kinds of fibers had excellent thermal stability, which meet the temperature requirement in the mixing and compaction process of asphalt mixture.

(b) The addition of the three kinds of fibers improved high temperature anti-rutting ability of asphalt mixture. In addition, those fibers increased flexural strength and flexural modulus, while slightly decreased flexural strain of asphalt mixture. Moisture stability of asphalt mixture containing the fiber had a slight decrease. In addition, the fibers could prolong fatigue life of asphalt mixture under the action of encapsulated rejuvenator. In short, the road performances of asphalt mixture containing the fiber meet the requirements.

(c) The self-healing ability of asphalt mixture with fiber was better than that of asphalt mixture without the fiber. It was worth noting that the synergistic action of microwave heating and rejuvenator could further significantly improve the self-healing ability of asphalt mixture.

This study provides the possibility of industrial production of self-healing asphalt mixtures. Different types of self-healing fibers can be fabricated in large scale by this method. Personally, I believe that the future trend is a combination of capsule method and induction method. Combining the advantages of capsules method and induction heating method to achieve a substantial improvement in the self-healing properties of asphalt concrete. As in this work, the composite Ca–alginate/GO fibers with crack response behavior and microwave heating response behavior combined the advantages of the capsules method and induction heating method. The fiber not only can rejuvenate the asphalt, but also can rapidly increase the temperature of the asphalt by microwave heating, which quickly increased the fluidity of asphalt, and thus the self-healing properties of the asphalt mixture can be quickly improved under the synergistic effect of the rejuvenator and the temperature.

Highlights

- Three kinds of fibers encapsulating asphalt rejuvenator were synthesized;
- Road performance of asphalt mixture containing the fiber met the requirements;
- Three kinds of fibers could improve self-healing ability of asphalt mixture.

Author Contributions: Conceptualization, B.S. and Q.L.; Data curation, B.S. and S.B.; Formal analysis, B.S.; Investigation, B.S.; Methodology, B.S., C.L., L.D. and Q.W.; Writing–original draft preparation, B.S.; Writing–review and editing, Q.L., X.Y. and J. N.; Project administration, S.W.; Supervision, S.W. and Q.L.

Funding: This research was funded by National Key R&D Program of China (No. 2018YFB1600200). This work presented in this paper was supported by National Natural Science Foundation of China (71961137010). The authors acknowledge the financial support by the Natural Science Foundation of China (No. 51778515) and National Key R&D Program of China (No. 2017YFE0111600). Finally, the sixth author thanks the financial support by the National Science Foundation of China (No. 5170081212).

Conflicts of Interest: There are no conflicts of interest regarding the publication of this paper.

References

1. Li, Y.; Wu, S.; Pang, L.; Liu, Q.; Wang, Z.; Zhang, A. Investigation of the effect of Mg-Al-LDH on pavement performance and aging resistance of styrene-butadiene-styrene modified asphalt. *Constr. Build. Mater.* **2018**, *172*, 584–596. [CrossRef]

2. Sreedhar, S.; Coleri, E. Effects of binder content, density, gradation, and polymer modification on cracking and rutting resistance of asphalt mixtures used in Oregon. *J. Mater. Civil Eng.* **2018**, *30*, 04018298. [CrossRef]

3. Garcia, A.; Jelfs, J.; Austin, C.J. Internal asphalt mixture rejuvenation using capsules. *Constr. Build. Mater.* **2015**, *101*, 309–316. [CrossRef]

4. Li, R.; Yu, Y.; Zhou, B.; Guo, Q.; Li, M.; Pei, J. Harvesting energy from pavement based on piezoelectric effects: Fabrication and electric properties of piezoelectric vibrator. *J. Renew. Sustain. Energy* **2018**, *10*, 54701. [CrossRef]

5. Li, R.; Guo, Q.; Shi, Z.; Pei, J. Effects of conductive carbon black on PZT/PVDF composites. *Ferroelectrics* **2018**, *526*, 176–186. [CrossRef]

6. Li, R.; Dai, Y.; Wang, P.; Sun, C.; Zhang, J.; Pei, J. Evaluation of Nano-ZnO dispersed state in bitumen with digital imaging processing technologies. *J. Test. Eval.* **2018**, *46*, 974–983. [CrossRef]

7. Chen, Z.; Wu, S.; Xiao, Y.; Zeng, W.; Yi, M.; Wan, J. Effect of hydration and silicone resin on basic oxygen furnace slag and its asphalt mixture. *J. Clean. Prod.* **2016**, *112*, 392–400. [CrossRef]

8. Xiao, Y.; Wang, F.; Cui, P.; Lei, L.; Lin, J.; Yi, M. Evaluation of fine aggregate morphology by image method and its effect on skid-resistance of micro-surfacing. *Materials* **2018**, *11*, 920. [CrossRef]

9. Kong, D.; Chen, M.; Xie, J.; Zhao, M.; Yang, C. Geometric characteristics of BOF slag coarse aggregate and its influence on asphalt concrete. *Materials* **2019**, *12*, 741. [CrossRef]

10. Ganjei, M.A.; Aflaki, E. Application of nano-silica and styrene-butadiene-styrene to improve asphalt mixture self healing. *Int. J. Pave. Eng.* **2019**, *20*, 89–99. [CrossRef]

11. Gupta, S.; Zhang, Q.; Emrick, T.; Balazs, A.C.; Russell, T.P.J.N.M. Entropy-driven segregation of nanoparticles to cracks in multilayered composite polymer structures. *Nat. Mater.* **2006**, *5*, 229–233. [CrossRef]

12. Liu, Q.; Yu, W.; Wu, S.; Schlangen, E.; Pan, P. A comparative study of the induction healing behaviors of hot and warm mix asphalt. *Constr. Build. Mater.* **2017**, *144*, 663–670. [CrossRef]

13. Norambuena-Contreras, J.; Garcia, A. Self-healing of asphalt mixture by microwave and induction heating. *Mater. Des.* **2016**, *106*, 404–414. [CrossRef]
14. Li, C.; Wu, S.; Chen, Z.; Tao, G.; Xiao, Y. Enhanced heat release and self-healing properties of steel slag filler based asphalt materials under microwave irradiation. *Constr. Build. Mater.* **2018**, *193*, 32–41. [CrossRef]
15. Norambuena-Contreras, J.; Gonzalez, A.; Concha, J.L.; Gonzalez-Torre, I.; Schlangen, E. Effect of metallic waste addition on the electrical, thermophysical and microwave crack-healing properties of asphalt mixtures. *Constr. Build. Mater.* **2018**, *187*, 1039–1050. [CrossRef]
16. Sun, D.; Sun, G.; Zhu, X.; Guarin, A.; Li, B.; Dai, Z.; Ling, J. A comprehensive review on self-healing of asphalt materials: Mechanism, model, characterization and enhancement. *Adv. Coll. Interface Sci.* **2018**, *256*, 65–93. [CrossRef] [PubMed]
17. Garcia, A.; Austin, C.J.; Jelfs, J. Mechanical properties of asphalt mixture containing sunflower oil capsules. *J. Clean. Prod.* **2016**, *118*, 124–132. [CrossRef]
18. Tabakovic, A.; Post, W.; Cantero, D.; Copuroglu, O.; Garcia, S.J.; Schlangen, E.J.S.M.S. The reinforcement and healing of asphalt mastic mixtures by rejuvenator encapsulation in alginate compartmented fibres. *Smart Mater. Struct.* **2016**, *25*, 084003. [CrossRef]
19. García, Á.; Schlangen, E.; van de Ven, M.; Sierra-Beltrán, G. Preparation of capsules containing rejuvenators for their use in asphalt concrete. *J. Hazard. Mater.* **2010**, *184*, 603–611. [CrossRef]
20. Al-Mansoori, T.; Norambuena-Contreras, J.; Micaelo, R.; Garcia, A. Self-healing of asphalt mastic by the action of polymeric capsules containing rejuvenators. *Constr. Build. Mater.* **2018**, *161*, 330–339. [CrossRef]
21. Shi, X.; Alvaro, G.; Junfeng, S.; Quantao, L.; Amir, T.; Erik, S.J.A.M.I. Self-Healing Asphalt Review: From Idea to Practice. *Adv. Mater. Interfaces* 2018. [CrossRef]
22. Chung, K.; Lee, S.; Cho, W.; Seo, J.; Hong, Y. Rheological analysis of self-healing property of microcapsule-containing asphalt. *J. Ind. Eng. Chem.* **2018**, *64*, 284–291. [CrossRef]
23. Sun, D.; Li, B.; Ye, F.; Zhu, X.; Lu, T.; Tian, Y. Fatigue behavior of microcapsule-induced self-healing asphalt concrete. *J. Clean. Prod.* **2018**, *188*, 466–476. [CrossRef]
24. Sun, D.; Lu, T.; Zhu, X.; Li, B.; Tian, Y. Optimization of synthesis technology to improve the design of asphalt self-healing microcapsules. *Constr. Build. Mater.* **2018**, *175*, 88–103. [CrossRef]
25. Li, R.; Zhou, T.; Pei, J. Design, preparation and properties of microcapsules containing rejuvenator for asphalt. *Constr. Build. Mater.* **2015**, *99*, 143–149. [CrossRef]
26. Hongliang, Z.; Yuzan, B.; Fenglei, C.J.C.; Materials, B. Rheological and self-healing properties of asphalt binder containing microcapsules. *Constr. Build. Mater.* **2018**, *187*, 138–148.
27. Norambuena-Contreras, J.; Yalcin, E.; Garcia, A.; Al-Mansoori, T.; Yilmaz, M.; Hudson-Griffiths, R. Effect of mixing and ageing on the mechanical and self-healing properties of asphalt mixtures containing polymeric capsules. *Constr. Build. Mater.* **2018**, *175*, 254–266. [CrossRef]
28. Shu, B.; Wu, S.; Dong, L.; Wang, Q.; Liu, Q.J.M. Microfluidic synthesis of Ca–alginate microcapsules for self-healing of bituminous binder. *Materials* **2018**, *11*, 630. [CrossRef]
29. Shu, B.; Zhang, L.; Wu, S.; Dong, L.; Liu, Q.; Wang, Q. Synthesis and characterization of compartmented Ca–alginate/silica self-healing fibers containing bituminous rejuvenator. *Constr. Build. Mater.* **2018**, *190*, 623–631. [CrossRef]

Article

The Effect of UV Irradiation on the Chemical Structure, Mechanical and Self-Healing Properties of Asphalt Mixture

Shaopeng Wu [1], Yong Ye [1], Yuanyuan Li [1,2,3,*], Chuangmin Li [2,4,*], Wei Song [1], Hechuan Li [1], Chao Li [1], Benan Shu [1] and Shuai Nie [1]

[1] State Key Laboratory of Silicate Materials for Architectures, Wuhan University of Technology, Wuhan 430070, China

[2] Engineering Laboratory of Spatial Information Technology of Highway Geological Disaster Early Warning in Hunan Province, Changsha University of Science & Technology, Changsha 410114, China

[3] Nottingham Transportation Engineering Centre, School of Civil Engineering, University of Nottingham, University Park, Nottingham NG7 2RD, UK

[4] School of Traffic and Transportation Engineering, Changsha University of Science & Technology, Changsha 410114, China

* Correspondence: liyuanyuan@whut.edu.cn (Y.L.); lichuangmin@csust.edu.cn (C.L.)

Received: 23 June 2019; Accepted: 27 July 2019; Published: 30 July 2019

Abstract: Although huge numbers of investigations have been conducted for the ultraviolet (UV) aging of asphalt binder, research rarely focuses on the asphalt mixture. In order to evaluate the aging effect of UV radiation on the asphalt mixture, a dense grade of asphalt mixture was designated and aged by UV radiation for 7, 14 and 28 days respectively. After that, the chemical functional groups of asphalt binder were tested by Fourier transform infrared spectrometer (FTIR). The semi-circular bending strength and fatigue resistance of asphalt concrete were tested to characterize the mechanical properties of the asphalt concrete. To evaluate the self-healing effect of the macro-structure continuity of asphalt concrete intuitively, the computed tomography (CT) scanning machine was used to characterize the crack size of asphalt concrete samples both before and after self-healing. The results show that, with the increase of UV irradiation time, the relative ratios of the C=O and S=O bands' areas of recovered asphalt binder increase significantly. UV radiation can significantly weaken the mechanical and self-healing properties of asphalt mixture, making the asphalt mixture to have worse macro-structure continuity, lower failure strength and worse fatigue resistance. Moreover, the longer the UV irradiation time is, the degradation effect of UV radiation on asphalt mixture becomes more obvious.

Keywords: UV radiation; asphalt mixture; chemical structure; mechanical performance; self-healing performance; CT scanning

---

1. Introduction

For the complex work condition of asphalt pavement, there are many factors that can affect the service span, such as the vehicle, temperature [1], water and UV radiation [2–4]. Therefore, in general, the service time of asphalt pavement is always shorter than its designated life. UV radiation has attracted more and more focus for its significant aging effects on the asphalt materials [5]. After UV aging, the elemental ratios and chemical components of asphalt binder are changed [6], which results in the deterioration of physical and rheological performance of asphalt binder and the pavement performance of its concrete [7–9]. Especially for the areas with high intensity of UV radiation, the aging effect is more obvious [10].

In the previous work [11], the researchers conducted much work on the UV aging mechanism and aging effects on the technical performance of asphalt materials. For instance, Vallerga [12] used UV radiation and infrared light to irradiate the asphalt film in a film oven respectively and the results indicated that UV radiation could significantly affect the softening point and ductility of asphalt binder, therefore the UV radiation was an important factor causing the aging phenomenon of asphalt binder. Zhang [13] studied the aging effect of three types of aging methods namely thin film oven test (TFOT), pressure aging vessel (PAV), and UV radiation, the results showed that the UV aging and PAV aging have more obvious effects on the degradation of styrene–butadiene–styrene (SBS) polymer. Glotova [14] studied the effect of UV aging on asphalt binder by testing the chemical properties, structures and composition contents of asphalt binder before and after aging, and found that UV radiation has obvious degeneration effects on these properties, and the type of light irradiation also has a close relationship with the photo-oxidation aging speed. Wu et al. [15] used a high-pressure mercury lamp to irradiate both a base asphalt binder and a polymer modified asphalt binder; the results of the Fourier infrared test and dynamic shear rheometer (DSR) test also showed that UV radiation can obviously age the asphalt binder, and different intensities of UV radiations caused different aging effects. The above research verifies the degradation effects of UV radiation on the technical performance of asphalt materials, and it is truly important to investigate the degradation behaviors of this.

Compared with the asphalt binder, there was less work focused on the UV aging effect of UV radiation on the pavement performance asphalt mixture. In fact, the UV radiation always directly affects the asphalt mixture during the service period of pavement [16], so conducting the UV aging test on the asphalt mixture is better to simulate the true service situation of the asphalt road. In this research, a dense grade of asphalt mixture with the nominal maximum aggregate size of 13.2 mm (AC-13) was designated, and aged by UV radiation for 7, 14 and 28 days respectively. After UV aging, the mechanical performance of asphalt concrete was tested by a semi-circle bending test and the anti-fatigue properties of asphalt mixture were investigated by three-point bending fatigue test.

On the other hand, basing on Reference [17], it can be found that the thermal oxidative aging (85 °C for 240 h) can decrease the mechanical performance and healing levels of asphalt concrete obviously. Although the mechanism of thermal oxidative aging of asphalt material is not at all the same with UV aging, the effect tendency on the performance is somehow the same [18], so that UV aging may also cause a significant effect on the self-healing performance of asphalt concrete. In this research, the chemical functional groups of asphalt binder were tested by Fourier transform infrared spectrometer (FTIR). The healing index of the mechanical performance of asphalt concrete was investigated to characterize the self-healing performance. Meanwhile, a computed tomography (CT) scanning machine was also used to quantitatively test the crack size of asphalt concrete before and after self-healing. The results can be used to express the degradation behaviors of the chemical structure, mechanical and self-healing properties of asphalt mixture exposed during UV irradiation.

2. Materials and Experimental Methods

2.1. Materials

The most commonly used asphalt mixture with the nominal maximum aggregate size of 13.2 mm (AC–13) was designated for the UV aging simulation test. The asphalt binder used in this research was base asphalt with penetration grade of 60/80. The physical performance and dynamic viscosity of asphalt binder are shown in Table 1.

Table 1. Physical performance and dynamic viscosity of asphalt binder.

Technical Parameter	25 °C Penetration (0.1 mm)	Softening Point (°C)	10 °C Ductility (cm)	60 °C Dynamic Viscosity (Pa·s)
Result	66.3	51.0	165	302.500
Standard	ASTM D5 [19]	ASTM D36 [20]	ASTM D113 [21]	ASTM D4402 [22]

The mix ratio of asphalt mixture was ensured with the Marshall method. The optimum oil–aggregate ratio of the asphalt mixture is 4.7%. The ratios of different grades of aggregate were 13.2–16.0 mm:9.5–13.2 mm:4.75–9.5 mm:2.36–4.75 mm:0–2.36 mm:filler = 3:15:33:13:32:4. The aggregate was basalt and the filler was the fine powder of limestone. The composite aggregate grading of the asphalt mixture is given in Figure 1. The volume parameters of asphalt concrete are shown in Table 2.

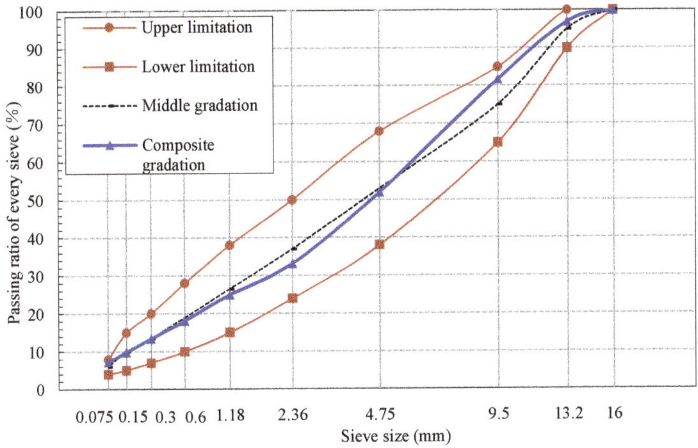

Figure 1. Composite aggregate grading of the asphalt mixture.

Table 2. Volume parameters of asphalt concrete.

Volume Parameters	Theoretical Maximum Specific Gravity	Bulk Volume Relative Density	VV(%)	VMA(%)	VFA(%)
Results	2.676	2.556	4.5	14.5	69.0

2.2. Experimental Methods

2.2.1. Parameter Design of UV Aging Test

The irradiation intensity of UV light used in the UV aging simulation test of asphalt mixture was 21 w/m^2, which was produced by four UV lamps, the power of every UV lamp was 500 W. The aging time was 7 days, 14 days and 28 days respectively. The wavelength range of the UV radiation was from 200 nm to 400 nm. The temperature of chamber of the UV aging simulation instrument was 25 °C, it was tested by the temperature sensor in the chamber. The surface temperature of the asphalt mixture was about 50 °C, it was tested by an infrared temperature tester (SMART SENSOR AR-300+, SMART SENSOR, Hongkong, China). According to the thermal oxidative aging test (short term aging) of asphalt mixture, the spread density of asphalt mixture was 22 kg/m^2. During the UV aging test, the asphalt mixture was mixed every 24 h. Figure 2 shows the UV aging test of loose asphalt mixtures.

Figure 2. UV aging test of loose asphalt mixtures.

2.2.2. Recovery of Asphalt Binder from Asphalt Mixtures

After different times of UV irradiation, the asphalt binders were recovered from the asphalt mixtures by dissolution-centrifugal separation method. The solvent was trichloroethylene (C_2HCl_3), the purity of which was higher than 99%. The detailed procedure was as follows. First, the asphalt mixture was placed into C_2HCl_3 for half an hour, and then the centrifugal separation method was used to get the asphalt–C_2HCl_3 solution; this step was repeated 2–3 times. Finally, the asphalt–C_2HCl_3 solution was placed in the fume hood and dried naturally for 72 h.

In order to detect whether the C_2HCl_3 would affect the Fourier transform infrared spectrometer (FTIR, Nexus, Thermo Nicolet, Columbus, OH, USA) result of recovered asphalt binder, after 72 h volatilization and drying, the chemical structures of the virgin asphalt binder, recovered 28 days aged asphalt binder and C_2HCl_3 were tested by FTIR respectively. The results are shown in Figure 3. From Figure 3, a special absorption band at the wavenumber of 932 cm^{-1} can be observed in the FTIR spectrum of C_2HCl_3, it is caused by the stretching vibration of C–Cl of C_2HCl_3. This absorption band was very obvious and could be found in the FTIR spectrum of the virgin asphalt binder. Therefore, it could be used to test whether the C_2HCl_3 volatilized completely by comparing the FTIR spectra differences of the virgin asphalt binder and recovered 28 days aged asphalt binder. It was found that, in the FTIR spectrum of 28 days aged asphalt binder, there was no absorption band at wavenumber 932 cm^{-1}, therefore the C_2HCl_3 volatilized completely after 72 h volatilization (at least, it did not obviously affect the FT-IR of recovered asphalt binder).

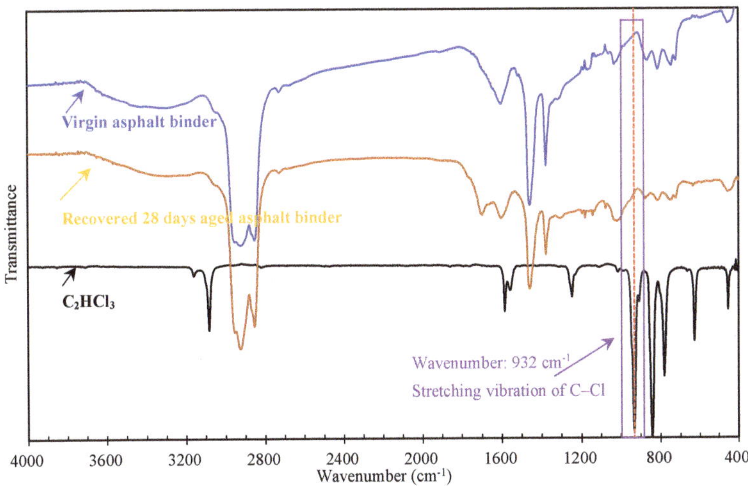

Figure 3. FTIR spectra of the pure asphalt binder, recovered 28 days aged asphalt binder and C_2HCl_3.

2.2.3. Chemical Functional Group Test of Recovered Asphalt Binders

The chemical structures of recovered asphalt binders were characterized by FTIR, the procedure of the FTIR test was as follows. The asphalt was dissolved in carbon disulfide (CS_2), the concentration of the asphalt binder was 5.0 wt%; then the asphalt–CS_2 solution was dropped on a potassium bromide (KBr) window, and the potassium bromide window was set under an incandescent light bulb until the CS_2 volatized completely. The sweep times was 32, the sweep range of wavenumber was from 4000 cm^{-1} to 400 cm^{-1}.

2.2.4. Mechanical Properties Tests of Asphalt Concrete

The mechanical properties tests of asphalt concretes before and after UV aging were investigated by both the semi-circular bending test and three-point bending fatigue test.

The machine used for the semi-circular bending test was UTM-25 (IPC, Sydney, Australia). The diameter of the sample was 100 mm, the thickness was 50 mm. In order to control the crack position during the test, a notch with depth of 10 mm and width of 2 mm was precut. The testing temperature was −10 °C, and the loading speed was 0.5 mm/min. The schematic diagram of semi-circular bending test of asphalt concrete is shown in Figure 4.

The fatigue tests of asphalt concretes were tested at both 0.4 and 0.6 stress ratios, which was conducted with UTM-25 machine as well. The testing temperature was 25 °C and the Poisson ratio was designated as 0.35. A semi-sine wave of loading was used, and the loading frequency was 1.0 Hz (loading time was 0.1 s, and interval was 0.9 s). To control the temperature of the samples, all of them were put in the UTM chamber under constant temperature (25 °C) for more than five h. For the same aging status, three samples of asphalt concretes were tested to calculate the average values and standard deviations of the fatigue lives.

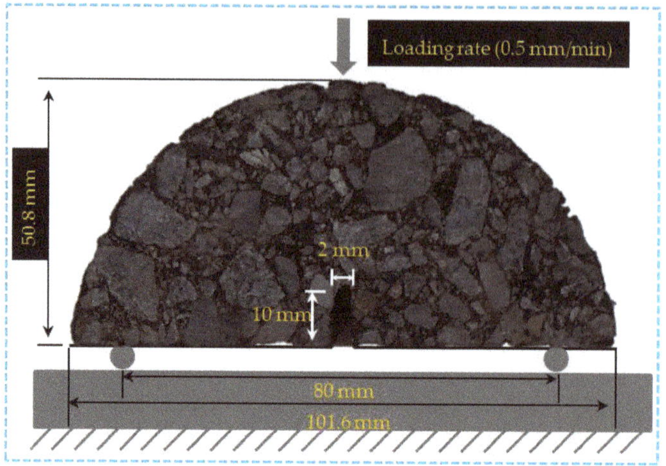

Figure 4. Schematic diagram of the semi-circular bending test of asphalt concrete.

2.2.5. Self-Healing Performance Tests of Asphalt Concrete

The semi-circular bending strength and fatigue life of initial asphalt concrete and asphalt concrete after self-healing were tested respectively, which was to obtain the healing index of asphalt concrete. A higher value of healing index expresses the better self-healing performance of asphalt concrete, and vise versa.

Meanwhile, a CT scanning machine (Xradia 510 Versa, ZEISS, Oberkochen, Germany) was used to characterize the crack size of asphalt concrete samples before and after self-healing. Figure 5 shows the diagram of this test. The real spatial resolution of the machine was 0.7 microns, and the voxel size was as low as 70 nanometers. The self-healing time of the samples was 72 h, and the temperature in the environmental chamber was 50 °C.

Figure 5. Experimental diagram of the computed tomography (CT) scanning test.

2.3. Logic Map of Experimental Design

The logic map of experimental design is shown in Figure 6.

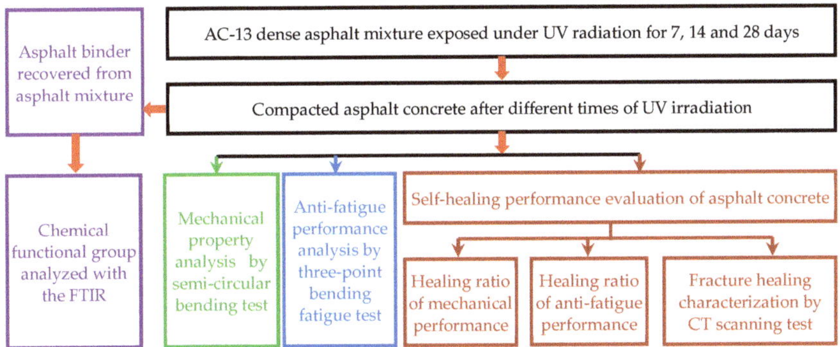

Figure 6. Logic map of the experimental design.

3. Results and Discussions

3.1. Characterization of the UV Aging Status of Recovered Asphalt Binder

The atoms that form the chemical bonds or functional group of organic molecules are always in a constant state of vibration, and the vibration frequency of that is equal to the vibration frequency of the infrared light [23]. When the infrared light irradiates on the organic molecules, the vibrational absorption of infrared light can be observed in the chemical bonds or functional groups in molecules [24]. The different chemical bonds or functional group have different absorption frequencies, the infrared spectrum will be reflected in different positions, thus the FTIR can be used to characterize the chemical bonds or functional molecules in organic molecules [25]. Because of the irradiation of UV light, the chemical groups such as C=C and C–H without oxygen can be oxidized, and form oxygen-containing functional groups, such as the carbonyl group (C=O) and sulfoxide functional group (S=O) [26]. Therefore, the content of C=O and S=O will increase after UV aging, the absorption band areas of these two chemical functional groups will increase as well, which can be used to characterize the aging status of asphalt binder. The bigger the absorption band areas of C=O and S=O, the more serious the UV aging status.

The FTIR spectra of asphalt binders recovered from aged mixtures are shown in Figure 7. From Figure 7, the absorption bands of the main chemical functional groups of asphalt binder were in the wavenumber range of 600–2000 cm^{-1}, the absorption bands at 1700 cm^{-1} and 1030 cm^{-1} were caused by the C=O and S=O respectively. However, the absorption bands of the C=O and S=O can also be observed in FTIR spectra of original asphalt binder. This is because thermo–oxidative aging happens during the production and storage of asphalt binder, and the mixing of asphalt mixture. From Figure 7, the absorption bands C=O and S=O increased with the increasing aging time, indicating the more serious aging status of asphalt binders recovered from asphalt mixtures under longer times of UV irradiation.

To quantificationally characterize the aging status of UV aging status of recovered asphalt binder, the relative ratios of the areas of C=O absorption band and S=O absorption band were calculated based on Equations (1) and (2) respectively.

$$\mathrm{CRR} = \frac{S_{1700\ \mathrm{cm}^{-1}}}{\sum S_{2000\ \mathrm{cm}^{-1} \sim 600\ \mathrm{cm}^{-1}}}, \tag{1}$$

$$\mathrm{SRR} = \frac{S_{1030\ \mathrm{cm}^{-1}}}{\sum S_{2000\ \mathrm{cm}^{-1} \sim 600\ \mathrm{cm}^{-1}}}, \tag{2}$$

where, the (CRR) and (SRR) are the C=O relative ratio and S=O relative ratio respectively, $S_{1700 \text{ cm}^{-1}}$ and $S_{1030 \text{ cm}^{-1}}$ are the areas of C=O absorption band and S=O absorption band respectively, $\sum S_{2000 \text{ cm}^{-1} \sim 600 \text{ cm}^{-1}}$ is the sum of areas of all bands in the wavenumber range of 2000 cm^{-1} ~600 cm^{-1}.

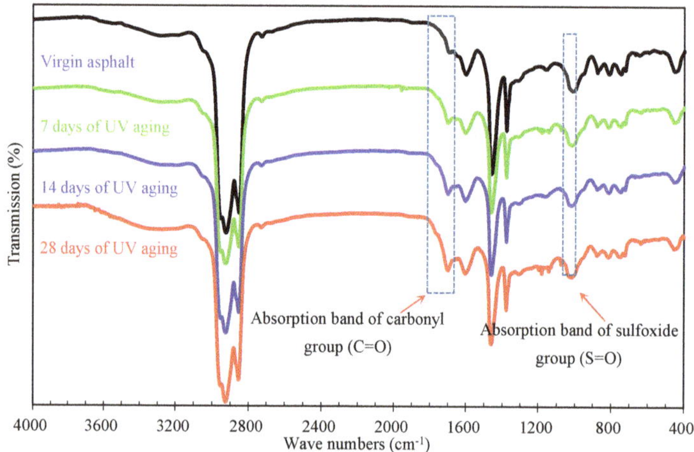

Figure 7. FTIR spectra of asphalt binders recovered from aged mixtures.

The relative ratios of the areas of C=O and S=O absorption bands are shown in Table 3. The bigger the CRR and SRR values are, the more serious the aging status of asphalt binder is [27]. From Table 3, after 7 days, 14 days and 28 days of UV irradiation, the CRR value of which increased by 128.6%, 226.2% and 358.3% respectively, the SRR value of which increased by 21.7%, 34.5% and 40.4% respectively. The CRR and SRR values of recovered asphalt binder increased significantly with the increase of UV irradiation time, the content of oxygen-containing functional groups in asphalt binder increased obviously as well, indicating that the aging status of asphalt binder was more serious with longer UV irradiation time.

Table 3. Carbonyl relative ratio (CRR) and sulfoxide relative ration (SRR) of asphalt recovered from aged asphalt mixtures.

Aging Time	C=O		S=O	
	CRR Value	**Relative Change to Original State (%)**	**SRR Value**	**Relative Change to Original State (%)**
Virgin asphalt	0.0084	-	0.0750	-
7 days	0.0192	128.6	0.0913	21.7
14 days	0.0274	226.2	0.1009	34.5
28 days	0.0385	358.3	0.1053	40.4

3.2. UV Radiation Effects on Mechanical Properties of Asphalt Concrete

3.2.1. UV Radiation Effects on the Semi-Circular Bending Strength of Asphalt Concrete

Asphalt concrete is granular in nature, and its macroscopic behavior is mainly a function the interactions of asphalt binder and aggregate and the UV aging of asphalt binder may significantly influence the mechanical property of asphalt concrete [28,29]. The semi-circular bending test was conducted on the asphalt concrete before and after UV aging. Three parallel tests were performed at each aging time, the bending strength of asphalt concrete was calculated according to Equation (3), the average bending strength of these three specimens were taken as the final result.

$$S_t = \frac{2 \cdot P}{\pi \cdot d \cdot h}$$

(3)

where, S_t is the semi-circular bending strength of asphalt concrete, MPa; P is the maximum loading of every specimen, N; d is the diameter of every specimen, mm; h is the thickness of every specimen, mm.

Figure 8 gives the average bending strengths of all asphalt concretes. From Figure 8, when after 7, 14 and 28 days of UV aging, the semi-circular bending strengths of asphalt concretes at −10 °C decreased by 10.8%, 15.6% and 31.4% respectively; while the semi-circular bending strengths of asphalt concretes at 25 °C increased by 97.6%, 110.0% and 118.6% respectively. Therefore, after different times of UV aging, the semi-circular bending strengths of asphalt concretes at −10 °C tended to decrease, while they showed an opposite tendency at 25 °C. With the extension of UV exposure time, the tendencies were more obvious. The reason is that, after aging the asphalt binder tends to be stiffer and harder, the asphalt binder is inherently brittle at low temperature condition (−10 °C); the comprehensive action of UV aging and low temperature causes the degradation of the three bending strengths of asphalt concretes. However, at relative high temperature (25 °C), the harder effect of UV aging on asphalt binder can partially offset the softening effect of high temperature, therefore the 25 °C semi-circular bending strengths of asphalt concretes were increased. This results also the same as other research work [30].

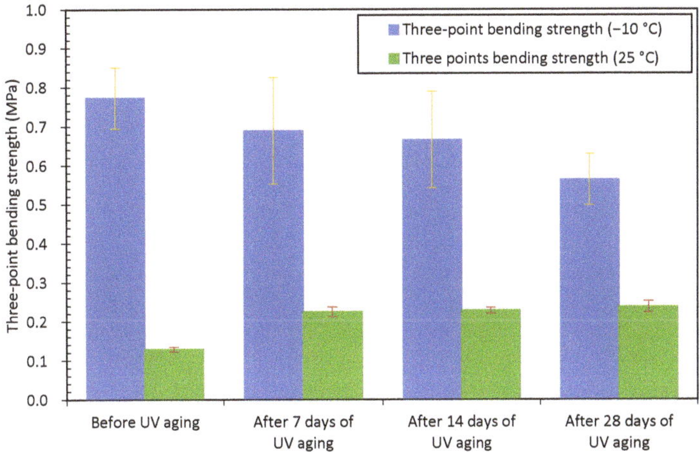

Figure 8. Semi-circular bending strengths of asphalt concretes before and after UV aging.

3.2.2. UV Radiation Effects on the Fatigue Resistance of Asphalt Concrete

The fatigue tests of asphalt concretes before and after UV aging were conducted at 0.4 and 0.6 stress ratios respectively. The results are shown in Figure 9.

It can be found from Figure 9 that, after UV aging, the fatigue lives of asphalt concretes at both 0.4 and 0.6 stress ratios decreased, and with prolonging UV irradiation time, the decreased range was more significant. In detail, after being aged for 7, 14 and 28 days, the fatigue lives of asphalt concrete under 0.4 stress ratio of loading decreased by 42.2%, 58.9% and 66.2% respectively, and under 0.6 stress ratio of loading decreased by 32.3%, 58.2% and 69.6% respectively. Therefore, UV radiation decreased the fatigue resistance of asphalt concretes; the longer the UV irradiation time was, the more significant the aging effect.

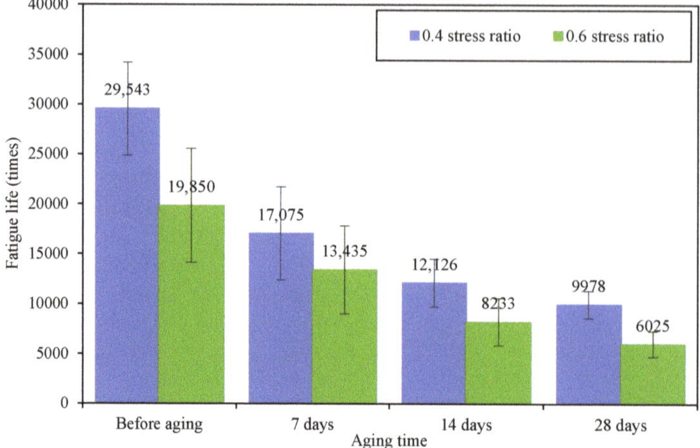

Figure 9. Fatigue lives of asphalt concrete under different stress ratios.

3.3. UV Aging Effects on the Self-Healing Performance of Asphalt Concrete

3.3.1. Healing Percentages of Semi-Circular Bending Strengths

The healing percentages of semi-circular bending strengths of asphalt concretes were calculated according to Equation (4). The healing percentages of semi-circular bending strengths (HPBS) values of asphalt concretes before and after UV irradiation are listed in Figures 10 and 11.

$$\text{HPBS} = \frac{BS_{after\ aging}}{BS_{before\ aging}} \times 100\%$$

(4)

where, the HPBS is the healing percentages of semi-circular bending strengths of asphalt concretes, %; $BS_{before\ aging}$ and $BS_{after\ aging}$ are the healing percentages of semi-circular bending strengths of asphalt concretes before and after UV aging respectively, MPa.

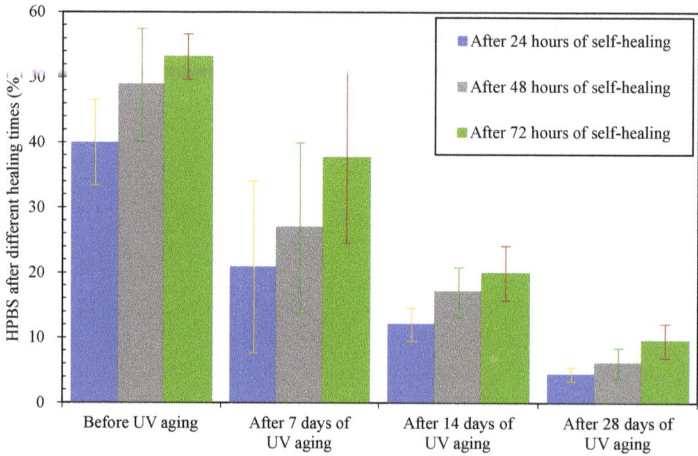

Figure 10. Healing percentages of −10 °C bending strengths after different healing times.

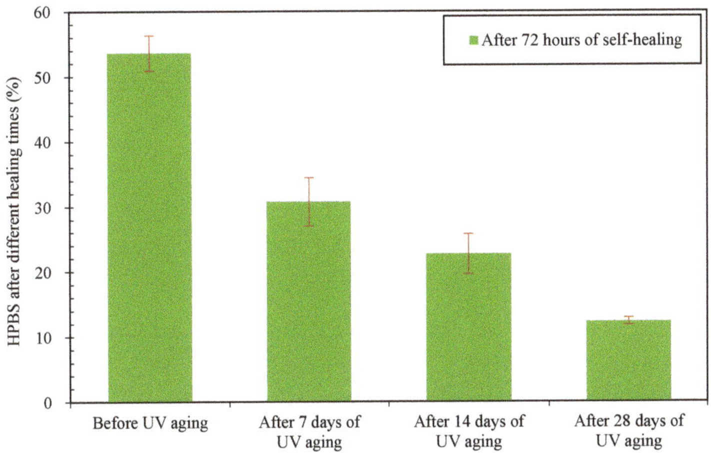

Figure 11. Healing percentages of 25 °C bending strengths after 72 h of self-healing.

From Figure 10, for the −10 °C bending strengths, the self-healing effects of asphalt concretes had time sensitivity: the HPBS values of asphalt concretes increased with the increase of healing time. When the healing times were 24 h and 72 h, the HPBS values of asphalt concretes before UV aging were 40.0% and 53.2% respectively. Compared with asphalt concrete without UV irradiation, after self-healing, the HPBS values of asphalt concretes after 7, 14 and 28 days UV irradiation decreased significantly; the HPBS decrement of asphalt concretes after different times of UV irradiation are listed in detail in Table 4. From Figure 11 and Table 4, after 72 h of self-healing, the same tendency could also be observed for the 25 °C bending strengths of asphalt concrete. Therefore, the UV irradiation could weaken the self-healing performance of asphalt concrete obviously, and the longer the irradiation time was, the worse the self-healing performance of asphalt concrete.

Table 4. HPBS decrement of asphalt concretes after different times of UV irradiation (%).

Aging Time	HPBS Decrement at −10 °C			HPBS Decrement at 25 °C
	24 h	**48 h**	**72 h**	**72 h**
7 days	47.7	44.8	29.6	42.5
14 days	70.0	64.9	62.5	57.6
28 days	89.0	87.5	82.1	76.8

3.3.2. Healing Percentages of Fatigue Performance

Initial fatigue lives and the fatigue lives after self-healing of asphalt concretes are shown in Table 5. From Table 5, after UV aging, the fatigue lives of all asphalt concretes before and after healing decreased obviously, with the increase of aging time, the reductions were more significant.

To quantificationally compare the abilities of the self-healing performance of asphalt concretes, the healing percentages of fatigue lives (HPFL) were calculated according to Equation (5).

$$\mathrm{HPFL} = \frac{FL_{after\ aging}}{FL_{before\ aging}} \times 100\% \tag{5}$$

where, the HPFL is the healing percentages of fatigue lives of asphalt concretes, %; $FL_{before\ aging}$ and $FL_{after\ aging}$ are the healing percentages of fatigue lives of asphalt concretes before and after UV aging respectively, times.

Table 5. Initial fatigue lives and the fatigue lives after self-healing of asphalt concretes.

Stress Ratio	Status	Before Aging	7 Days	14 Days	28 Days
0.4 stress ratio	Initial fatigue life	29543	17075	12126	9978
	After self-healing	19352	8703	2156	819
0.6 stress ratio	Initial fatigue life	19850	13435	8233	6025
	After self-healing	13413	5360	1611	379

The HPFL values of asphalt concrete after different times of UV irradiation are listed in Figure 12. From Figure 12, for the 0.4 stress ratio, the HPFL values of asphalt concretes before UV aging was 65.7%, it decreased to 21.4%, 72.9% and 87.5% after 7, 14, and 28 days of UV irradiation respectively; for the 0.6 stress ratio, the HPFL values of asphalt concretes also showed a decreasing tendency. The lower HPFL values meant a worse self-healing effect, so the self-healing effect of asphalt concretes after UV irradiation decreased, the self-healing performance of asphalt concrete decreased with increasing UV irradiation time. The reason is that the self-healing of asphalt concrete is accomplished mainly based on two patterns, firstly, due to the drain of asphalt binder into the cracks [31], it can fill the cracks and recover the structural continuity of asphalt concrete [32]; secondly, the thermal expansion of asphalt binder also plays an important role in the self-healing of asphalt concrete [33]. However, after UV aging, the viscosity of asphalt binder in its concrete is much higher than that of before aging [26]. At the same temperature, the asphalt binder in asphalt concrete after UV aging flows more slowly than that of initial asphalt concrete, the relative high viscosity limits the flow of asphalt binder [34–36], therefore results in a lower self-healing ratio.

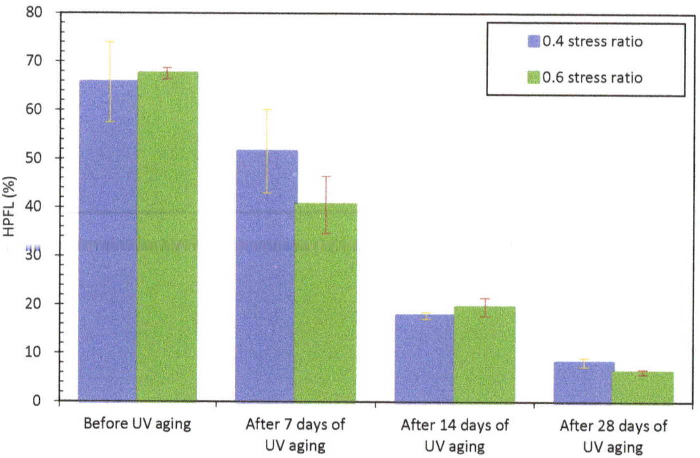

Figure 12. HPFL values of asphalt concretes after different aging times.

3.3.3. Fracture Healing Characteristics Analysis by CT Scanning Test

The CT scanning figures of asphalt concretes before and after UV irradiation are shown in Figures 13 and 14 respectively, where the self-healing effect of the cracks in the asphalt concrete intuitively can be observed. From Figure 13, the asphalt concrete had an obvious crack before self-healing, while, after self-healing, the crack almost disappeared, and the crack could not be found unless we observed very carefully, which indicated that the crack healed very well. From Figure 14, after self-healing, the crack of asphalt concrete after 28 days of UV irradiation also decreased obviously, especially the lower part of sample almost healed completely, but a small crack could still be observed in the upper part (marked in Figure 14 by the red arrow), the continuity of macro structure of asphalt concrete was not completely repaired. Therefore, despite being under the same self-healing condition, the self-healing effect of asphalt concrete before UV irradiation was much better than that of asphalt concrete after 28 days of UV irradiation. The results indicated that the asphalt concrete after UV irradiation still had a certain self-healing ability, but the self-healing ability of which was reduced obviously during UV irradiation, therefore UV radiation can significantly decrease the self-healing performance of asphalt concrete.

Figure 13. CT scanning figures of initial asphalt concretes (**a**) before self-healing; (**b**) after self-healing.

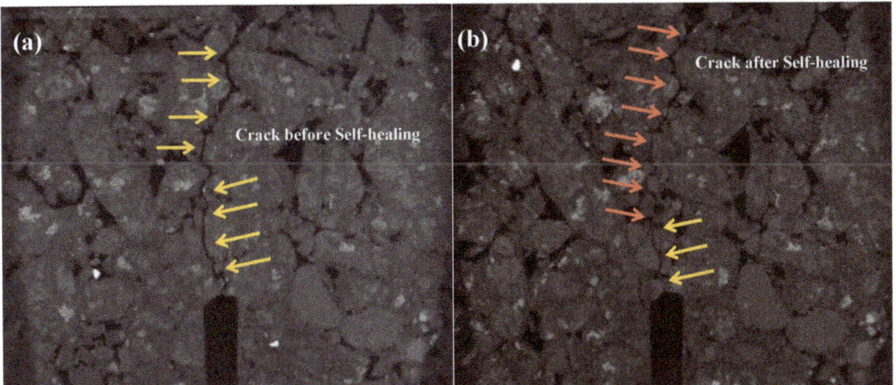

Figure 14. CT scanning figures of asphalt concretes after UV irradiation (**a**) before self-healing; (**b**) after self-healing.

4. Conclusions

In this research, the AC-13 asphalt mixture was exposed under UV radiation for 7, 14 and 28 days to investigate the degradation behaviors of the chemical structure, mechanical and self-healing properties of asphalt mixture exposed under the UV radiation, the flowing conclusions can be obtained.

(1) With the increase of UV irradiation time, the CRR and SRR values of recovered asphalt binder increased significantly, meanwhile, the fatigue life of asphalt concrete under both the 0.4 and 0.6

stress ratios decreased gradually, indicating that the UV radiation could weaken the chemical structure and fatigue resistance of asphalt concrete.

(2)   UV radiation significantly affected the asphalt semi-circular bending strength of asphalt concrete, it decreased the −10 °C bending strength of asphalt concrete, while increased the 25 °C bending strength, and the longer the exposure duration was, the more obvious the aging effect.

(3)   After UV aging, the HPBS and HPFL values of asphalt concrete decreased obviously; the CT scanning figures also showed the same tendency with HPBS and HPFL values, under the same self-healing condition. The self-healing effect of the crack in the asphalt concrete after UV irradiation was worse than that of initial asphalt concrete—a crack could still be observed in the asphalt concrete after UV irradiation. The results indicated the UV radiation could significantly reduce the self-healing performance of asphalt concrete, causing a worse macro-structure continuity, lower failure strength and worse fatigue resistance.

Author Contributions: S.W., Y.Y. and Y.L. conceived and designed the experiments. Y.L., B.S., H.L. and S.N. performed the experiments. S.W., B.S. and Y.Y. analyzed the data. C.L. (Chao Li) contributed reagents/materials/analysis tools. Y.Y., W.S. and Y.L. wrote the paper. S.W. and C.L. (Chuangmin Li) reviewed the paper.

Funding: The authors acknowledge the financial supported by the National Key Research and Development Program of China (No. 2017YFE0111600), the Natural Science Foundation of China (No. 51778515), and the Open Fund of Engineering Laboratory of Spatial Information Technology of Highway Geological Disaster Early Warning in Hunan Province (Changsha University of Science and Technology) of China (No. kfj180601). The authors also thank the test support from the Shiyanjia lab (www.shiyanjia.com), and Wuhan University of Technology for their materials and experimental instruments support.

Conflicts of Interest: The authors declare no conflict of interest.

References

1.   Bazzaz, M.; Darabi, M.K.; Little, D.N.; Garg, N. A straightforward procedure to characterize nonlinear viscoelastic response of asphalt concrete at high temperatures. *Transp. Res. Rec.* **2018**, *2672*, 481–492. [CrossRef]

2.   Zeng, W.; Wu, S.; Wen, J.; Chen, Z. The temperature effects in aging index of asphalt during UV aging process. *Constr. Build. Mater.* **2015**, *93*, 1125–1131. [CrossRef]

3.   de Sá, M.D.; Lins, V.D.; Pasa, V.M.; Leite, L.F. Weathering aging of modified asphalt binders. *Fuel Process. Technol.* **2013**, *115*, 19–25. [CrossRef]

4.   Fernández-Gómez, W.D.; Rondón Quintana, H.; Reyes Lizcano, F. A review of asphalt and asphalt mixture aging: Una revisión. *Ing. Investig.* **2013**, *33*, 5–12.

5.   Li, R.; Xiao, F.; Amirkhanian, S.; You, Z.; Huang, J. Developments of nano materials and technologies on asphalt materials–A review. *Constr. Build. Mater.* **2017**, *143*, 633–648. [CrossRef]

6.   Li, Y.; Wu, S.; Liu, Q.; Xie, J.; Li, H.; Dai, Y.; Li, C.; Nie, S.; Song, W. Aging effects of ultraviolet lights with same dominant wavelength and different wavelength ranges on a hydrocarbon-based polymer (asphalt). *Polym. Test.* **2019**, *75*, 64–75. [CrossRef]

7.   Liu, X.; Wu, S.; Liu, G.; Li, L. Effect of ultraviolet aging on rheology and chemistry of LDH-modified bitumen. *Materials* **2015**, *8*, 5238–5249. [CrossRef] [PubMed]

8.   Zhang, C.; Yu, J.; Xu, S.; Xue, L.; Cao, Z. Influence of UV aging on the rheological properties of bitumen modified with surface organic layered double hydroxides. *Constr. Build. Mater.* **2016**, *123*, 574–580. [CrossRef]

9.   Zhang, H.; Zhu, C.; Kuang, D. Physical, rheological, and aging properties of bitumen containing organic expanded vermiculite and nano-zinc oxide. *J. Mater. Civ. Eng.* **2015**, *28*, 04015203. [CrossRef]

10.   Xiao, F.; Amirkhanian, S.N.; Karakouzian, M.; Khalili, M. Rheology evaluations of WMA binders using ultraviolet and PAV aging procedures. *Constr. Build. Mater.* **2015**, *79*, 56–64. [CrossRef]

11.   Lins, V.; Araújo, M.; Yoshida, M.; Ferraz, V.; Andrada, D.; Lameiras, F. Photodegradation of hot-mix asphalt. *Fuel* **2008**, *87*, 3254–3261. [CrossRef]

12. Vallerga, B.; Monismith, C.; Granthem, K. A study of some factors influencing the weathering of paving asphalts. In *Assoc Asphalt Paving Technol Proc*; Association of Asphalt Paving Technologists: Lino Lake, MN, USA, 1957.

13. Zhang, D.; Zhang, H.; Shi, C. Investigation of aging performance of SBS modified asphalt with various aging methods. *Constr. Build. Mater.* **2017**, *145*, 445–451. [CrossRef]

14. Glotova, N.; Kats, B.; Gorshkov, V. Photooxidation of asphalts in thin films. *Chem. Technol. Fuels Oils.* **1974**, *10*, 876–879. [CrossRef]

15. Wu, S.; Pang, L.; Liu, G.; Zhu, J. Laboratory study on ultraviolet radiation aging of bitumen. *J. Mater. Civ. Eng.* **2010**, *22*, 767–772. [CrossRef]

16. Li, Y.; Wu, S.; Dai, Y.; Pang, L.; Liu, Q.; Nie, S.; Li, H.; Wang, Z. Laboratory and field evaluation of sodium stearate organically modified LDHs effect on the anti aging performance of asphalt mixtures. *Constr. Build. Mater.* **2018**, *189*, 366–374. [CrossRef]

17. Norambuena-Contreras, J.; Yalcin, E.; Garcia, A.; Al-Mansoori, T.; Yilmaz, M.; Hudson-Griffiths, R. Effect of mixing and ageing on the mechanical and self-healing properties of asphalt mixtures containing polymeric capsules. *Constr. Build. Mater.* **2018**, *175*, 254–266. [CrossRef]

18. Wu, S.; Zhao, Z.; Li, Y.; Pang, L.; Amirkhanian, S.; Riara, M. Evaluation of aging resistance of graphene oxide modified asphalt. *Appl. Sci.* **2017**, *7*, 702. [CrossRef]

19. ASTM, D. Standard test method for penetration of bituminous materials. In *Annual Book of ASTM Standards USA*; ASTM International: West Conshohocken, PA, USA, 1992.

20. Standard, A. D36. In *Standard Test Method for Softening Point of Bitumen (Ring-and-Ball Apparatus)*; ASTM International: West Conshohocken, PA, USA, 2009.

21. ASTM. *Standard Test Method for Ductility of Bituminous Materials*; Annu Book ASTM Stand; ASTM International: West Conshohocken, PA, USA, 1979.

22. ASTM. *Standard Test Method for Viscosity Determination of Asphalt at Elevated Temperatures Using a Rotational Viscometer*; ASTM International: West Conshohocken, PA, USA, 2012.

23. Zhong, K.; Cao, D.W.; Luo, S. Determination the modifier content in SBS modified asphalt based on infrared spectroscopy technique. *Appl. Mech. Mater.* **2010**, *34*, 1129–1134. [CrossRef]

24. Hou, X.; Lv, S.; Chen, Z.; Xiao, F. Applications of Fourier transform infrared spectroscopy technologies on asphalt materials. *Measurement* **2018**, *121*, 304–316. [CrossRef]

25. Sun, D.Q.; Zhang, L.W.; Zhang, X.L. Quantification of SBS content in SBS polymer modified asphalt by FTIR. *Adv. Mater. Res.* **2011**, *287*, 953–960. [CrossRef]

26. Li, Y.; Wu, S.; Pang, L.; Liu, Q.; Wang, Z.; Zhang, A. Investigation of the effect of Mg-Al-LDH on pavement performance and aging resistance of styrene-butadiene-styrene modified asphalt. *Constr. Build. Mater.* **2018**, *172*, 584–596. [CrossRef]

27. Xue, Y.; Hu, Z.; Wang, C.; Xiao, Y. Evaluation of dissolved organic carbon released from aged asphalt binder in aqueous solution. *Constr. Build. Mater.* **2019**, *218*, 465–476. [CrossRef]

28. Misra, A.; Poorsolhjouy, P. Granular micromechanics based micromorphic model predicts frequency band gaps. *Contin. Mech. Thermodyn.* **2016**, *28*, 215–234. [CrossRef]

29. Nejadsadeghi, N.; Placidi, L.; Romeo, M.; Misra, A. Frequency band gaps in dielectric granular metamaterials modulated by electric field. *Mech. Res. Commun.* **2019**, *95*, 96–103. [CrossRef]

30. Li, Y.; Wu, S.; Liu, Q.; Nie, S.; Li, H.; Dai, Y.; Pang, L.; Li, C.; Zhang, A. Field evaluation of LDHs effect on the aging resistance of asphalt concrete after four years of road service. *Constr. Build. Mater.* **2019**, *208*, 192–203. [CrossRef]

31. García, Á. Self-healing of open cracks in asphalt mastic. *Fuel* **2012**, *93*, 264–272. [CrossRef]

32. Ayar, P.; Moreno-Navarro, F.; Rubio-Gámez, M.C. The healing capability of asphalt pavements: A state of the art review. *J. Clean. Prod.* **2016**, *113*, 28–40. [CrossRef]

33. Grossegger, D.; Garcia, A. Influence of the thermal expansion of bitumen on asphalt self-healing. *Appl. Therm. Eng.* **2019**, *156*, 23–33. [CrossRef]

34. Liu, Q.; García, Á.; Schlangen, E.; van de Ven, M. Induction healing of asphalt mastic and porous asphalt concrete. *Constr. Build. Mater.* **2011**, *25*, 3746–3752. [CrossRef]

35. Little, D.N.; Prapnnachari, S.; Letton, A.; Kim, Y. *Investigation of the Microstructural Mechanism of Relaxation and Fracture Healing in Asphalt*; Texas Transportation Inst College Station: College Station, TA, USA, 1993.

36. Lv, Q.; Huang, W.; Zhu, X.; Xiao, F. On the investigation of self-healing behavior of bitumen and its influencing factors. *Mater. Des.* **2017**, *117*, 7–17. [CrossRef]

Article

Investigation of the Effect of Induction Heating on Asphalt Binder Aging in Steel Fibers Modified Asphalt Concrete

Hechuan Li [1], Jianying Yu [1], Shaopeng Wu [1], Quantao Liu [1,*], Yuanyuan Li [1], Yaqi Wu [2] and Haiqin Xu [1]

[1]   State Key Laboratory of Silicate Materials for Architectures, Wuhan University of Technology, Wuhan 430070, China; lihc@whut.edu.cn (H.L.); jyyu@whut.edu.cn (J.Y.); wusp@whut.edu.cn (S.W.); liyuanyuan@whut.edu.cn (Y.L.); xuhaiqin@whut.edu.cn (H.X.)

[2]   School of Foreign languages, China University of Geosciences (Wuhan), Wuhan 430074, China; Vilcky2012@163.com

*   Correspondence: liuqt@whut.edu.cn

Received: 15 March 2019; Accepted: 29 March 2019; Published: 1 April 2019

Abstract: Induction heating is a valuable technology to repair asphalt concrete damage inside. However, in the process of induction heating, induced particles will release a large amount of heat to act on asphalt binder in a short time. The purpose of this paper was to study the effect of induction heating on asphalt binder aging in steel fibers modified asphalt concrete. The experiments were divided into two parts: induction heating of Dramix steel fibers coated with asphalt binder (DA) and steel wool fibers modified asphalt concrete. After induction heating, the asphalt binders in the samples were extracted for testing aging indices with Fourier Transform Infrared (FTIR), Dynamic Shear Rheometer (DSR), and Four-Components Analysis (FCA) tests. The aging of asphalt binder was analyzed identifying the change of chemical structure, the diversification of rheological properties, and the difference of component. The experiments showed that the binder inside asphalt concrete began aging during induction heating due to thermal oxygen reaction and volatilization of light components. However, there was no peak value of the carbonyl index after induction heating of ten cycles, and the carbonyl index of DA was equivalent to that of binder in asphalt concrete after three induction heating cycles, which indicated the relatively closed environment inside asphalt concrete can inhibit the occurrence of the aging reaction.

Keywords: asphalt concrete; induction heating; asphalt binder aging; FTIR; DSR; FCA

---

1. Introduction

It has been widely proven that asphalt concrete is a self-healing material and induction heating can magnify the healing ability to extend the pavement service life [1–4]. The healing mechanisms of asphalt concrete have been reported by many researchers. Bitumen is traditionally regarded as a colloidal system consisting of high molecular weight asphaltene micelles dispersed or dissolved in the lower molecular weight oily maltenes [5]. Castro and Sánchez explained the healing of asphalt mixes during rest periods by sol–gel theories [6]. Phillips proposed a three steps diffusion model to explain the healing of bitumen: (1) surface approach due to consolidating stresses and bitumen flow, (2) wetting (adhesion of two cracked surfaces to each other driven by surface energy density), and (3) diffusion and randomization of asphaltene structures. The first two steps cause the recovery of the modulus (stiffness), and the third step causes recovery of the strength [7,8]. Kringos et al. used a chemo-mechanical model to simulate healing of bitumen [9]. Garcia et al. applied capillary flow theory and flow behavior factor index to explain the healing mechanism [10,11]. All these studies provide some basis for the study of self/induction healing behavior of asphalt concrete.

The success of the test section has further accelerated the process of wide application of this technology [1,12]. Subsequently, a series of improvement studies focused on induction heating technology were conducted, among which improving induction heating efficiency became the key factor to optimize the healing ability of asphalt concrete [13–19]. High temperature is conducive to the occurrence and rapid progress of the healing process, which results in a higher healing rate of asphalt concrete. Liu et al. found that the optimal heating temperature was 85 °C to obtain the best healing effect for the asphalt concrete designed in his research [15]. Menozzi et al. pointed out that the total lifetime of asphalt mixture under fatigue could be increased at approximately 55 °C [19]. García et al. stated that the mechanical resistance of the test samples could be recovered up to 60% at around 100 °C, and the effect of multiple healing cycles was not compromised [20].

The above-mentioned advantages of induction heating technology, such as high induction heating efficiency, high healing rate, and multiple induction healings, emphasize the importance of temperature, especially temperature which is high enough. High temperature makes the asphalt binder to obtain better fluidity to enhance healing performance. However, it brings a significant potential damage to the asphalt material. During the process of induction heating, induced particles will produce a large amount of heat to act on asphalt binder in a short time, which may result in the possibility of thermal oxygen aging of asphalt binder. Researchers also hope that asphalt concrete can be repaired for multiple times, which greatly increases the likelihood of aging happening. Garcia et al. demonstrated there was no aging phenomenon of the binder in induction heating of asphalt mastic [18]. But Menozzi et al. [19] stated the change of flow behavior factor of the binder in asphalt mixture, which proved that the binder was aged during induction heating. These studies were not focused on the aging of asphalt binder during induction heating. Thus, there were obvious shortcomings. Garcia only studied aging of asphalt mastic, whose void volume was much smaller than that of actual pavement, and the volume of asphalt binder was too large to react with less oxygen inside asphalt mastic adequately. While Menozzi did not analyze the changes of oxygen functional groups in asphalt binders after induction heating, which is considered to be an important evidence of asphalt binder aging. In addition, their studies both concerned the aging of asphalt binders after primary induction heating, without mentioning the case after multiple induction heating.

In view of the differences and problems in the above researches, it is necessary to redesign the experiments to study the effect of induction heating on asphalt binder aging, which will provide an important basis for the application of electromagnetic induction technology in asphalt concrete. In this paper, the experiments were divided into two parts: induction heating test of Dramix steel fibers coated with asphalt (DA) and induction heating test of steel wool fibers modified asphalt concrete. Dramix steel fiber is a tough steel fiber, and it was coated with asphalt binder film to make an induction heating binder-fiber sample, while the asphalt concrete sample was cut from the rutting plate. After induction heating test of different times under different oxygen concentration, the internal asphalt binders were extracted for testing the aging indices with Fourier Transform Infrared (FTIR) test, Dynamic Shear Rheometer (DSR) test, and Four-Components Analysis (FCA) test [21–23].

2. Materials and Experiments

2.1. Materials

AH-70# base asphalt obtained from Hubei Guochuang Hi-tech Material Co., Ltd., of China (Yichang, China) was used in this paper. Steel wool fibers provided by Jiangsu Golden Torch Metal products Co., Ltd (Yancheng, China) and Dramix steel fibers provided by Bekaert Corp. (Brussels, Belgium) were used as the heating units for asphalt mixtures and asphalt binders via induction heating, respectively. The morphology of the two kinds of fibers are shown in Figure 1, and the properties of bitumen, steel wool fiber, and Dramix steel fiber are shown in Table 1. The optimal content of fibers was 6% by the volume of asphalt according to previous researches [4]. Asphalt mixture with 6%

steel wool fibers had the best mechanical properties (highest strength and particle loss resistance) and acceptable induction heating speed. Basalt aggregate and limestone filler were used in this study.

Figure 1. Morphology of steel wool fibers (**a**) and Dramix steel fibers (**b**).

Table 1. The properties of bitumen and steel fibers.

Materials	Properties	Values	Specifications
Bitumen	Penetration (25 °C, 100 g, 5 s, 0.1 mm)	68	60–80
	Ductility (15 °C, cm)	>100	100
	Softening point (°C)	47.5	47
	Density (g/cm^3)	1.034	-
Steel wool fiber	Average length (mm)	4.2	-
	Equivalent diameter (µm)	70–130	-
	Density (g/cm^3)	7.8	-
	Average heating rate (°C/s)	12.2	
Dramix steel fiber	Average length (mm)	13	-
	Equivalent diameter (µm)	200	-
	Density (g/cm^3)	7.8	-
	Minimum tensile strength (N/mm^2)	2	-
	Average heating rate (°C/s)	8.5	

2.2. Specimen Preparation

2.2.1. Dramix Steel Fiber Covered with Asphalt Binder Film (DA)

To better study the effect of high temperature produced via induction heating on asphalt binder aging, Dramix steel fibers were immersed into the heated asphalt to obtain the asphalt-fiber samples wrapped by 0.5 mm asphalt film, which were prepared for an FTIR test and FCA test, as shown in Figure 2a. For DSR test, it needed more asphalt, 3.9 g asphalt was poured into a glass dish with a diameter of 9.35 cm to obtain 0.57 mm asphalt film. The number of Dramix steel fibers was 100 to ensure adequate induction heating ability, as shown in Figure 2b.

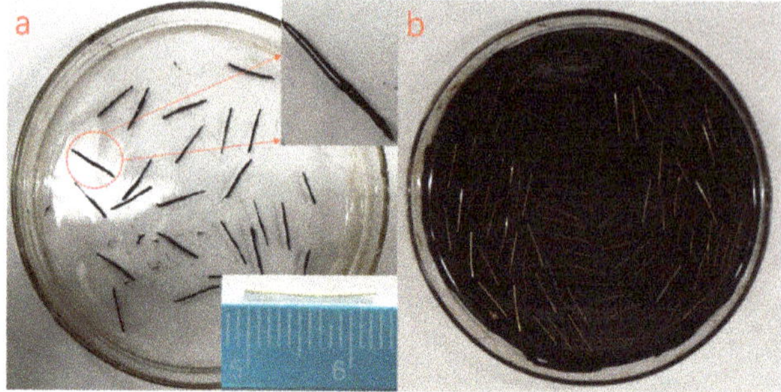

Figure 2. Dramix steel fibers coated with asphalt binder (DA) for (**a**) Fourier Transform Infrared (FTIR) and Four-Components Analysis (FCA) tests and (**b**) Dynamic Shear Rheometer (DSR) test.

2.2.2. Asphalt Concrete Sample

AC-13 basalt asphalt concrete including steel wool fibers was designed in this paper according to Marshall design method, and the asphalt/aggregate ratio was 4.7%. The aggregate grading curves of asphalt mixtures are shown in Figure 3. In the induction heating test, the specimen was a rectangular beam cut from rutting plate in a size of 85 mm × 50 mm × 10 mm. It is important to note that in our design concept, the thinner the sample is, the better the accuracy of the test results is. There were three considerations for making the asphalt concrete specimen this size: First, in the application of induction heating, there exists gradient heating and healing, and the temperature of the surface layer is the most significant, which has been proved in previous studies [4,24], so the asphalt binders aging at the surface layer during induction heating are more representative. Second, thinner samples can eliminate as much ununiformity as possible in the binder aging of different layers due to gradient heating, which is required because we need to extract the asphalt from the mixture after induction heating. Thicker asphalt concrete sample may result in inaccurate aging testing due to an excessive gradient. Finally, the asphalt concrete gradation causes the above size to be the limit of asphalt concrete sample. Otherwise, there will be a lot of broken stones. The cut samples are shown in Figure 4.

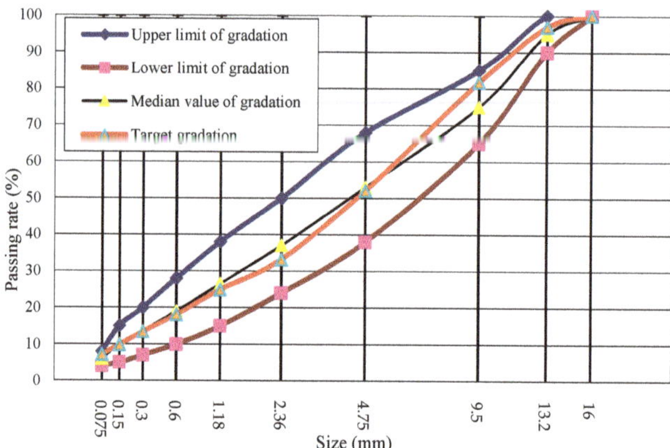

Figure 3. Grading curve of AC-13 steel wool fibers modified basalt asphalt mixture.

Figure 4. Induction heating test samples.

2.3. Induction Heating Test

The power needed for induction heating is dependent on the conductivity and size of the specimen used for the test. Figure 5 shows the induction heating test of Dramix asphalt-fibers and asphalt concrete, corresponding to power 8.8 kW and 8.4 kW, respectively. The frequency of the induction heating apparatus was always 123 kHz. The distance between the coil and the surface of the specimen has a significant influence on the heating speed. The distance used in this paper for all samples was 10 mm according to the preliminary study [4], as shown in Figure 5. The induction heating times were 10 s for Dramix asphalt-fibers and 40 s for asphalt concrete while the corresponding temperature was 85 °C, which had proven to be the best temperature for induction healing in previous studies [15]. An infrared image camera with a resolution of 320 × 240 pixels was used to detect the temperature of the samples during induction heating, as shown in Figure 6. After each induction heating, the samples had enough time to recover to ambient temperature before starting the next induction heating test. Additionally, the phenomenon of the gradient temperature during induction heating had been demonstrated in previous studies [11,24], so in this paper, the gradient aging during induction heating was studied, and the three beams were stacked together, as shown in Figure 7. For all induction heating tests, once the surface average temperature reached 85 °C, the induction heating was stopped. For the three-layer beams induction heating test, to amplify the effect of gradient on the asphalt binder aging, the sample was heated 10 times in all, still giving the sample a full time to return to ambient temperature after each induction heating. In this paper, all Dramix steel fibers samples were induction heated 1 time. Pure asphalt (PA), as a contrast sample, was also conducted with FTIR, DSR, and FCA test. The test summary is shown in Table 2.

Figure 5. Induction heating test of Dramix asphalt-fibers (**a**) and asphalt concrete (**b**).

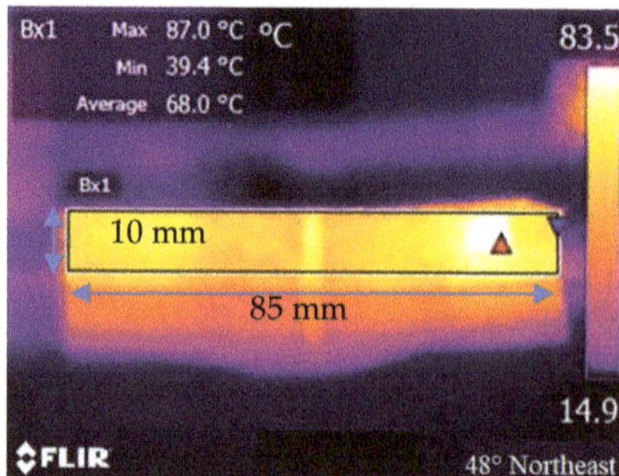
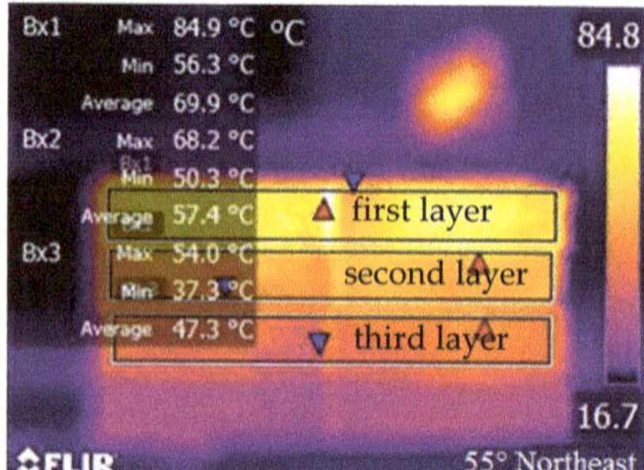

Figure 6. Infrared imaging of samples during induction heating.

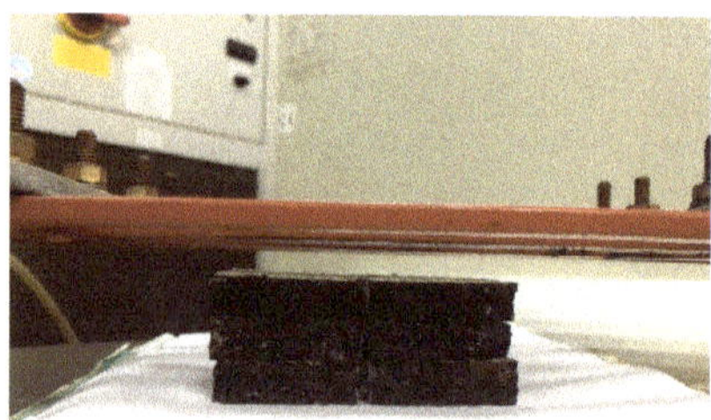

Figure 7. Induction heating of three-layer beams.

Table 2. The test summary.

Specimen	Number	Induction Heating Times	Test
PA	–	0	
Single-layer beam	1	1	FTIR, DSR & FCA
	1	2	
	1	3	
	1	5	
	1	10	
DM	25 & 100	1	
Three-layer beams	1	10	FTIR

2.4. Extraction of Asphalt Binder

After induction heating, the asphalt binder was extracted by dissolution–filtration when the samples were restored to ambient temperature. For the Dramix asphalt-fibers, trichloroethylene was poured into the glass dish containing the Dramix asphalt-fibers [25]. After the asphalt coated on the Dramix steel fibers was dissolved, the fibers were removed, and the binder was stored in a closed tube for future testing. The experimental operation is shown in Figure 8. Regarding asphalt concrete, after the asphalt binder was dissolved, the solid solution mixture was filtered to separate the asphalt binder and other substances. To improve the dissolution effect and prevent the volatilization of trichloroethylene, it was necessary to dissolve the binder in a sealed state. And it is important to note in particular that the particle size of limestone filler was less than 0.075 mm. Inadequate filtration may result in the retention of limestone filler in asphalt binder, which in turn may have an impact on the test results. Therefore, the particle size of limestone fillers was determined by laser particle size analyzer (Mastersizer 2000), as shown in Figure 9. The minimum particle size of the limestone fillers was 0.363 μm, so the slow speed filter paper (pore size is 1~3 μm) was enough to filter out the most limestone filler from the asphalt mastic to obtain the pure asphalt binder. The extraction process of asphalt binder in asphalt concrete is shown in Figure 10.

Figure 8. Extraction and storage of asphalt binder coated on Dramix steel fibers.

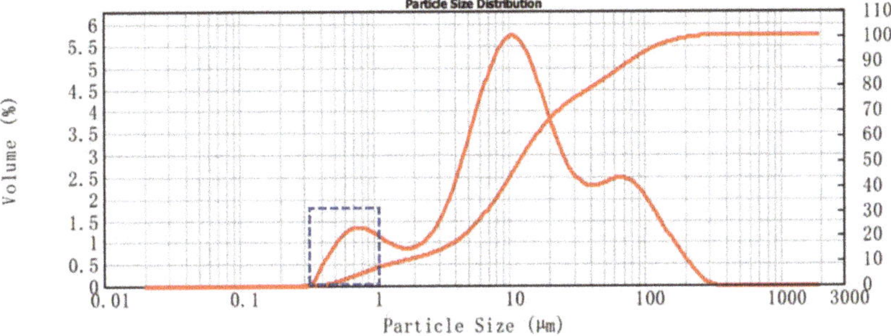

Figure 9. Laser particle size analysis of limestone fillers.

Figure 10. Extraction process of asphalt binder in asphalt concrete: dissolution (**a**), filtration (**b**), and volatilization (**c**).

2.5. Fourier Transform Infrared (FTIR) Test

The chemical structures of the virgin asphalt and extractive asphalt binders before and after induction heating were studied by a Fourier transform infrared (FTIR, Nexus, ThermoNicolet Crop., Waltham, MA, USA). In this paper, FTIR tests were performed under a scanned area between 4000 and 400 cm^{-1}. And the scanning resolution was 4 cm^{-1}. The carbonyl functions C=O (centered around 1700 cm^{-1}) were monitored by studying their changes in spectra. The carbonyl functions C=O can provide the information about oxidation of asphalt. The carbonyl functions C=O index could be calculated by the area of its bands by the following equation [26]:

$$I_{C=O} = \frac{\text{Area of carbonyel band centered around 1700 cm}^{-1}}{\sum \text{Area of spectral bands between 2000 and 600 cm}^{-1}} \tag{1}$$

2.6. Dynamic Shear Rheometer (DSR) Test

DSR (MCR101, Anton Paar, Graz, Austria) was applied to investigate the rheological properties of extracted asphalt binder previously. Frequency sweep test was performed in the range of 40 to 76 °C. Seven temperature levels were measured at intervals of 6 °C. At each temperature level, the frequency scanning range was 0.1~100 Hz. To ensure the uniform temperature of the specimen, the temperature equilibrium time of the asphalt specimen at each sweeping temperature was 3 min. A plate with 25 mm diameter and 1 mm gap were used. According to Liu et al.'s research [27], the master curve was simulated through the logPen. model.

2.7. Four-Component Analysis (FCA) Test

Asphalt consists of various molecular weights of hydrocarbons and their derivatives and based on the relative molecular size and polarity of asphalt, it can be divided into four components, namely saturates, aromatics, resins, and asphaltenes. To test the effects of induction heating on these four components, asphalt binders induced heated in different conditions and times were tested through TLC-FID (Iatron Laboratories Inc., Tokyo, Japan). According to [23], two percent (*w/v*) solutions of asphalt binders were prepared in dichloromethane, and 1 µL sample solution was spotted on chromarods. There was a three-stage process for the separation of asphalt fractions. The first stage was in n-heptane (70 mL) and expanded to 100 mm of the chromarods, the second stage in toluene/n-heptane (70 mL, 4/1 by volume) was developed to 50 mm of the chromarods, and the last development was in toluene/ethanol (70 mL, 11/9 by volume) and expanded to 25 mm of the chromarods. The solvent was dried in an oven at 80 °C after each stage. Then, the chromarods were scanned in the TLC-FID analyzer. Four chromarods were tested for each sample, and finally, the average values were used as the results. In this paper, 10 parallel tests were performed on each sample.

3. Results and Discussion

3.1. DA Induction Heating

Figure 11 shows the infrared images of Dramix steel fibers and steel wool fibers after induction heating for 10 s and 5 s, respectively. It was found that the maximum temperature reached to 142 °C and 87.1 °C, with a heating rate 8.5 °C/s and 12.2 °C/s. Predictably, if the heating time is prolonged, the temperature of the fibers will be higher. Because of the rapid temperature rising of asphalt wrapped on the surface of fibers, asphalt binder is subjected to very high thermal radiation directly, which is enough to cause the asphalt binder aging.

Dramix asphalt-fibers were completely exposed to air during induction heating. In this condition, the amount of oxygen was sufficient for binder aging, which could rule out that there would be no binder aging in asphalt concrete due to lack of oxygen. The purpose of the experiment was to verify whether asphalt binders would age in the case of sufficient oxygen during induction heating.

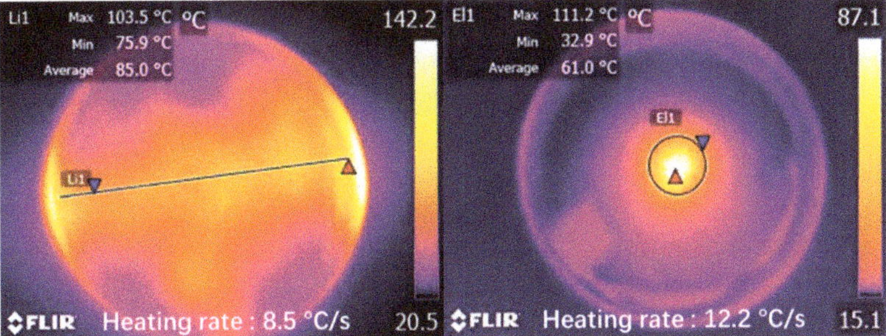

Figure 11. Infrared images of Dramix steel fibers and steel wool fibers after induction heating.

3.2. Rheological Properties Analysis

The master curves of complex moduli and phase angles of asphalt binders after induction heating in different condition are shown in Figures 12 and 13, respectively. The reference temperature for establishing the master curves was 46 °C. From Figures 12 and 13, the complex moduli and phase angle of all asphalt binders changed obviously after induction heating, and the change was aggravated with the increase of induction heating times. After induction heating, the complex moduli and phase angle of asphalt binders increased and decreased severally. Compared with the complex moduli, the change and trend of phase angle were more irregular, but it still proved that the asphalt binder was aging during induction heating. The increase tendency of complex moduli indicated that the resistance of asphalt to deformation under repeated shear loading increased. Meanwhile, the decrease tendency of phase angle demonstrated that the ratio of elastic modulus (or storage modulus) to complex modulus increased after induction heating. These were the aging characteristics of asphalt binders. These changes can be explained in two aspects: On the one hand [28], some asphalt molecules produced oxygen-containing functional groups with higher molecular weights as a result of the oxidation reaction. On the other hand [29], due to the volatilization of the light components in the asphalt at high temperature, the proportion of the light component fraction decreased, while the proportion of the weight component fraction increased. This imbalance of components can lead to the deterioration of physical and rheological properties of asphalt binders, which will be validated and discussed according to FTIR test and FCA test in the following two chapters.

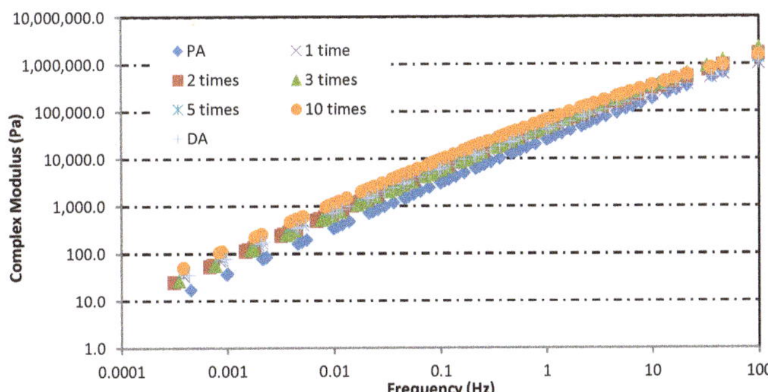

Figure 12. Complex modulus master curve of asphalt binders in different induction heating conditions.

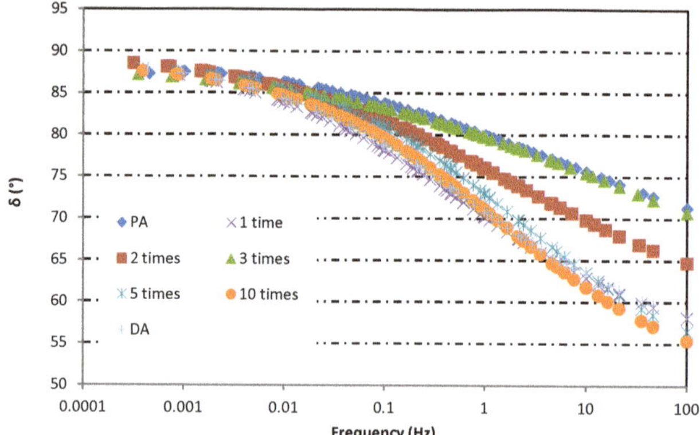

Figure 13. Phase angle master curve of asphalt binders in different induction heating conditions.

3.3. Chemical Structure Analysis

3.3.1. Multiple Induction Heating

Figure 14 shows the changing areas of carbonyl functions and Figure 15 presents the changes in FTIR index of asphalt binders obtained from different induction heating conditions. From Figure 14, it can be observed that with the increase of induction heating times, the carbonyl index in asphalt binder increased gradually, but there was no peak value. For pure asphalt, $I_{C=O}$ was 0.000531, which was far below the value of other samples, indicating that slow aging occurred during production and storage. After induction heating for 10 cycles, $I_{C=O}$ reached to 0.006231, which demonstrated that the asphalt binder aging was very obvious at this time. However, for DA, its carbonyl index was equivalent to the value of the asphalt binder inside the asphalt concrete after inducting heating for three cycles, which was more serious than that after induction heating for one and two cycles. This showed that in the case of more oxygen, the asphalt binder did incur more serious thermal oxygen aging during induction heating, and in the meantime, the relatively closed space in asphalt concrete could restrain the aging.

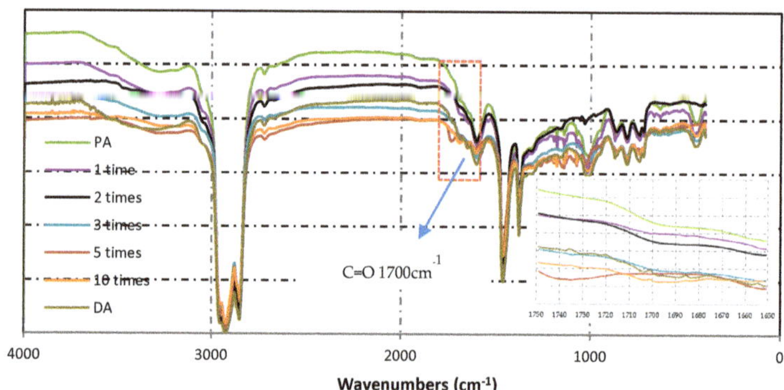

Figure 14. FTIR spectra of asphalt binders before and after induction heating.

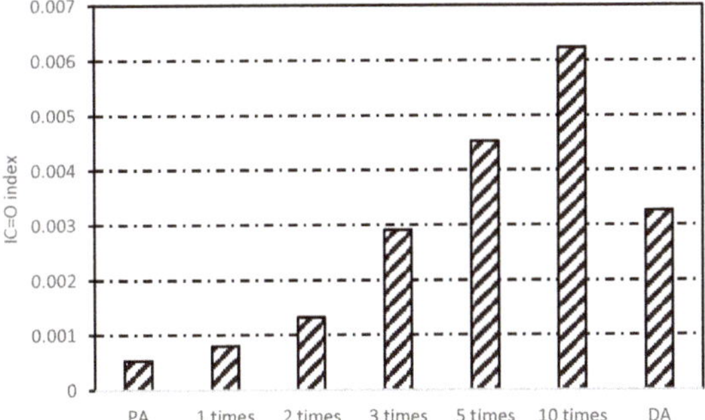

Figure 15. $I_{C=O}$ variation of asphalt binders before and after induction heating.

3.3.2. Gradient Heating of Three-Layer Beams

Figure 16 shows the $I_{C=O}$ index of asphalt binder after induction heating in different layers. The $I_{C=O}$ of the one, two and three layers were 0.00619, 0.00516, and 0.00384 in turn, which was the same gradient phenomenon with the induction heating temperature (40 s, 8.6 kW) in Figure 6. The carbonyl index of asphalt binder in the surface layer was similar to that in asphalt concrete which was induced heated for 10 cycles in the preceding section. With the decrease of induction temperature, the carbonyl functional groups produced inside the asphalt binder gradually decreased, indicating that the gradient aging of binder happened during induction heating. The surface average heating temperature was 85 °C, which was not enough to cause the asphalt aging in theory, but the actual situation was that even if the average temperature of the third layer was only 47.3 °C, the asphalt binder was still aged. This was easy to understand, because the so-called average temperature was the average temperature of asphalt concrete, and from the above study, it was found that steel fibers gave off a lot of heat in a short period during induction heating, which was enough for the asphalt binders aging. From FTIR tests, it was found that one of the reasons for the asphalt binders aging was the thermal oxygen aging reaction of asphalt molecules with oxygen at high temperature, and the other mechanism of asphalt aging will be explained in the next chapter.

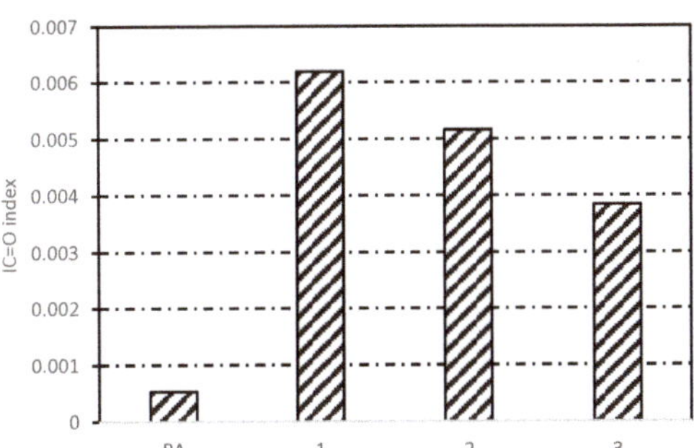

Figure 16. $I_{C=O}$ index of asphalt binder after induction heating in different layers.

3.4. Four-Component Analysis

Figure 17 shows the component fractions of asphalt binders after induction heating in different conditions. Significant changes of four component fractions of different asphalt binders took place. After induction heating for 10 cycles, the fractions of saturates decreased from 20.29% to 13.69%, and the fractions of aromatic decreased from 37.53% to 30.16%, while the fractions of resins increased from 29.08% to 35.82%, and the fractions of asphaltenes increased from 13.10% to 20.33%. From Figure 17, the trend of components changes showed that the number of induction heating cycles was the decisive factor. The change of component fractions may be due to volatilization of light components (saturates and aromatic) or transition of light components to weight components at high temperature. In addition, the component fractions of DA were similar to that of the asphalt binder after induction heating for 3 cycles. This may be because open space was more conducive to the volatilization of light components, thus, avoiding the reattachment of volatile light components to the interior of asphalt concrete.

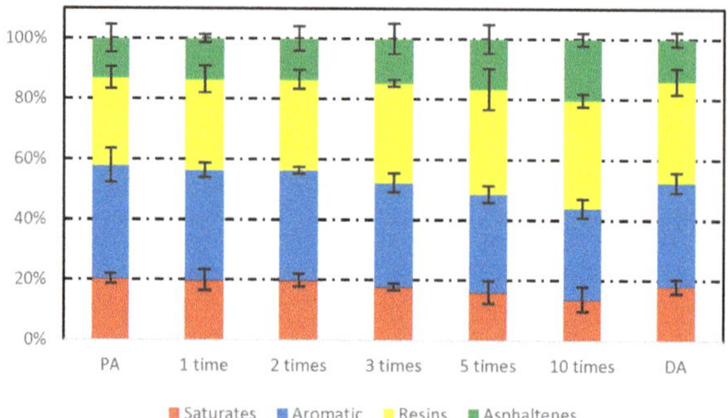

Figure 17. Component fractions of asphalt binders before and after induction heating in different conditions.

4. Conclusions

In this research, the effect of induction heating on asphalt binder aging in steel fibers modified asphalt concrete was investigated. Based on the results discussed above, the following conclusions could be drawn:

- It was demonstrated that the asphalt binder inside asphalt concrete began aging during induction heating due to the rapid temperature rise of asphalt wrapped on the surface of fibers, whose aging mechanisms were thermal oxygen aging and volatilization of light components or transition of light components to weight components.
- According to DSR, the complex moduli and phase angle of asphalt binders increased and decreased severally after induction heating, indicating that the rheological properties of asphalt binders changed.
- For the binder inside asphalt concrete, there was no peak value of carbonyl index after ten cycles of induction heating, and the carbonyl index of DA was equivalent to that of asphalt binder after three cycles induction heating, indicating that relatively closed environment inside the asphalt concrete could restrain the thermal oxygen aging.
- The number of induction heating was the decisive factor to influence the change of asphalt binder component fractions, and the binder component fractions changed more slowly compared to DA.
- Although the asphalt binder aging inside asphalt concrete was slower, it was still necessary to study the effect of binders aging on the healing performance of asphalt concrete.

Further research: The test will be conducted in two aspects: On the one hand, the change in healing performance of the asphalt binder itself will be investigated, and we plan to evaluate it by the flow behavior factor index, which is the most recognized method now. On the other hand, we will design experiments to verify whether the incomplete strength recovery of asphalt mixture after induction healing is related to the aging of asphalt binder.

Author Contributions: Conceptualization, H.L. and Q.L.; Data curation, H.L. and H.X.; Formal analysis, H.L.; Funding acquisition, S.W.; Investigation, H.L.; Methodology, H.L. and Y.L.; Writing–original draft preparation, H.L.; Writing–review and editing, Q.L.; Project administration, S.W.; Resources, Y.W.; Software, H.L.; Supervision, J.Y. and Q.L.

Funding: This research was funded by National Key R&D Program of China (No. 2017YFE0111600) National Natural Science Foundation of China (No. 51778515 and No. 51508433).

Conflicts of Interest: The authors declare no conflict of interest.

References

1. Liu, Q.; García, Á.; Schlangen, E.; van de Ven, M. Induction healing of asphalt mastic and porous asphalt concrete. *Constr. Build. Mater.* **2011**, *25*, 3746–3752. [CrossRef]

2. Liu, Q.; García, Á.; Schlangen, E.; van de Ven, M. Induction heating of electrically conductive porous asphalt concrete. *Constr. Build. Mater.* **2010**, *24*, 1207–1213. [CrossRef]

3. Norambuena-Contreras, J.; Garcia, A. Self-healing of asphalt mixture by microwave and induction heating. *Mater. Des.* **2016**, *106*, 404–414. [CrossRef]

4. Liu, Q.; Chen, C.; Li, B.; Sun, Y.; Li, H. Heating Characteristics and Induced Healing Efficiencies of Asphalt Mixture via Induction and Microwave Heating. *Materials* **2018**, *11*, 913. [CrossRef] [PubMed]

5. Read, J.; Whiteoak, D. *The Shell Bitumen Handbook*, 5th ed.; Thomas Telford Ltd.: London, UK, 2003.

6. Castro, M.; Sánchez, J.A. Fatigue and Healing of Asphalt Mixtures: Discriminate Analysis of Fatigue Curves. *Transp. Eng.* **2006**, *132*, 168–174. [CrossRef]

7. Phillips, M.C. Multi-step models for fatigue and healing, and binder properties involved in healing. In Proceedings of the Eurobitume Workshop on Performance Related Properties for Bituminous Binders, Kirchberg, Luxembourg, 3–6 May 1999.

8. Little, D.N.; Bhasin, A. Exploring mechanisms of healing in asphalt mixtures and quantifying its impact. In *Self Healing Materials*; Springer: Dordrecht, The Netherlands, 2007; Volume 100, pp. 205–218.

9. Kringos, N.; Schmets, A.; Pauli, T.; Scarpas, T. A finite element base chemo-mechanical model to similate healing in bitumen. In Proceedings of the International Workshop on Chemo-mechanics of Bituminous materials, Delft, The Netherlands, 9–11 June 2009.

10. García, Á. Self-healing of open cracks in asphalt mastic. *Fuel* **2011**, *93*, 264–272. [CrossRef]

11. García, Á.; Schlangen, E.; van de Ven, M.; Liu, Q. A simple model to define induction heating in asphalt mastic. *Constr. Build. Mater.* **2012**, *31*, 38–46. [CrossRef]

12. Liu, Q.; Schlangen, E.; van de Ven, M. Characterization of the material from the induction healing porous asphalt concrete trial section. *Mater. Struct.* **2013**, *46*, 831–839. [CrossRef]

13. Liu, Q.; Wu, S.; Schlangen, E. Induction heating of asphalt mastic for crack control. *Constr. Build. Mater.* **2013**, *4*, 345–351. [CrossRef]

14. Liu, Q.; Yu, W.; Schlangen, E.; van Bochove, G. Unravelling Porous Asphalt Concrete with Induction Heating. *Constr. Build. Mater.* **2014**, *71*, 152–157. [CrossRef]

15. Liu, Q.; Schlangen, E.; van de Ven, M.; van Bochove, G.; van Montfort, J. Evaluation of the induction healing effect of porous asphalt concrete through four-point bending fatigue test. *Constr. Build. Mater.* **2012**, *29*, 403–409. [CrossRef]

16. Garcia, A.; Schlangen, E.; Van de Ven, M. Two Ways of Closing Cracks on Asphalt Concrete Pavements: Microcapsules and Induction Heating. *Key Eng. Mater.* **2009**, *417*, 573–576. [CrossRef]

17. García, A.; Norambuena-Contreras, J.; Partl, M.N. Experimental evaluation of dense asphalt concrete properties for induction heating purposes. *Constr. Build. Mater.* **2013**, *46*, 48–54. [CrossRef]

18. García, A.; Schlangen, E.; Van de Ven, M.; van Vliet, D. Induction heating of mastic containing conductive fibers and fillers. *Mater. Struct.* **2011**, *44*, 499–508. [CrossRef]

19. Menozzi, A.; Garcia, A.; Partl, M.N.; Tebaldi, G.; Schuetz, P. Induction healing of fatigue damage in asphalt test samples. *Constr. Build. Mater.* **2015**, *74*, 162–168. [CrossRef]

20. García, A.; Bueno, M.; Norambuena-Contreras, J.; Partl, M.N. Induction healing of dense asphalt concrete. Constr. *Build. Mater.* **2013**, *49*, 1–7. [CrossRef]

21. Huo, K.-F.; Zhai, Y.-C.; Liao, K.J.; Yang, P.; Yan, F.; Wei, Y. A study on change of family composition and properties of Liaoshu paving asphalt on aging. *J. Petroleum Sci. Technol.* **2001**, *19*, 651–660.

22. Liu, X.; Wu, S.; Liu, G.; Li, L. Effect of ultraviolet aging on rheology and chemistry of LDH-modified bitumen. *Materials* **2015**, *8*, 5238–5249. [CrossRef]

23. Pang, L.; Zhang, X.; Wu, S.; Ye, Y.; Li, Y. Influence of water solute exposure on the chemical evolution and rheological properties of asphalt. *Materials* **2018**, *11*, 983. [CrossRef]

24. Li, H.; Yu, J.; Wu, S.; Liu, Q.; Li, B.; Li, Y.; Wu, Y. Study on the gradient heating and healing behaviors of asphalt concrete induced by induction heating. *Constr. Build. Mater.* **2019**, *208*, 638–645. [CrossRef]

25. Zeng, W.; Wu, S.; Wen, J.; Chen, Z. The temperature effects in aging index of asphalt during UV aging process. *Constr. Build. Mater.* **2015**, *93*, 1125–1131. [CrossRef]

26. Lamontagne, J.; Dumas, P.; Mouillet, V.; Kister, J. Comparison by Fourier transform infrared (FTIR) spectroscopy of different ageing techniques: Application to road bitumens. *Fuel* **2001**, *80*, 483–488. [CrossRef]

27. Liu, G.; Leegwater, G.; Nielsen, E.; Komacka, J.; van de Ven, M. Evaluating the rheological properties of PMB-containing RA binders from surface-layer asphalt mixtures to be recycled. *Constr. Build. Mater.* **2013**, *49*, 8–14. [CrossRef]

28. Li, Y.; Wu, S.; Dai, Y.; Pang, L.; Liu, Q.; Xie, J.; Kong, D. Investigation of sodium stearate organically modified LDHs effect on the anti-aging properties of asphalt binder. *Constr. Build. Mater.* **2018**, *172*, 509–518. [CrossRef]

29. Li, Y.; Wu, S.; Pang, L.; Liu, Q.; Wang, Z.; Zhang, A. Investigation of the effect of Mg-Al-LDH on pavement performance and aging resistance of styrene-butadiene-styrene modified asphalt. *Constr. Build. Mater.* **2018**, *172*, 584–596. [CrossRef]

Article

Study of the Self-Healing Performance of Semi-Flexible Pavement Materials Grouted with Engineered Cementitious Composites Mortar based on a Non-Standard Test

Xu Cai, Wenke Huang and Kuanghuai Wu *

School of Civil Engineering, Guangzhou University, Guangzhou 510006, China; cx_caixu@163.com (X.C.); hwk_gzu@163.com (W.H.)
* Correspondence: wukuanghuai@163.com; Tel.: +86-020-39366661

Received: 12 August 2019; Accepted: 23 October 2019; Published: 24 October 2019

Abstract: Semi-flexible pavement (SFP) materials, with their characteristics of good high temperature stability, strong durability, and lower cost, are suitable for heavy-duty roads, but their cracking problem has hindered the development and popularization of this kind of pavement to a certain extent. In this study, engineered cementitious composites (ECC) were used to form ECC-SFP materials. The self-healing properties of ECC-SFP materials with three kinds of voids of matrix asphalt mixtures were studied. The test results showed that the fluidity and strength of the ECC mortars met the specification requirements when the water–cement ratio was 0.23 and the ECC fiber dosage was 1–2%. The flexural strength of ECC mortar is better than that of ordinary mortar. The higher the ECC fiber dosage, the higher the flexural strength. Increasing the void of the matrix asphalt mixture and the amount of ECC mortar increased the toughness of the ECC-SFP material, which was seen as an increase of the flow value. Curing conditions are key factor affecting the self-healing properties of ECC mortar and ECC-SFP materials. The self-healing effect of materials in 60 °C water is the best. When an ECC fiber dosage of 1% was used, the HI_{mor} of ECC mortar and HI_{mix} of ECC-SFP material were 27.5% and 24.8%, respectively. With the addition of ECC material, ECC-SFP material achieved a certain degree of self-healing, but this still needs to be further optimized. Studies of grouting process optimization and increasing the ECC fiber dosage are feasible directions to explore in order to improve the self-healing properties of ECC-SFP materials in the future.

Keywords: self-healing pavement materials; semi-flexible pavement; ECC mortar; crack resistance; curing conditions

---

1. Introduction

Semi-flexible pavement (SFP) material is a kind of pavement material which is composed of a special cement mortar infused in the large voids of an open-graded matrix asphalt mixture [1,2]. Its high temperature stability is better than that of asphalt pavement, and it has superior deformation resistance, water damage resistance, and skid resistance. Therefore, SFP material is mainly used in long steep slope sections, tunnels, small radius curve roads, toll crossings, urban road intersections, and at bus stops [3–7].

However, due to the presence of asphalt, cement mortar and aggregate, SFP materials contain interface areas composed of different materials. The differences in the physical and mechanical properties of these materials makes it easy to produce stress concentration inside the SFP materials, which makes them prone to crack damage. At present, research on the cracking of SFP materials usually focuses on the properties of asphalt, aggregate, and cement mortar, and their influences on cracking resistance. For example, Husain [8] studied a cementing material's aggregate grading, durability and strength through statistical analysis, and obtained the influence of gradations in the performance of SFP materials. Pei [9] added different kinds and amounts of additives into the mortar, and found that water reducer, expansion agents, and air-entraining agents had different effects on fluidity, strength, and shrinkage. Wang [10] tested nine mixtures to find the relationship between compressive strength, rupture strength, the water–cement ratio, and the sand-cement ratio. The test indicted that both the cement mortar and the asphalt skeleton affected the performance of the SFP materials. For the cement mortar, the water–cement ratio and sand-cement ratio had great influences on its performance, such as the workability, mechanical strength, and volume stability, and therefore in the performance of the SFP materials. Ling [5] applied SFP materials with asphalt-rubber to a heavy-duty test road, and both laboratory tests and the test road showed that the SFP with asphalt-rubber achieved an excellent performance. Suhana [11] investigated the mechanical properties of cement-bitumen composites as an alternative SFP surfacing material. The findings showed that by replacing 5% of the cement with silica fume there was an improvement in compressive strength and the tensile stiffness modulus. Yang [12] conducted a runway test bench by using eight groups of SFP specimens. The results showed that the maternal asphalt mixture void content was the most important factor that affected durability to cyclic wheel load. When the maternal asphalt mixture void content was 26%, the mechanical performance of the semi-flexible material was superior. Zhang [13] studied the effects of composition and formulation on grouting material. The results showed that cement paste, with its optimal formulation, was more suitable as a grouting material and achieved better performance. The optimal ratio of water to cement was 0.58, content of coal ash was 10%, and content of mineral powder was 10%. Hong [14] evaluated the freeze–thaw durability of SFP mixes grouted with early-strength (ES) and high-strength (HS) grouting materials. The results showed that enhancement of the freeze-thaw resistance of an SFP mix can be achieved by incorporating a high-strength grouting material.

Although most studies have revealed the influencing factors of cracking of SFP materials, it is still difficult to effectively avoid cracking. In fact, the asphalt component in SFP material has certain self-healing properties, meeting the basic conditions of self-healing design. Recently, research into pavement self-healing has been paid more attention. Sun [15] pointed out that self-healing capability for repairing micro-crack damage can restore functionality, at least to some extent, and considerably reduce maintenance costs, as well as extend the pavement's service life and eventually decrease the emissions of greenhouse gas from pavement production. The self-healing of asphalt materials can be described as the partial restoration of the intrinsic asphalt structure across adjacent crack surfaces [16]. Mohammad [17] evaluated the effects of a combination of nano-silica and styrene-butadiene-styrene (SBS) on the self-healing ability of hot mix asphalt (HMA) by applying an indirect tension test (IDT). The main reason for crack healing was the flow of bitumen mortar into micro-cracks under gravity forces. Giorgia [18] evaluated the self-healing potential and thixotropy of bituminous mastics. The results showed that the self-healing properties of the material were not significantly related to the degree of aging of the asphalt. An aged bitumen content of up to 45% (45R) improved the overall fatigue response without hindering self-healing capability. Xu [19] explored the potential use of calcium alginate capsules in porous asphalt. The results showed that samples with capsules were able to achieve a healing index that was 6% higher than the reference samples. Wang [20] estimated the fatigue and healing characteristics of asphalt binder by newly developed linear amplitude sweep (LAS) and LAS-Based Healing (LASH) protocols. The combined use of LAS and LASH tests is recommended for effectively distinguishing and designing the fatigue-healing performance of neat and modified asphalt binders. Fan [21] evaluated the fracture resistance and self-healing performance of asphalt concrete at low temperatures using a semicircular bending (SCB) test. The results showed that the optimum healing temperature was 60 degrees at a healing time of 8 h.

To broadly define the healing phenomenon, healing acts as long as the material does not completely fail and there is still contact between the crack faces, since an external force can be applied to make two fractured surfaces come into contact [16,22]. Due to the existence of asphalt mortar, the cracking of

SFP materials is less than that of cement pavement, and the crack width is generally smaller [2,23]. Therefore, SFP material has good self-healing design potential. Since the matrix asphalt mixture has a good self-healing property, the emphasis of the self-healing design of SFP material is to control the crack width and the self-healing design of cement paste.

In fact, each failure and self-healing recombination is the concentration and dissipation of internal stress in the material. After repeated failure and healing cycles, the rigid and flexible components in SFP materials will have better collaborative deformation capacity. Based on this concept, this study used engineered cementitious composites (ECC) mortar to replace the original cement mortar. The mixing ratio of ECC mortar was determined through experiments, and the self-healing properties of ECC mortar and SFP materials grouted with ECC mortar under different curing conditions were studied.

2. Experimental Design

According to the predetermined research objectives, we developed the following experimental program.

2.1. Materials

Materials used in this study include asphalt, aggregates, powder mortar and engineered cementitious composites.

2.1.1. Asphalt and Aggregates

Shell 70# asphalt was selected for this test. The technical indexes of asphalt are shown in Table 1. The aggregates were formed from granite from the Guangdong Furong quarry (Shenzhen, China), and the technical indicators are shown in Table 2.

Table 1. Technical indexes of 70# asphalt.

Technical Indexes	Unit	Test Results
Penetration at 25 °C	0.1 mm	70.2
Softening point	°C	49
Ductility at 10 °C	cm	51.3
Flash point	°C	335
Bitumen solubility (trichloroethylene)	%	99.8
Density at 15 °C	g/cm^3	1.037

Table 2. Technical indexes of the aggregates.

Aggregate Types	Technical Indexes	Unit	Test Results
Coarse aggregate	Crushing value	%	18
	Los Angeles abrasion	%	25
	Water absorption	%	0.17
	Apparent relative density	-	2.679
	Content of particles smaller than 0.075 mm	%	0.55
	Needle flake	%	12
Fine aggregate	Robustness	%	15
	Sand equivalent	%	68
	Apparent relative density	-	2.568

2.1.2. Powder Mortar

The powder mortar used in this study is based on the research results obtained by our research team [24–26], and is composed of cement, fine sand, fly ash, mineral powder, poly-carboxylic acid water-reducing agent, and other additives. The composition of the mortar is shown in Table 3.

Table 3. The components of the powder mortar.

Component	Cement	Sand	Fly ash	Rubber Powder	Water Reducer	Shrinkage Reducing Agent	Other Additives
Dosage %	38	25	25	3	0.5	0.7	7.8

2.1.3. Engineered Cementitious Composites

The performance of fiber in a cement mortar mainly depends on its strength, elongation, elastic modulus, and other mechanical properties [27–29]. Considering the economic technology and feasibility of the test, PVA-ECC fiber was used in the preparation of the ECC mortar in this paper. The physical parameters of the ECC are shown in Table 4.

Table 4. Physical properties of the PVA fiber.

Density (g/cm^3)	Elasticity Modulus (Gpa)	Tensile Strength (MPa)	Diameter (μm)	Elongation (%)	Length (mm)
1.30	30–50	1600–2500	15	6	5–10

2.2. Matrix Asphalt Mixture

The matrix asphalt mixture not only needs to have the required strength, but also have enough connected and semi-connected voids to ensure the grouting of the cement mortar when forming a semi-flexible structure. In this study, the volume method was used to design three matrix asphalt mixtures with voids of 20%, 25%, and 28%. The asphalt mixtures were expressed by MA20, MA25, and MA28, respectively. The optimal asphalt contents were determined by the Cantabro Test and Shellenburg leakage test. The design results are shown in Table 5.

Table 5. Design of the matrix asphalt mixture.

Granular Composition	MA20	MA25	MA28
Voids (%)	20	25	28
Asphalt aggregate ratio (%)	4.0	3.6	3.3
16 mm	100	100	100
13.2 mm	94.6	94.6	94.6
9.5 mm	63.5	64.1	64.4
4.75 mm	14.7	12.4	11.9
2.36 mm	5.9	4.7	4.3
0.075 mm	3.5	3.0	3.0

2.3. Engineered Cementitious Composites(ECC) Mortar

Cement mortar used for SFP materials should have better fluidity, higher strength, and greater volume stability. According to the existing research results [30–32], this study used a cement mortar mixer to prepare the ECC mortar, as shown in Figure 1. The ECC was soaked in water before mixing to ensure that the ECC fully absorbed the water. There were four steps to prepare the ECC mortar: (1) the dry powder mortar was blended at a low speed for 1 min; (2) 2/3 water was added and blended at high speed for 4 min; (3) another 1/3 water was added and blended at high speed for 4 min; (4) the ECC was added and blended for 2 min to disperse the fiber evenly in the mortar.

Figure 1. Cement mortar mixer.

2.3.1. Test of Fluidity

The fluidity was tested by a special v-shaped funnel. The upper diameter of the funnel was 178 mm, the lower diameter was 13 mm, the outflow pipe length was 38 mm, and the effective volume was 1000 mL, as shown in Figure 2. When the discharge time of 1000 mL mortar from the funnel was 12–18 s, it was considered that the mortar met the requirement for fluidity [24]. Mortars with a water–cement ratio of 0.20, 0.21, 0.22, 0.23 and 0.24 were tested, and 0%, 1%, 2% and 3% ECC fibers were added into the mortar. Three parallel experiments were conducted for each proportion, and the average value was used as the final test result.

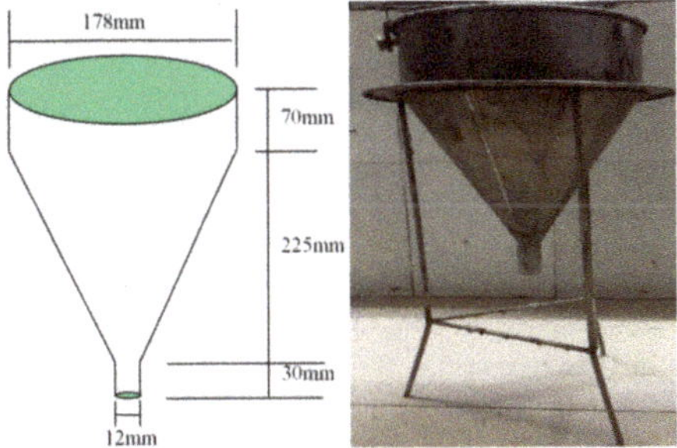

Figure 2. The funnel used for the test of fluidity.

2.3.2. Strength Test for Mortar

According to the fluidity test results, the ECC mortar ratio with the desired fluidity performance was selected, and the flexural and compressive performance test [33] was carried out according to the specifications. Specimens of 40 mm × 40 mm × 160 mm were used in the flexural strength test, as shown in Figure 3. Under the same water–cement ratio, molding mode and curing environment, the flexural strength of the ECC mortar and ordinary mortar with different fiber contents of 3 days, 7 days, 14 days and 28 days was measured.

Figure 3. Cement mortar specimens.

According to the Test Methods of Cement and Concrete for Highway Engineering, the specimen used in the compression test was the specimen selected after the flexural test, and the compression surface was the side of the specimen during molding. The loading rate of the press was set at 2400 N/s ± 200 N/s until the specimen was damaged. The press and test fixtures used are shown in Figure 4.

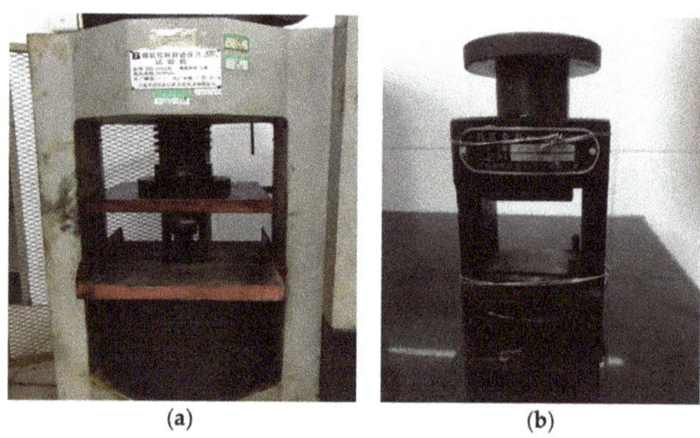

(a)　　　　　　　　　　　　　　　　(b)

Figure 4. Settings of the compressive strength test. (**a**) Setup of compression test; (**b**) Test fixture.

2.3.3. Self-Healing Test of Mortar

According to the strength test results, the optimal mix ratio of ECC mortar was selected. The self-healing properties of the ECC mortar were evaluated based on the shear strength test. Firstly, specimens of 40 mm × 40 mm × 160 mm were prepared and maintained by conventional means for seven days. Secondly, according to the shear strength test method, the sample was preloaded at the rate of 1 mm/min. The load was controlled to reach 50% of the material's 7 day flexural strength, and the preloading specimen was obtained, as shown in Figure 5.

Three different curing conditions were adopted for the specimens with cracks, and the curing time was 21 days. The first condition was dry air curing. The specimen was placed in a dry environment with a humidity of 10% at 20 °C. The second condition was wet air curing, in which the specimen was placed in an environment of 20 °C with 100% humidity. The last was hot and humid curing. The specimen was placed in a 60 °C constant temperature water tank. After curing, the flexural strength of the specimen was measured.

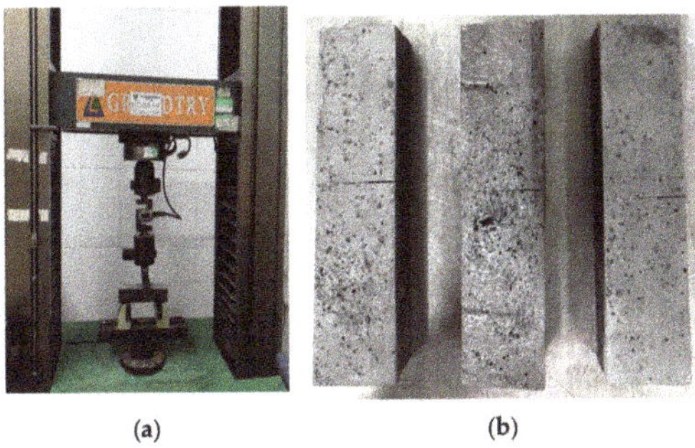

(a)   (b)

Figure 5. Preloading treatment of the specimens. (**a**) Preloading setup; (**b**) Preloaded specimens.

During the test we found the strength of cement mortar varies from 30% to 50% between 7 days and 28 days. Considering that the strength of cement mortar will increase with the increase of curing time, and the self-healing ratio calculated using the ratio of the strength of preload specimens at 28 days and 7 days cannot eliminate the contribution of cement hydration. Therefore, the self-healing property of the material was characterized by the ratio of the 28 days flexural strength of the preloaded specimen to the 28 days flexural strength of a standard specimen in this study, as shown in Equation (1).

$$HI_{\mathrm{mor}} = \frac{R_f{}'}{R_f} \tag{1}$$

where HI_{mor} is the self-healing ratio of flexural strength; $R_f{}'$ is the 28 days flexural strength of the preloaded specimen; and R_f is the 28 days flexural strength of the standard specimen.

2.4. The Semi-Flexible Pavement (SFP) Materials

According to the test results of the ECC mortar, the optimal mix ratio of the ECC mortar was selected to carry out the self-healing performance test of the SFP material, including the Marshall stability test and asphalt mixture three-point bending test.

2.4.1. Specimen Grouting

The matrix asphalt mixture specimen was first formed and the bottom and sides of the specimen were sealed with transparent tape. Then, mortar was poured onto the surface of the specimen, and a vibration table was used to assist the grouting. The vibration time was generally controlled at about 90 s. During grouting, the surface of the specimen was observed until there was no bubbling. Excess grout was scraped off the surface after grouting. The grouting effect is shown in Figure 6.

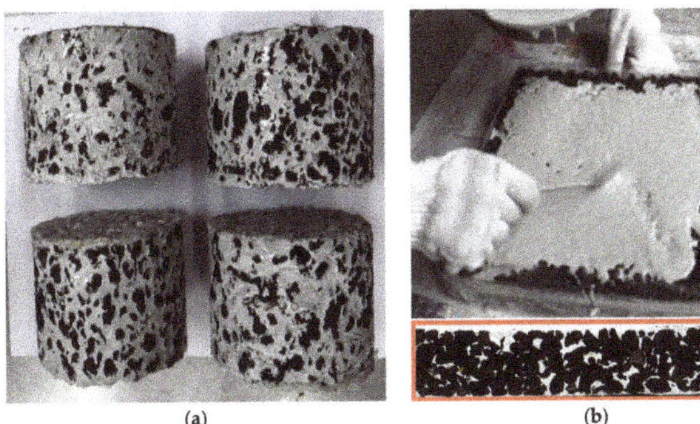

(a)   (b)

Figure 6. Grouting of the SFP material. (**a**) Marshall specimen grouting; (**b**) Rutting specimen grouting.

2.4.2. Marshall Stability Test

A batch of matrix asphalt mixture was formed according to the designed voids of 20%, 25%, and 28%. The ECC mortar and ordinary mortar were respectively grouted, and the specimens were cured for 3 days, 7 days and 28 days, respectively. Four Marshall specimens of each mixing ratio were tested under the condition of 60 °C. The test results are expressed as averages.

2.4.3. Three-Point Bending Test

According to the Standard Test Methods of Bitumen and Mixtures for Highway Engineering [34], the 300 mm × 300 mm × 60 mm rutting specimens of matrix asphalt mixtures with voids of 20%, 25%, 28% were formed, and the ECC mortar and ordinary mortar were respectively grouted. After 7 days of curing time, the rutting specimens were cut into 250 mm × 30 mm × 35 mm beam specimens, as shown in Figure 7. The beam specimens were divided into two groups. One group was maintained for 7 days, and the other group was maintained for 28 days. The three-point bending test was carried out after curing.

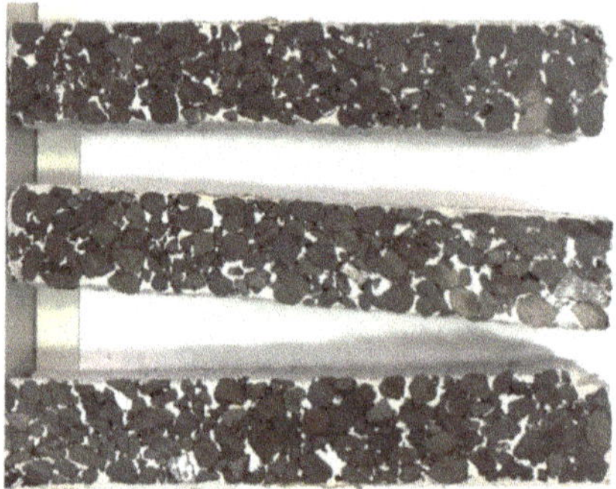

Figure 7. The beam specimens of the SFP materials.

In the three-point bending test, the beam specimen was placed on a bracket with a span of 200 mm, and vertical load was applied at the central point of the beam. The test temperature was −10 °C and the loading rate was 50 mm/min. The calculation of the three-point bending strength, R_B, the maximum bending strain, ε_B, and the stiffness modulus, S_B, are shown in Equations (2)–(4).

$$R_B = \frac{3 \times L \times P_B}{2 \times b \times h^2} \tag{2}$$

$$\varepsilon_B = \frac{6 \times h \times d}{L^2} \tag{3}$$

$$S_B = \frac{R_B}{\varepsilon_B} \tag{4}$$

where R_B is the three point bending strength of the specimen; ε_B is the maximum bending strain; S_B is the stiffness modulus of the materials; b is the width of the specimen; h is the height of the specimen; L is the span of the specimen; P_B is the maximum load when the specimen fails; and d is the mid-span deflection of the specimen when it fails.

2.4.4. Self-Healing Test of the Semi-Flexible Pavement (SFP) Materials

The self-healing properties of the SFP materials were evaluated based on the three-point bending test. Beam specimens were formed according to the method outlined in Section 2.4.3, and the specimens were preloaded at the rate of 1 mm/min under the condition of −10 °C for 7 days of curing time. The load was controlled to reach 50% of the material's 7 days three-point bending strength, and preloaded specimens were obtained. After that, the three curing conditions mentioned in Section 2.3.3 were used for curing for 21 days. After curing, the three-point bending strength of the SFP beam specimens at 10 °C was measured. Based on the same considerations as the calculation of self-healing ratio of mortar, the self-healing properties of the SFP materials were characterized by the ratio of the 28 days three-point bending strength of the preloaded specimens to the 28 days three-point bending strength of standard specimens, as shown in Equation (5).

$$HI_{\text{mix}} = \frac{R_B{'}}{R_B} \tag{5}$$

where HI_{mix} is the self-healing ratio of the SFP materials; $R_B{'}$ is the 28 days three-point bending strength of the preloaded SFP beam specimens; and R_B is the 28 days three-point bending strength with the standard test.

3. Results and Discussion

According to the proposed tests, the following results can be obtained.

3.1. Mortar

3.1.1. Fluidity

According to the designed fluidity test, mortars with water–cement ratios of 0.20, 0.21, 0.22, 0.23, and 0.24 were tested separately, and ECC fiber dosages of 1%, 2%, and 3% was added to the mortar for the fluidity test. The results are shown in Table 6.

Table 6. Result of fluidity test of mortar.

Water-Cement Ratio	Dosage (%)	Fluidity (s)	Deviation
0.2	0	15.1	0.81
	1	16.3	0.65
	2	18.5	1.05
	3	21.6	1.56
0.21	0	13.6	0.41
	1	15.5	1.01
	2	17.3	0.81
	3	20.5	1.29
0.22	0	12.4	0.93
	1	14.8	1.25
	2	16.0	1.14
	3	19.6	1.26
0.23	0	10.8	0.53
	1	13.2	0.87
	2	15.1	0.85
	3	18.7	1.53
0.24	0	8.9	0.77
	1	11.2	0.71
	2	14.5	1.07
	3	17.5	1.24

According to Table 6, the fluidity of the mortar deteriorates after the addition of the ECC, and the flow time increases. Under the same water–cement ratio, the fluidity decreases with the increase of ECC content. Secondly, the fluidity increases as the water–cement ratio increases. When the ECC content reaches 3%, there is too much flocculent in the mortar, and the water–cement ratio needs to reach 0.24 to meet the minimum requirement of fluidity. When 1% and 2% ECC are mixed in, the fluidity value of the mortar basically meets the requirements of 12–18 s flow time. The flow state of the mortar is shown in Figure 8.

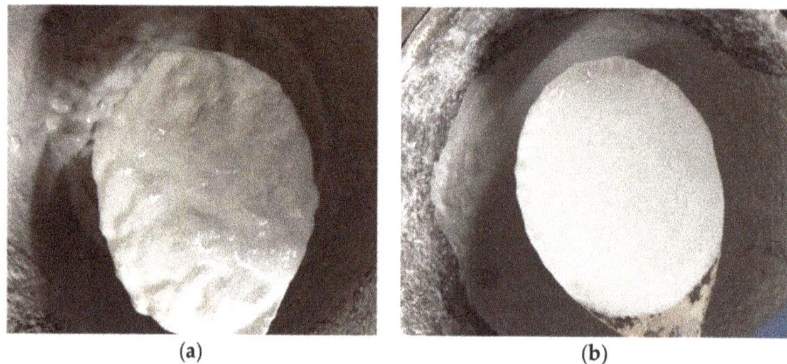

Figure 8. Flow state of ECC mortar. (**a**) Dosage 3%, w/c 0.24; (**b**) Dosage 2%, w/c 0.23.

According to the fluidity test results, the dosage of ECC should not be more than 2%. At the same time, in order to increase the fluidity of the ECC mortar, a larger water–cement ratio should be used. The fluidity of 2% ECC mortar is less than 12 s when the water–cement ratio is 0.24, which does not meet the requirements of the specification. Therefore, a water–cement ratio of 0.23 was adopted in subsequent studies to carry out mechanical tests on mortars with ECC fiber dosages of 0%, 1%, and 2%.

3.1.2. Flexural Strength and Compressive Strength of Mortar

According to ECC fiber dosages of 0%, 1% and 2%, four groups of 40 mm × 40 mm × 160 mm test pieces were formed. There were three specimens in each group, for a total of 36 specimens. Four groups of specimens with the same ECC fiber dosage were cured for 3 days, 7 days, 14 days, and 28 days. The flexural strength and compressive strength were tested after curing. The test results are shown in Table 7.

Table 7. Test results of the flexural strength and compressive strength of mortar.

Dosage/%	Curing Time/d	Flexural Strength (MPa)		Compressive Strength (MPa)	
		Mean	Deviation	Mean	Deviation
0	3	8	0.21	32	0.57
	7	10.6	0.18	52.3	0.55
	14	11.2	0.19	65.6	0.42
	28	13.3	0.14	75.9	0.68
1	3	8.3	0.25	29.8	0.44
	7	10.2	0.24	53.4	0.58
	14	12.8	0.64	65.9	0.75
	28	14.9	0.53	74.3	0.69
2	3	7.9	0.54	30.3	0.87
	7	10.3	0.49	51.7	0.74
	14	13.2	0.35	66.1	0.88
	28	15.5	0.47	73.6	0.61

According to Table 7, under the same ECC fiber dosage, the flexural strength and compressive strength of the cement mortar increased with the increase of curing time. The flexural strength and compressive strength after curing for 7 days were 65% and 70% of that found after curing for 28 days, respectively. Secondly, with the increase of ECC fiber dosage, the flexural strength of the cement mortar increased but the compressive strength decreased. The 28 days flexural strengths of 1% and 2% ECC mortar were 12% and 16% higher than that of mortar without ECC, and the compressive strengths were 98% and 97% of the mortar without adding ECC. The addition of ECC caused the specimen show a certain degree of toughness and increased the crack resistance of the material. Failure modes of the specimens are shown in Figure 9.

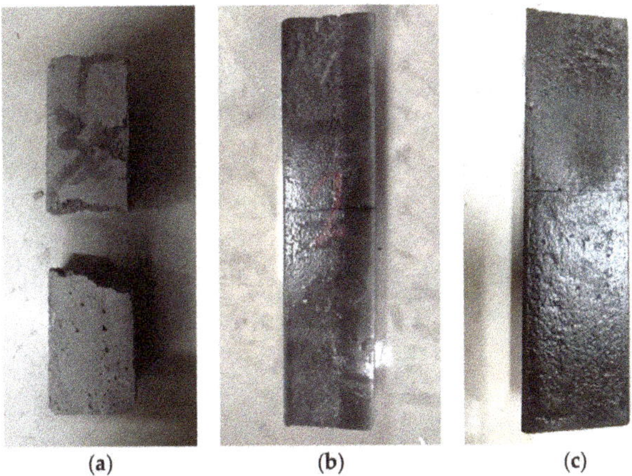

Figure 9. Failure modes of specimens with different ECC fiber dosages. (**a**) 0% Dosage; (**b**) 1% Dosage; (**c**) 2% Dosage.

Figure 9 indicates that the ECC performed as a bridge-connection effect when the specimen was damaged, so that the specimen could still be connected through the ECC after becoming cracked. The ECC mortar exhibited stronger tensile resistance, higher toughness, and greater energy absorption capacity than ordinary mortar.

3.1.3. Self-Healing Evaluation of Mortar

Nine specimens of 40 mm × 40 mm × 160 mm were prepared with the ECC fiber dosages of 0%, 1% and 2%. The mortar self-healing test was carried out according to the method mentioned in Section 2.3.3, and the test results are shown in Table 8.

The mortar without ECC were completely fractured at the preloading stage, so further test data could not be obtained. As can be seen from Table 8, curing conditions and ECC content were the main factors affecting the self-healing performance of ECC mortar. The recovery of specimens under the three curing conditions varied. The HI_{mor} for curing in dry air was 5.5–6.5%. The HI_{mor} for curing in moist air was 13.5–16.0%. However, in a hot and humid environment, the maximum value of HI_{mor} was obtained, of 27.5–30.0%. The increase of ECC fiber dosage slightly improved the self-healing property of ECC mortar. Under the same curing conditions, the self-healing performance of ECC mortar with 2% dosage was slightly better than that of ECC mortar with 1% dosage.

In conclusion, the self-healing performance of ECC mortar is related to environmental temperature, environmental humidity, and ECC fiber dosage. At 20 °C, increasing ambient humidity increases HI_{mor} by about 8.5%. Raising the ambient temperature to 60 °C in wet conditions increases the HI_{mor} by about 14%. Increasing the dosage of ECC can increase HI_{mor} by 1.3–3.3%. Therefore, the curing condition is the main factor affecting the self-healing effect of ECC mortar. The recommended dosage of ECC is 1%.

Table 8. Results of the mortar self-healing test.

Dosage (%)	Curing Condition	Specimen Code	R_f' (MPa)	HI_{mor}		
				Single Value (%)	Mean (%)	Deviation (%)
1	temp. 20 °C, humidity 10%	1-1	0.5	3.4		
		1-2	1.1	7.4	5.4	8.0
		1-3	0.8	5.4		
	temp. 20 °C, humidity 100%	1-4	2.1	14.1		
		1-5	2.3	15.4	13.7	5.5
		1-6	1.8	12.1		
	temp. 60 °C, in water	1-7	4.2	28.2		
		1-8	4.3	28.8	27.5	6.2
		1-9	3.8	25.5		
2	temp. 20 °C, humidity 10%	2-1	1.1	7.1		
		2-2	0.8	5.2	6.7	3.4
		2-3	1.2	7.7		
	temp. 20 °C, humidity 100%	2-4	2.8	18.1		
		2-5	2.5	16.1	16.1	7.6
		2-6	2.2	14.2		
	temp. 60 °C, in water	2-7	4.4	28.4		
		2-8	4.9	31.6	30.3	5.8
		2-9	4.8	31.0		

3.2. Semi-Flexible Pavement (SFP) Materials

According to the self-healing test results of ECC mortar, 0% and 1% ECC fiber dosage were selected to prepare the cement mortar. Marshall specimens and three-point bending specimens of SFP materials were prepared with matrix asphalt mixture with voids of 20%, 25%, and 28%.

3.2.1. Marshall Test

Twenty-four Marshall specimens of matrix asphalt mixture for each void ratio were filled with cement mortar with ECC fiber dosages of 0% and 1%, respectively (called SFP material and ECC-SFP material, respectively) to test the Marshall stability and flow value. The test results are shown in Figure 10.

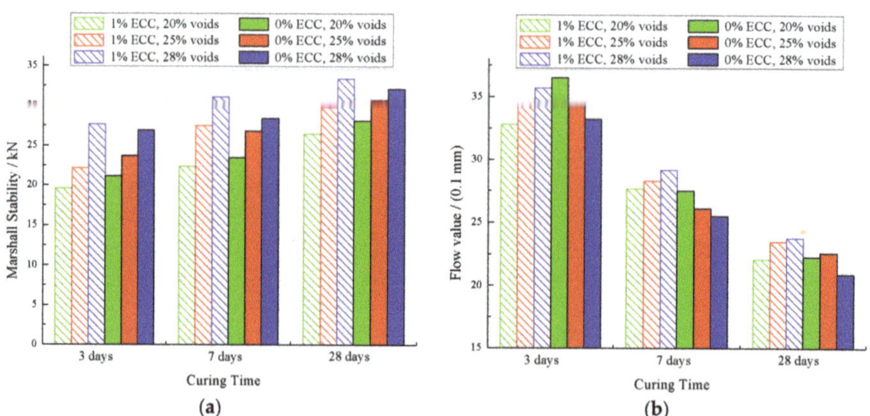

Figure 10. Marshall test results of SFP materials. (**a**) Results of the Marshall stability test; (**b**) Results of the flow value.

According to Figure 10a, the void ratio of matrix asphalt mixture was the main factor affecting the stability of the material. The Marshall stability increased with the increase of the void ratio of the matrix asphalt mixture. Secondly, the Marshall stability increased with the increase of curing time, and the Marshall stability at 3 and 7 days reached 45% and 70% respectively, as it was at 28 days. However, ECC has little effect on the Marshall stability. Under the same ratio of air voids and curing time, the difference in the Marshall stability between ECC-SFP materials and SFP materials was not obvious.

According to Figure 10b, the flow value decreased gradually with the increase of curing time. The void ratio of the matrix asphalt mixture also affected the trend of the flow value. The flow value of ECC-SFP materials increased with the increase of the void ratio, while that of SFP materials decreased with the increase of the void ratio. Meanwhile, the flow value of the ECC-SFP material was higher than that of the SFP material, which indicated that ECC can increase the toughness of SFP materials and give them a better deformation capacity.

3.2.2. Three-Point Bending Test

Four rutting specimens were molded for each void ratio of matrix asphalt mixture, and the ECC mortar and ordinary mortar were grouted. After 7 days of curing, the rutting specimens were cut to obtain beam specimens. The three-point bending test was carried out according to the method in Section 2.4.3. The results of the three-point bending test at 7 days and 28 days are shown in Figure 11.

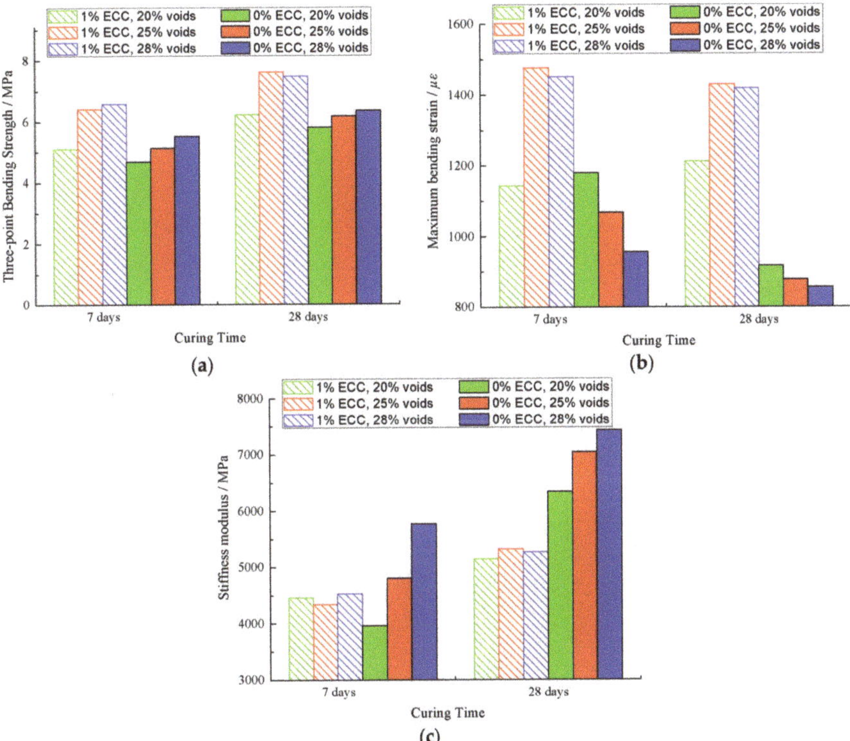

Figure 11. Results of the three-point bending test. (**a**) Results of three-point bending strength; (**b**) Results of maximum bending strain; (**c**) Results of stiffness modulus.

According to Figure 11a, the three-point bending strength increased with the increase of the void ratio of the matrix asphalt mixture. Secondly, the use of ECC mortar can improve the three-point bending strength. Under the condition of 28 days of curing time, the bending modulus values of the ECC-SFP materials with three voids were 7.2%, 23.1% and 17.8% higher than for the SFP materials, respectively.

According to Figure 11b, the void ratio of the matrix asphalt mixture affected the maximum bending strain. The maximum bending strain of the ECC-SFP material increased with the increase of the void ratio, reaching the maximum value at 25% void. Meanwhile, the maximum bending strain of SFP materials decreased with the increase of the void ratio. Secondly, the grouted amount of ECC mortar significantly increased the maximum bending. The larger the void ratio of the matrix asphalt mixture, the more ECC mortar was grouted, and the larger the maximum bending strain. Under the condition of 28 days of curing time, the maximum bending strain values of the ECC-SFP materials with three voids were 32.2%, 62.9% and 65.9% higher than for the SFP materials, respectively.

According to Figure 11c, the addition of ECC reduced the stiffness modulus. The stiffness modulus of the ECC-SFP materials was smaller than that of the SFP materials with the same void ratio of matrix asphalt mixture. Under the condition of 28 days of curing time, the stiffness modulus values of ECC-SFP were 18.9%, 24.4% and 28.9% smaller than for the SFP materials respectively.

In conclusion, adding ECC into cement mortar can increase the toughness of the SFP material, giving it a better deformation ability and resistance to cracking.

3.2.3. Self-Healing Evaluation of ECC-SFP Materials

Two rutting specimens were molded for each void ratio of the matrix asphalt mixture, and the ECC mortar and ordinary mortar were grouted. After 7 days of curing, the rutting specimens were cut to obtain beam specimens. The beam specimens were divided into three groups for three curing conditions. The self-healing evaluation test was carried out according to the method in Section 2.4.4, and the test results are shown in Table 9.

Table 9. Self-healing test results of the ECC-SFP materials.

AC Type	Curing Condition	Specimen Code	$R_B{'}$ (MPa)	HI_{mix} Single Value (%)	HI_{mix} Mean (%)	HI_{mix} Deviation (%)
MA20	temp. 20 °C, humidity 10%	1-1	0.46	7.4	7.3	2.5
		1-2	0.52	8.3		
		1-3	0.38	6.1		
	temp. 20 °C, humidity 100%	1-4	1.08	17.3	15.5	5.9
		1-5	0.87	14.0		
		1-6	0.94	15.1		
	temp. 60 °C, in water	1-7	1.53	24.6	24.8	8.8
		1-8	1.42	22.8		
		1-9	1.68	27.0		
MA25	temp. 20 °C, humidity 10%	2-1	0.53	7.0	5.7	2.6
		2-2	0.36	4.7		
		2-3	0.42	5.5		
	temp. 20 °C, humidity 100%	2-4	1.25	16.4	16.1	2.1
		2-5	1.14	15.0		
		2-6	1.29	17.0		
	temp. 60 °C, in water	2-7	1.86	24.4	25.9	6.8
		2-8	2.13	28.0		
		2-9	1.93	25.4		
MA28	temp. 20 °C, humidity 10%	3-1	0.38	5.1	4.4	0.9
		3-2	0.28	3.7		
		3-3	0.33	4.4		
	temp. 20 °C, humidity 100 %	3-4	1.28	17.1	16.6	4.3
		3-5	1.12	15.0		
		3-6	1.33	17.8		
	temp. 60 °C, in water	3-7	2.02	27.0	26.2	5.0
		3-8	1.82	24.3		
		3-9	2.03	27.1		

It can be seen from Table 9 that curing conditions were the main factors affecting the self-healing properties of ECC-SFP materials. Consistent with the test results of ECC mortar, the recovery of specimens under three different curing conditions was different. The HI_{mix} of specimens cured in dry air were 4.4–7.3%. The HI_{mix} of specimens cured in moist air were 15.5–16.6%. The HI_{mix} of specimens cured in high temperature water were 24.8–26.2%. Secondly, as the void ratio of the matrix asphalt mixture increased, the HI_{mix} of specimens cured under dry conditions decreased, while the HI_{mix} of specimens cured under hot and humid conditions increased. Combined with the mix design results in Table 5, it can be inferred that the higher the amount of asphalt, the higher the HI_{mix} will be when cured in a dry environment. Asphalt is a major contributor to the self-healing properties of ECC-SFP materials in dry environments.

The self-healing properties of ECC-SFP materials are related to ambient temperature and humidity. HI_{mix} increased by 8.2–12.2% at 20 °C when the ambient humidity increased from 10% to 100%. We can increase HI_{mix} by 9.3–9.6% under 100% humidity by raising the temperature from 20 °C to 60 °C. In hot and humid conditions, HI_{mix} can be increased by increasing the void ratio of the matrix asphalt mixture, but the effect is not obvious. Meanwhile, it can be seen from the deviation of the data that increasing the void ratio can increase the material uniformity. In conclusion, the void ratio of the matrix asphalt mixture should be 25–28%.

4. Conclusions

This paper studied the influence of Engineered Cementitious Composites (ECC) on the fluidity and strength of cement mortar, and studied the self-healing properties of ECC mortar and Semi-Flexible Pavement (SFP) materials grouted with ECC under three different curing conditions. According to the experimental results, the following conclusions can be drawn.

Firstly, the addition of ECC will reduce the fluidity of mortar. The fluidity of mortar decreases with the increase of ECC fiber dosage. To prepare ECC mortar which meets grouting requirements, a larger water–cement ratio should be used and the dosage of ECC should be controlled. In this study, cement mortar with a 0.23 water–cement ratio and less than 2% ECC fiber dosage was found to ensure fluidity.

Secondly, the addition of ECC gives the mortar better toughness. The flexural strength of ECC mortar is better than that of ordinary mortar, and the higher the ECC fiber dosage, the more significant the improvement of flexural strength.

Third, the grouted amount of ECC mortar affects the Marshall stability and flow value of the ECC-SFP material. As the void ratio of the matrix asphalt mixture increases, the amount of ECC mortar increases, and the Marshall stability and flow value increases.

Finally, curing conditions are the key factor affecting the self-healing property of ECC mortar and ECC-SFP materials. The materials have the best self-healing effect under a high temperature and humidity. When an ECC fiber dosage of 1% is used, HI_{mor} and HI_{mix} can reach 27.5% and 24.8%, respectively.

From the above conclusions, we can see that the use of ECC does give SFP materials certain self-healing properties, but the self-healing effect is still relatively poor. Improving the self-healing property of ECC-SFP materials should start with increasing the dosage of ECC, but this conflicts with the fluidity of the mortar. Therefore, ECC mortar is not suitable for using with the existing grouting methods. It is more important at this stage to optimize the grouting method and develop the ECC mortar grouting equipment for ECC-SFP materials. In terms of curing conditions, ECC-SFP materials should be used in areas with high temperatures and humidity. For example, the Guangdong-Hong Kong-Macao greater bay area in south China is a prime area for use of these materials. This bay area has a large amount of heavy traffic and perennial high temperatures and rain, which would maximize the usefulness of the self-healing properties of ECC-SFP materials. However, ECC-SFP materials are not suitable for cold and dry areas at middle and high latitudes.

In this study, some preliminary results regarding the self-healing properties of ECC-SFP materials were obtained; however, the fatigue durability at the macro-scale and the interface physicochemical properties of the materials at the micro-scale still need to be further studied.

Author Contributions: Conceptualization, X.C. and K.W.; Validation, W.H.; Investigation, X.C.; Resources, K.W.; Data curation, X.C.; writing—original draft preparation, X.C.; Writing—review and editing, W.H.; Supervision, K.W.

Funding: This research was funded by the National Natural Science Foundation of China, grant number 51708144 and 51878193.

Conflicts of Interest: The authors declare no conflict of interest.

References

1. Yang, Y.; Huang, S.L.; Ding, Q.J. The property research on interfacial medicated semi-flexible pavement material. *Appl. Mech. Mater.* **2011**, *71–78*, 1090–1098.
2. Ding, Q.J.; Sun, Z.; Shen, F.; Huang, S.L. The performance analysis of semi-flexible pavement by the volume parameter of matrix asphalt mixture. *Adv. Mater. Res.* **2011**, *168*, 351–356. [CrossRef]
3. Ding, Q.J.; Zhao, M.Y.; Shen, F. Mechanical behavior and failure mechanism of recycled semi-flexible pavement material. *J. Wuhan Univ. Technol. Mater. Sci. Ed.* **2015**, *30*, 981–988. [CrossRef]
4. Setyawan, A. Assessing the Compressive Strength Properties of Semi-Flexible Pavements. *Proced. Eng.* **2013**, *54*, 863–874. [CrossRef]
5. Ling, T.Q.; Zhao, Z.J.; Xiong, C.H. The application of semi-flexible pavement on heavy traffic roads. *Int. J. Pavement Res. Technol.* **2009**, *2*, 211–217.
6. Husain, N.M.; Karim, M.R.; Mahmud, H.B.; Koting, S. Effects of aggregate gradation on the physical properties of semiflexible pavement. *Adv. Mater. Sci. Eng.* **2014**. [CrossRef]
7. Vila-Cortavitarte, M.; Jato-Espino, D.; Castro-Fresno, D.; Calzada-Pérez, M.Á. Self-healing capacity of asphalt mixtures including by-products both as aggregates and heating inductors. *Materials* **2018**, *11*, 800. [CrossRef]
8. Tran, T.N.; Nguyen, H.T.T.; Nguyen, K.S.; Nguyen, H.N.T. Semi-flexible Material: The Sustainable Alternative for the Use of Conventional Road Materials in Heavy-Duty Pavement. In Proceedings of the 4th Congrès International de Géotechnique-Ouvrages-Structures, Singapore, 26–27 October 2017.
9. Pei, J.Z.; Cai, J.; Zou, D. Design and performance validation of high-performance cement paste as a grouting material for semi-flexible pavement. *Constr. Build. Mater.* **2016**, *126*, 206–217. [CrossRef]
10. Wang, Y.J.; Guo, C.Y.; Tian, Y.F.; Wang, J.J. Design of mix proportion of cement mortar with high-performance composite semi-flexible pavement. *Adv. Mater. Res.* **2013**, *641–642*, 342–345. [CrossRef]
11. Suhana, K.; Mohamed, R.K.; Hilmibin, M.; Norjidah, A.A.H. Mechanical properties of cement-bitumen composites for semi-flexible pavement surfacing. *Baltic J. Road Bridge Eng.* **2014**, *9*, 191–199.
12. Yang, B.; Weng, X. The influence on the durability of semi-flexible airport pavement materials to cyclic wheel load test. *Constr. Build. Mater.* **2015**, *98*, 171–175. [CrossRef]
13. Zhang, J.; Cai, J.; Pei, J.Z. Formulation and performance comparison of grouting materials for semi-flexible pavement. *Constr. Build. Mater.* **2016**, *115*, 582–592. [CrossRef]
14. Hong, J.X.; Wang, K.J.; Xiong, Z.J.; Gong, M.H.; Deng, C.; Peng, G.; Zhu, H.R. Investigation into the freeze–thaw durability of semi-flexible pavement mixtures. *Road Mater. Pavement Design* **2019**, *20*, 1–17. [CrossRef]
15. Sun, D.Q.; Sun, G.Q.; Zhu, X.Y.; Guarin, A.; Li, B.; Dai, Z.W.; Ling, J.M. A comprehensive review of self-healing of asphalt materials: Mechanism, model, characterization and enhancement. *Adv. Colloid Interface Sci.* **2018**, *256*, 65–93. [CrossRef]
16. Little, D.N.; Lytton, R.L.; Chairl, B.; Williams, D.; Texas, A. An analysis of the mechanism of micro damage healing based on the application of micromechanics first principles of fracture and healing. *J. Assoc. Asphalt Paving Technol.* **1999**, *68*, 501–542.
17. Mohammad, A.G.; Esmail, A. Application of nano-silica and styrene-butadiene-styrene to improve asphalt mixture self healing. *Int. J. Pavement Eng.* **2019**, *20*, 89–99.
18. Giorgia, M.; Amedeo, V.; Francesco, C. Influence of different fillers and SBS modified bituminous blends on fatigue, self-healing and thixotropic performance of mastics. *Road Mater. Pavement Design* **2019**, *20*, 656–670.
19. Xu, S.; Liu, X.Y.; Amir, T.; Erik, S. Investigation of the potential use of calcium alginate capsules for self-healing in porous asphalt concrete. *Materials* **2019**, *12*, 168. [CrossRef]

20. Wang, C.; Xie, W.; Underwood, B.S. Fatigue and healing performance assessment of asphalt binder from rheological and chemical characteristics. *Mater. Struct.* **2018**, *51*. [CrossRef]

21. Fan, S.P.; Wang, H.; Zhu, H.Z.; Sun, W. Evaluation of self-healing performance of asphalt concrete for low-temperature fracture using semicircular bending test. *J. Mater. Civil Eng.* **2018**, *30*, 04018218. [CrossRef]

22. Sun, D.Q.; Li, B.; Ye, F.Y.; Zhu, X.Y.; Lu, T.; Tian, Y. Fatigue behavior of microcapsule-induced self-healing asphalt concrete. *J. Clean. Prod.* **2018**, *188*, 466–476. [CrossRef]

23. Wu, G.X.; Wang, Y.; Wan, R.Y. A study of the strength mechanism of semi-flexible composite pavement material and a modified method for the determination of its compressive resilient modulus. In Proceedings of the 10th International Conference of Chinese Transportation Professionals (ICCTP), Beijing, China, 4–8 August 2010.

24. *Specifications for Special Mortar Semi-Flexible Pavement Application Technology*; DB44/T 1296; Administration of Quality and Technology Supervision of Guangdong Province: Guangzhou, China, 2014.

25. Wang, W.M.; Gao, D.; Wu, K.H. The research of performance on semi-flexible pavement materials. *Highw. Eng.* **2014**, *39*, 78–82.

26. Wang, F.H.; Wang, W.M.; Wu, K.H. Study on the design and performance of large void basic asphalt mixture for semi-flexible pavement. *Highw. Eng.* **2015**, *40*, 72–76.

27. Wu, M.; Johannesson, B.; Geiker, M. A review self-healing in cementitious material and engineered cementitious composites as a self-healing material. *Constr. Build. Mater.* **2012**, *28*, 571–583. [CrossRef]

28. Ali, M.A.E.M.; Nehdi, M.L. Innovation crack-healing hybrid fiber reinforced engineered cementitious composites. *Constr. Build. Mater.* **2017**, *150*, 689–702. [CrossRef]

29. Liu, H.Z.; Zhang, Q.; Gu, C.S. Self-healing of microcracks in Engineered cementitious composites under sulfate and chloride environment. *Constr. Build. Mater.* **2017**, *153*, 948–956. [CrossRef]

30. Wu, C.; Pan, Y.; Ueda, T. Characterization of the abrasion resistance and the acoustic wave attenuation of the engineered cementitious composites for runway pavement. *Constr. Build. Mater.* **2018**, *174*, 537–546. [CrossRef]

31. Zeng, M.L.; Nan, D.; Wu, C.F.; Zhu, T. Research on self-healing performance of PVA-ECC stabilized crushed-stone mixture. *J. Highw. Transp. Res. Dev.* **2015**, *32*, 42–45.

32. Jiang, J.W.; Ni, F.J.; Wu, F.; Husam, S.; Lv, Q. Evaluation of the healing potential of asphalt mixtures based on a modified semi-circular bending test. *Constr. Build. Mater.* **2019**, *196*, 284–294. [CrossRef]

33. *Test Methods of Cement and Concrete for Highway Engineering*; JTG E30; China Communications Press: Beijing, China, 2005.

34. *Standard Test Methods of Bitumen and Mixtures for Highway Engineering*; JTG E20; China Communications Press: Beijing, China, 2011.

 *materials* 

Article

Laboratory and Numerical Investigation of Microwave Heating Properties of Asphalt Mixture

Haopeng Wang [1,*], Yue Zhang [2], Yi Zhang [3], Shuyin Feng [4], Guoyang Lu [2] and Lintao Cao [5,*]

[1] Section of Pavement Engineering, Faculty of Civil Engineering and Geosciences, Delft University of Technology, Stevinweg 1, 2628 CN Delft, The Netherlands

[2] Institute of Highway Engineering, RWTH Aachen University, Mies-van-der-Rohe-Street 1, 52074 Aachen, Germany; 25yuezhang@web.de (Y.Z.); lu@isac.rwth-aachen.de (G.L.)

[3] School of Highway, Chang'an University, Xi'an 710064, Shaanxi, China; yizhang@chd.edu.cn

[4] Department of Civil Engineering, University of Bristol, BS8 1TR Bristol, UK; shuyin.feng@bristol.ac.uk

[5] School of Civil Engineering and Architecture, Hubei University of Arts and Science, Xiangyang 441053, China

* Correspondence: haopeng.wang@tudelft.nl (H.W.); 30caolintao@tongji.edu.cn (L.C.); Tel.: +31-062-536-1801 (H.W.)

Received: 12 December 2018; Accepted: 28 December 2018; Published: 4 January 2019

Abstract: Microwave heating is an encouraging heating technology for the maintenance, recycling, and deicing of asphalt pavement. To investigate the microwave heating properties of asphalt mixture, laboratory tests and numerical simulations were done and compared. Two types of Stone Mastic Asphalt (SMA) mixture samples (with basalt aggregates and steel slag aggregates) were heated using a microwave oven for different times. Numerical simulation models of microwave heating of asphalt mixture were developed with finite element software COMSOL Multiphysics. The main thermal and electromagnetic properties of asphalt mixture, served as the model input parameters, were measured through a series of laboratory tests. Both laboratory-measured and numerical simulated surface temperatures were recorded and analyzed. Results show that the replacement of basalt aggregates with steel slag aggregates can significantly increase the microwave heating efficiency of asphalt mixture. Numerical simulation results have a good correlation with laboratory test results. It is feasible to use the developed model coupling electromagnetic waves with heat transfer to simulate the microwave heating process of asphalt mixture.

Keywords: asphalt mixture; microwave heating; steel slag; dielectric loss; electromagnetic; numerical simulation

---

1. Introduction

Microwave heating has been widely applied in various industrial fields, such as drying, material preparation, food processing, healthcare, etc. [1] Microwaves have the potential to provide rapid, uniform, high efficient, safe, and environment-friendly heating of materials [2]. Due to the above advantages of microwave heating, there have been increased interests in utilizing microwave heating in the asphalt paving industry. Specifically, three main applications in pavement engineering include asphalt pavement maintenance (such as crack healing, pothole patching, rut repair, etc.) [3–5], recycling of the old asphalt pavement (heating of reclaimed asphalt pavement using a microwave power unit) [6] and snow melting or deicing [7,8]. The main mechanism of microwave heating is the dielectric loss of a material under the microwave filed, including polarized relaxation loss and conductive loss [2]. Asphalt mixture usually consists of about 5% of asphalt, about 95% of coarse aggregate, fine aggregate and other mineral powders [9]. When asphalt mixture is exposed to microwave radiation, heat is generated through conversion of the energy of the electromagnetic field. In the conventional heating methods, such as hot-air heating and infrared heating, energy is transferred from the surfaces of the material internally by convection, conduction, and radiation [10]. In contrast, microwave heating is achieved by molecular excitation inside the material without relying on the temperature gradient. Therefore, microwave heating is a direct energy conversion process rather than heat transfer from external heat sources [3]. This fundamental difference in transferring energy endows microwave heating many exclusive advantages. More recently, induction heating was introduced in asphalt pavement. It is based on the Faraday's electromagnetic induction theory and only applicable to the conductive asphalt materials [11,12]. However, the skin effect caused by the induced eddy currents causes high surface temperature and low internal temperature, thus producing a large temperature gradient through the system [13]. Therefore, microwave heating is a promising, competitive, and effective heating technology.

Nevertheless, the microwave heating efficiency of ordinary asphalt mixture is relatively low due to the low microwave absorbing properties. The capability of a material in absorbing microwave energy can be described by its dielectric properties [14]. The dielectric property of a material is usually expressed by the dielectric permittivity ε^* in Equation (1).

$$\varepsilon^*(f) = \varepsilon\prime(f) - i\varepsilon''(f) \tag{1}$$

where $\varepsilon\prime$ is the dielectric constant; ε'' is the dielectric loss factor; f is the frequency of the external electric field, and $i = \sqrt{-1}$. The dielectric property is highly dependent on the frequency. The dielectric constant determines the amount of storable energy in the material in the form of an electric field. The dielectric loss factor indicates how much of that energy can dissipate in the form of heat. The loss tangent *tanδ*, defined as $\varepsilon''/\varepsilon\prime$, reflects the material's ability of transforming microwave energy into heat. The complex permittivity of asphalt mixtures are influenced not only by frequency and temperature, but also by other properties such as density, asphalt type and content, aggregate type and size, void ratio, and moisture content [15]. It was reported that asphalt has a very low loss tangent of about 0.001. Most of the conventional mineral aggregates, such as albite, marble, orthoclase, and quartz, have poor microwave absorbing characteristics [16]. Therefore, efforts were put into improving the microwave absorbing efficiency of asphalt mixture by adding microwave absorbers or using magnetite-bearing aggregates, such as taconite aggregate mineral and steel slag. Other attempts to improve microwave-absorbing capability included the addition of graphite, carbonyl iron powders (CIPs), carbon nanotubes, steel wool, and ferrite particles [10,13,17,18]. The magnetism of asphalt mixture introduced by ferrite additives is responsible for the magnetic loss during microwave heating. Permeability is the parameter to describe the degree of magnetization that a material experiences under the influence of an external magnetic field. Similar to the complex permittivity, the real part of permeability ($\mu\prime$) is related to energy storage, and the imaginary part (μ'') implies the magnetic loss in particles [10]. Therefore, by improving the permittivity and permeability of asphalt mixture, the microwave heating properties will be enhanced.

Through various time and material consuming laboratory tests, it can be found microwave heating is a promising technology for asphalt pavement recycling and maintenance. However, fewer studies applied numerical modelling to investigate the microwave heating process and mechanism of asphaltic materials [10,13,19]. The aim of this study was to investigate the microwave heating properties of different types of asphalt mixtures through both laboratory test and numerical simulation.

2. Materials and Methods

2.1. Materials and Mix Design

In this study, the used asphalt binder was neat Pen-70 asphalt from Shell. Table 1 presents the basic properties of the binder. Basalt aggregates, steel slag aggregates and limestone fillers were used to produce asphalt mixture samples. Various properties of both aggregates are shown in Table 2. Polyester fiber was added as the drain-down stabilizer at the dosage of 0.3% by the total weight of the mix. As industrial waste, steel slag contains some metal oxides, especially transition metal such as ferric oxide. It was reported that steel slag can significantly influence the thermal and electromagnetic properties of asphalt mixture [10,20]. Steel slag was used as a partial substitute for basalt [21]. Due to the strong absorption of asphalt, fine steel slag aggregates were not chosen to substitute for fine basalt aggregates smaller than 2.36 mm. Since the specific gravity of steel slag is different from basalt aggregate, the equivalent-volume method was used to replace the coarse basalt aggregates above 2.36 mm with steel slag.

Table 1. Basic properties of asphalt binder.

Properties	Value
Penetration (25 °C, 100 g, 5 s, 0.1 mm)	71
Ductility (5 cm/min, 5 °C, cm)	32.2
Softening point (R&B, °C)	47.5
Flash point (°C)	272
Rotational viscosity (60 °C, Pa·s)	203
Wax content (%)	1.6
Density (15 °C, g/cm^3)	1.032

Table 2. Basic properties of aggregates.

Aggregate	Specific Gravity (g/cm^3)	Water Absorption (%)	Crushing Value (%)	Asphalt Affinity (%)	Abrasion Loss (%) (Los Angeles)
Basalt	2.82	0.72	12.8	>85	14.6
Steel slag	3.47	1.26	12.2	>95	13.8

Stone Mastic Asphalt (SMA) with 13.2-mm nominal maximum aggregate size, designated as SMA-13, was used in this study. Gradation SMA-13 shown in Figure 1 was designed in accordance with the standard Marshall Design method (ASTM D6927) [22]. Two types of asphalt mixture samples, control mix with basalt aggregates (SMA-B) and mix with steel slag substitution (SMA-S), were prepared. To avoid the influence of varying grain sizes, both aggregates were sieved into different sieve sizes and then mixed into the specific gradation. The optimum asphalt content for SMA-B was 6.2%. The same asphalt content was chosen for SMA-S to minimize the control variable. The air void for both types of the mix was around 4.0%. Standard Marshall cylindrical specimens (101.6 mm in diameter and 63.5 mm in height) were fabricated for microwave heating test.

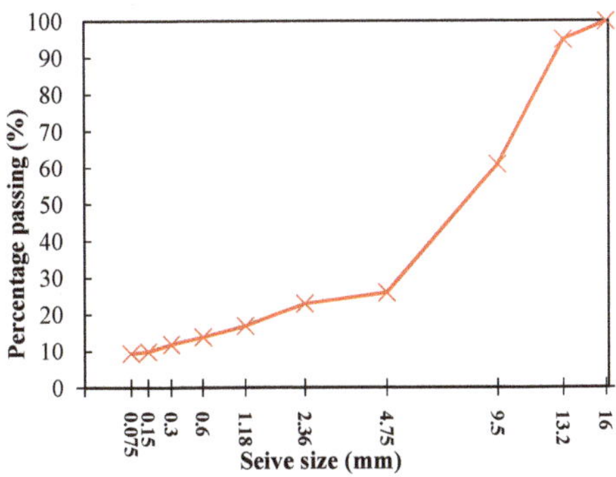

Figure 1. Mix gradation of Stone Mastic Asphalt-13 (SMA-13).

2.2. Thermal Properties Measurement

Thermal conductivity, thermal diffusivity and specific heat capacity are the three most important factors that affect the microwave heating process of materials, which refers to heat transfer phenomena [23]. The thermal conductivity was measured through a steady-state method using a heat flow meter (HFM 446, NETZSCH Group, Selb, Germany) according to ASTM C518 [24]. A slab specimen (15 cm × 15 cm × 4 cm) was placed between two plates with temperature gradient, and the heat flow created by the well-defined temperature difference is measured with a heat flux sensor. In this case, 5 °C and 35 °C were set as the constant temperatures for the cold plate and the hot plate respectively. The thermal conductivity can be calculated based on the acquired data according to the Fourier's Law for heat conduction (Equation (2)).

$$\mathbf{q} = -k\nabla T \tag{2}$$

where $\mathbf{q}$ is the conductive heat flux; k is the thermal conductivity; T is the transient temperature. The specific heat capacity C_p was also measured by the heat flow meter. With the total heat consumption required to heat the sample and temperature development, the specific heat capacity can be determined at a certain temperature. Thermal diffusivity (α) is the coefficient that characterizes the rate of heat energy diffusion throughout a material when it is exposed to a fluctuating thermal environment. Thermal diffusivity is calculated as thermal conductivity divided by density (ρ) and specific heat capacity at a constant pressure.

$$\alpha = \frac{k}{\rho \times c_p} \tag{3}$$

2.3. Electromagnetic Properties Measurement

Electromagnetic parameters, including complex permittivity and complex permeability, are the main indicators to quantify the microwave absorbing efficiency of a material. To obtain these above parameters of asphalt mixture specimens, measurements were carried out with an Agilent E5071C vector network analyzer (Santa Clara, CA, USA) using the free-space method [25]. The detailed measurement system and calculation process can be found in Reference [22]. The electrical conductivity of asphalt mixture was measured by the simple two-probe method [9,26].

2.4. Temperature Measurement under Microwave Heating

Asphalt mixture samples were heated using a commercial microwave oven (Galanz P100M25ASL-H4, Guangdong Galanz Enterprise Co, ltd., Foshan, China) with an input of 1200 W and a 220 V, 50 Hz power supply. The oven can generate microwaves of up to 1000 W at an excitation frequency of 2.45 GHz, which corresponds to a wavelength of 122.4 mm. Each type of asphalt mixture sample has two replicates due to the potential variation of test results. The cylindrical specimen (Φ101.6 mm × 63.5 mm) was placed on the center of the glass plate in the microwave oven. The surface temperature was measured every 20 s by swiftly opening the door using a thermal infrared camera as shown in Figure 2 [19]. The total heating time was 120 s. The average temperature value of six randomly selected points from the specimen surface was calculated as the experimental temperature.

Figure 2. Surface temperature measurement of the specimen in the microwave oven.

3. Numerical Simulation

As discussed before, microwave heating is a multiphysics phenomenon that involves the physics of electromagnetic waves and heat transfer. The rapidly varying electric and magnetic fields lead to four sources of heating. First, any electric field applied to a conductive material will generate eddy currents. In addition, a time-varying electric field will cause dipolar molecules within the material to oscillate back and forth to generate molecular friction. A time-varying magnetic field applied to a conductive material will also induce current flow. For certain types of magnetic materials, the hysteresis losses also make contribution to the heating [27]. To simulate the electro-magneto-thermal phenomenon in a real-time mode, the finite element software COMSOL Multiphysics (Version 5.3, COMSOL BV, Zoetermeer, The Netherlands) has been utilized for modelling microwave heating of asphalt mixture.

3.1. Electromagnetic Waves

Electromagnetic analysis of asphalt mixture on a macroscopic level involves solving Maxwell's equations subject to certain boundary conditions. These equations can be formulated in partial differential form, which can be handled by the finite element method.

$$\nabla \times \mathbf{H} = \mathbf{J} + \frac{\partial \mathbf{D}}{\partial t} \tag{4a}$$

$$\nabla \times \mathbf{E} = -\frac{\partial \mathbf{B}}{\partial t} \tag{4b}$$

$$\nabla \cdot \mathbf{D} = \rho_e \tag{4c}$$

$$\nabla \cdot \mathbf{B} = 0 \tag{4d}$$

To apply the Maxwell equations self-consistently, the constitutive relations describing the macroscopic behaviors of matter under the influence of fields need to be obtained. Assuming asphalt mixture is an isotropic and linear material, the constitutive equations can be formulated as follows.

$$\mathbf{J} = \sigma\mathbf{E} \tag{5a}$$

$$\mathbf{D} = \varepsilon\mathbf{E} \tag{5b}$$

$$\mathbf{B} = \mu\mathbf{H} \tag{5c}$$

where $\mathbf{H}$ is the magnetic field intensity; $\mathbf{J}$ is the electric current density; $\mathbf{D}$ is the electric displacement or electric flux density; $\mathbf{E}$ is the electric field intensity; $\mathbf{B}$ is the magnetic flux density; ρ_e is the electric charge density; σ is the material electrical conductivity; ε is the material permittivity; and μ is the material permeability.

3.2. Heat Transfer

Applied microwave energy is transformed into power based on the electromagnetic field distribution at a particular location. The absorbed power term is considered a source term in heat transfer equations to calculate transient temperature profile. The diffusion of heat into continua is governed by:

$$\rho C_p \frac{\partial T}{\partial t} = \nabla \cdot (k\nabla T) + Q_e \tag{6}$$

where ρ is the density; C_p is the specific heat at constant pressure; k is the thermal conductivity; T is the temperature at time t; and Q_e is the internal heat source (absorbed power). The surface of the matter exchanges heat with surrounding air by convection expressed as:

$$-\mathbf{n} \cdot \mathbf{q} = h(T - T_a) \tag{7}$$

where **q** is the conductive heat flux, which is proportional to the temperature gradient in Equation (2); h is the surface convective coefficient; **n** is the normal vector on the boundary; T is the transient temperature; and T_a is the ambient temperature.

3.3. Multiphysics Coupling

The electro-magneto-thermal phenomenon often encountered in microwave heating is usually solved in a coupled manner because the power dissipation calculated from electromagnetic fields influences other physical phenomenon, such as heat transfer, component evaporation, and microstructural change in heated materials. These complex physical situations result in rapid changes in material properties, which in turn makes the problem highly nonlinear [27]. However, the nonlinearity in this study was not considered because of the difficulty to measure the input material parameters of such an in homogenous material. The process of coupling electromagnetic waves and heat transfer in microwave heating is shown in Figure 3. The distributed heat source, which includes resistive heating (ohmic heating) and magnetic losses in Equation (8) [28], is computed from a stationary electromagnetic analysis in the frequency domain. Then a transient heat transfer simulation showing how the heat redistributes in the asphalt mixture samples was followed. In the software, the frequency domain study is only used for the electromagnetics interface, whereas the time-dependent study is only applicable to the heat transfer interface. Notice that the electromagnetic heat source will be computed first, and then used in the time-dependent heat transfer study step.

$$Q_e = Q_{rh} + Q_{ml} \tag{8a}$$

$$Q_{rh} = \frac{1}{2}\mathrm{Re}(\mathbf{J}\cdot\mathbf{E}) \tag{8b}$$

$$Q_{ml} = \frac{1}{2}\mathrm{Re}(i\omega\mathbf{B}\cdot\mathbf{H}) \tag{8c}$$

where Q_{rh} is the resistive heating of dielectric material due to the electric current; Q_{ml} is the magnetic loss of magnetic material interacting with the magnetic field component of microwave. $\mathrm{Re}()$ is the real part of the variable.

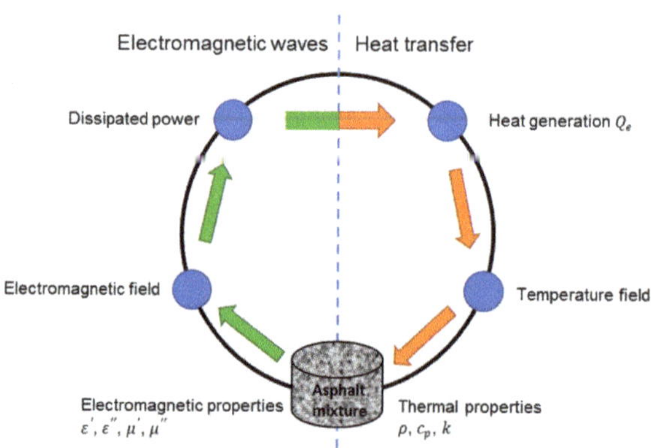

Figure 3. Schematic flow chart of coupling electromagnetic and thermal fields [28].

3.4. Model Definition

The microwave oven is a metallic box connected to a 2.45 GHz microwave source via a rectangular waveguide. The dimensions of the oven are 267 mm (width) × 270 mm (depth) × 188 mm (height). The size of the waveguide is 50 mm (width) × 78 mm (depth) × 18 mm (height). There is a cylindrical glass plate near the bottom of the oven. A cylindrical asphalt mixture sample was placed on top of the glass plate. The microwave operates at 1000 W, but because the symmetrical model was built to reduce the model size by one half, only 500 W was input in the simulation. The symmetry cut is applied vertically through the oven, waveguide, asphalt mixture sample, and plate. The symmetrical geometry and 3D mesh are shown in Figures 4 and 5, respectively. Copper was applied for the walls of the oven and waveguide in this model. The applied impedance boundary condition on these walls ensures the small resistive metals losses get accounted for. The symmetry cut has mirror symmetry for the electric field and is represented by the boundary condition as shown in Equation (9).

$$\mathbf{n} \times \mathbf{H} = 0 \tag{9}$$

where $\mathbf{n}$ is the outward unit normal vector to the port boundary; $\mathbf{H}$ is the magnetic field vector.

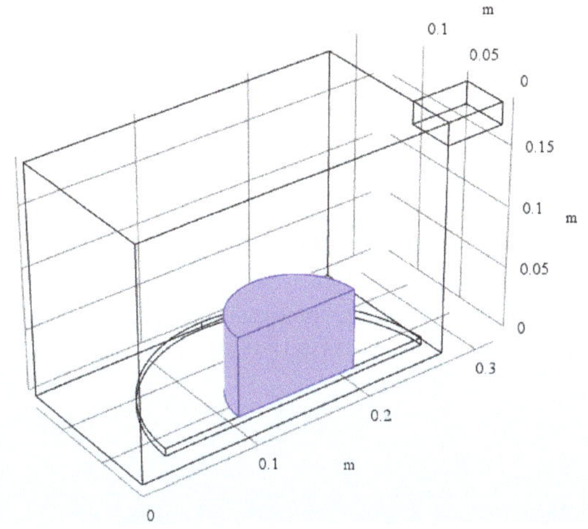

Figure 4. Geometry of microwave oven, asphalt mixture sample, and waveguide feed.

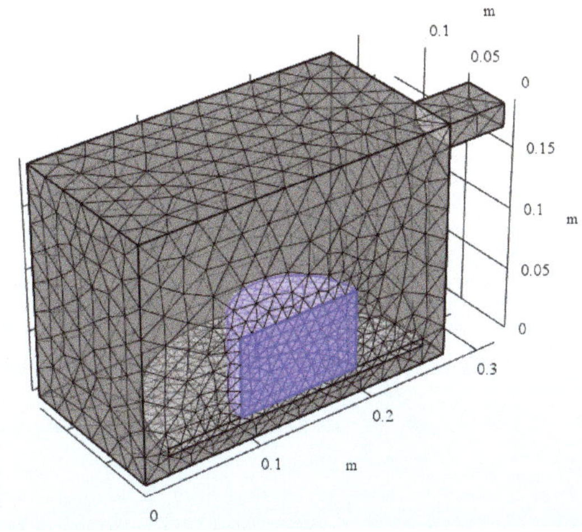

Figure 5. Mesh of microwave oven, asphalt mixture sample, and waveguide feed.

The rectangular port is excited by a transverse electric (TE) wave, which is a wave that has no electric field component in the propagating direction. At an excitation frequency of 2.45 GHz, TE_{10} mode is the only mode of propagation through the rectangular waveguide. The propagation constant β required for the port mode settings is frequency (v) dependent:

$$\beta = \frac{2\pi}{c}\sqrt{v^2 - v_c^2} \tag{10}$$

where c is the speed of light and v_c is the cutoff frequency.

3.5. Material Properties

As discussed before, to implement the finite element model of microwave heating on asphalt mixtures, their material properties need to be obtained as the input parameters. Specifically, the thermal and electromagnetic parameters of both SMA-B and SMA-S mixtures were presented in Tables 3 and 4. The presented data were the averaged values from test results of three replicates. It can be noted that the replacement of basalt with steel slag decreased the thermal conductivity and specific heat capacity of asphalt mixture, while the thermal diffusivity was increased. Steel slag has a porous inter-structure. Many small pores within the porous steel slag obstruct the heat transfer process, which is accounted for the decrease of the thermal conductivity of asphalt mixture. The high porosity of steel slag also contributes to the heat retention characteristics, which is responsible for the decrease of heat capacity [29]. In terms of electromagnetic properties, SMA-S has higher electrical conductivity than SMA-B. The addition of steel slag also increases both permittivity and permeability of asphalt mixture as shown in Table 4. The improvement of the electromagnetic properties of SMA-S is due to the ferric components and other metal elements in steel slag.

Table 3. Thermal parameters of asphalt mixtures.

Mixture Type	Density (kg/m^3)	Thermal Conductivity (W/(m·K))	Specific Heat Capacity (J/(kg·K))	Thermal Diffusivity (m^2/s)
SMA-B	2530	1.508	918.5	6.49×10^{-7}
SMA-S	2632	1.446	756.5	7.26×10^{-7}

Table 4. Electromagnetic parameters of asphalt mixtures at 2.45 GHz.

Mixture Type	σ	ε'	ε''	μ'	μ''
SMA-B	4.26×10^{-9}	5.34	0.49	1.0	0
SMA-S	3.85×10^{-7}	5.68	0.52	1.03	0.006

4. Results and Discussions

4.1. Numerical Simulation Results

4.1.1. Microwave Heat Source Distribution

The numerical analysis using the current model took several minutes with common personal computer configuration. The distributed microwave heat source as a slice plot through the center of asphalt mixture sample SMA-B and SMA-S are shown in Figures 6 and 7, respectively. It indicates that the resistive loss distribution shows a complicated oscillating pattern, which has several strong peaks inside the sample. Since sample SMA-B is a non-magnetic material, there is no magnetic loss during the microwave heating process. On the contrary, sample SMA-S has both resistive loss and magnetic loss. Through a volume integration of the microwave heating, the amount of resistive loss and magnetic loss, as well as the total power loss during the heating process were calculated in Table 5. The total microwave energy absorbed by the asphalt mixtures is more than 90% of the input microwave power (500 W). It is interesting to note that the resistive loss of SMA-S after microwave heating is lower than that of SMA-B. However, from the total heat source, SMA-S with steel slag aggregates has a higher microwave absorbing efficiency than SMA-B with basalt aggregates.

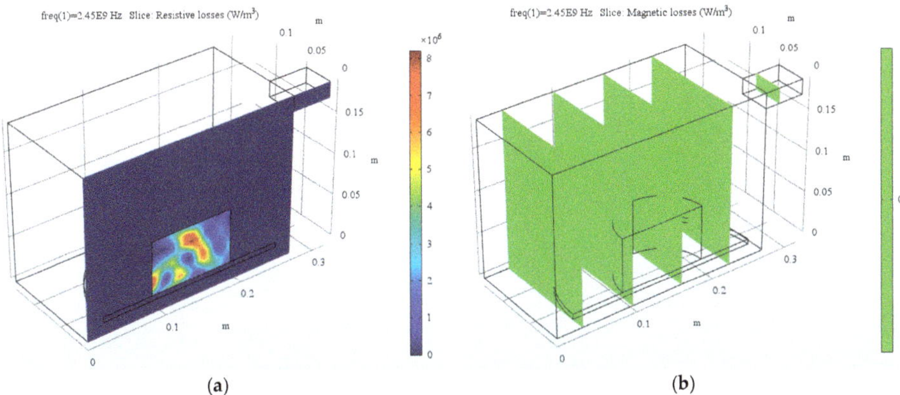

Figure 6. The dissipated microwave power distribution of asphalt mixture SMA-B: (**a**) Resistive loss; (**b**) Magnetic loss.

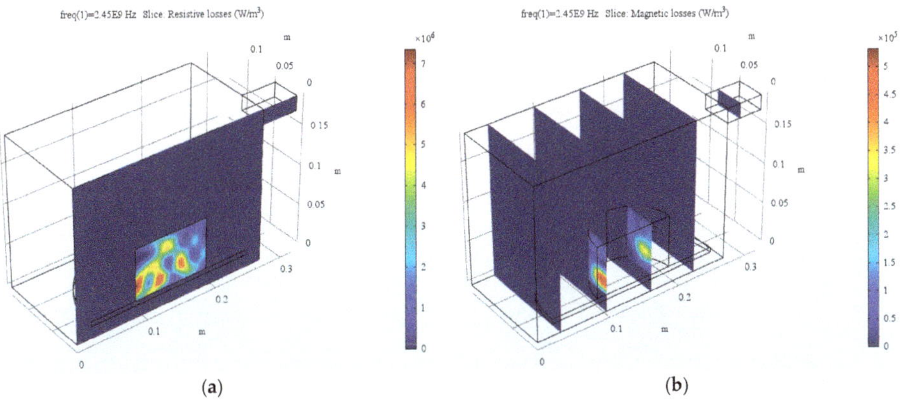

Figure 7. The dissipated microwave power distribution of asphalt mixture SMA-S: (**a**) Resistive loss; (**b**) Magnetic loss.

Table 5. The distributed heat source of asphalt mixtures at 2.45 GHz.

Mixture Type	Resistive Losses (W)	Magnetic Losses (W)	Total Heat Source (W)
SMA-B	455.19	0	455.19
SMA-S	442.06	28.37	470.43

4.1.2. Temperature Distribution of Test Samples

The temperature distribution of two types of asphalt mixture after 120 s simulative microwave heating are presented in Figure 8. It looks like the surface temperature of SMA-S is higher than that of SMA-B from the temperature contour plot. Quantitative analysis will be conducted in the following part. It should be emphasized that the rectangular cross section of the cylindrical sample seen as the surface area is actually the internal part of the sample due to the symmetric treatment of the model. It is obvious that the temperature distribution of the asphalt mixture specimen during heating was not uniform. The internal temperatures were higher the surface ones. This is possibly due to the fact that heat dissipation on the surface of a specimen is greater than its interior. This simulation results coincide with the laboratory results [5,24,30]. When heating the asphalt mixture to certain temperatures, the inside water contents start boiling and transporting heat as steam to outer layers. Asphalt may start flowing due to softening, resulting in the change of air voids and skeleton structure. These physio-chemical changes of the mix constituents also affect the electromagnetic properties of the asphalt mixture. The simple microwave absorption and heat conduction model used here does not capture these nonlinear effects. However, the model can serve as a starting point for a more advanced analysis.

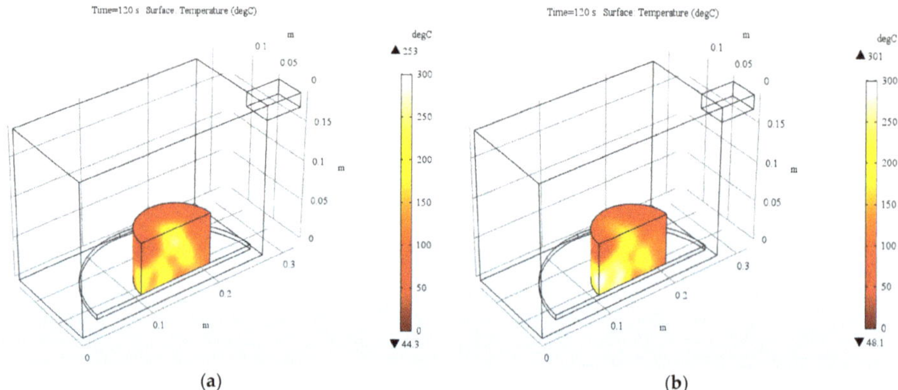

(a)   (b)

Figure 8. Surface temperature distribution of asphalt mixture: (**a**) SMA-B; (**b**) SMA-S.

4.2. Comparison between Laboratory and Simulation Results

The lateral surface temperatures of laboratory test and numerical simulation are compared in Figure 9. Here the simulative surface temperature values were averaged through area integral. It is obvious that numerical simulation results have a good correlation with the experimental results for both types of asphalt mixtures. However, the numerically simulated temperatures are somewhat higher than the laboratory test results. This is possibly due to the temperature loss during the laboratory measurement for several seconds. In addition, the nonlinear effects during the heating process can be the reason for this, which needs to be further included in the model. Nevertheless, it is feasible to use a numerical method to simulate the microwave heating process of asphalt mixture.

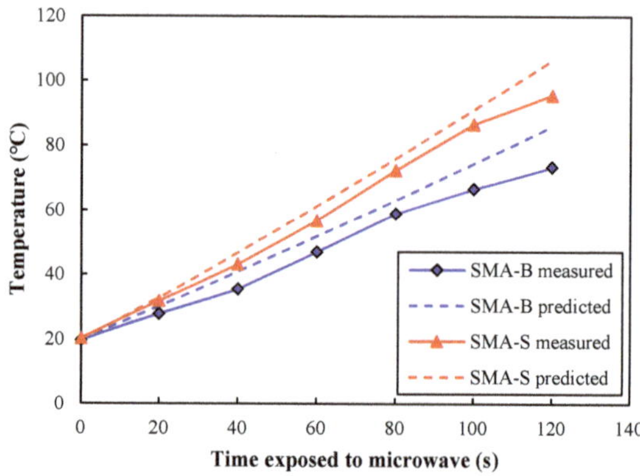

Figure 9. Surface temperature comparison of asphalt mixture between laboratory test and numerical simulation.

More precisely, the initial temperatures, final temperatures and heating rates of asphalt mixture samples during microwave heating are summarized in Table 6. SMA-S has a better microwave heating performance than SMA-B. The final lateral surface temperature of SMA-S reached 95.6 °C while that of SMA-B was only 73.4 °C. The higher heating rate of SMA-S than SMA-B confirms the higher microwave absorbing efficiency of SMA-S due to the addition of steel slag.

Table 6. Microwave heating performance of asphalt mixture samples on the lateral surface.

Mixture Type		Initial Temperature (°C)	Final Temperature (°C)	Heating Rate (°C/s)
SMA-B	Experiment	19.7	73.4	0.448
	Simulation	20.0	86.1	0.551
SMA-S	Experiment	20.3	95.6	0.623
	Simulation	20.0	106.5	0.721

5. Conclusions

This study investigated the microwave heating properties of two types of asphalt mixture through both laboratory test and numerical simulation. The main thermal and electromagnetic properties of used asphalt mixtures were explored through laboratory tests. The following conclusions can be drawn based on the study:

(1) The partial replacement of basalt aggregates with steel slag aggregates improve the electromagnetic properties of asphalt mixture. Microwave heating of asphalt mixture sample containing steel slag includes both resistive heating and magnetic heating due to the altered permeability of the sample.

(2) Asphalt mixture sample containing steel slag aggregates has a higher microwave heating efficiency than ordinary asphalt mixture sample with basalt aggregates.

(3) There is a good correlation between laboratory measured temperatures and numerically simulated temperatures of asphalt mixture samples.

(4) It is feasible to use the developed FEM model, which coupled electromagnetic waves with heat transfer, to simulate the microwave heating process of asphalt mixture.

For further research, the size effect of test samples and specific material parameters, such as moisture content, air voids, asphalt content, aggregate properties, etc. should be considered. Developing more advanced numerical models which consider the nonlinear effects and time discretization will be a challenge. In addition, the microstructural changes and mechanical performance of asphalt mixtures after microwave heating will be further investigated.

Author Contributions: Conceptualization, H.W.; Formal analysis, Y.Z. (Yue Zhang); Funding acquisition, L.C. and G.L.; Investigation, Y.Z. (Yue Zhang); Methodology, H.W. and S.F.; Project administration, G.L.; Supervision, S.F.; Writing—original draft, H.W. and Y.Z. (Yi Zhang); Writing—review and editing, H.W., Y.Z. (Yi Zhang), and L.C.

Funding: This research received no external funding.

Acknowledgments: The first author would like to acknowledge the scholarship from the China Scholarship Council.

Conflicts of Interest: The authors declare no conflict of interest.

References

1. Jones, D.A.; Lelyveld, T.P.; Mavrofidis, S.D.; Kingman, S.W.; Miles, N.J. Microwave heating applications in environmental engineering—A review. *Resour. Conserv. Recycl.* **2002**, *34*, 75–90. [CrossRef]

2. Metaxas, A.A.; Meredith, R.J. *Industrial Microwave Heating*; IET: Stevenage, UK, 1983.

3. Benedetto, A.; Calvi, A. A pilot study on microwave heating for production and recycling of road pavement materials. *Constr. Build. Mater.* **2013**, *44*, 351–359. [CrossRef]

4. Bosisio, R.; Spooner, J.; Granger, J. Asphalt road maintenance with a mobile microwave power unit. *J. Microw. Power* **1974**, *9*, 381–386. [CrossRef]

5. Gallego, J.; del Val, M.A.; Contreras, V.; Paez, A. Heating asphalt mixtures with microwaves to promote self-healing. *Constr. Build. Mater.* **2013**, *42*, 1–4. [CrossRef]

6. Sun, T.; Chen, L. Dielectric loss model for asphalt mixture based on microwave heating. *Electromagnetics* **2017**, *37*, 49–63. [CrossRef]

7. Wang, Z.J.; Gao, J.; Ai, T.; Zhao, P. Laboratory investigation on microwave deicing function of micro surfacing asphalt mixtures reinforced by carbon fiber. *J. Test. Eval.* **2014**, *42*, 498–507. [CrossRef]

8. Sun, Y.; Wu, S.; Liu, Q.; Hu, J.; Yuan, Y.; Ye, Q. Snow and ice melting properties of self-healing asphalt mixtures with induction heating and microwave heating. *Appl. Therm. Eng.* **2018**, *129*, 871–883. [CrossRef]

9. Wang, H.; Yang, J.; Liao, H.; Chen, X. Electrical and mechanical properties of asphalt concrete containing conductive fibers and fillers. *Constr. Build. Mater.* **2016**, *122*, 184–190. [CrossRef]

10. Liu, W.; Miao, P.; Wang, S.-Y. Increasing microwave heating efficiency of asphalt-coated aggregates mixed with modified steel slag particles. *J. Mater. Civ. Eng.* **2017**, *29*, 04017171. [CrossRef]

11. Apostolidis, P.; Liu, X.; Scarpas, A.; Kasbergen, C.; van de Ven, M.F.C. Advanced evaluation of asphalt mortar for induction healing purposes. *Constr. Build. Mater.* **2016**, *126*, 9–25. [CrossRef]

12. Liu, Q.; Wu, S.; Schlangen, E. Induction heating of asphalt mastic for crack control. *Constr. Build. Mater.* **2013**, *41*, 345–351. [CrossRef]

13. Miao, P.; Liu, W.; Wang, S. Improving microwave absorption efficiency of asphalt mixture by enriching Fe_3O_4 on the surface of steel slag particles. *Mater. Struct.* **2017**, *50*, 134. [CrossRef]

14. Al-Ohaly, A.A.; Terrel, R.L. *Effect of Microwave Heating on Adhesion and Moisture Damage of Asphalt Mixtures*; National Research Council, Transportation Research Board: Washington, DC, USA, 1988; pp. 27–36.

15. Jaselskis, E.J.; Grigas, J.; Brilingas, A. Dielectric properties of asphalt pavement. *J. Mater. Civ. Eng.* **2003**, *15*, 427–434. [CrossRef]

16. Hopstock, M.D.; Zanko, M.L. *Minnesota Taconite as a Microwave-Absorbing Road Aggregate Material for Deicing and Pothole Patching Applications*; Natural Resources Research Institute, University of Minnesota Duluth: Duluth, MN, USA, 2005.

17. Zhao, H.D.; Zhong, S.; Zhu, X.Y.; Chen, H.Q. High-efficiency heating characteristics of ferrite-filled asphalt-based composites under microwave irradiation. *J. Mater. Civ. Eng.* **2017**, *29*, 04017007. [CrossRef]

18. Wang, H.; Yang, J.; Lu, G.; Liu, X. Accelerated healing in asphalt concrete via laboratory microwave heating. *J. Test. Eval.* **2018**, *48*. [CrossRef]

19. Wang, H.; Apostolidis, P.; Liu, X.; Scarpas, T.; Yang, J.; Xu, L. Laboratory test and numerical simulation of microwave heating properties of asphalt mixture. In Proceedings of the 10th International Conference on the Bearing Capacity of Roads, Railways and Airfields (BCRRA 2017), Athens, Greece, 28–30 June 2017.

20. Liu, Q.T.; Li, B.; Schlangen, E.; Sun, Y.H.; Wu, S.P. Research on the mechanical, thermal, induction heating and healing properties of steel slag/steel fibers composite asphalt mixture. *Appl. Sci.* **2017**, *7*, 1088. [CrossRef]

21. Wu, S.P.; Xue, Y.J.; Ye, Q.S.; Chen, Y.C. Utilization of steel slag as aggregates for stone mastic asphalt (SMA) mixtures. *Build. Environ.* **2007**, *42*, 2580–2585. [CrossRef]

22. ASTM. *ASTM D6927-15 Standard Test Method for Marshall Stability and Flow of Asphalt Mixtures*; ASTM International: West Conshohocken, PA, USA, 2015.

23. Nguyen, Q.T.; Di Benedetto, H.; Sauzeat, C. Determination of thermal properties of asphalt mixtures as another output from cyclic tension-compression test. *Road Mater. Pavement Des.* **2012**, *13*, 85–103. [CrossRef]

24. Wang, H. Design and Evaluation of Conductive Asphalt Concrete for Self-Healing. Master's Thesis, Southeast University, Nanjing, China, 2016.

25. Liu, W.; Wang, S.; Gu, X. Improving microwave heating efficiency of asphalt concrete by increasing surface magnetic loss of aggregates. *Road Mater. Pavement Des.* **2018**, 1–15. [CrossRef]

26. Wu, S.; Mo, L.; Shui, Z.; Chen, Z. Investigation of the conductivity of asphalt concrete containing conductive fillers. *Carbon* **2005**, *43*, 1358–1363. [CrossRef]

27. Pitchai, K. *Electromagnetic and Heat Transfer Modeling of Microwave Heating in Domestic Ovens*; University of Nebraska at Lincoln: Lincoln, NE, USA, 2011.

28. Kopyt, P.; Celuch, M. Coupled electromagnetic-thermodynamic simulations of microwave heating problems using the fdtd algorithm. *J. Microw. Power Electromagn. Energy* **2007**, *41*, 18–29. [CrossRef]

29. Ahmedzade, P.; Sengoz, B. Evaluation of steel slag coarse aggregate in hot mix asphalt concrete. *J. Hazard. Mater.* **2009**, *165*, 300–305. [CrossRef]
30. Sun, Y.; Wu, S.; Liu, Q.; Zeng, W.; Chen, Z.; Ye, Q.; Pan, P. Self-healing performance of asphalt mixtures through heating fibers or aggregate. *Constr. Build. Mater.* **2017**, *150*, 673–680. [CrossRef]

Article

Preparation and Characteristics of Ethylene Bis(Stearamide)-Based Graphene-Modified Asphalt

Xia Zhang [1], Jun-Xi He [1], Gang Huang [1,*], Chao Zhou [1], Man-Man Feng [1] and Yan Li [2]

[1]   National and Local Joint Engineering Laboratory of Traffic Civil Engineering Materials, Chongqing Jiaotong University, Chongqing 400074, China; 990020050886@cqjtu.edu.cn (X.Z.); 622160951066@mails.cqjtu.edu.cn (J.-X.H.); zczc6@cqjtu.edu.cn (C.Z.); 622160111017@mails.cqjtu.edu.cn (M.-M.F.)

[2]   Chongqing Tongli Expressway Maintenance Engineering Co., Ltd., Chongqing 400074, China; 622150951030@mails.cqjtu.edu.cn

*   Correspondence: 990000000659@cqjtu.edu.cn; Tel.: +86-23-6265-2574

Received: 7 February 2019; Accepted: 1 March 2019; Published: 5 March 2019

Abstract: In this study, graphene-modified asphalt (GMA) was prepared from SK-70# matrix asphalt and ethylene bis(stearamide) (EBS). Based on the uniform design method, a model was created using Data Processing System (DPS) software and First Optimization (1stOpt) software using the graphene mixing amount, EBS mixing amount, shear rate, shear time, and shear temperature as factors and using the asphalt penetration, softening point, force ductility, SHRP-PG test, and multistress creep recovery data as indices. Calculations and analysis showed that the optimal composition and preparation parameters of GMA are as follows: the graphene proportion is 20‰, the EBS proportion is 1%, the shear rate is 6000 r.p.m., the shear time is 180 min, and the shear temperature is 140 °C. The prepared GMA had a significantly improved softening point, low-temperature fracture energy, antirutting factor, and creep recovery rate, indicating that adding graphene can improve the high- and low-temperature performance of asphalt. The prepared GMA was characterized by X-ray diffraction (XRD). The dispersibility of graphene in asphalt was evaluated by fluorescence microscopy and Image-Pro Plus imaging software. The results show that graphene can exist in asphalt in a stable form, which increases the loose-layered structure of stacked asphalt or gum. The intense adsorption effect of graphene strengthens the ordered structure of asphalt. However, due to its dispersibility characteristics, some graphene exists in asphalt in clustered form. When the graphene-to-dispersant ratio approaches the optimal value, the dispersant changes the form of graphene in asphalt from irregular clusters to regular clusters and from large, distinct clusters to small, indistinct clusters. When dispersant cannot uniformly disperse graphene in asphalt, graphene clusters primarily form medium-sized grains.

Keywords: graphene-modified asphalt; ethylene bis(stearamide); uniform design; dispersibility; modification

1. Introduction

Graphene, formed by carbon atoms via sp2 electron orbital hybridization, is a beehive-shaped, two-dimensional carbon nanometer inorganic material with various superior properties. In recent years, graphene has become a focus area of scientific research [1–11]. Asphalt pavement is the primary form of pavement in road engineering. Based on the characteristics of the basic chemical structures of graphene and asphalt (components), graphene and asphalt share similar structures and the two should have excellent affinity [12,13]. Graphene has an enormous specific surface area and can have an intense physical adsorption effect with asphalt. Additionally, graphene is capable of physical adsorption and nonpolarized adsorption with light components and polycyclic aromatic hydrocarbons released by asphalt when heated. Under high temperature, graphene effectively suppresses the release of poisonous, harmful asphalt fumes and is environmentally friendly [14,15]. Therefore, graphene-modified asphalt (GMA) has numerous excellent properties and multiple functional groups which can significantly improve asphalt performance (such as its viscoelasticity), reduce or eliminate various asphalt pavement hazards, such as ruts, fractures, and surface wear, and reduce the cost of the entire pavement life cycle. GMA has important scientific and application value for promoting the development of high-performance and durable long-life asphalt pavement [16]. In recent years, Wang Z. et al. [17–19] showed that expanded graphite nanoplatelet composite-modified asphalt materials can effectively enhance the fracture recovery energy, strength, and healing capabilities of an asphalt mixture. Yao H. et al. [20] found that graphite nanoplatelet-modified asphalt can improve asphalt's high- and low-temperature performance, its complex shear modulus, and the antirutting and waterproof capabilities of the asphalt mixture. Li Y. et al. [21] showed that when graphene oxide (GO) and asphalt were mixed, CO_2 gas was released during GO decomposition; the GO structure was completely stripped and was scattered in asphalt to form a single layer. Huang Gang et al. [14,22,23] used expanded graphite to suppress asphalt fumes and proved that expanded graphite was infiltrated by asphalt and was stripped to form graphene platelets that were partially scattered in asphalt. Cheng I. F. et al. [24] developed a technique to produce large graphene flakes on an asphalt surface, which proved that graphene can stably exist in an asphalt medium in a single layer. The existing studies primarily focus on the modification of pavement asphalt using graphene oxide or graphene nanoplatelets to improve asphalt performance [25–34]. There is no report of research on pavement asphalt modification using graphene.

Based on the uniform design method and using the asphalt penetration index, softening point, force ductility, SHRP-PG test, and multistress creep recovery test data as indices, this paper employed Data Processing System (DPS) and First Optimization (1stOpt) software to establish a mathematical model to investigate the material composition and preparation parameters of GMA. In addition, a microscopic analysis method and Image-Pro Plus software were applied to evaluate the dispersibility of graphene in asphalt.

2. Experimental Method and Performance Evaluation

2.1. Materials

The matrix asphalt used in this study was SK-70# asphalt (PG64-22). Each index was tested based on the Standard Test Method of Bitumen and Bituminous Mixture for Highway Engineering (JTG E20-2011) by the Chinese Ministry of Communications [35]. The graphene was NK-1 graphene produced by the Sichuan Deyang Graphene Carbon Technology Co., Ltd., Deyang, Sichuan, China. The dispersant was ethylene bis(stearamide) (EBS) from the Malaysia Kao Company, Petaling Jaya, Malaysia. The basic solvent was trichloroethylene. The technical parameters of SK-70# asphalt, graphene, and EBS are listed in Tables 1–3, respectively.

Table 1. Parameters of SK-70# asphalt.

	Test Item	Test Result	Technology Index	Test Method
	penetration (25 °C, 5 s, 100 g)/0.1 mm	64.70	60.0~80.0	T0604
	ductility (15 °C, 5 cm/min)/cm	103.00	≥100.0	T0605
	softening point/°C	48.10	≥45.0	T0606
	density (15 °C)g/cm^3	1.21	actual measurement	T0603
	wax content/%	2.04	≤2.2	T0615
	dynamic viscosity(60 °C)/Pa·s	197	≥180	T0620
	flash point/°C	315	≥260	T0611
after RTFOT	mass change/%	−0.18	≤±0.8	T0610
	residual penetration ratio/%	63.50	≥61.0	T0604
	10 °C ductility/cm	8.60	≥6.0	T0605

Table 2. Parameters of graphene NK-1.

Parameter	Index
graphene layers/thickness	1–3, monolayer rate >80%
ash content/%	<3.0
specific surface area/m^2/g	110.0
film electrical conductivity/S/cm	550.0
flake diameter (D50)/um	7.0~12.0
flake diameter (D90)/um	11.0~15.0
appearance	Black-grey powder
bulk density/g/mL	0.01~0.02
water content/%	<2.0

Table 3. Parameters of dispersant ethylene bis(stearamide).

Parameter	Index
appearance	White powder
initial melting point/°C	141.0~146.0
total amine/mg KOH/g	$\leq$3.0
color value	$\leq$5.0
acid value/mg KOH/g	$\leq$7.0
fineness degree/mesh	600
heating decrement/%	$\leq$0.5
flash point/°C	$\geq$28.0

2.2. Equipment and Characterization

The shear processing of the modified asphalt was performed using a BME200L intense shear and mix emulsion machine (motor power 0.4 kw, rotational speed range 0–10,000 r.p.m.) from the Shanghai Weikang Machine Manufacturing Co., Ltd., Shanghai, China. The ultrasonic separation of the graphene mixture solution was performed using JP-040 ultrasonic equipment (ultrasonic wave power: 240 W, ultrasonic wave frequency: 40 kHz) from the Skymen Cleaning Equipment Shenzhen Co., Ltd., Shenzhen, China. The asphalt penetration index, softening point, and force ductility were measured using an SYD-2801D penetration index tester, an SYD-2806E softening point tester, and an SYD-45DBF ductility/tension tester with temperature and speed regulation from the Shanghai Changji geological instrument Co., Ltd., Shanghai, China. The asphalt rheological performance was tested with a Bohlin DSR I dynamic shear rheometer from the Malvern Panalytical Instrument Co., Ltd., Malvern, UK. The structure characterization of the modified asphalt was performed with a D8-Advance X-ray diffractometer (copper/palladium, voltage: 40 kV, current: 40 MA, test rate: 0.1 sec/step, wavelength: 1.5418 angstrom) from the Bruker Corporation, Karlsruhe, Germany. The graphene dispersion in GMA was observed using a DM6 M microscope from the Leica Microsystems Inc. Co., Ltd., Buffalo Grove, IL, USA.

2.3. GMA Preparation

The GMA preparation process was as follows:

(1) The graphene and EBS were measured using an electric analytical balance (resolution: 0.0001 g, SHIMADZU Co., Tokyo, Japan) and were placed in a 1000 mL beaker. A total of 350 mL of trichloroethylene was added and mixed with a glass bar to produce a mixed solution. The mixed solution was heated in a constant temperature (80 °C) hot water bath for 15 min. Then, the opening was covered with preservative film. The mixed solution was ultrasonically processed for 2.0 h with a 5-min break every 30 min.

(2) First, 350 g of matrix asphalt was prepared. Then, the mixed solution (after ultrasonic processing) was poured into the container filled with 350 g of matrix asphalt. The container opening was sealed with 3–4 layers of preservative film and cultured for 12 h so that the asphalt was completely dissolved in the mixed solution. The trichloroethylene in asphalt was completely removed using a

rotary evaporator (from the Büchi Labortechnik AG, Uster, Switzerland) with the following parameters: oil bath temperature: 110 °C, rotational speed: 85~90 r.p.m., and evaporation time: 60 min. After the trichloroethylene was removed, the asphalt was poured into a container for the shear test to prepare the GMA with importing nitrogen into the bottom of the container continually. The GMA preparation process is shown in Figure 1.

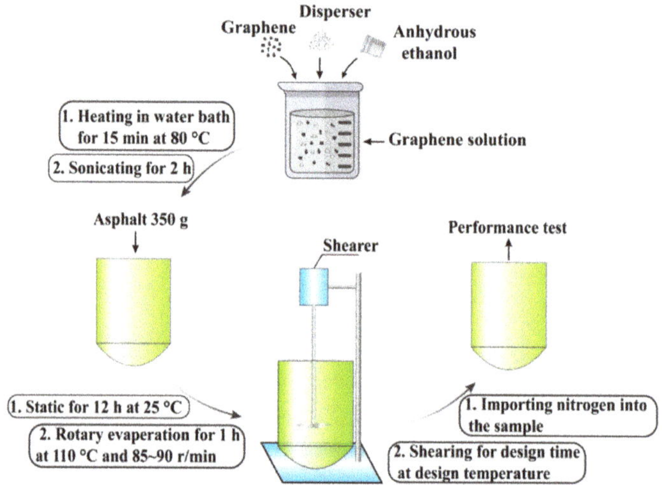

Figure 1. Graphene-modified asphalt preparation process.

2.4. Experimental Design

Uniform design is an application of the "pseudo-Monte Carlo method" in number theory. Uniform design can select a subset of typical test points from the entire set of test points, ensure the uniform distribution of test points in a test range, and reflect major features of the test system. The uniform design method is widely employed to investigate material composition and demonstrates excellent applicability and accuracy [36,37]. Therefore, in this paper, the uniform design method was employed for GMA composition design. In test design, each uniform design table has a code $U_n(q^s)$. "U" represents uniform design; "n" represents n tests; "q" indicates that each factor has q levels; "s" means the table has s columns [38–40].

Five factors with significant impact (X_1, X_2, X_3, X_4, X_5) were selected to investigate GMA composition and preparation parameters [41]. The details of these factors follow: X_1 is the shear rate (r.p.m.); X_2 is the shear time (min); X_3 is the graphene proportion (‰) (mass fraction of matrix asphalt); X_4 is the EBS proportion (%) (mass fraction of graphene); and X_5 is the shear temperature (°C). Each factor has 10 levels, as listed in Table 4.

Table 4. Test design factor levels.

Factor	Level									
	1	2	3	4	5	6	7	8	9	10
X_1/r.p.m.	2000	2500	3000	3500	4000	4500	5000	5500	6000	7000
X_2/min	30	30	60	60	90	90	120	120	180	180
X_3/‰	2	4	6	8	10	12	14	16	18	20
X_4/%	1	2	3	4	5	6	7	8	9	10
X_5/°C	110	110	120	120	130	130	140	140	150	150

Based on the factor levels in Table 4, a corresponding uniform design table and a usage table were generated to design combinations of test parameters. The obtained test parameter combinations are listed in Table 5.

Table 5. Test combinations design table.

Test #	Factor				
	X_1/r.p.m.	X_2/min	X_3/‰	X_4/%	X_5/°C
1#	2000	60	8	5	150
2#	2500	90	16	10	140
3#	3000	180	2	4	130
4#	3500	30	10	9	120
5#	4000	60	18	3	110
6#	4500	120	4	8	150
7#	5000	180	12	2	140
8#	5500	30	20	7	130
9#	6000	90	6	1	120
10#	7000	120	14	6	110

Based on the preparation parameters of each test group in Table 5, the GMA was prepared and subsequent performance tests were performed.

2.5. Performance Evaluation and Microanalysis

The GMA pavement performance was analyzed via its penetration index, softening point, and force ductility index. An SHRP-PG test and a multistress creep recovery test were performed to analyze the viscoelasticity of GMA. The GMA structure was characterized via XRD (from the Bruker Corporation, Karlsruhe, Germany) and a fluorescence microscope (from the Leica Microsystems Inc. Co., Ltd., Buffalo Grove, IL, USA).

3. Results and Discussion

3.1. Indices Data Analysis

The penetration index represents the asphalt thickness at the test temperature, which reflects asphalt's rheological performance to some extent [42,43]. The test conditions were as follows: the water bath was at a constant temperature of 25 °C, the standard penetration load was 100 g, and the penetration time was 5 s. The softening point is the critical temperature at which asphalt changes from a solid state to a liquid, which reflects the temperature response performance of the asphalt material [44]. The ductility reflects asphalt's deformation capability at a specified temperature and its stretch rate before it is stretched to rupture [45,46]. In this study, the force ductility test environment was as follows: the water bath was at a constant temperature of 5 °C, and the stretch rate was 5 cm/min. Three indices were obtained during the asphalt specimen tensile process: force, ductility, and fracture energy (the integral of force and ductility). The test results for the asphalt indices are shown in Figure 2a,b.

The penetration test result in Figure 2 shows that after graphene was added, except for test groups 1 and 8, the asphalt penetration indices decreased. Test group 6 had the minimum penetration at 5.02 mm. Test groups 1 and 8 had the maximum penetration indices, at 6.54 mm. The penetration test results indicate that the graphene addition hardened the asphalt overall, improving its high-temperature performance. The softening point test results show that after adding graphene, the asphalt softening points in all test groups increased. Test group 7 had the maximum softening point at 51.7 °C. The softening point test results suggest that adding graphene improves asphalt's high-temperature performance. The force ductility test results show that after adding graphene, the maximum ductility force, ductility, and fracture energy improved significantly. Test group 7 had the maximum ductility force at 150.0 N; test group 6 had the maximum ductility at 46.70 cm; and test group 9 had the maximum fracture energy at 3633.0 N·mm. The force ductility test results demonstrate that adding graphene significantly improves asphalt's low-temperature performance. To summarize, graphene addition improves both the high- and low-temperature performance of asphalt. The optimal material composition and preparation parameters for preparing GMA are similar to the design parameters of test groups 6–9.

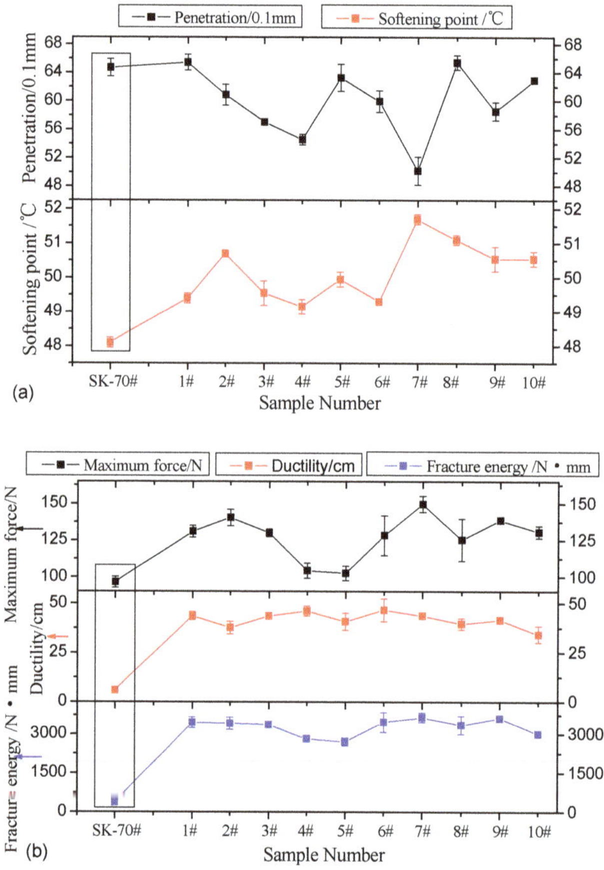

Figure 2. (**a**,**b**) Conventional asphalt performance index test results.

3.2. DSR Test

The rheological parameter of graphene asphalt was tested using the Dynamic Shear Rheological test (DSR) proposed by the Strategic Highway Research Project (SHRP) in the United States to characterize viscoelastic energy and evaluate the high- and low-temperature performance and the antifatigue performance of asphalt [47].

3.2.1. SHRP-PG Test

The SHRP-PG evaluates the high-temperature performance indices of asphalt cement material. The test reflects two important parameters of asphalt's viscoelasticity: the complex shear modulus G^* and the phase angle δ. The complex shear modulus G^* is the ratio of the maximum shear stress and the maximum shear strain in the SHRP-PG classification test. The complex shear modulus G^* represents the overall resistance of a material under repeated shear deformation, which includes the elastic modulus G′ and the viscous modulus G″. The elastic modulus is given by G′ = G*cosδ, which reflects the asphalt energy stored and released during shear deformation. The viscous modulus is given by G″ = G*sinδ, which represents the dissipated energy in the form of heat due to internal friction during the asphalt shear process. G*sinδ is defined as the antirutting factor, which represents the capability of asphalt cement material to resist permanent deformation under high temperature [48,49]. In this study, the test temperature was 64 °C; the diameter of the smooth metal plate was 25 ± 0.05 mm; the gap between the test plate and the roof was 1 ± 0.05 mm; and the test frequency was 10 rad/s. The test results are shown in Figure 3.

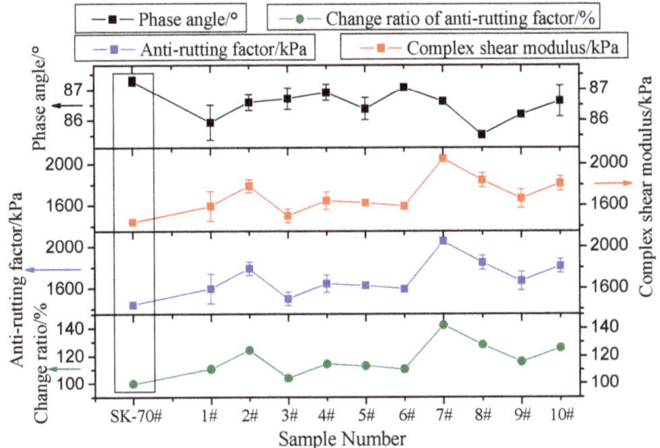

Figure 3. SHRP-PG test results (64 °C): The changing trends of phase angle, complex shear modulus, antirutting factor, and change ratio of antirutting factor are shown separately. The change ratio of antirutting factor is that the antirutting factor of each test group is divided by the antirutting factor of SK-70# base asphalt.

Figure 3 shows that after graphene is added, the GMA phase angle decreases to some extent, while the complex shear modulus and the antirutting factor improve to some extent. Test groups 7, 8, and 10 had the most significant improvement in antirutting factor (42.4%, 28.2%, and 25.9%, respectively). It can be inferred that adding graphene improves asphalt's high-temperature stability. The optimal proportions of graphene and dispersant for graphene asphalt preparation is similar to the material design parameters for test groups 7, 8, and 10.

3.2.2. Multistress Creep Recovery (MSCR) Test

Repeated multistress creep recovery tests were performed to further evaluate GMA's high-temperature stability. The test temperature was based on the SHRP-PG classification test result and the AASHTO T350-14 specification [50–52]. First, a 100 Pa shear stress was applied for 100 s. Then, while the 100 Pa shear stress was applied, cyclic loading (1 s loading and 9 s unloading) was repeated 10 times. Next, a 3200 Pa shear stress was applied to repeat the above process. The entire test included 30 cycles and took 300 s. The delayed elasticity recovery capability of GMA was evaluated via the recovery rate R and the unrecoverable creep compliance Jnr. The test results are shown in Figure 4a–c.

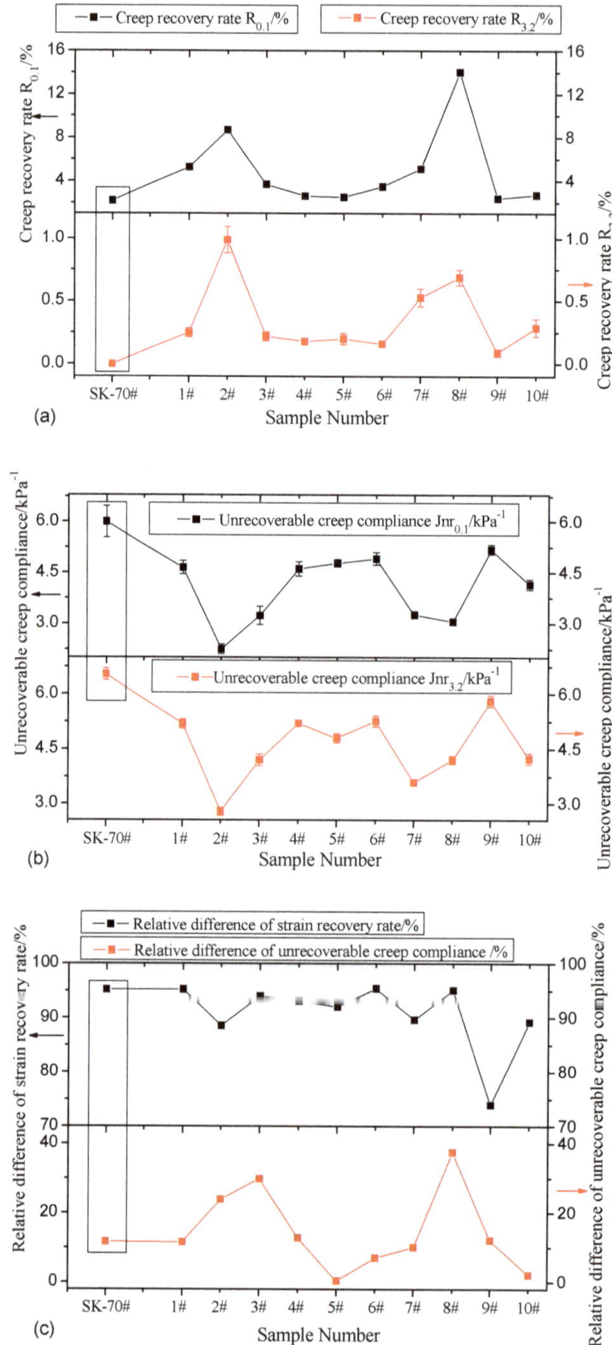

Figure 4. (**a**–**c**) Creep recovery test result.

Figure 4 shows that compared with matrix asphalt, GMA's creep recovery rate under 0.1 kPa shear stress and its creep recovery rate under 3.2 kPa of shear stress improve to some extent, indicating that the addition of graphene improves the asphalt's viscoelastic recovery capability. Test groups 8, 2, and 7 have superior creep recovery rates at 14.04%, 8.68%, and 5.12%, respectively, which are 6.41 times, 3.96 times, and 2.34 times greater than those for matrix asphalt. In the 3.2 kPa creep recovery test, matrix asphalt has almost no creep recovery, while groups 2, 8, and 7 have improved creep recovery rates at 0.99%, 0.69%, and 0.53%, respectively. The optimal parameters for GMA are similar to the parameters for groups 2, 7, and 8.

To summarize, based on a test of three major indices and the DSR test result, the optimal material composition and parameters for GMA preparation are similar to the design parameters for test groups 7 and 8.

3.3. Determining the Optimum Mixing Ratio

In this paper, Data Processing System (DPS) analysis software (Version DPSv17.10) and First Optimization (1stOpt) software (Version 7.0) are employed to calculate the optimal material composition for GMA preparation. DPS is a data processing system that integrates functions such as numeric calculation, statistical analysis, model simulation, drawing, and table generation [53,54]. 1stOpt is general-purpose numerical optimization simulation software with various classical and modern optimization algorithms that produce accurate solutions for nonlinear optimization problems [55,56]. Because conventional least square multiple linear regression and progressive regression analysis methods cannot meet the requirement of multiparameter and nonlinear test design modeling, three regression models, "partial least square quadratic polynomial", "partial least square quadratic term", and "partial least square interaction term", are employed in this paper. DPS software and 1stOpt software are employed to find the optimal GMA material composition.

The interdependency of three force ductility test parameters (force, ductility, and fracture energy) in modeling results in multiple colinearity and an unstable calculation result, which impacts the model creation significantly. Therefore, five indices (penetration Y_1, fracture energy Y_2, softening point Y_3, 64 °C antirutting factor Y_4, and 0.1 kPa creep recovery rate Y_5) are selected to create the regression model for calculation and analysis. During modeling, based on the PRESS statistics after data standardization and a declining trend in the sum of the squared errors, the determinant coefficient R^2 is defined as the major criterion to evaluate the regression model's effectiveness. A larger determinant coefficient indicates better equation fitting. The relationship between the number of latent variables and the determinant coefficient in three regression models calculated by DPS software is given in Table 6.

Table 6. The number of latent variables versus the determinant coefficient.

The Number of Latent Variables	Partial Least Square Quadratic Polynomial Regression Determinant Coefficient R^2					Partial Least Square Quadratic Term Regression Determinant Coefficient R^2					Partial Least Square Interaction Term Regression Determinant Coefficient R^2				
	Y_1	Y_2	Y_3	Y_4	Y_5	Y_1	Y_2	Y_3	Y_4	Y_5	Y_1	Y_2	Y_3	Y_4	Y_5
1	0.720	0.274	0.262	0.294	0.001	0.740	0.387	0.278	0.216	0.009	0.694	0.346	0.326	0.363	0.001
2	0.923	0.336	0.401	0.317	0.591	0.777	0.391	0.795	0.714	0.424	0.911	0.374	0.559	0.470	0.646
3	0.944	0.424	0.658	0.764	0.631	0.813	0.760	0.822	0.732	0.882	0.912	0.651	0.663	0.733	0.652
4	0.961	0.688	0.701	0.787	0.699	0.843	0.921	0.921	0.825	0.941	0.973	0.789	0.820	0.849	0.669
5	0.965	0.864	0.796	0.835	0.749	0.976	0.941	0.943	0.876	0.973	0.977	0.922	0.881	0.906	0.879

Table 6 shows that as the number of latent variables increases, the determinant coefficient R^2 gradually increases. When the number of latent variables is 5, the determinant coefficient R^2 reaches its maximum level. This means the regression method created using the partial least square method has a higher degree of fitting, and the model is closer to the actual situation and reliable. The coupling of five factors in the model leads to significant changes in GMA performance. The equation groups of three regression models are given in Table 7.

Table 7. Equation data of regression fitting model.

Regression Model	Partial Least Square Quadratic Polynomial Regression Model	Partial Least Square Quadratic Term Regression Model	Partial Least Square Interaction Term Regression Model
regression equation of penetration	$Y_1 = 69.065 + 5.02 \times 10^{-4} \times X_1 + 0.248 \times X_2 + 0.236 \times X_3 - 2.019 \times X_4 - 0.291 \times X_5 - 2.58 \times 10^{-4} \times X_2{}^2 + 1.99 \times 10^{-2} \times X_3{}^2 - 8.66 \times 10^{-2} \times X_4{}^2 + 2.21 \times 10^{-3} \times X_5{}^2 - 5 \times 10^{-6} \times X_1 \times X_2 + 6.3 \times 10^{-5} \times X_1 \times X_3 + 1.36 \times 10^{-4} X_1 \times X_4 - 3.2 \times 10^{-5} \times X_1 \times X_5 - 1.24 \times 10^{-3} \times X_2 \times X_3 + 6.94 \times 10^{-3} \times X_2 \times X_4 - 1.7 \times 10^{-3} \times X_2 \times X_5 + 1.5 \times 10^{-2} \times X_3 \times X_4 - 5.54 \times 10^{-3} \times X_3 \times X_5 + 1.33 \times 10^{-2} \times X_4 \times X_5$	$Y_1 = 145.630 - 6.78 \times 10^{-3} \times X_1 + 9.45 \times 10^{-2} \times X_2 - 1.488 \times X_3 + 2.012 \times X_4 - 1.153 \times X_5 + 1 \times 10^{-6} \times X_1{}^2 - 6.22 \times 10^{-4} \times X_2{}^2 + 7.31 \times 10^{-2} \times X_3{}^2 - 0.189 \times X_4{}^2 + 4.47 \times 10^{-3} \times X_5{}^2$	$Y_1 = 12.794 + 5.05 \times 10^{-3} \times X_1 + 0.266 \times X_2 + 1.397 \times X_3 - 4.966 \times X_4 + 0.434 \times X_5 - 5 \times 10^{-6} \times X_1 \times X_2 + 6.9 \times 10^{-5} \times X_1 \times X_3 + 1.91 \times 10^{-4} \times X_1 \times X_4 - 4.9 \times 10^{-5} \times X_1 \times X_5 - 1.39 \times 10^{-3} \times X_2 \times X_3 + 1.13 \times 10^{-2} \times X_2 \times X_4 - 2.39 \times 10^{-3} \times X_2 \times X_5 + 9.34 \times 10^{-3} \times X_3 \times X_4 - 1.09 \times 10^{-2} \times X_3 \times X_5 + 2.39 \times 10^{-2} \times X_4 \times X_5$
regression equation of fracture energy	$Y_2 = -7.782 + 1.16 \times 10^{-2} \times X_1 + 1.081 \times X_2 - 8.445 \times X_3 - 15.546 \times X_4 + 4.022 \times X_5 + 2 \times 10^{-6} \times X_1{}^2 + 5.57 \times 10^{-4} \times X_2{}^2 + 4.82 \times 10^{-2} \times X_3{}^2 + 0.233 \times X_4{}^2 - 9.80 \times 10^{-3} \times X_5{}^2 - 8 \times 10^{-6} \times X_1 \times X_2 - 1.67 \times 10^{-4} \times X_1 \times X_3 - 7.05 \times 10^{-4} \times X_1 \times X_4 - 1.37 \times 10^{-4} \times X_1 \times X_5 + 1.92 \times 10^{-3} \times X_2 \times X_3 - 5.54 \times 10^{-3} \times X_2 \times X_4 - 6.86 \times 10^{-3} \times X_2 \times X_5 + 0.349 \times X_3 \times X_4 + 4.28 \times 10^{-2} \times X_3 \times X_5 + 7.85 \times 10^{-2} \times X_4 \times X_5$	$Y_2 = -465.825 - 5.36 \times 10^{-2} \times X_1 + 0.209 \times X_2 + 0.885 \times X_3 - 9.127 \times X_4 + 12.342 \times X_5 + 7 \times 10^{-6} \times X_1{}^2 + 5.2 \times 10^{-5} \times X_2{}^2 - 1.56 \times 10^{-2} \times X_3{}^2 + 0.616 \times X_4{}^2 - 4.19 \times 10^{-2} \times X_5{}^2$	$Y_2 = 112.752 + 2.48 \times 10^{-2} \times X_1 + 0.944 \times X_2 - 8.115 \times X_3 - 10.189 \times X_4 + 1.410 \times X_5 + 1.3 \times 10^{-5} \times X_1 \times X_2 - 2.61 \times 10^{-4} \times X_1 \times X_3 - 9.33 \times 10^{-4} \times X_1 \times X_4 - 1.16 \times 10^{-4} \times X_1 \times X_5 + 5.59 \times 10^{-3} \times X_2 \times X_3 - 1.51 \times 10^{-2} \times X_2 \times X_4 - 5.56 \times 10^{-3} \times X_2 \times X_5 + 0.361 \times X_3 \times X_4 + 5.25 \times 10^{-2} \times X_3 \times X_5 + 7.08 \times 10^{-2} \times X_4 \times X_5$
regression equation of softening point	$Y_3 = 41.772 + 1.71 \times 10^{-4} \times X_1 + 1.13 \times 10^{-2} \times X_2 - 0.106 \times X_3 - 0.142 \times X_4 + 8.69 \times 10^{-2} \times X_5 + 1.2 \times 10^{-5} \times X_2{}^2 + 1.88 \times 10^{-3} \times X_3{}^2 - 2.17 \times 10^{-3} \times X_4{}^2 - 2.02 \times 10^{-4} \times X_5{}^2 + 1 \times 10^{-6} \times X_1 \times X_2 + 4 \times 10^{-6} \times X_1 \times X_3 - 3 \times 10^{-6} \times X_1 \times X_5 + 8.6 \times 10^{-5} \times X_2 \times X_3 - 4.69 \times 10^{-4} \times X_2 \times X_4 - 8 \times 10^{-5} \times X_2 \times X_5 + 6.23 \times 10^{-3} \times X_3 \times X_4 + 6.17 \times 10^{-4} \times X_3 \times X_5 + 7.55 \times 10^{-4} \times X_4 \times X_5$	$Y_3 = 19.483 - 8.74 \times 10^{-4} \times X_1 - 8.32 \times 10^{-4} \times X_2 + 0.129 \times X_3 - 0.348 \times X_4 + 0.467 \times X_5 + 2.4 \times 10^{-5} \times X_2{}^2 - 2.08 \times 10^{-3} \times X_3{}^2 + 2.6 \times 10^{-2} \times X_4{}^2 - 1.72 \times 10^{-3} \times X_5{}^2$	$Y_3 = 45.499 + 3.88 \times 10^{-4} \times X_1 + 1.39 \times 10^{-3} \times X_2 - 8.17 \times 10^{-2} \times X_3 - 3.67 \times 10^{-2} \times X_4 + 2.26 \times 10^{-2} \times X_5 + 1 \times 10^{-6} \times X_1 \times X_2 + 1 \times 10^{-6} \times X_1 \times X_3 - 8 \times 10^{-6} \times X_1 \times X_4 - 2 \times 10^{-6} \times X_1 \times X_5 + 1.98 \times 10^{-4} \times X_2 \times X_3 - 7.9 \times 10^{-4} \times X_2 \times X_4 + 2 \times 10^{-6} \times X_2 \times X_5 + 5.77 \times 10^{-3} \times X_3 \times X_4 + 8.89 \times 10^{-4} \times X_3 \times X_5 + 2.43 \times 10^{-4} \times X_4 \times X_5$
regression equation of anti-rutting factor	$Y_4 = 903.586 + 2.12 \times 10^{-4} \times X_1 - 1.517 \times X_2 + 1.279 \times X_3 + 4.146 \times X_4 + 5.787 \times X_5 + 3 \times 10^{-6} \times X_1{}^2 + 2.48 \times 10^{-3} \times X_2{}^2 + 5.2 \times 10^{-2} \times X_3{}^2 - 0.291 \times X_4{}^2 - 7.68 \times 10^{-3} \times X_5{}^2 + 2.44 \times 10^{-4} \times X_1 \times X_2 + 4.46 \times 10^{-4} \times X_1 \times X_3 + 1.98 \times 10^{-4} \times X_1 \times X_4 - 2.54 \times 10^{-4} \times X_1 \times X_5 + 3.84 \times 10^{-2} \times X_2 \times X_3 - 0.114 \times X_2 \times X_4 + 6.66 \times 10^{-3} \times X_2 \times X_5 + 0.361 \times X_3 \times X_4 + 3.92 \times 10^{-2} \times X_3 \times X_5 - 2.03 \times 10^{-2} \times X_4 \times X_5$	$Y_4 = -2748.009 - 0.116 \times X_1 - 0.855 \times X_2 + 55.224 \times X_3 - 50.766 \times X_4 + 63.364 \times X_5 + 1.8 \times 10^{-5} \times X_1{}^2 + 7.32 \times 10^{-3} \times X_2{}^2 - 1.742 \times X_3{}^2 + 4.757 \times X_4{}^2 - 0.228 \times X_5{}^2$	$Y_4 = 1256.714 + 5.72 \times 10^{-3} \times X_1 - 3.325 \times X_2 + 1.769 \times X_3 + 21.236 \times X_4 + 0.849 \times X_5 + 2.97 \times 10^{-4} \times X_1 \times X_2 - 1.4 \times 10^{-5} \times X_1 \times X_3 - 6.69 \times 10^{-4} \times X_1 \times X_4 + 1.25 \times 10^{-4} \times X_1 \times X_5 + 4.61 \times 10^{-2} \times X_2 \times X_3 - 0.184 \times X_2 \times X_4 + 2.57 \times 10^{-2} \times X_2 \times X_5 + 0.196 \times X_3 \times X_4 + 7.24 \times 10^{-2} \times X_3 \times X_5 - 9.25 \times 10^{-2} \times X_4 \times X_5$
regression equation of creep recovery rate	$Y_5 = -50.358 - 1.12 \times 10^{-4} \times X_1 + 0.160 \times X_2 - 1.083 \times X_3 - 0.878 \times X_4 + 0.748 \times X_5 + 2.8 \times 10^{-4} \times X_2{}^2 + 2.52 \times 10^{-2} \times X_3{}^2 - 3.78 \times 10^{-2} \times X_4{}^2 - 2.43 \times 10^{-3} \times X_5{}^2 - 8 \times 10^{-6} \times X_1 \times X_2 + 6.5 \times 10^{-5} \times X_1 \times X_3 + 6.1 \times 10^{-5} \times X_1 \times X_4 - 7 \times 10^{-6} \times X_1 \times X_5 - 1.72 \times 10^{-3} \times X_2 \times X_3 - 2.55 \times 10^{-4} \times X_2 \times X_4 - 8.69 \times 10^{-4} \times X_2 \times X_5 + 4.20 \times 10^{-2} \times X_3 \times X_4 + 3.57 \times 10^{-3} \times X_3 \times X_5 + 5.04 \times 10^{-3} \times X_4 \times X_5$	$Y_5 = -146.464 - 4.88 \times 10^{-3} \times X_1 - 1.68 \times 10^{-2} \times X_2 - 0.597 \times X_3 + 0.508 \times X_4 + 2.342 \times X_5 + 1 \times 10^{-6} \times X_1{}^2 + 3.9 \times 10^{-5} \times X_2{}^2 + 4.6 \times 10^{-2} \times X_3{}^2 - 2.84 \times 10^{-2} \times X_4{}^2 - 8.52 \times 10^{-3} \times X_5{}^2$	$Y_5 = -19.031 - 3.4 \times 10^{-5} \times X_1 + 0.221 \times X_2 - 0.521 \times X_3 - 1.787 \times X_4 + 0.151 \times X_5 - 1.2 \times 10^{-5} \times X_1 \times X_2 + 9.8 \times 10^{-5} \times X_1 \times X_3 + 1.15 \times 10^{-4} \times X_1 \times X_4 - 1 \times 10^{-6} \times X_1 \times X_5 - 2.75 \times 10^{-3} \times X_2 \times X_3 - 7.99 \times 10^{-4} \times X_2 \times X_4 - 1.06 \times 10^{-3} \times X_2 \times X_5 + 5.01 \times 10^{-2} \times X_3 \times X_4 + 3.51 \times 10^{-3} \times X_3 \times X_5 + 7.24 \times 10^{-3} \times X_4 \times X_5$

Based on Table 6, a comparison of the determinant coefficients in the three regression models shows that the partial least square quadratic polynomial regression model has small Y_3 and Y_5 determinant coefficients and a relatively low degree of fitting. Therefore, this model is excluded. By comparison, partial least square interaction term regression and partial least square quadratic term regression have large determinant coefficients and regression models with higher degrees of fitting.

Therefore, these two models are employed to find the optimal solution for GMA material composition and preparation parameters.

Table 7 shows that all three regression models are nonlinear. Considering that there are multiple solutions in the actual calculation, 1stOpt software is employed in the regression model for optimization. The results are listed in Table 8.

Table 8. Optimization solution and corresponding dependent variable in each regression model.

Regression Model and Calculation Method		Partial Least Square Quadratic Term Regression Model	Partial Least Square Interaction Term Regression Model	
		B-1	B-2	B-3
optimization solution	shear rate X_1/r.p.m.	6500	7000	6500
	shear time X_2/min	180	200	30
	graphene mixing amount X_3/‰	20	20	20
	stearic amide mixing amount X_4/%	1.00	8.26	10.00
	shear temperature X_5/°C	140	160	150
value of dependent variable	penetration index Y_1/0.1 mm	88.15	51.27	66.22
	fracture energy Y_2/N·mm	4301.6	3927.4	3541.9
	softening point Y_3/	47.51	52.68	51.39
	64 °C antirutting factor Y_4/kPa	2099.27	2338.77	1909.48
	0.1 kPa creep recovery rate Y_5/%	30.13	9.25	23.43

Based on Table 8, the optimal graphene asphalt material composition and preparation parameters are obtained to prepare GMA for performance tests and verification. The results are listed in Table 9.

Table 9. Performance test results of optimal formula.

Item		SK70# Matrix Asphalt	Partial Least Square Quadratic Term Regression Model		Partial Least Square Interaction Term Regression Model			
			B-1	Change Rate/%	B-2	Change Rate/%	B-3	Change Rate/%
	penetration/0.1 mm	64.7	61.5	−4.95	62.3	−3.71	58.6	−9.43
	softening point/°C	48.1	58.6	21.83	52.3	8.73	54.3	12.89
5 °C force ductility	maximum force/N	96.6	168.0	73.91	136.0	40.79	123.0	27.33
	ductility/mm	6.11	42.54	596.24	44.21	623.57	48.39	691.98
	fracture energy/N·mm	387.7	4035.7	940.93	3542.4	813.70	3358.3	766.21
	64 °C antirutting factor/Pa	1442.22	2099	45.54	1643	13.92	1443	0.05
	0.1 kPa creep recovery rate/%	2.19	20.24	824.20	8.75	299.54	7.93	262.10

In Table 9, change rate is that the test value of GMA is divided by that of SK-70# matrix asphalt in the same test item. Table 9 shows that compared with matrix asphalt, the prepared GMA has a smaller penetration and a significantly higher softening point, force ductility force, ductility, fracture energy, 64 °C anti-rutting factor, and 0.1 kPa creep recovery rate. In tests B-1, B-2, and B-3, compared with matrix asphalt, fracture energy values at low temperature improve by 940.93%, 813.70%, and 766.21%, respectively; 64 °C anti-rutting factors improve by 45.54%, 13.92%, and 0.05%, respectively; and creep recovery rates improve by 824.20%, 299.54%, and 262.10%, respectively.

To summarize, in three optimal formulae, compared with matrix asphalt, the prepared GMA has a smaller penetration index, and the asphalt is hardened. Additionally, high- and low-temperature performance and delayed elasticity recovery improve significantly. This is likely because some of the graphene has intercalated in the asphalt, which causes a strengthening effect. Test group B-1 had the most significant performance improvement; hence, test B-1 parameters are selected as optimal GMA mix parameters: the high-speed shear rate is 6500 r.p.m.; the shear time is 180 min; the graphene proportion is 20‰; the EBS proportion is 1%; and the shear temperature is 140 °C.

3.4. Textural Characterization

3.4.1. XRD Test

In the XRD test, the material under analysis undergoes X-ray diffraction to obtain a diffraction spectrum, which is used to investigate useful material characteristics such as crystal structure and elemental composition [57]. XRD analysis was performed on the SK-70# matrix asphalt and the BEST-1 GMA; the results are shown in Figure 5.

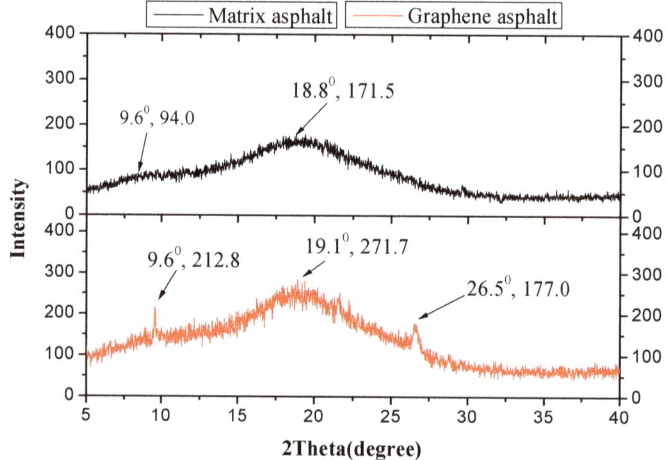

Figure 5. XRD test spectrum of matrix asphalt and GMA.

Based on Figure 5, the matrix asphalt spectrum shows the most intense peak is at approximately $2\theta = 18.8°$. Based on Bragg's law, $2d\sin\theta = n\lambda$, the interplanar spacing is $d_1 = 0.472$ nm and there is an extremely weak peak at $2\theta = 9.6°$; the interplanar spacing is $d_2 = 0.921$ nm, which is a loose-layered structure of stacked asphalt or gum. The GMA spectrum shows peaks at $2\theta = 9.6°$ and $2\theta = 19.1°$; the interplanar spacing values are $d_3 = 0.921$ nm and $d_4 = 0.467$ nm, respectively; there is a new peak at approximately $2\theta = 26.5°$, which is the graphene characteristic peak [57] with a strength of 177 cps and an interplanar spacing of $d_5 = 0.336$ nm. The spectrum demonstrates the existence of graphene in asphalt. After graphene is added, the strength of the asphalt or gum characteristic peak increases to some extent, which means that graphene increases its loose-layered structure of stacked asphalt or gum. Peak spacing decreases to some extent, indicating that the intense adsorption effect of graphene enhances the ordered structure of asphalt.

3.4.2. Microscope Test

Due to its advantages, including convenient operation and easy sample preparation, the fluorescence microscope has become a widely used tool to observe micromorphology of materials, and has been used in asphalt characterization [58]. In this paper, SK-70# matrix asphalt, GMA in uniform design test groups 1–10, and GMA in the three groups with optimal admixtures were observed using a fluorescence microscope. The test results are shown in Figure 6.

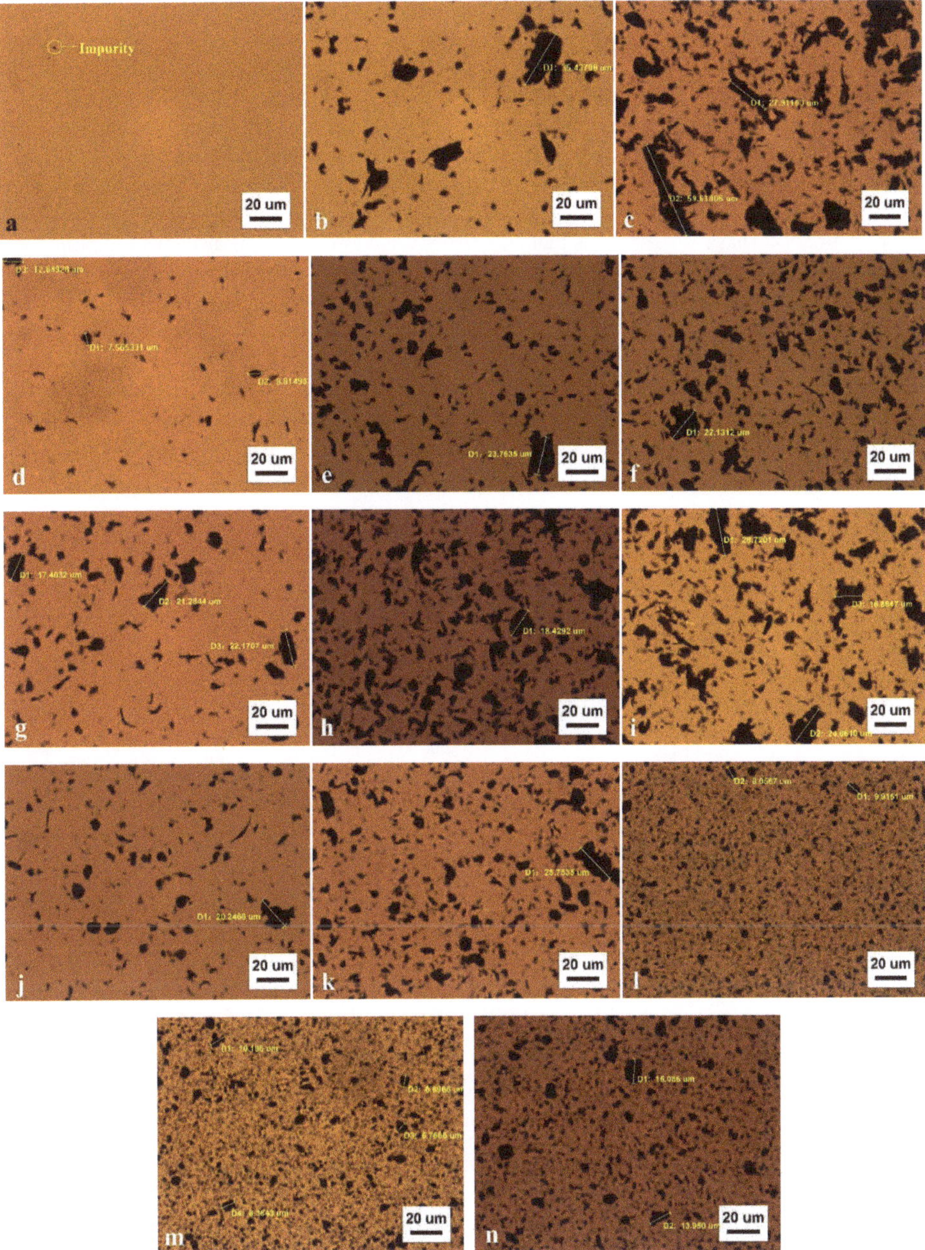

Figure 6. Microscopy test results ((**a**) is the test result of SK-70# matrix asphalt; (**b–k**) are the test results of GMA uniform design groups 1–10; (**l–n**) are the test results of B-1~B-3 GMA).

A comparison of matrix asphalt in Figure 6a and GMA in Figure 6b–k shows that various forms of black substances are observed in all graphene asphalt samples. As graphene is a nanometer material, observation under a normal fluorescence microscopy condition is very difficult. If graphene is distributed evenly in asphalt under the effect of stearic amide dispersant, then graphene asphalt topography observed in a fluorescence microscopic image with 500× magnification should essentially be identical to matrix asphalt topography. However, the actual observation shows that graphene asphalt contains a large amount of a black substance. Graphene has an extremely large specific surface area and a strong interlayer force, and therefore is very difficult to distribute completely uniformly [59–61]. Because the XRD test proves the stable existence of graphene in asphalt, this black substance should be graphene clusters. EBS cannot distribute graphene evenly in asphalt.

Figure 6l–n show that compared with materials with other compositions, the graphene clusters in the GMA prepared with the optimal material composition obtained from modeling have more regular, spherical shapes. This means that with the optimal graphene and dispersant mixture ratio, the dispersant changes the graphene topography in asphalt; the graphene clusters evolve from large, distinct, irregular shapes to small, indistinct, regular shapes.

Image-Pro Plus is widely used microscopy image analysis software with accurate and reliable image analysis results. In recent years, image analysis has been applied extensively in civil engineering research [62–64]. In this paper, Image-Pro Plus software is employed to analyze GMA images and obtain test group parameters, such as the number of graphene clusters, the maximum area, minimum area, total area, cluster average area, total area, and ratio to maximum cluster area. The results are shown in Figures 7–9.

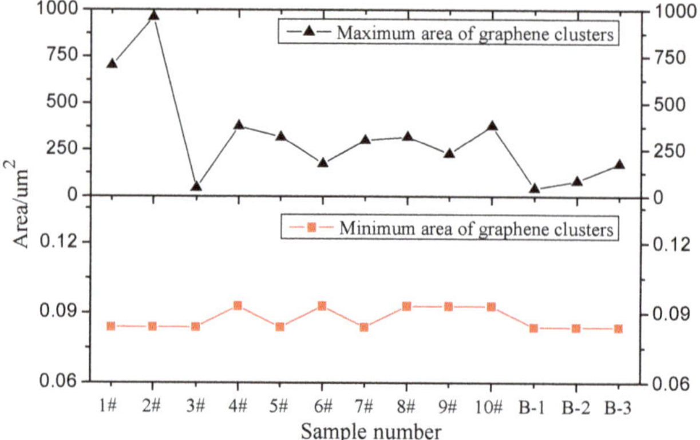

Figure 7. Variation trend of graphene cluster max and min areas.

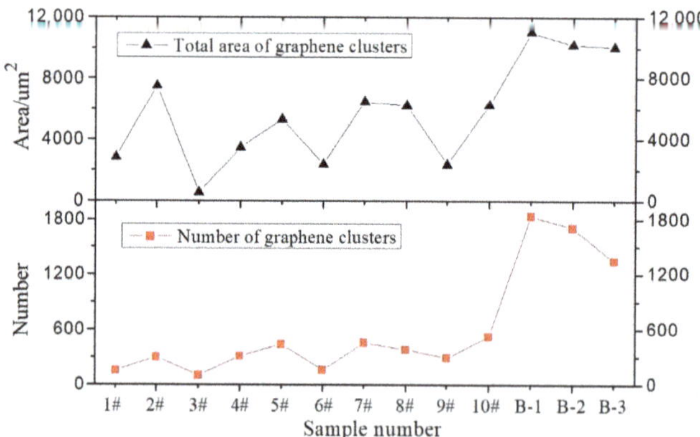

Figure 8. Variation trend of graphene cluster total area and quantity.

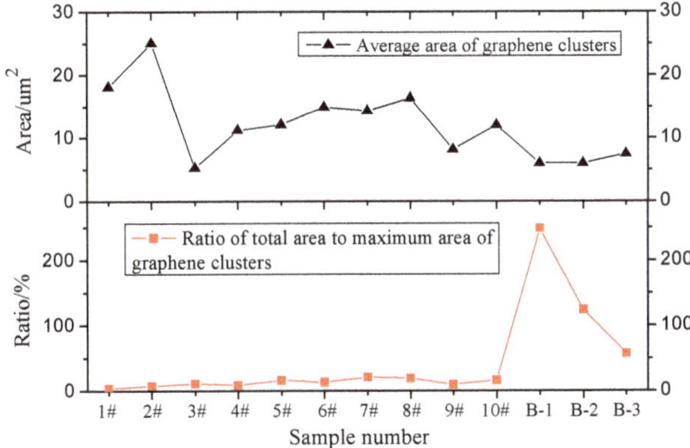

Figure 9. Variation trend of graphene cluster average area and ratio of total area to maximum area of graphene clusters.

Figures 7 and 8 show that the dispersant has significantly different graphene dispersion effects in different test groups (i.e., the graphene distribution in asphalt is affected by differences in parameters including the dispersant and graphene mix ratio, shear rotating speed, shear time, and shear temperature). In different test groups, the graphene clusters have similar minimum areas. Although Figure 7 shows the maximum graphene cluster areas in different test groups differ significantly, such differences reflect differences between individual graphene cluster areas and cannot represent the general variation pattern of graphene clusters in the test groups. Therefore, the maximum and minimum graphene cluster areas have no comparative significance. In the optimal parameter solution, the optimal graphene asphalt material composition and preparation parameters are based on test group B-1. Figure 8 shows a larger graphene cluster total area and more clusters. Figure 9 shows a small graphene cluster average area with a significantly larger total area and maximum area ratio than other test groups. This means this test group has properly distributed graphene in asphalt.

Based on the above graphene cluster characteristics, clusters in images are divided into three categories based on dimension: fine clusters, medium clusters, and coarse clusters. The fine cluster area is less than 1 μm^2; the medium cluster area is between 1 μm^2 and 10 μm^2; the coarse cluster area is larger than 10 μm^2. Based on these categories, the graphene cluster distribution patterns in various test groups scattered by EBS are shown in Figures 10 and 11.

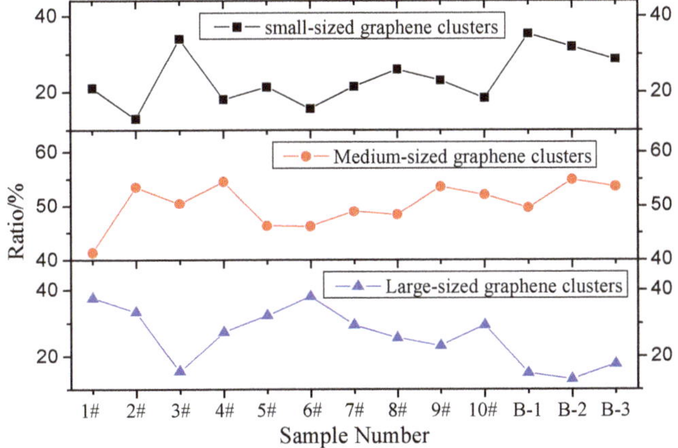

Figure 10. Variation trend of different categories of graphene clusters.

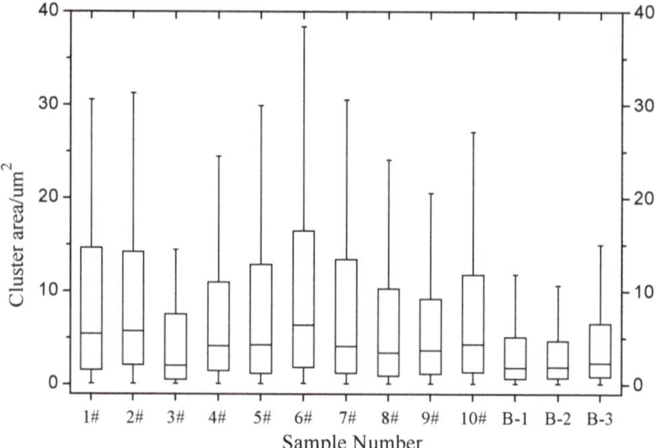

Figure 11. Box plot of graphene cluster area.

Figures 10 and 11 show that in different test groups, the coarse- and fine-grain proportions of graphene clusters in graphene asphalt vary significantly. In contrast, the medium grain ratio has a small variation and is essentially stable. This means when dispersant cannot distribute graphene evenly in asphalt, the majority of graphene clusters in asphalt are medium-sized.

The performance comparison shows that test group B-1 had the smallest quartile and median among all test groups. Test group B-1 had the highest proportion of fine graphene clusters, a small proportion of coarse clusters, the largest total cluster area, clusters with small dimension, and the maximum softening point, low temperature ductility fracture energy, antirutting factor, and 0.1 kPa creep recovery rate at 58.6 °C, 4035.7 N·mm, 2099.00 kPa, and 20.24%, respectively. Again, this means graphene in this test group is distributed properly in asphalt, resulting in a significant improvement in macroscopic asphalt performance.

To summarize, in EBS-based GMA, when the graphene and dispersant proportions and corresponding preparation parameters are optimal, graphene is distributed properly in asphalt, which significantly improves the softening point, low-temperature ductility fracture energy, antirutting factor, and creep recovery rate of the material.

4. Conclusions

(1) A method for calculating the optimal parameters of GMA and a process to prepare GMA were proposed. For EBS-based GMA, the optimal parameters are as follows: the graphene proportion is 20‰; the EBS proportion is 1%; the high-speed shear rate is 6000 r.p.m.; the shear time is 180 min; the shear temperature is 140 °C. The prepared GMA had a significantly improved softening point, low temperature fracture energy, antirutting factor, and creep recovery rate.

(2) The prepared GMA had a softening point of 58.6 °C, a low-temperature ductility force of 168.0 N, low-temperature ductility of 42.54 mm, low-temperature fracture energy of 2099 N·mm, and a 0.1 kPa creep recovery rate of 20.24%. Compared with SK-70# matrix asphalt, the performance of GMA was significantly improved.

(3) Graphene can exist in an asphalt medium in a stable form, and some graphene in asphalt is in the form of clusters. When the graphene and dispersant composition is close to the optimal ratio, the dispersant changes the form of graphene in asphalt from irregular clusters to regular clusters and from distinct, large clusters to indistinct, small clusters. When the graphene distribution in asphalt is closer to the ideal situation, graphene asphalt has improved high- and low-temperature performance. When the dispersant cannot distribute graphene evenly in asphalt, the majority of graphene clusters in asphalt are medium-sized.

(4)   Although EBS is used in this study, graphene is still not distributed evenly in asphalt in the form of flakes but is in the form of small clusters. Methods to ideally disperse or intercalate graphene in asphalt to substantially improve asphalt performance require further investigation.

Author Contributions: Conceptualization, X.Z., G.H., and J.-X.H.; Methodology, X.Z., G.H., and J.-X.H.; Software, J.-X.H. and M.-M.F.; Validation, X.Z., J.-X.H., and Y.L.; Formal Analysis, J.-X.H.; Investigation; G.H. and C.Z.; Resources, G.H. and C.Z.; Data Curation, J.-X.H. and M.-M.F.; Writing—Original Draft Preparation, J.-X.H.; Writing—Review & Editing, X.Z. and G.H.; Visualization, X.Z. and G.H.; Supervision, G.H.

Funding: This research was funded by National Natural Science Foundation of China (No. 51402030 & No. 51778096) and Municipal Foundation Project of CQEC (No. yjgl33018) and Natural Science Foundation Project of CQ CSTC (No. cstc2016jcyjA0119 & No. cstc2017jcyjB0028).

Acknowledgments: The authors would like to gratefully acknowledge the financial supports from National Natural Science Foundation of China and Municipal Foundation Project of CQEC and Natural Science Foundation Project of CQ CSTC.

Conflicts of Interest: The authors declare no conflict of interest.

References

1.   Geim, A.K. Graphene: Status and prospects. *Science* **2009**, *324*, 1530–1534. [CrossRef] [PubMed]

2.   Wu, J.; Pisula, W.; Müllen, K. Graphenes as potential material for electronics. *Chem. Rev.* **2007**, *107*, 718–747. [CrossRef] [PubMed]

3.   Rao, C.N.R.; Sood, A.K.; Voggu, R.; Subrahmanyam, K.S. Some novel attributes of graphene. *J. Phys. Chem. Lett.* **2010**, *1*, 572–580. [CrossRef]

4.   Zhang, Y.; Tan, Y.W.; Stormer, H.L.; Kim, P. Experimental observation of the quantum Hall effect and Berry's phase in graphene. *Nature* **2005**, *438*, 201–204. [CrossRef] [PubMed]

5.   Bolotin, K.I.; Sikes, K.J.; Jiang, Z.; Klima, M.; Fudenberg, G.; Hone, J.; Kim, P.; Stormer, H.L. Ultrahigh electron mobility in suspended graphene. *Solid State Commun.* **2008**, *146*, 351–355. [CrossRef]

6.   Schadler, L.S.; Giannaris, S.C.; Ajayan, P.M. Load transfer in carbon nanotube epoxy composites. *Appl. Phys. Lett.* **1998**, *73*, 3842–3844. [CrossRef]

7.   Balandin, A.A.; Ghosh, S.; Bao, W.; Calizo, I.; Teweldebrhan, D.; Miao, F.; Lau, C.N. Superior thermal conductivity of single-layer graphene. *Nano Lett.* **2008**, *8*, 902–907. [CrossRef] [PubMed]

8.   Chae, H.K.; Siberio-Pérez, D.Y.; Kim, J.; Go, Y.; Eddaoudi, M.; Matzger, A.J.; O'Keeffe, M.; Yaghi, O.M. A route to high surface area, porosity and inclusion of large molecules in crystals. *Nature* **2004**, *427*, 523. [CrossRef] [PubMed]

9.   Lee, C.; Wei, X.; Kysar, J.W.; Hone, J. Measurement of the elastic properties and intrinsic strength of monolayer graphene. *Science* **2008**, *321*, 385–388. [CrossRef] [PubMed]

10.   Van den Brink, J. From strength to strength. *Nat. Nanotechnol.* **2007**, *2*, 199–201. [CrossRef] [PubMed]

11.   Weitz, R.T.; Yacoby, A. Graphene rests easy. *Nat. Nanotechnol.* **2010**, *5*, 699–700. [CrossRef] [PubMed]

12.   Wu, J.; Xu, H.; Zhang, J. Raman spectroscopy of graphene. *Acta Chim. Sin.* **2014**, *72*, 301–318. [CrossRef]

13.   Qi-Sen, Z.; Xin, X. Research review on constitutive model and microstructure of asphalt and asphalt mixture. *China J. Highw. Transp.* **2016**, *29*, 26–33.

14.   Huang, G. Exploitation of Modified Asphalt of Fume Suppression and Study on Performance of Its Mixture under the Elevated Temperture. Ph.D. Thesis, Chongqing Jiaotong University, Chongqing, China, 2013.

15.   Zhang, B. The Effect of Modified Graphene on the Flame Retardancy and Smoke Suppression Properties of Polymer. Master's Thesis, Anhui Jianzhu University, Hefei, China, 2017.

16.   Liu, Y.; Han, M.; Yin, Y.; Fenglei, Z.; Liu, L.; Jing, L. Research progress and prospective exploration of graphene in intelligent highway. *Mater. Rev.* **2017**, *31*, 169–172.

17.   Wang, Z.; Dai, Q.; Guo, S. Microwave-healing performance of modified asphalt mixtures with flake graphite and exfoliated graphite nanoplatelet. *Constr. Build. Mater.* **2018**, *187*, 865–875. [CrossRef]

18.   Wang, Z.; Dai, Q.; Guo, S.; Wang, R.; Ye, M.; Yap, Y.K. Experimental investigation of physical properties and accelerated sunlight-healing performance of flake graphite and exfoliated graphite nanoplatelet modified asphalt materials. *Constr. Build. Mater.* **2017**, *134*, 412–423. [CrossRef]

19.   Wang, Z.; Dai, Q.; Guo, S. Laboratory performance evaluation of both flake graphite and exfoliated graphite nanoplatelet modified asphalt composites. *Constr. Build. Mater.* **2017**, *149*, 515–524. [CrossRef]

20. Yao, H.; Dai, Q.; You, Z.; Ye, M.; Yap, Y.K. Rheological properties, low-temperature cracking resistance, and optical performance of exfoliated graphite nanoplatelets modified asphalt binder. *Constr. Build. Mater.* **2016**, *113*, 988–996. [CrossRef]

21. Li, Y.; Wu, S.; Amirkhanian, S. Investigation of the graphene oxide and asphalt interaction and its effect on asphalt pavement performance. *Constr. Build. Mater.* **2018**, *165*, 572–584. [CrossRef]

22. Huang, G. Suppression mechanism of expanded graphite for asphalt fume and dynamic performance of asphalt mixture of fume suppression. *China J. Highw. Transp.* **2015**, *28*, 1–10.

23. He, Z. *Study on Mechanism and Suppression Technology of Modified Asphalt Smoke at High Temperature*; National Natural Science Foundation: Beijing, China, 2014; pp. 85–93.

24. Cheng, I.F.; Xie, Y.; Allen Gonzales, R.; Brejna, P.R.; Sundararajan, J.P.; Fouetio Kengne, B.A.; Eric Aston, D.; McIlroy, D.N.; Foutch, J.D.; Griffiths, P.R. Synthesis of graphene paper from pyrolyzed asphalt. *Carbon* **2011**, *49*, 2852–2861. [CrossRef]

25. Amirkhanian, A.; Xiao, F.; Amirkhanian, S. Evaluation of high temperature rheological characteristics of asphalt binder with carbon nano particles. *J. Test. Eval.* **2011**, *39*, 583–591.

26. Moreno-Navarro, F.; Sol-Sánchez, M.; Gámiz, F.; Rubio-Gámez, M.C. Mechanical and thermal properties of graphene modified asphalt binders. *Constr. Build. Mater.* **2018**, *180*, 265–274. [CrossRef]

27. Le, J.; Marasteanu, M.; Turos, M. *Graphene Nanoplatelet (GNP) Reinforced Asphalt Mixtures: A Novel Multifunctional Pavement Material*; NCHRP IDEA Project 173; University of Minnesota: Minneapolis, MN, USA, 2016; pp. 1–29.

28. Zhou, X.; Zhang, X.; Xu, S.; Wu, S.; Liu, Q.; Fan, Z. Evaluation of thermo-mechanical properties of graphene/carbon-nanotubes modified asphalt with molecular simulation. *Mol. Simulat.* **2017**, *43*, 312–319. [CrossRef]

29. Liu, X.; Wu, S. Study on the graphite and carbon fiber modified asphalt concrete. *Constr. Build. Mater.* **2011**, *25*, 1807–1811. [CrossRef]

30. Liu, K.; Zhang, K.; Wu, J.; Muhunthan, B.; Shi, X. Evaluation of mechanical performance and modification mechanism of asphalt modified with graphene oxide and warm mix additives. *J. Clean. Prod.* **2018**, *193*, 87–96. [CrossRef]

31. Han, M.; Li, J.; Muhammad, Y.; Yin, Y.; Yang, J.; Yang, S.; Duan, S. Studies on the secondary modification of SBS modified asphalt by the application of octadecyl amine grafted graphene nanoplatelets as modifier. *Diam. Relat. Mater.* **2018**, *89*, 140–150. [CrossRef]

32. Han, M.; Li, J.; Muhammad, Y.; Hou, D.; Zhang, F.; Yin, Y.; Duan, S. Effect of polystyrene grafted graphene nanoplatelets on the physical and chemical properties of asphalt binder. *Constr. Build. Mater.* **2018**, *174*, 108–119. [CrossRef]

33. Zeng, W.; Wu, S.; Pang, L.; Sun, Y.; Chen, Z. The utilization of graphene oxide in traditional construction materials: Asphalt. *Mater.* **2017**, *10*, 48. [CrossRef] [PubMed]

34. Habib, N.Z.; Aun, N.C.; Zoorob, S.E.; Lee, P.I. Use of graphene oxide as a bitumen modifier: An innovative process optimization study. *Adv. Mater. Res.* **2015**, *1105*, 365–369. [CrossRef]

35. *Standard Test Methods of Asphalt and Asphalt Mixtures for Highway Engineering: JTG E20-2011*; Ministry of Communications of the People's Republic of China: Beijing, China, 2011.

36. Xin, C.; Lu, Q.; Ai, C.; Rahman, A.; Qiu, Y. Optimization of hard modified asphalt formula for gussasphalt based on uniform experimental design. *Constr. Build. Mater.* **2017**, *136*, 556–564. [CrossRef]

37. Yu, J.; Zhang, X. Evaluation of environmental factors to fatigue performance of asphalt mixes based on uniform design. *Highway* **2010**, *21*, 82–86.

38. Li, T.Z.; Yang, X.L. An efficient uniform design for *Kriging-based* response surface method and its application. *Comput. Geotech.* **2019**, *109*, 12–22. [CrossRef]

39. Fang, K.; Lin, D.K.J.; Winker, P.; Zhang, Y. Uniform Design: Theory and Application. *Technometrics* **2000**, *42*, 237–248. [CrossRef]

40. Zhang, G.; Wang, W. A citation review on the uniform experimental design. *J. Appl. Stat. Mgmt.* **2013**, *32*, 89–99.

41. Li, Y. The Preparation Technology of Graphene Modified Asphalt with High-Performance Road. Master's Thesis, Chongqing Jiaotong University, Chongqing, China, 2018.

42. Bao, D.; Yu, Y.; Zhao, Q. Evaluation of the chemical composition and rheological properties of bio-asphalt from different biomass sources. *Road Mater. Pavement* **2019**, *1*, 1–15. [CrossRef]

43. Hadiwardoyo, S.P.; Sinaga, E.S.; Fikri, H. The influence of Buton asphalt additive on skid resistance based on penetration index and temperature. *Constr. Build. Mater.* **2013**, *42*, 5–10. [CrossRef]

44. Wang, R.; Fan, H.; Jiang, W.; Ni, G.; Qu, S. Amino-functionalized graphene quantum dots prepared using high-softening point asphalt and their application in Fe^{3+} detection. *Appl. Surf. Sci.* **2019**, *467–468*, 446–455. [CrossRef]

45. Sun, D.; Lv, W. Evaluation of low temperature performances of polymer modified asphalts by force-ductility test. *J. Build. Mater.* **2007**, *10*, 37–42.

46. Gu, M. Asphalt and Asphalt Mixture at Low Temperature Performance Evaluation Methods of Research. Master's Thesis, Shandong Jianzhu University, Jinan, China, 2017.

47. Jahanbakhsh, H.; Karimi, M.; Moghadas Nejad, F.; Jahangiri, B. Viscoelastic-based approach to evaluate low temperature performance of asphalt binders. *Constr. Build. Mater.* **2016**, *128*, 384–398. [CrossRef]

48. Huang, G.; He, Z.; Hu, C.; Huang, T. Performance evaluation of rock modified asphalt based on analysis of microstructure and rheological property. *J. Southeast Univ.* **2010**, *40*, 367–372.

49. Xu, O.; Xiao, F.; Han, S.; Amirkhanian, S.N.; Wang, Z. High temperature rheological properties of crumb rubber modified asphalt binders with various modifiers. *Constr. Build. Mater.* **2016**, *112*, 49–58. [CrossRef]

50. Ali, A.W.; Kim, H.H.; Mazumder, M.; Lee, M.-S.; Lee, S.-J. Multiple Stress Creep Recovery (MSCR) characterization of polymer modified asphalt binder containing wax additives. *Int. J. Pavement Res. Technol.* **2018**, *1*, 1–15. [CrossRef]

51. Huang, W.; Tang, N. Characterizing SBS modified asphalt with sulfur using multiple stress creep recovery test. *Constr. Build. Mater.* **2015**, *93*, 514–521. [CrossRef]

52. Yang, X.; You, Z. High temperature performance evaluation of bio-oil modified asphalt binders using the DSR and MSCR tests. *Constr. Build. Mater.* **2015**, *76*, 380–387. [CrossRef]

53. Liu, Y. Response Surface Modeling by Local Kernel Partial Least Squares. Master's Thesis, Tsinghua University, Beijing, China, 2013.

54. Tang, Q.Y.; Zhang, C.X. Data Processing System (DPS) software with experimental design, statistical analysis and data mining developed for use in entomological research. *Insect Sci.* **2013**, *20*, 254–260. [CrossRef] [PubMed]

55. Wang, Y.; Wang, T. Comparison of the interpolation data produced by CSI and 1stopt in sedimentation prediction. *J. Ludong Univ.* **2017**, *33*, 374–378.

56. Chen, X. *Optimizing, Fitting and Modeling 1stopt Application*; China Building Material Press: Beijing, China, 2012.

57. Zhang, Y. Graphene and High Quality Graphene: Controllable Synthesis, Characterization, Properties and Application. Ph.D. Thesis, Wuhan University, Wuhan, China, 2014.

58. Li, Y.; Li, J.; Ding, S.; Sun, X. Characterization of remaining oil after polymer flooding by laser scanning confocal fluorescence microscopy. *J. Disper. Sci. Technol.* **2014**, *35*, 898–906. [CrossRef]

59. Liu, P.; Yang, C.; Ling, Z.; Zhu, E.; Shi, Q. Technological routes toward homogeneous dispersion of graphene: A review. *Mater. Rev.* **2016**, *30*, 39–45.

60. Bai, M. Preparation and Application of Graphene and Graphene Dispersions. Ph.D. Thesis, Beijing University of Chemical Technology, Beijing, China, 2016.

61. Poorsargol, M.; Alimohammadian, M.; Sohrabi, B.; Dehestani, M. Dispersion of graphene using surfactant mixtures: Experimental and molecular dynamics simulation studies. *Appl. Surf. Sci.* **2019**, *464*, 440–450. [CrossRef]

62. Zhang, X.; Huang, T.; Zhang, Y.; Gao, H.; Jiang, M. Image-pro plus analysis of pore structure of concrete. *J. Build. Mater.* **2015**, *18*, 177–182.

63. Salemi, M.; Wang, H. Image-aided random aggregate packing for computational modeling of asphalt concrete microstructure. *Constr. Build. Mater.* **2018**, *177*, 467–476. [CrossRef]

64. Xing, C.; Tan, Y.; Liu, X.; Anupam, K.; Scarpas, T. Research on local deformation property of asphalt mixture using digital image correlation. *Constr. Build. Mater.* **2017**, *140*, 416–423. [CrossRef]

Article

Rheological Properties, Compatibility, and Storage Stability of SBS Latex-Modified Asphalt

Shisong Ren [1], Xueyan Liu [1], Weiyu Fan [2], Haopeng Wang [1,*] and Sandra Erkens [1]

[1]   Section of Pavement Engineering, Faculty of Civil Engineering and Geosciences, Delft University of Technology, Stevinweg 1, 2628 CN Delft, The Netherlands; Shisong.Ren@tudelft.nl (S.R.); x.liu@tudelft.nl (X.L.); S.M.J.G.Erkens@tudelft.nl (S.E.)

[2]   State Key Laboratory of Heavy Oil Processing, China University of Petroleum, Qingdao 266580, Shandong, China; Weiyu.Fan@upc.edu.cn

*   Correspondence: haopengwang@tudelft.nl; Tel.: +31-062-536-1801

Received: 5 October 2019; Accepted: 6 November 2019; Published: 8 November 2019

Abstract: A styrene-butadiene-styrene (SBS) latex modifier can be used for asphalt modification due to the fact of its energy-saving, construction convenience, and economic advantages. The objective of this study was to investigate the influence of asphalt type and SBS latex dosage on the rheological properties, compatibility, and storage stability of asphalt through temperature and frequency sweep, steady-state flow, multiple stress creep and recovery (MSCR) tests, Cole-Cole plots and thermal storage tests. The results indicated that high SBS latex content is beneficial for improving anti-rutting, anti-fatigue, viscous flow resistance, and elastic recovery abilities of modified asphalt. The chemical composition of asphalt had a significant effect on the properties of the SBS latex-modified asphalt. High asphaltenes and low resins were favorable to enhancing anti-rutting and recovery properties but weakened the anti-fatigue, compatibility, and storage stability of modified asphalt. Furthermore, compared to SBS particle-modified asphalt, SBS latex-modified asphalt had greater rutting and fatigue resistance. However, SBS latex-modified asphalt had some disadvantages in compatibility and storage stability. Comprehensively considering the balance of viscoelastic properties, compatibility, and storage stability of SBS latex-modified asphalt, the mixing dosage of SBS latex modifier is recommended at 4.0 wt% which could feasibly replace SBS particle in asphalt modification.

Keywords: SBS-modified asphalt; SBS latex; rheological properties; compatibility; storage stability

---

1. Introduction

Asphalt is always used as binder material for road construction, and it is obtained from petroleum refining processes [1]. In terms of chemical composition, asphalt is composed of saturates, aromatics, resins, and asphaltenes. The main elements of bitumen is C, H, S, N, and O which is similar to petroleum products. Currently, with the increase in traffic loading and temperature conditions, asphalt roads are easily damaged including rutting, cracking, flacking, etc. [2,3]. Therefore, it is urgently needed to improve the performance of asphalt pavement and prolong its service life.

It is clear that base asphalt has many disadvantages, such as high-temperature flow, low-temperature cracking, and temperature susceptibility, that make it incapable of meeting the performance requirements of high-grade pavement [4]. Thus, in order to guarantee the pavement properties of an asphalt mixture under complex environments, additives have been used to modify asphalt. The main modifiers for asphalt binder include styrene-butadiene-styrene (SBS), crumb rubber (CR), polythene (PE), ethylene-vinyl acetate (EVA), styrene-butadiene rubber (SBR), polyphosphoric acid (PPA), gilsonite, and other nanomaterials [1,5,6]. According to previous studies [2,4], the addition of a polymer modifier can effectively improve the high-temperature rutting and low-temperature cracking resistance of asphalt and prolong the service life of the pavement. Zhang et al. [7] studied the effect of SBS on the rheological and aging properties of asphalt, and the results showed that SBS could remarkably enhance the viscoelastic performance and anti-aging capacity of asphalt. Meanwhile, Liang et al. [8] investigated the influence of sulfur and SBS type on both the properties and mechanism of SBS-modified asphalt, showing that there existed a chemical reaction between the SBS copolymer and the base asphalt. The SBS polymer molecule in asphalt can crosslink which results in the formation of a three-dimensional polymer network structure. This is why SBS-modified asphalt possesses better high- and low-temperature properties.

On the other hand, it is obvious that SBS-modified asphalt has disadvantages in terms of compatibility and storage stability [9]. Another problem is that modified asphalt with high-content SBS has poor workability and low economic efficiency which further limits the application of SBS copolymer [10,11]. Recently, researchers proposed that other additives can be added into SBS-modified asphalt to prepare complex modified asphalt which can alleviate the stability issue and further improve the pavement performance of SBS-modified asphalt. Qian et al. [12] investigated and concluded that CR/SBS composite-modified asphalt produced via a high-cured method had better performance, and it was a cost-effective approach to producing asphalt binder. In addition, Li et al. [13] prepared a modified heavy calcium carbonate and SBS composite-modified asphalt and found that the addition of calcium carbonate could strengthen the modulus and better enhance the viscoelastic properties than pristine SBS-modified asphalt. Importantly, SBS-modified asphalt is always manufactured by the asphalt industry because of the high-speed shear process used, and then it is transported to the construction site where asphalt binder and aggregate are mixed and compacted. Thus, separation between asphalt and SBS could easily happen during thermal storage and transportation [2,14]. Moreover, the preparation process for SBS-modified asphalt is complicated, including high-speed shear and mixing, and is energy intensive and costly.

Both the storage stability and the compatibility of SBS-modified asphalt are important to its engineering application and performance in asphalt roads. Clearly, a strong phase separation of the SBS copolymer from bitumen during storage and transportation is not expected [15]. Many researchers have studied the effects of SBS dosage on the compatibility and storage stability of modified asphalt. Lu et al. [16] investigated the compatibility and storage stability of SBS-modified bitumen using fluorescence microscopy and dynamic mechanical analysis. The results showed that the storage stability of modified asphalt decreased with an increase in SBS content; in addition, the degree of SBS dispersion in bitumen influenced the storage stability and rheological properties of modified binders as well. Meanwhile, Fu et al. [17] added SBS-g-M grafted with vinyl monomer into SBS-modified asphalt and found that the compatibility of the SBS-modified asphalt improved significantly. Therefore, the content of SBS modifier needs to be controlled to prevent the modified asphalt from phase separation.

The objective of this study was to explore the possibility of using an SBS latex modifier in asphalt modification and to investigate the effects of SBS latex dosage and asphalt components on the rheological properties, compatibility, and storage stability of modified asphalt. The experimental work in this study is shown in Figure 1 and described in the subsequent sections.

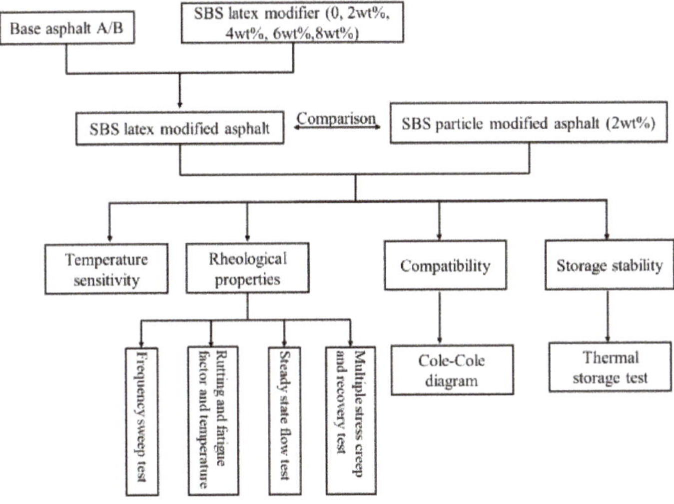

Figure 1. Flow chart of the experimental projects.

2. Materials and Methods

2.1. Raw Materials

In this paper, two base asphalts (PetroChina Fuel Oil Co. Ltd., Beijing, China) (coded as A and B) with a 60/80 penetration grade were used. The conventional properties and chemical compositions of these two base asphalts are displayed in Table 1. It is clear that, although the two asphalts have a similar penetration grade, their components have significant differences. The SBS latex modifier was provided by Shandong Dashan Road&Bridge Engineering Co. Ltd., Shandong Province, China. The apparent condition of the SBS-latex modifier showed milky liquid. The SBS solid concentration in latex modifier was 33.55 wt%. The SBS latex is the product prior to the preparation of the SBS particle. Owing to the lack of concentration and prilling procedure, a large amount of solvent exists in SBS latex. Therefore, the preparation of SBS latex-modified asphalt can not only omit the solidification and off-solvent operational processes, but it can also be prepared with only the use of a mixing stirrer.

Table 1. Conventional properties and chemical compositions of base asphalt.

Test Properties	Asphalt A	Asphalt B	Methods
25 °C Penetration (0.1 mm)	68	71	ASTM D5 [18]
Softening point (°C)	49.8	48.6	ASTM D36 [19]
15 °C Ductility (cm)	>150	>150	ASTM D113 [20]
Saturates (wt%)	23.4	27.1	ASTM D4124 [21]
Aromatics (wt%)	41.2	32.9	
Resins (wt%)	26.8	39.1	
Asphaltenes (wt%)	8.6	0.9	
Colloidal index (CI) *	0.47	0.39	

* Colloidal index (CI) = (saturates + asphaltenes)/(aromatics + resins).

2.2. Preparation of SBS Latex-Modified Asphalt

In this study, the preparation process of the SBS-latex-modified asphalt is shown in Figure 2. This process can be separated into three parts: the mixing section, condensation part, and recovery device. The dosage of the SBS latex modifier was determined in accordance with previous studies [22,23]. Firstly, the base asphalt was preheated in a temperature-controlled oven to a flow state. Then, it was poured into a three-necked flask which was heated in a cylindrical container to 150 °C and mixed with a certain amount of SBS latex modifier (0, 2 wt%, 4 wt%, 6 wt%, and 8 wt% of neat asphalt) with a speed of 1000 rpm for 30 min using a variable-speed blender. After that, 0.2 wt% sulfur powder was added into the above blend which was then mixed at 1000 rpm and 150 °C for 30 min. For simplicity, the modified asphalt A (B) containing various SBS latex dosages are abbreviated as A0 (B0), A2 (B2), A4 (B4), A6 (B6), and A8 (B8). Owing to the existence of solvent in SBS latex, the preparation device for SBS latex-modified asphalt was closed and equipped with a solvent condensation and recovery unit which can be seen in Figure 2. In order to evaluate and compare the performance grades of the SBS latex-modified asphalt, a conventional SBS particle-modified asphalt with the same SBS solid content was also prepared. The base asphalt was preheated to 175 °C and then the SBS particle modifier (the same solid content with 6 wt% SBS latex) was added and was sheared at 4000 rpm for 30 min using a high shear machine. Then, 0.2 wt% sulfur powder was added into the above blend and mixed at 1000 rpm and 150 °C for 30 min. Finally, the conventional SBS particle-modified asphalt was obtained and named ASBS (BSBS).

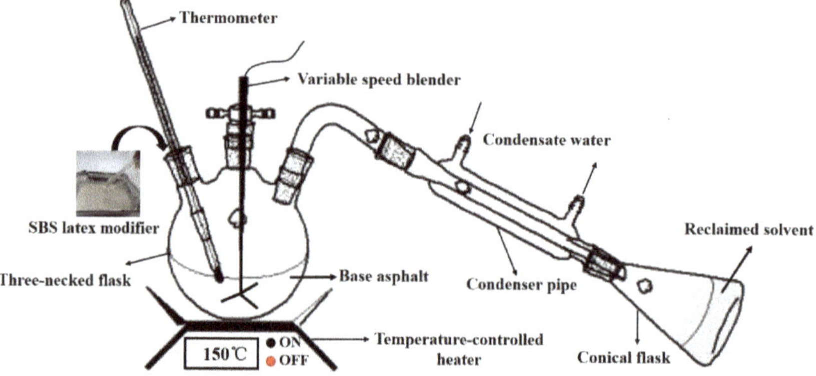

Figure 2. The preparation process of styrene-butadiene-styrene (SBS) latex-modified asphalt.

2.3. Test Methods

In this study, SBS latex-modified asphalts with various SBS latex dosages were prepared via distilling and condensing device. Then, dynamic shear rheometer (DSR) (Anton Par, Graz, Austrian) was conducted to analyze the viscoelastic properties of the modified asphalts. In addition, Cole Cole diagrams and thermal storage tests were performed to evaluate the compatibility and storage stability of the SBS latex-modified asphalts, respectively. Finally, the performance of the SBS latex-modified asphalt was compared with the conventional SBS particle-modified asphalt, and the optimum content of SBS latex modifier was obtained.

Rheological characterization of the asphalt samples was performed using the DSR device [24]. Frequency sweep tests were conducted at 60 °C with the frequency increasing from 0.1 rad/s to 100 rad/s. In addition, temperature sweep tests were performed at 10 rad/s from 48 °C to 84 °C with a parallel plate geometry of 25 mm and a gap setting of 1 mm to measure the moduli and rutting factor of the asphalts. Meanwhile, fatigue factors were also obtained by temperature sweep tests at 10 rad/s with the testing temperature increasing from 10 °C to 50 °C. To guarantee the linear viscoelastic ranges of the asphalt, train sweep tests were conducted to determine the loaded strain before these two tests.

In addition, a steady-state flow method was performed to obtain the influence of SBS latex dosage on the zero-shear viscosity (ZSV) and deformation resistance abilities of modified asphalt [20]. The shear rate regions were chosen from 10^{-3} s^{-1} to 10^{2} s^{-1}. Furthermore, multiple stress creep and recovery tests (MSCR), employed under two loading levels of 0.1 KPa and 3.2 KPa, were conducted at 60 °C. For both the loading levels, the asphalt samples underwent ten cycles of one-second creep and nine seconds recovery [25,26]. The applied parallel plates' geometry and gap were determined according to the relevant literatures [27–29].

3. Results and Discussion

3.1. Frequency Sweep Tests

It is clear that road performance depends, to some extent, on the viscoelastic and mechanical properties of asphalt binder [30,31]. Characterization of the linear viscoelastic behavior of the asphalt binder was performed to evaluate the effect of the SBS latex dosage on the rheology of the modified asphalts. Asphalt is a viscoelastic material that behaves very differently with various temperature and loading times. Therefore, a frequency sweep test was performed in the frequency region of 0.1–100 rad/s at 60 °C, and the experimental results are shown in Figure 3.

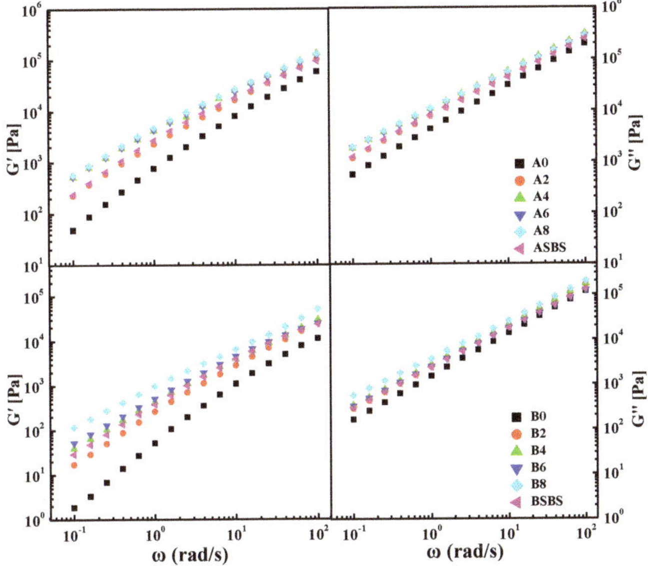

Figure 3. Effects of loading frequency on the storage and loss modulus of the base and the SBS latex-modified asphalts.

From Figure 3, clear increasing trends in the storage modulus G′ and loss modulus G″ can be seen with the frequency increasing gradually. It is well known that storage modulus G′ represents the energy recovery and elastic section of asphalt, while loss modulus G″ shows the energy loss and viscous part. However, the increasing trends of the two moduli were very heterogeneous. With regard to all asphalt samples, storage modulus G′ showed lower values than loss modulus G″ in the whole testing frequency region, indicating that the viscoelastic performance of the base and modified asphalts were primarily dominated by viscous properties.

This phenomenon was more obvious in the case of neat asphalt which is attributed to its poor viscoelastic properties at high temperature. Meanwhile, the gap between G′ and G″ became smaller as the frequency declined. After adding SBS latex, the G′ and G″ values of the modified asphalt both underwent an increasing trend compared with the neat asphalt. With the increase in the SBS latex dosage, the G′ and G″ values of asphalt samples both continued to increased, while the increasing trend weakened when the SBS latex content exceeds 6 wt%. The results also reveal that adding SBS latex increased the proportion of elastic components in the binder, which contributes to improving the stiffness of asphalt at high temperature.

Furthermore, asphalt composition also has a remarkable influence on the modulus and stiffness of SBS latex-modified asphalt. Both the storage and loss modulus of asphalt A were higher than that of asphalt B with the same SBS latex concentration; this is attributed to the differences in asphalt composition. This phenomenon suggests that there was a stronger reaction between the SBS latex modifier and base asphalt A with a higher asphaltenes content. As a consequence, asphalt type and composition are very crucial for SBS latex modification, and a relatively high asphaltenes dosage is beneficial to obtaining a satisfactory modification result. Additionally, the SBS latex-modified asphalt had better moduli than that of the conventional SBS particle-modified asphalt when the SBS concentration was the same, indicating that the SBS latex-modified asphalt possessed better deformation resistance performance. In addition, the moduli difference between the SBS latex-modified asphalt and the conventional SBS particle-modified asphalt was more distinct for asphalt A.

3.2. Temperature Sweep Tests

It is well known that rutting resistance factor $G^*/\sin\delta$ is an important index to evaluate the high-temperature stability of asphalt. The rutting resistance factor $G^*/\sin\delta$ of asphalt was obtained by temperature sweep tests with the temperature increasing from 48 °C to 84 °C and with a frequency of 10 rad/s. The rutting resistance factors $G^*/\sin\delta$ of the base and modified asphalts with various SBS latex dosages are presented in Figure 4. There was an obvious linear relationship between the testing temperature and rutting resistance factors of the asphalts. In order to evaluate the impact of SBS latex content on the anti-rutting properties of the modified asphalts, the linear fitting formula was used and the failure temperatures of the asphalts were calculated when the rutting resistance factor $G^*/\sin\delta$ was equal to 1.0 KPa.

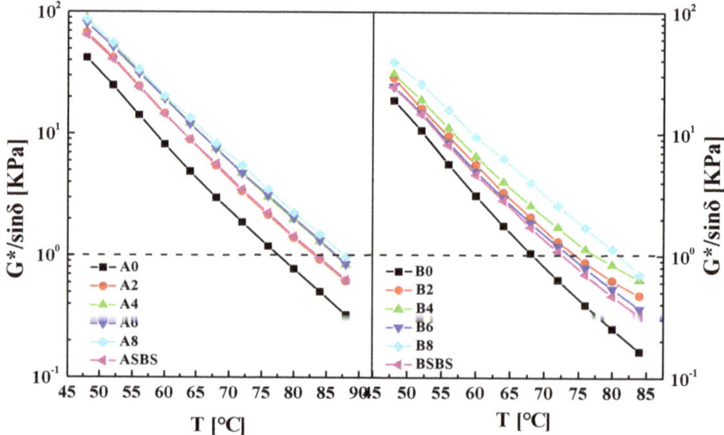

Figure 4. Effects of temperature on the rutting factor of the base and SBS-latex-modified asphalts.

Table 2 shows the failure temperatures of the base and modified asphalts. It can be seen that the addition of SBS latex increased the failure temperature of the base asphalt dramatically, indicating SBS latex can improve the anti-rutting ability of asphalt. However, the SBS latex dosage had a different effect on the improvement of the failure temperature of the asphalts. For asphalt A, the failure temperature increased by 7.01%, 10.74%, 10.79%, and 12.25% when the SBS latex content was 2 wt%, 4 wt%, 6 wt%, and 8 wt%, respectively, while the failure temperature of asphalt B increased by 9.15%, 14.28%, 15.27%, and 18.42%. Clearly, the greater influence of the SBS latex concentration on asphalt B over asphalt A is obvious. Meanwhile, the degree of influence became smaller when the SBS latex dosage exceeded 4 wt%.

Table 2. Failure temperature and fatigue temperature of all asphalt samples.

SBS Latex Contents	Failure Temperature/°C		Fatigue Temperature/°C	
	Asphalt A	**Asphalt B**	**Asphalt A**	**Asphalt B**
0	78.13	68.12	10.12	4.79
2	83.61	74.35	9.45	1.80
4	86.52	77.85	7.33	−0.19
6	86.56	78.52	7.71	0.58
8	87.70	80.67	8.27	3.10
SBS particle	83.89	72.37	7.74	2.16

In addition, the failure temperature of the SBS latex-modified asphalt was higher than that of the SBS particle-modified asphalt with the same SBS solid content. That is to say, SBS latex-modified asphalt had a better anti-rutting ability than the SBS particle-modified asphalt which is the main binder used in pavement construction. Moreover, for asphalt A, the failure temperature of the SBS latex-modified asphalt was 2.67 °C larger than that of the SBS particle-modified asphalt, while the difference for asphalt B was 6.15 °C. Meanwhile, the failure temperature values of the modified asphalts A were all higher than that of asphalts B, showing that the SBS latex-modified asphalt A had greater rutting and deformation resistance because of the higher asphaltenes concentration.

Fatigue cracking is another defect of asphalt pavement, thus the fatigue resistance factor is an important and essential index to evaluate the long-term pavement performance and service life of asphalt roads which has been receiving increasing attention [32,33]. In this study, temperature sweep tests were performed with temperature changes from 10 °C to 50 °C with 5 °C increments to investigate the effect of SBS latex dosage on the fatigue resistance of modified asphalt. The relationship between testing temperature and fatigue factors G*sinδ in SBS-latex-modified asphalt is plotted in Figure 5. It was found that the fatigue resistance factor of modified asphalt increased as the temperature gradually increased, illustrating that a high-temperature environment has an adverse effect on the fatigue resistance performance of SBS latex-modified asphalt.

In order to assess the fatigue property of modified asphalt intuitively, the ultimate fatigue temperature was calculated and used for analysis [34]. The relationship between the temperature and fatigue resistance of asphalt can be demonstrated by a linear formula, which is also shown in Figure 5. There was a clear linear relationship between the temperature and fatigue factor of asphalt, and the correlation coefficient values R^2 were all more than 0.99. Besides, the fatigue temperature of of base and modified asphalts with various SBS latex concentrations are displayed in Table 2. The SBS latex dosage has great influence on the fatigue property of modified asphalt. The addition of SBS latex remarkably decreased the fatigue temperature of base asphalt, showing the positive effect of SBS latex on improving the fatigue resistance of asphalt. However, as SBS latex dosage increased, the fatigue temperature of modified asphalt first declined and then increased, indicating that a higher SBS latex dosage leads to a loss in fatigue resistance for SBS latex-modified asphalt, and that there exists an optimal SBS latex content to ensure the best fatigue resistance performance. Regardless of the asphalt components, the modified asphalt with an SBS latex content of 4 wt% had the lowest fatigue temperature and best fatigue resistance ability. That is to say, the optimal dosage of SBS latex in this study was 4 wt% when considering the anti-fatigue performance of modified asphalt.

Furthermore, asphalt type and chemical composition also affected the fatigue resistance of SBS-latex-modified asphalt dramatically [35,36]. It was found that, with the same SBS latex dosage, the fatigue temperature values of modified asphalts A were all higher than that of modified asphalts B, indicating that modified asphalts B had better fatigue resistance performance. This is due to the difference in asphalt composition, and the lower asphaltene concentration of asphalt B contributed to the improvement of fatigue resistance performance. It is interesting that the SBS latex-modified asphalt had a lower fatigue temperature than the SBS particle-modified asphalt, indicating that the SBS latex-modified asphalt had better fatigue resistance. Thus, in regard to the fatigue resistance properties, it is feasible and superior to use SBS latex as a modifier of asphalt to enhance its properties.

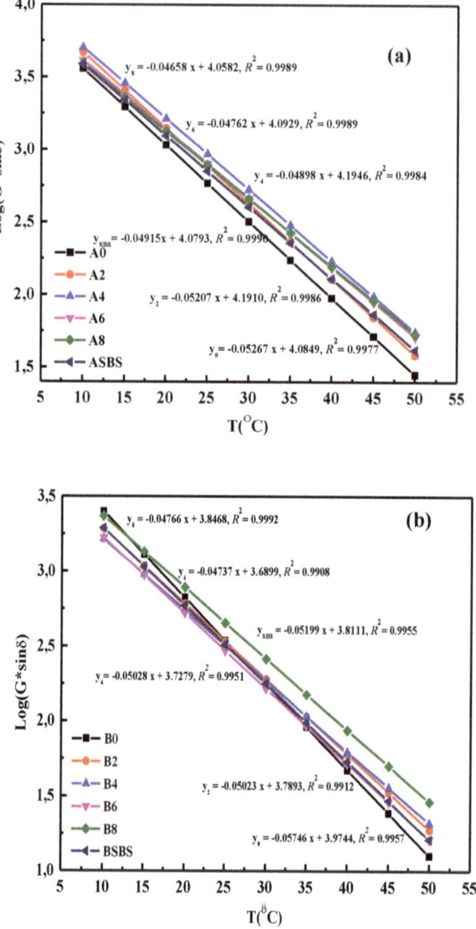

Figure 5. Effects of temperature on the fatigue factor of (**a**) the base and (**b**) the SBS latex-modified asphalts.

3.3. Steady-State Flow Tests

Characterization under large deformation always gives a very distinct viscoelastic response in comparison with those in small deformation, especially for polymer-modified asphalt [37,38]. Thus, the 60 °C flow behavior of asphalt was characterized by steady-state flow tests with the shear rate increasing from 10^{-3} s^{-1} to 10^2 s^{-1}. The flow curves of base and SBS-latex-modified asphalts are plotted in Figure 6. On the whole, the viscous flow behavior of asphalt shows a dependence on shear rate, which is extraordinarily distinct for modified asphalt. The shear viscosity of asphalt keeps constant when the shear rate is in the low range, representing the typical Newtonian fluid characteristic [39]. When the shear rate exceeds this value, the shear viscosity of the asphalt decreases remarkably which presents an apparent shear-thinning behavior. Neat asphalt has the widest Newtonian behavior region, which is shortened after adding SBS latex. The non-Newtonian fluid characteristics of asphalt with a higher SBS latex dosage were more obvious, which is attributed to the complicated entanglement between asphalt and polymer. On the other hand, the viscosity of the base asphalt was the lowest, increasing gradually with the increase in SBS latex content, indicating that adding SBS latex can improve the flow resistance of asphalt dramatically. Overall, base asphalt presented primarily Newtonian fluid characteristics, while SBS latex-modified asphalts showed non-Newtonian behavior in the studied region of shear rates. In addition, it is clear that the type and composition of base asphalt had great influence on the flow behavior of SBS-latex-modified asphalt. The Newtonian fluid region of asphalt B was larger than that of asphalt B, which is attributed to the lower amount of asphaltenes in asphalt B. Meanwhile, it was found that the viscosity values of SBS-latex-modified asphalt B were all lower than that of modified asphalt A. That is to say, asphalt B with lower asphaltenes had the advantages of workability and inferior flow resistance ability.

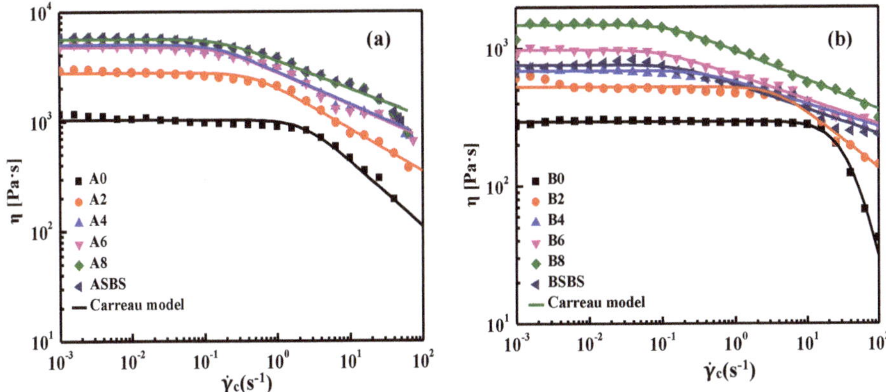

Figure 6. Flow curves of (**a**) the base and (**b**)SBS-latex-modified asphalts.

In order to evaluate the effects of SBS latex dosage on the viscous properties of asphalt quantitatively, the relationship between shear rates and viscosity of asphalt can be fitted using Carreau model [8], which is as follows:

$$\frac{\eta_0}{\eta} = \left[1 + \left(\frac{\dot{\gamma}}{\dot{\gamma}_c} \right)^2 \right]^s \tag{1}$$

where η shows the viscosity of asphalt, η_0 is the ultimate viscosity when the shear viscosity approaches zero which is also called the zero-shear viscosity (ZSV), s represents the slope of the fitting curve in the non-Newtonian region, while $\dot{\gamma}$ is shear rate, and $\dot{\gamma}_c$ represents the critical shear rate value when the viscoelastic characteristic of asphalt binder changes from Newtonian behavior to the non-Newtonian region [40,41]. From Figure 6, the flow curve of asphalt fits the Carreau model fairly well, showing the parameters obtained from the Carreau model are very reliable. Table 3 shows the results of all the studied samples calculated from the Carreau model including zero-shear viscosity η_0, critical shear rate value $\dot{\gamma}_c$, and fitting curve slope s. It is clear that the neat asphalt has the lowest η_0 value (ZSV), showing the poor flow resistance ability of the base asphalt at high temperature. From Table 3, it can be seen that the zero-shear viscosity of asphalt B was lower than that of asphalt A, while the $\dot{\gamma}_c$ value of asphalt B was much higher than asphalt A. The results show that compared to asphalt A, asphalt B had a wider Newtonian fuild range and better workability, which is attributed to its higher content of light oil fractions and lower dosage asphaltenes. Meanwhile, it can also be found that the s value of asphalt B is higher than asphalt A, which means asphalt B has higher shear susceptibility than asphalt B, which is also connected to its chemical component characteristics.

Table 3. The results calculated from the Carreau model at 60 °C of all studied samples.

Items	SBS Latex Contents/wt%					
	0	**2**	**4**	**6**	**8**	**SBS Particle**
Asphalt A	A0	A2	A4	A6	A8	ASBS
$\eta_0 \times 10^{-3}$ (Pa·s)	1.032	2.773	4.733	5.026	5.629	5.718
$\dot{\gamma}_c$ (s^{-1})	2.416	0.471	0.142	0.169	0.165	0.155
s	0.300	0.193	0.146	0.149	0.128	0.126
Asphalt B	B0	B2	B4	B6	B8	BSBS
$\eta_0 \times 10^{-3}$ (Pa·s)	0.295	0.527	0.692	0.977	1.494	0.762
$\dot{\gamma}_c$ (s^{-1})	40.207	3.743	0.321	0.092	0.128	0.164
s	1.141	0.206	0.082	0.089	1.107	0.090

Meanwhile, SBS latex can remarkably enhance the ZSV value of asphalt and this improvement was more obvious for modified asphalt A. Interestingly, the zero shear viscosity value of modified asphalt A was higher than that of modified asphalt B, indicating the viscous flow behavior was also significantly affected by the chemical composition of base asphalt. In terms of viscoelastic properties, high aromatic content and moderate asphaltenes dosage were beneficial for enhancing the flow resistance of asphalt. Furthermore, SBS particle-modified asphalt A had a higher η_0 value than that of SBS-latex-modified asphalt A, even though the SBS latex content increased to 8 wt%, while the η_0 value of SBS particle-modified asphalt B was lower than that of 6 wt% SBS particle-modified asphalt B. On the other hand, adding SBS latex results in a decrease in the critical shear rate $\dot{\gamma}_c$ of the asphalt, indicating that the shear-thinning behavior of modified asphalt became more distinct. This is related to the presence of macro-molecules due to the incorporation of polymers and the complicated structure of modified asphalt being prone to disruption by shear stress. Moreover, the addition of SBS latex decreases the s value and improves the resistance to shear thinning of the modified asphalt.

The relationship between the SBS latex dosage and η_0 value as well as the critical shear rate $\dot{\gamma}_c$ was fitted by a linear formula as shown in Figure 7. With the increase in SBS latex, the η_0 value of modified asphalt increased dramatically, while the critical shear rate $\dot{\gamma}_c$ decreased, showing that SBS latex can improve the deformation resistance and shear-thinning behavior of asphalt. However, the influence of SBS latex content on the ZSV value of modified asphalt A was more obvious than that of modified asphalt B, while the effect on the $\dot{\gamma}_c$ value of modified asphalt B was more distinct than that of modified asphalt B. Furthermore, the asphalt component can also affect the ZSV and $\dot{\gamma}_c$ values of modified asphalt. It was found that the ZSV value of modified asphalt A was higher than that of modified asphalt B, while the $\dot{\gamma}_c$ value of modified asphalt A was lower than that of modified asphalts B, indicating that asphalt A had better deformation resistance ability and was closer to the non-Newtonian fluid characteristic.

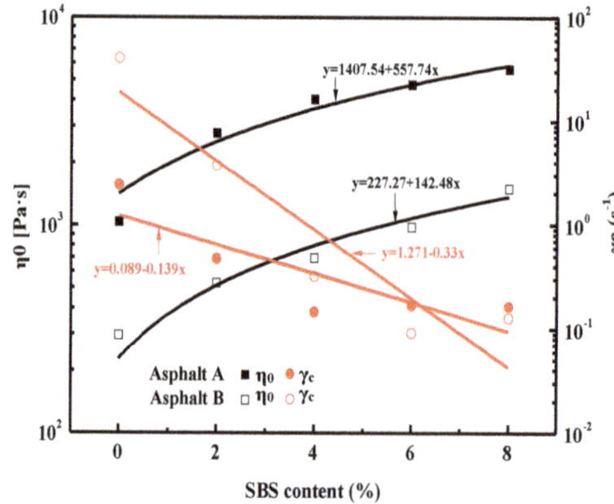

Figure 7. Effects of SBS latex content on the Carreau model parameters of modified asphalts.

3.4. Creep and Recovery Behavior

In order to further evaluate the effects of SBS latex dosage on the high-temperature properties and recovery behavior of modified asphalt, MSCR tests were conducted with respect to AASHTO TP70 [25]. Figure 8 shows the MSCR results of asphalt samples at stress levels of 0.1 KPa and 3.2 KPa, respectively. It is well known that the percent recovery (R%) is an elastic response indicator of asphalt, and the increased value of R% indicates the improvement in the elastic component in asphalt. Meanwhile, non-recoverable creep compliance (Jnr) is well related to the rutting performance of asphalt, and the lower Jnr value represents the better resistance to permanent deformation of binder.

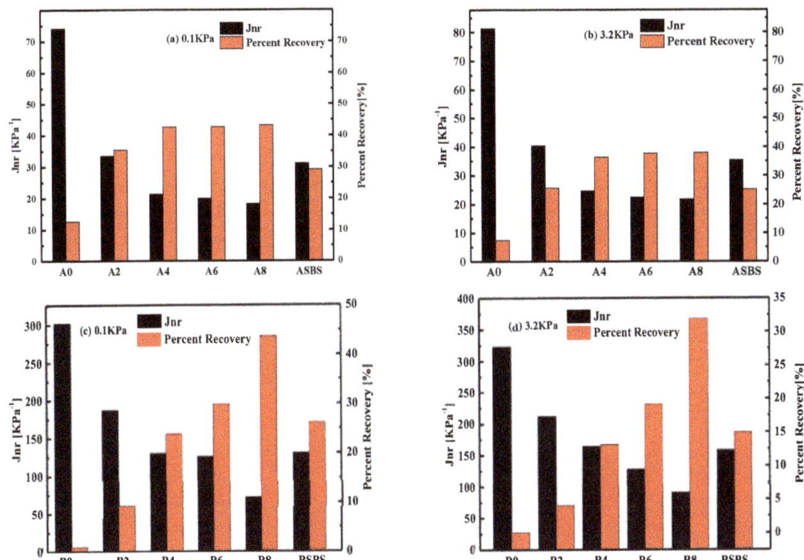

Figure 8. Effects of SBS latex content on the percent recovery (R%) and non-recovery compliance (Jnr) of modified asphalts.

As expected, base asphalt had the lowest R% value and highest Jnr value regardless of the loading stress level. The R% value of asphalt was damaged when the loading stress increased from 0.1 KPa to 3.2 KPa, while the non-recoverable creep compliance Jnr increased dramatically. Meanwhile, adding SBS latex can remarkably increase the R% value and decrease the Jnr value of asphalt. The R% and Jnr value of modified asphalt increased and declined with the increase in SBS latex dosage, respectively. That is to say, SBS latex can improve the elastic recovery and deformation resistance of asphalt. However, the degree of influence of SBS latex content on the R% and Jnr value of modified asphalt depends on the chemical composition of the base asphalt. The effect of SBS latex dosage on the elastic properties of asphalt B was more obvious than that of asphalt A, which means the positive effect of SBS latex on the resistance to permanent deformation was more significant for asphalt B.

Additionally, modified asphalt A always had a higher R% and lower Jnr value than those of modified asphalt B, indicating modified asphalt A had better elastic response and deformation resistance performance than asphalt B. On the other hand, it was found that the R% value of SBS latex-modified asphalt was higher than that of SBS particle-modified asphalt, while the Jnr value of the former was lower than the latter, regardless of loading stress level when the SBS particle content was the same as the latex. Obviously, the SBS latex-modified asphalt possessed better elastic properties and anti-deformation abilities than conventional SBS particle-modified asphalt. As to asphalt A, both R% and Jnr values of SBS particle-modified asphalt were similar to that of 2 wt% SBS latex-modified asphalt, while SBS particle-modified asphalt B had the same grade of anti-rutting and elastic response performance with 4 wt% SBS latex-modified asphalt B. Furthermore, it can be found that the influence of SBS latex dosage on improving the permanent deformation resistance of asphalt A became less obvious when the SBS latex dosage exceeded to 4 wt%. However, the influence of SBS latex dosage on the recovery and creep compliance of asphalt B was more obvious. Therefore, asphalt B was more suitable to be modified with SBS latex which is consistent with the above results.

3.5. Compatibility

As is well known, compatibility between polymer and base asphalt is very essential, which can result in phase separation and performance inhomogeneity of polymer-modified asphalt [12]. According to previous studies [9,22,42], a Cole-Cole diagram is the most efficient method to investigate the compatibility of polymer-modified asphalts. In this study, the Cole-Cole plot was selected to assess the compatibility differences during the preparation of the SBS latex-modified asphalts. It is well known that the compatibility of modified asphalt is related to the shape of Cole-Cole diagrams, and a symmetrical parabola shows great compatibility, while a conclusion of incompatibility between the asphalt and modifier is drawn when the shape deviates from symmetry [12].

Figure 9 displays the Cole-Cole diagrams of the base asphalt and the SBS latex-modified asphalts at 60 °C. For modified asphalt A, when the SBS latex dosage was lower than 4 wt%, the η'' value first increased and then descended with the increase in η'. However, when the SBS latex dosage was larger than 4 wt%, the η'' value of the modified asphalt rapidly increased with the increase in η'. This means that the compatibility of SBS latex-modified asphalt A became worse with the increase in the SBS latex dose; this is attributed to the concentretion limitation of resin in asphalt A. On the other hand, Cole-Cole plots of modified asphalt B present symmetrical parabolas when the SBS latex dosage was less than 8 wt%. With the increase in the SBS latex modifier, the shape of the Cole-Cole diagram deviated from this symmetry gradually, indicating the SBS latex dosage had a passive influence on the compatibility of modified asphalt.

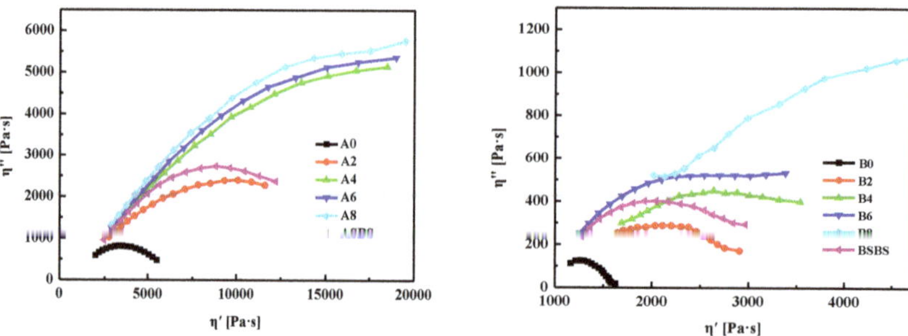

Figure 9. Cole-Cole plots of the base and SBS latex-modified asphalts.

Furthermore, the compatibility between asphalt and polymer also depended on the composition of the base asphalt. It was clear that the shape of the Cole-Cole plot for modified asphalt B was closer to symmetrical parabolas than that of modified asphalt A under the same SBS latex dosage [42–44]. This result showed that SBS latex-modified asphalt B had better compatibility than asphalt A, which is attributed to the composition differences of the base asphalt. From Table 1, it can be seen that asphalt B had more resins and a lower asphaltene concentration than asphalt A which is beneficial to the formation of a stable colloid structure and enhanced the compatibility between asphalt and SBS latex modifier. In conclusion, neat asphalt with a higher resins dosage and lower asphaltenes content is suitable to be used to prepare stable SBS latex-modified asphalt with satisfactory compatibility. Furthermore, it was found that the shape of the Cole-Cole plot for the SBS particle-modified asphalt trended to symmetrical parabolas better than that of the SBS latex-modified asphalt, regardless of the type and composition of the base asphalt. This indicates that the SBS latex-modified asphalt had worse compatibility than the conventional SBS particle-modified asphalt.

3.6. Storage Stability

Polymer-modified asphalts are multiphase systems and the difference in the solubility and density between polymer and asphalt can result in phase separation of the modified asphalt during high-temperature storage processes [45]. On the basis of this, a storage stability test was performed on SBS latex-modified asphalt according to the aforementioned testing method. To reveal the effect of SBS latex dosage on storage stability, thermal storage tests were conducted for all modified asphalts at 163 °C for 2 d. Figure 10 shows the results of storage stability tests for the SBS latex-modified asphalts. The results indicate that, with the increase in the SBS latex dosage, the softening point difference of the modified asphalts became gradually larger, showing that the higher SBS latex content had a passive influence on the storage stability and compatibility of modified asphalt. Meanwhile, the softening point difference value of modified asphalt B was lower than that of modified asphalt A. Hence, it was concluded that all SBS latex-modified asphalts suffered from phase separation during thermal storage, and the storage stability of modified asphalt was associated with the base asphalt composition as well as the SBS latex dosage.

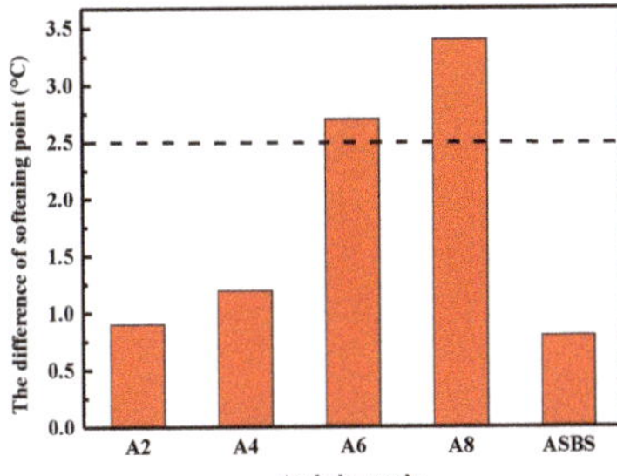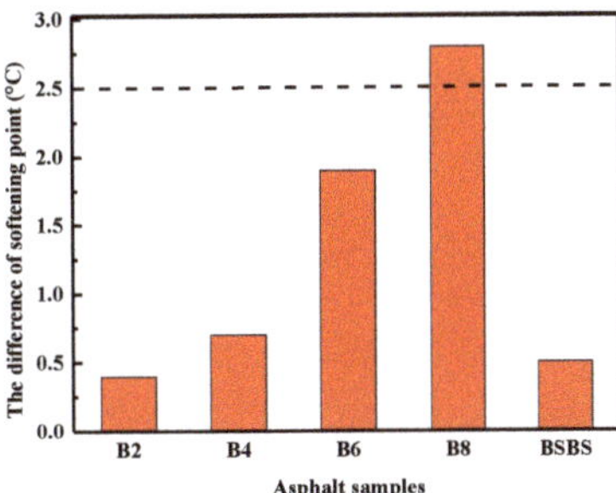

Figure 10. Effect of SBS latex content on the storage stability of modified asphalts.

It was found that the softening point difference of the modified asphalt increased as the SBS latex dosage increased. When the SBS dosage content reached 8 wt%, the softening point difference of modified asphalt A and B increased to 3.4 °C and 2.7 °C, respectively, which obviously exceeded the specification that polymer-modified asphalt is storage-stable with a softening point difference lower than 2.5 °C. These results demonstrate that a high SBS latex dosage has an adverse effect on the storage stability of modified asphalt. Furthermore, the thermal stability of modified asphalt also depends on the components of the base asphalt. It was clear that SBS latex-modified asphalt B had a smaller softening point difference value and greater storage stability than modified asphalt A which is associated with the asphalt components. From Table 1, asphalt A contained more asphaltenes and a lower amount of resins than asphalt B. Therefore, the difference in storage stability can be interpreted by the difference in asphalt compositions. Furthermore, the softening point difference of SBS latex-modified asphalt A was 1.9 °C larger than that of SBS particle-modified asphalt, while modified asphalt B had a 1.4 °C higher softening point difference. That is to say, SBS latex-modified asphalt had worse storage stability than conventional SBS particle-modified asphalt. Thus, it can be concluded that SBS latex-modified asphalt is suitable to be prepared and used at the construction site immediately. Certainly, SBS latex-modified asphalt with a lower content SBS latex can be prepared by the asphalt industry and transported to the site of the construction of an asphalt road.

4. Conclusions

This paper prepared and evaluated the rheological properties, compatibility, and storage stability of SBS latex-modified asphalt. The effects of SBS latex dosage and asphalt composition on the moduli, rutting and fatigue temperature, ZSV, R%, Jnr, compatibility, and storage stability of modified asphalt were investigated which were also compared with conventional SBS particle-modified asphalt. On the basis of the experimental results and discussion, the following conclusions were drawn:

(1) The SBS latex dosage had a great effect on the rheological properties of modified asphalt. With an increase in the SBS latex content, both the viscoelastic and high-temperature performance of modified asphalt improved including the moduli, rutting resistance, anti-fatigue, zero shear viscosity, and elastic recovery abilities.

(2) The chemical composition of neat asphalt also had a significant influence on the properties of SBS latex-modified asphalt. High asphaltenes and low resins are beneficial for anti-rutting and viscoelastic properties, while low asphaltenes and high resins are favorable to fatigue resistance ability, compatibility, and storage stability of modified asphalt.

(3) Compared to conventional SBS particle-modified asphalt, SBS latex-modified asphalt possesses higher moduli, greater rutting and fatigue resistance as well as superior viscoelastic performance which, however, has some disadvantages in terms of compatibility and storage stability.

(4) In this study, in comprehensively considering the balance of viscoelastic properties, compatibility, and storage stability of SBS latex-modified asphalt, the mixing content of SBS latex is recommended at 4.0 wt% which can be prepared at a construction site with a lower temperature. It is feasible for the SBS latex modifier to take the place of SBS particle and be used in the modification of asphalt.

Author Contributions: Conceptualization, S.R. and H.W.; Formal analysis, X.L.; Investigation, S.R. and L.X.; Resources, W.F.; Writing-original draft preparation, S.R.; Writing-review and editing, S.E.

Funding: This research received no external funding.

Acknowledgments: The first author would like to acknowledge the scholarship from the China Scholarship Council.

References

1. Polacco, G.; Filippi, S.; Merusi, F.; Stastna, G. A review of the fundamental of polymer-modified asphalts: Asphalt/polymer interactions and principle of compatibility. *Adv. Colloid Interface Sci.* **2015**, *224*, 72–112. [CrossRef]

2. Bai, M. Investigation of low-temperature properties of recycling of aged SBS modified asphalt binder. *Constr. Build. Mater.* **2017**, *150*, 766–773. [CrossRef]

3. Shan, L.; Xie, R.; Wagner, N.; He, H.; Liu, Y. Microstructure of neat and SBS modified asphalt binder by small-angle neutron scattering. *Fuel* **2019**, *253*, 1589–1596. [CrossRef]

4. Wang, H.; Liu, X.; Apostolidis, P.; Tom, S. Rheological behavior and its chemical interpretation of crumb rubber modified asphalt containing warm-mix additives. *Transp. Res. Rec. J. Transp. Res. Board* **2018**, *2672*, 337–348. [CrossRef]

5. Liang, M.; Ren, S.; Fan, W.; Xin, X.; Shi, J.; Luo, H. Rheological property and stability of polymer modified asphalt: Effect of various vinyl-acetate structures in EVA copolymers. *Constr. Build. Mater.* **2017**, *137*, 367–380. [CrossRef]

6. Lin, P.; Huang, W.; Tang, N.; Xiao, F. Performance characteristics of terminal blend rubberized asphalt with SBS and polyphosphoric. *Constr. Build. Mater.* **2017**, *141*, 171–182. [CrossRef]

7. Zhang, D.; Zhang, H.; Shi, C. Investigation of aging performance of SBS modified asphalt with various aging methods. *Constr. Build. Mater.* **2017**, *145*, 445–451. [CrossRef]

8. Liang, M.; Xin, X.; Fan, W.; Wang, H.; Ren, S.; Shi, J. Effects of polymerized sulfur on rheological properties, morphology and stability of SBS modified asphalt. *Constr. Build. Mater.* **2017**, *150*, 860–871. [CrossRef]

9.   Liang, M.; Liang, P.; Fan, W.; Qian, C.; Xin, X.; Shi, J.; Nan, G. Thermo-rheological behavior and compatibility of modified asphalt with various styrene-butadiene structures in SBS copolymers. *Mater. Des.* **2015**, *88*, 177–185. [CrossRef]

10. Zhang, F.; Hu, C. The research of SBS and SBR compound modified asphalts with polyphosphoric acid and sulfur. *Constr. Build. Mater.* **2013**, *43*, 461–468. [CrossRef]

11. Lin, P.; Huang, W.; Li, Y.; Tang, N.; Xiao, F. Investigation of influence factors on low temperature properties of SBS modified asphalt. *Constr. Build. Mater.* **2017**, *154*, 609–622. [CrossRef]

12. Qian, C.; Fan, W.; Liang, M.; He, Y.; Ren, S.; Lv, X.; Nan, G.; Luo, H. Rheological properties, storage stability and morphology of CR/SBS composite modified asphalt by high-cured method. *Constr. Build. Mater.* **2018**, *193*, 312–322. [CrossRef]

13. Li, J.; Yang, S.; Liu, Y.; Muhammad, Y.; Su, Z.; Yang, J. Studies on the properties of modified heavy calcium carbonate and SBS composite modified asphalt. *Constr. Build. Mater.* **2019**, *218*, 413–423. [CrossRef]

14. Dong, F.; Zhao, W.; Zhang, Y.; Wei, J.; Fan, W.; Yu, Y.; Wang, Z. Influence of SBS and asphalt on SBS dispersion and the performance of modified asphalt. *Constr. Build. Mater.* **2014**, *62*, 1–7. [CrossRef]

15. Cong, Y.; Huang, W.; Liao, K.; Zhai, Y. Study on storage stability of SBS modified asphalt. *Pet. Sci. Technol.* **2005**, *23*, 39–46. [CrossRef]

16. Lu, X.; Isacsson, U. Compatibility and storage stability of styrene-butadiene-styrene copolymer modified bitumens. *Mater. Struct.* **1997**, *30*, 618–626. [CrossRef]

17. Fu, H.; Xie, D.; Li, L.; Yu, M.; Yao, S. Storage stability and compatibility of asphalt binder modified by SBS graft copolymer. *Constr. Build. Mater.* **2007**, *21*, 1528–1533. [CrossRef]

18. ASTM D5-06. *Standard Test Method for Penetration of Bituminous Materials*; ASTM: West Conshohocken, PA, USA, 2006.

19. ASTM D36-06. *Standard Test Method for Softening Point of Bitumen (Ring and Ball Apparatus)*; ASTM: West Conshohocken, PA, USA, 2006.

20. ASTM D113-99. *Standard Test Method for Ductility of Bituminous Materials*; ASTM: West Conshohocken, PA, USA, 1999.

21. ASTM D4124-01. *Standard Test Methods for Separation of Asphalt into Four Fractions*; ASTM: West Conshohocken, PA, USA, 2001.

22. Ren, S.; Liang, M.; Fan, W.; Zhang, Y.; Qian, C.; He, Y.; Shi, J. Investigating the effects of SBR on the properties of gilsonite modified asphalt. *Constr. Build. Mater.* **2018**, *190*, 1103–1116. [CrossRef]

23. Yan, C.; Huang, W.; Lin, P.; Zhang, Y.; Lv, Q. Chemical and rheological evaluation of aging properties of high content SBS polymer modified asphalt. *Fuel* **2019**, *252*, 417–426. [CrossRef]

24. AASHTO M 320. *Standard Specification for Performance-Graded Asphalt Binder*; AASHTO: Washington, DC, USA, 2017.

25. AASHTO TP70. *Standard Method of Test for Multiple Stress Creep Recovery (MSCR) Test of Asphalt Binder Using a Dynamic SHEAR Rheometer (DSR)*; American Association of State Highway and Transportation Officials: Washington, DC, USA, 2009.

26. Wang, H.; Liu, X.; Zhang, H.; Apostolidis, P.; Scarpas, T.; Erkens, S. Asphalt-rubber interaction and performance evaluation of rubberised asphalt binders containing non-foaming warm-mix additives. *Road Mater. Pavement Des.* **2018**, 1–22. [CrossRef]

27. Zhao, X.; Wang, S.; Wang, Q.; Yao, H. Rheological and structural evolution of SBS modified asphalts under natural weathering. *Fuel* **2016**, *184*, 242–247. [CrossRef]

28. Zhang, P.; Guo, Q.; Tao, J.; Ma, D.; Wang, Y. Aging mechanism of a diatomite-modified asphalt binder using Fourier-Transform Infrared (FTIR) Spectroscopy analysis. *Materials* **2019**, *12*, 988. [CrossRef]

29. Huang, M.; Wen, X. Experimental study on photocatalytic effect of nano TiO_2 expoxy emulsified asphalt mixture. *Appl. Sci.* **2019**, *9*, 2464. [CrossRef]

30. Du, P.; Chen, Z.; Zhang, H. Rheological and aging behaviors of base and SBS modified asphalt with thermocheromic microcapsule. *Constr. Build. Mater.* **2019**, *200*, 1–9. [CrossRef]

31. Li, B.; Zhou, J.; Zhang, Z.; Yang, X.; Wu, Y. Effect of short-term aging on asphalt modified using microwave activation crumb rubber. *Materials* **2019**, *12*, 1039. [CrossRef]

32. Wang, P.; Dong, Z.; Liu, Z. Influence of carbon nanotubes on morphology of asphalts modified with styrene-butadiene-styrene. *Transp. Res. Rec. J. Transp. Res. Board* **2017**, *2632*, 130–139. [CrossRef]

33. Li, Y.; Lyv, Y.; Fan, L.; Zhang, Y. Effects of cement and emulsified asphalt on properties of mastics and 100% cold recycled asphalt mixtures. *Materials* **2019**, *12*, 754. [CrossRef]

34. Salim, R.; Gundia, A.; Underwood, B.S.; Kaloush, K.E. Effect of MSCR percent recovery on performance of polymer modified asphalt mixtures. *Transp. Res. Rec. J. Transp. Res. Board* **2019**, *2673*, 320–328. [CrossRef]

35. Mandi, S.; Nader, T.; Soheil, N. Performance evaluation of slurry seals containing reclaimed asphalt pavement. *Transp. Res. Rec. J. Transp. Res. Board.* **2019**, *2673*, 358–368.

36. Zhang, X.; He, J.; Huang, G.; Zhou, C.; Feng, M.; Li, Y. Preparation and characteristics of ethylene bis(stearamide)-based graphene-modified asphalt. *Materials* **2019**, *12*, 757. [CrossRef]

37. Jing, R.; Varveri, A.; Liu, X.; Scarpas, A.; Erkens, S. Laboratory and field aging effect on bitumen chemistry and rheology in porous asphalt mixture. *Transp. Res. Rec. J. Transp. Res. Board* **2019**, *2673*, 365–374. [CrossRef]

38. Liu, Y.; Wang, H.; Tighe, S.; Zhao, G.; You, Z. Effects of preheating on the rheological properties of rejuvenated asphalt binder. *Transp. Res. Rec. J. Transp. Res. Board* **2019**, *2673*, 546–557. [CrossRef]

39. Wang, H.; Liu, X.; Apostolidis, P.; Scarpas, T. Non-Newtonian behaviors of crumb rubber-modified bituminous binders. *Appl. Sci.* **2018**, *8*, 1760. [CrossRef]

40. Anderson, D.; Lapalu, L.; Marasteranu, M.; LeHir, Y.; Planche, J.; Martin, D. Low-temperature thermal cracking of asphalt binders as ranked by strength and fracture properties. *Transp. Res. Rec.* **2001**, *1766*, 1–6. [CrossRef]

41. Mohammad, J.; Abbas, B.; Sassan, A. Effects of stress levels on creep and recovery behavior of modified asphalt binders with the same continuous performance grades. *Transp. Res. Rec. J. Transp. Res. Board* **2015**, *2507*, 57–66.

42. Mirsayar, M. On the low-temperature mixed-mode fracture analysis of asphalt binder–theories and experiments. *Eng. Fract. Mech.* **2017**, *186*, 181–194. [CrossRef]

43. Wu, S.; Zhao, Z.; Li, Y.; Pang, L.; Amirkhanian, S.; Riara, M. Evaluation of aging resistance of graphene oxide modified asphalt. *Appl. Sci.* **2017**, *7*, 702. [CrossRef]

44. Hou, Y.; Wang, L.; Yue, P.; Sun, W. Fracture failure in crack interaction of asphalt binder by using a phase-field approach. *Mater. Struct.* **2015**, *48*, 2997–3008. [CrossRef]

45. Kim, B.; Roque, R.; Birgission, B. Effect of styrene butadiene styrene modifier on cracking resistance of asphalt mixture. *Transp. Res. Rec. J. Transp. Res. Board* **2003**, *1829*, 8–15. [CrossRef]

Article

Improvement of Low-Temperature Performance of Buton Rock Asphalt Composite Modified Asphalt by Adding Styrene-Butadiene Rubber

Xiyan Fan, Weiwei Lu, Songtao Lv * and Fangwei He

National Engineering Laboratory of Highway Maintenance Technology, Changsha University of Science & Technology, Changsha 410114, China
* Correspondence: lst@csust.edu.cn

Received: 1 July 2019; Accepted: 23 July 2019; Published: 24 July 2019

Abstract: To improve the low-temperature performance of the Buton rock asphalt (BRA)-modified asphalt, styrene-butadiene rubber (SBR) was added to it. The BRA-modified asphalt and SBR-BRA composite modified asphalt were prepared by high-speed shearing method. The penetration, softening point, ductility, and Brookfield viscosity of the two kinds of asphalt were measured. The dynamic shear rheometer (DSR) and the beam bending rheometer (BBR) were employed to research the performance of BRA-modified asphalt by adding SBR. The results showed that the pure asphalt in BRA was the main reason to reduce the low-temperature performance of neat asphalt when the content of BRA was 19%. However, the ash in BRA was the main factor to reduce the low-temperature performance when its content was more than 39.8%. When the BRA content was 59.8%, the SBR-BRA composite modified asphalt with SBR contents of 2%, 4%, 6%, and 8%, and it shows that the penetration and ductility of the BRA-modified asphalt are increased by the addition of SBR. The equivalent brittle point was reduced, the stiffness modulus was decreased, and the creep rate was increased. At the same time, the Brookfield viscosity was reduced and the rutting factor was increased. The stiffness modulus of the SBR-BRA composite modified asphalt mixture was increased. That is to say, when SBR was mixed into the BRA-modified asphalt, the low-temperature performance could be remarkably improved based on ensuring high-temperature performance. The low-temperature index of composite modified asphalt was analyzed. It was recommended to apply the equivalent brittle point to evaluate the low-temperature performance of SBR-BRA composite modified asphalt.

Keywords: road engineering; low-temperature performance; BRA-modified asphalt; BRA-SBR composite modified asphalt

---

1. Introduction

By the end of 2018, the total mileage on China's highways was 4.85 million kilometers, including 142,600 km of expressways. Most of the pavement structures are asphalt pavement [1–5]. However, with the growth of the traffic volume, neat asphalt binder is difficult to meet modern transport development demands and needs to be modified. Buton rock asphalt (BRA) is a kind of asphalt with excellent stability formed by the action of different natural environmental factors [6]. It has excellent compatibility with asphalt, low production cost, and convenient transportation [7]. Therefore, rock asphalt as a modifier to modify neat asphalt has prominent advantages [8,9]. Engineering practice shows that the application of rock asphalt-modified asphalt can improve the rutting and other diseases caused by an overload on the road surface, improve the service performance of the road surface, and extend the service life of it. It is also reported that the added BRA can enhance the skid resistance of the asphalt pavement [10]. It is indeed an excellent road asphalt modifier [11,12]. Other researches are shown that BRA-modified asphalt has excellent high-temperature stability and anti-aging performance, but its addition to neat asphalt destroys the low-temperature cracking resistance [13–15]. To prevent low-temperature shrinkage cracks and improve the service capacity of the highway, it is necessary to seek other kinds of modifiers to modify the neat asphalt [16], so that the composite-modified asphalt can improve the high-temperature performance as well as the low-temperature performance.

SBR is a modifier for polymer-modified asphalt. Because of its good compatibility with asphalt and its rich content of polycyclic aromatic hydrocarbons, it can significantly improve the low-temperature performance of neat asphalt. Besides, SBR modification technology is mature and has been widely used in asphalt modification [17–19]. The results have shown that SBR is a linear polymer material with high molecular weight (100,000–1.5 million). SBR will increase the average molecular weight of modified asphalt so that the modified asphalt forms a mosaic structure with a large surface area and has high surface energy. When the temperature decreases, SBR particles can play a role in toughening and plasticizing, offset part of the load effect, and hinder the further expansion of micro-cracks [20,21]. It has also been shown that SBR could reduce the hardening of asphalt during oxidative aging [22–25]. Therefore, the performance of SBR-modified asphalt is relatively good, and especially at a lower temperature, can show good flexibility, ductility, and crack resistance. Previous studies have shown that the comprehensive performance of BRA–SBR composite modified asphalt was better than styrene-butadiene-styrene (SBS)-modified asphalt and SBR-modified asphalt [26,27]. The study [28] has also shown that as the amount of BRA increases, the adhesion of the composite-modified asphalt to the aggregate and the Brookfield viscosity increased.

Although many studies have shown that the low-temperature performance of BRA-modified asphalt deteriorated, most of the studies have employed mechanism analysis [29,30]. It has not been found which component of BRA has a negative influence on the low-temperature performance of neat asphalt. Moreover, the studies have only shown that SBR could improve the low-temperature properties of neat asphalt [31], but its performance as a composite modifier currently lacks systematic research. Aiming at these issues, this study first characterizes the asphalt modified with BRA and BRA-ash, respectively. The performance between the BRA- and BRA-ash-modified asphalt binder was compared to determine whether the BRA-asphalt or BRA-ash content play a major role in asphalt modification. Then, based on the characterization results, the low-temperature performance of the BRA-modified asphalt was further improved with the added SBR content. The sensitivity of the low-temperature index to the content of SBR was analyzed.

2. Materials Preparation and Test Method

2.1. Materials Preparation

The material used in this paper is AH-70# neat asphalt, produced from Indonesia's Buton rock asphalt, SBR latex. Technical indicators are shown in Tables 1–3. The results show that the technical indicators of all raw materials are in line with the norms.

Table 1. Technical properties of 70# neat asphalt.

Technical Indicators	Industry Standard	Test Results
Penetration (25 °C,100 g, 5 s) (0.1 mm)	60–80	68.2
Ductility (5 cm/min, 15 °C) (cm)	≥100	>100
Softening point (°C)	≥46	49.1
Density (15 °C) (g/cm^{-3})	-	1.03

Table 2. Main technical indexes of Buton rock asphalt.

Technical Indicators		Industry Standard	Test Results
Appearance		Brown Powder	Brown Powder
Ash content (%)		<75	74.14
Solubility (%)		>25	25.86
Water content (%)		<2	0.97
Particle size range (%)	4.75 mm	100	100
	2.36 mm	90–100	100
	0.6 mm	10–60	100

Table 3. Technical performance of styrene-butadiene rubber (SBR) latex.

Property	Test Result	Specification of Experimental Methods
Appearance	White Latex	-
Molecular weight	50,000	GB/T12005.10-1992
Mooney viscosity (mPa·s)	4000	GB/T1231.1

Based on the research of other scholars [32–34], the external blending method was used to determine that the blending amount of BRA was 19%, 39%, 58%, 77%, and 97%. This reflects the mass ratio of BRA to modified asphalt. According to the principle that the ratio of ash to pure asphalt is the same, the amount of Buton rock ash mortar is shown in Table 4. For example, when the amount of BRA is 19%, pure asphalt accounts for 5% and ash accounts for 14% in the BRA-modified asphalt. It means that the percentage of pure asphalt is the same in BRA-modified asphalt and BRA ash mortar. The percentage of ash is the same in BRA-modified asphalt and BRA ash mortar, as shown in Figure 1. The amount of SBR is also determined according to the research of other scholars [35].

Table 4. Buton rock asphalt (BRA) and BRA ash content comparison table.

BRA Content	BRA Ash Content
0.19	0.14
0.39	0.29
0.58	0.43
0.77	0.57
0.97	0.72

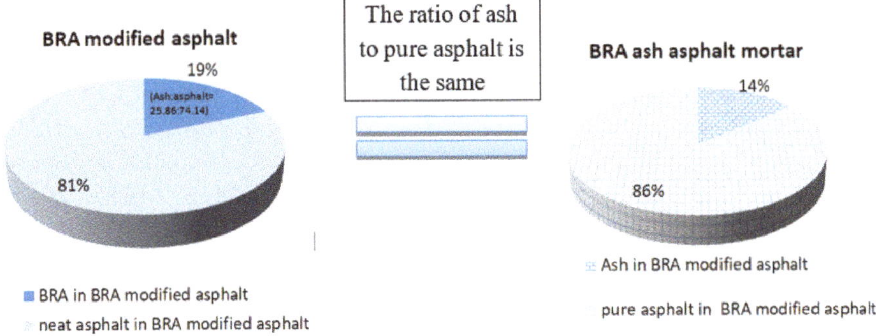

Figure 1. The ratio of ash to pure asphalt is the same.

Studies [36] have shown that when the BRA dosage range is between 40% and 80% (mass ratio), the BRA-modified asphalt has the same rutting resistance as the SBS-modified asphalt and has excellent high-temperature performance. However, according to the BBR test data of BRA-modified asphalt in this paper, 58% of the amount of BRA-modified asphalt reached the limit of low-temperature performance of asphalt at −6 °C. BRA content of 58% of was selected to prepare BRA-SBR composite modified asphalt.

The high-speed shear and induction cooker were used to heat the AH-70# neat asphalt binder to 140–145 °C before mixing of modified asphalt binder. The BRA with different proportions was added to the asphalt binder. The high-speed shear was turned for 30 min to mix the BRA and neat asphalt when the temperature was controlled at 165 °C, and then different amounts of SBR were added. The BRA-SBR composite modified asphalt was prepared [37]. The flow chart is shown in Figure 2.

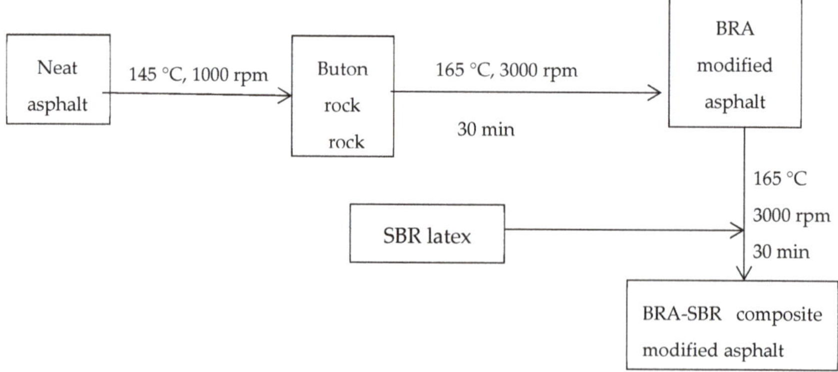

Figure 2. Flow chart for preparing modified asphalt.

The Buton rock asphalt ash is obtained by burning the Buton rock asphalt at a high temperature. Its main component is $CaCO_3$, and its decomposition temperature is 825–896.6 °C, and the melting point is 1339 °C. To ensure that the microstructure of the Buton rock asphalt is not resolved, the muffle furnace's combustion temperature is set to 482 °C to burn the BRA. The BRA ash mortar is prepared in a similar manner to the BRA-modified asphalt.

2.2. Test Method

The penetration test is a commonly used method for determining the consistency of asphalt. Penetration tests of 3 temperatures were carried out, and the penetration index (PI) and the equivalent brittle point ($T_{1.2}$) were calculated. The penetration index is an indicator of the temperature sensitivity of asphalt. $T_{1.2}$ means the corresponding temperature at which the penetration of the asphalt is 1.2. It reflects the low-temperature properties of asphalt.

The ductility test was carried out at 10 °C. The relation curve of load and ductility of asphalt can be obtained from the force ductility test, and the area enclosed by the curve and X-axis is usually called the abruption power. The abruption power (A) represents the work required by the external force in the process of stretching to the breaking of asphalt. This index takes into account the deformation and tension in the whole test process, and can better evaluate the viscosity and toughness of asphalt at low-temperature compared with the ductility. The ratio of the ductility to the tension is taken as the compliance in extension, and the value is used to measure the low-temperature viscosity and toughness of asphalt, which can better reflect the low-temperature performance of asphalt [38].

The low-temperature bending beam rheology (BBR) test was specified by Superpave as a test to evaluate the low-temperature properties of asphalt. Strategic Highway Research Program (SHRP) believes that the cracking of the road surface is related to the stiffness of the asphalt mixture at 7200 s. If it is less than or equal to 200 MPa, the cracking is small. It is difficult to control the temperature to be stable when the loading time is 7200 s. According to the time–temperature equivalent principle, the creep stiffness of 7200 s is equivalent to the test result of the BBR test for 60 s. The test results of BBR are expressed as creep stiffness and creep rate at 60 s.

In order to further observe the improvement of the low-temperature performance of BRA-modified asphalt by SBR, the low-temperature creep bending test was carried out in this paper. The composite modified asphalt was prepared by a BRA content of 58% and an SBR content of 5%. The failure strain index of the low-temperature bending test was used to evaluate the low-temperature performance of the modified asphalt mixture. The bending strength and the bending strain at the time of fracture of the test piece are used to calculate the stiffness modulus at the time of failure of the test piece [39]. The test conditions are shown in Table 5.

Table 5. Test conditions of low-temperature creep bending test.

Test Conditions	Specimen Size	Temperature	Loading Frequency
	250 mm × 30 mm × 35 mm	−10 °C	50 mm/min

To make the study more complete, the high-temperature performance of the BRA-SBR composite modified asphalt was also observed through the Brookfield rotary viscosity test and the DSR (dynamic shear rheometer) test. The Brookfield viscosity test temperatures were 135 °C, 155 °C, and 175 °C. The speed is set to 10 r/min. The initial recording temperature of the asphalt dynamic shear rheological test is 46 °C, which is recorded every 6 °C to obtain the complex modulus, phase angle, and rutting factor of the modified asphalt.

3. Test Result

3.1. Penetration Test

The penetration at 15 °C was related to the low-temperature performance of asphalt. The higher the penetration at 15 °C, the better the low-temperature performance of asphalt [40]. Table 6 can conclude that when BRA content was 19%, 39%, 58%, 77%, and 97%, compared with the neat asphalt, the penetration of BRA-modified asphalt (15 °C) was reduced by 18.3%, 25.5%, 37.8%, 45.4%, and 55.4%. As the temperature increased, the penetration increased. However, as the BRA dosage increases, the amplitude decreases. BRA is a granule. Its addition can reduce the rheological properties of asphalt. At the same time, the equivalent brittle point $T_{1.2}$ was increased by 0.75 °C, 1.51 °C, 2.93 °C, 3.61 °C, and 4.42 °C. It indicates that the low-temperature crack resistance of BRA-modified asphalt decreases, that is, the hardness of BRA-modified asphalt increases at a low temperature, and brittle fracture is likely to occur when the BRA-modified asphalt is stressed at a low temperature [41]. As the content of BRA increased from 0% to 97%, the penetration index of BRA-modified asphalt increased from −0.602 to 0.346, indicating that the thermal sensitivity of BRA-modified asphalt was improved.

Table 6. Test results of BRA-modified asphalt.

Property		Unit	BRA (%)					
			0	19	39	58	77	97
Penetration	15 °C	0.1 mm	25.1	20.5	18.7	15.6	13.7	11.2
	25 °C		68.2	53.7	46.9	39.0	33.9	28.2
	30 °C		114.4	87.5	79.4	64.5	54.5	41.1
PI			−0.602	−0.321	−0.258	−0.149	0.019	0.346
$T_{1.2}$		°C	−15.09	−14.34	−13.58	−12.16	−11.48	−10.67

Table 7 can conclude that when the ash content of BRA was 14%, 29%, 43%,57%, and 72%, the penetration (15 °C) of the BRA ash asphalt mortar was reduced by 4%, 12%, 20%, 28%, and 36%;

moreover, the equivalent brittle point $T_{1.2}$ increased by 0.21 °C, 0.92 °C, 1.26 °C, 1.85 °C, and 2.33 °C compared to the neat asphalt. It is shown that with the increase of BRA ash content, the hardness of BRA ash asphalt mortar increased at low temperatures, and the low-temperature crack resistance decreased. As the BRA ash content increased from 0% to 72%, the penetration index of BRA ash asphalt mortar increased from −0.662 to −0.174, indicating that the temperature sensitivity of the BRA ash asphalt mortar was improved.

Table 7. Penetration test results of the BRA ash asphalt mortar.

Property		Unit	The Content of BRA Ash (%)					
			0	14	29	43	57	72
Penetration	15 °C	0.1 mm	25.1	23.9	21.7	19.8	18.2	16.8
	25 °C		68.2	63.9	58.2	51.5	46.8	40.8
	30 °C		114.4	107.5	96.2	85.3	77.3	70.4
PI			−0.602	−0.539	−0.488	−0.35	−0.284	−0.174
$T_{1.2}$		°C	−15.09	−14.88	−14.17	−13.83	−13.24	−12.76

From Tables 8 and 9, it can be found that the addition of BRA and BRA ash reduced the penetration of neat asphalt. When the amount was small, pure asphalt in BRA played a major role. When the amount was high, ash in BRA was the main one. It can be seen from Table 9 that the addition of BRA and BRA ash increased the equivalent brittle point of the neat asphalt. When the dosage was low, pure asphalt in BRA played a significant role. When the dosage was high, ash in BRA played the main character. That is, as the amount of BRA is increased, the weakening effect of ash in BRA on the low-temperature performance of neat asphalt is more pronounced.

Table 8. Percentage of penetration of different components in BRA.

BRA Content (%)	BRA Ash Content (%)	Penetration (15 °C)		The Proportion of Ash and Pure Asphalt in the Difference between the Penetration of BRA Modified Asphalt and Neat Asphalt	
		BRA Modified Asphalt	BRA Ash Asphalt Mortar	Ash (%)	Pure Asphalt (%)
0	0	25.1	25.1	0	0
19	14	20.5	22.5	57	43
39	29	18.7	21.7	53	47
58	43	15.6	19.8	56	44
77	57	13.7	18.2	61	39
97	72	11.2	16.8	60	40

As can be seen from Table 10, with the increase of SBR, the penetration and penetration index of BRA-SBR composite modified asphalt increased. The greater the penetration, the softer the asphalt. The greater the penetration index, the lower the temperature sensitivity of the asphalt. It reflects that the addition of SBR improved the low-temperature performance of the BRA-SBR composite modified asphalt. When the SBR content was 2%, 4%, 5%, 6%, and 8%, the equivalent brittle point of the BRA-SBR composite modified asphalt as 2.42 °C, 3.19 °C, 4.35 °C, 5.96 °C, and 7.99 °C lower than that of the BRA-modified asphalt. The equivalent brittle point ($T_{1.2}$) is a low-temperature indicator. The smaller it is, the better the low temperature crack resistance of the composite-modified asphalt. This shows that the BRA–SBR composite modified asphalt has the best low-temperature performance, followed by the neat asphalt and the BRA-modified asphalt. With the increase of SBR, the equivalent brittle point of BRA-SBR composite modified asphalt increased first and then decreased.

Table 9. Percentage of equivalent brittle point of different components in BRA.

BRA Content (%)	BRA Ash Content (%)	$T_{1.2}$ (°C)		The Proportion of Ash and Pure Asphalt in the Difference between the Equivalent Brittle Point of BRA Modified Asphalt and Neat Asphalt	
		BRA Modified Asphalt	BRA Ash Asphalt Mortar	Ash (%)	Pure Asphalt (%)
0	0	−15.09	−15.09	0	0
19	14	−14.34	−14.88	28	72
39	29	−13.58	−14.57	66	34
58	43	−12.16	−13.83	43	57
77	57	−11.48	−13.64	40	60
97	72	−10.67	−12.98	48	52

Table 10. Penetration of BRA-SBR composite modified asphalt.

Property		Unit	The Content of SBR (%) (The Content of BRA is 58%)					
			0	2	4	5	6	8
Penetration	15 °C	0.1 mm	15.6	17.6	18.5	20.4	22.5	24.8
	25 °C		39	41.4	43.4	48	57.5	63.8
	30 °C		64.5	69.1	71.9	79	84.3	89.3
PI			−0.149	0.127	0.173	0.185	0.239	0.406
$T_{1.2}$		°C	−12.16	−14.58	−15.35	−16.51	−18.12	−20.15

3.2. Force Ductility Test

It can be seen from Table 11 and Figure 3 that the 10 °C ductility of the BRA-modified asphalt gradually decreased as the amount of BRA increased. After blending 19% and 39% of BRA, the variation of BRA-modified asphalt and neat asphalt is the same. The tensile force increased first and then gradually decreased to zero. The specimen did not suddenly break, which indicates that the material also has toughness and tenacity. However, the peak tensile strength of BRA-modified asphalt increased, indicating that the viscoelasticity of BRA-modified asphalt increased. When 58%, 77%, and 97% BRA were added, the tensile force of BRA-modified asphalt changed abruptly. Especially when 97% BRA was added, the tensile force reached its peak value and breaks suddenly. Compliance in extension is an elastic constant equal to the ratio of strain to stress. The greater the compliance, the easier it is to deform. As shown in Table 11, the tensile compliance decreased as the BRA content increased. It shows that the low-temperature performance of BRA-modified asphalt was reduced. The larger power was correlated to the higher the energy absorption during stretching. That is, asphalt had good toughness and strong fatigue resistance. As the content of BRA increased, the power decreased and the toughness deteriorated.

The curve of Figure 4 can be seen as a stress-strain curve, and as the ductility increased, the force also increased. When the peak was reached, the ductility continued to increase and the force began to decrease. That is, the first was the elastic deformation process, followed by the plastic deformation process. It can be seen from Table 12 and Figure 4 that the 10 °C ductility of the BRA ash asphalt mortar also exhibited a gradually decreasing change as the BRA content increased. When the ash content of BRA was 14% and 29%, the variation of BRA ash asphalt mortar was consistent with that of neat asphalt. The tensile force decreased gradually to zero with the increase of ductility, which indicates that the material has plastic characteristics. After the incorporation of 43%, 57%, and 72% BRA ash, the curve of the BRA ash asphalt mortar did not decay to 0, which indicates that the material has brittleness characteristics. With the addition of BRA ash, the compliance and power of BRA ash asphalt mortar were also reduced. The reason for the smaller amplitude than BRA-modified asphalt is due to the pure asphalt in BRA. It can improve the rheological properties of asphalt, making the asphalt softer, resulting in poor asphalt low-temperature performance.

Table 11. Force ductility test result of BRA-modified asphalt.

BRA Content (%)	Ductility (cm)	F_{Max} (N)	Compliance in Extension	A (J)
0	24.34	52.48	0.464	1265.708
19	7.05	84.56	0.083	269.8131
39	6.51	105.74	0.062	247.9926
58	5.14	119.96	0.043	170.0564
77	3.07	177.5	0.017	73.96904
97	0.60	298.56	0.002	1.63564

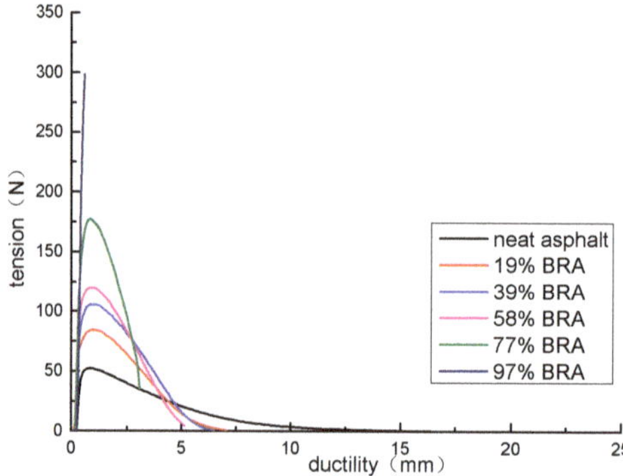

Figure 3. Relationship between tensile force and ductility of BRA-modified asphalt.

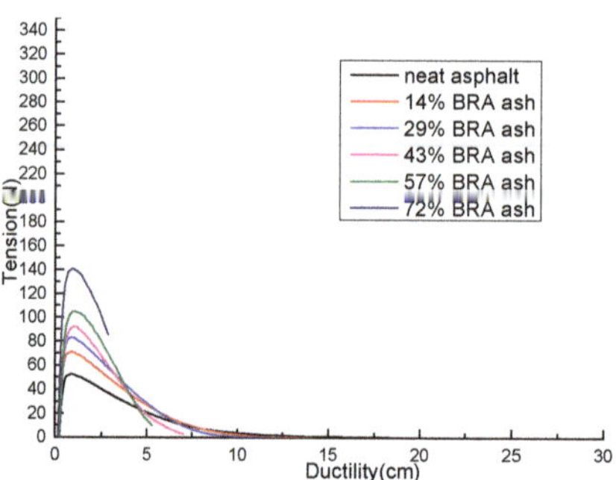

Figure 4. Comparison between tensile force and ductility of asphalt mortar with different BRA ash content.

Table 12. Force ductility test result of BRA ash asphalt mortar.

BRA Ash Content (%)	Ductility (cm)	F_{Max} (N)	Compliance in Extension	A (J)
0	24.34	52.48	0.464	1265.708
14	14.09	71.36	0.197	685.8082
29	12.28	83.3	0.147	553.6637
43	7.02	92.42	0.076	263.8006
57	5.28	105.12	0.05	185.625
72	2.85	140.8	0.02	73.196

Based on the above analysis, the BRA-modified asphalt contains a large number of irregular BRA particles, which quickly causes stress concentration inside the modified asphalt. As shown in Table 13, when the ductility test was carried out, BRA caused the modified asphalt to decrease in ductility, which is the central role of BRA ash. Besides, the pure asphalt in BRA improved the cohesive properties of the asphalt, resulting in an increase in the tensile strength of the BRA-modified asphalt and a decrease in flexibility. Under the combined effect of these two factors, the low-temperature performance of BRA-modified asphalt was inferior to that of neat asphalt.

Table 13. Percentage of ductility of different components in BRA.

BRA Content (%)	BRA Ash Content (%)	Ductility (cm)		The Proportion of Ash and Pure Asphalt in the Difference between the Ductility of BRA Modified Asphalt and Neat Asphalt	
		BRA Modified Asphalt	BRA Ash Asphalt Mortar	Ash (%)	Pure Asphalt (%)
0	0	24.34	24.34	0	0
19	14	7.05	14.09	59	41
39	29	6.51	12.28	68	32
58	43	5.14	7.02	90	10
77	57	3.07	5.28	90	10
97	72	0.60	2.85	91	9

It can be seen from Table 14 that when SBR output was 2%, 4%, 5%, 6%, and 8%, the ductility increased by 4.59 cm, 8.04 cm, 11.73 cm, 14.31 cm, and 15.47 cm, respectively. The low-temperature performance of asphalt was effectively improved by adding SBR, and the effect was better with the increase of SBR content. With the addition of SBR, the compliance and power of BRA-SBR composite modified asphalt increased. It indicates that the SBR improve the toughness and fatigue resistance of BRA-modified asphalt.

Table 14. Results of 10 °C ductility test of BRA-SBR composite modified asphalt.

SBR Content (%)	Ductility (cm)	F_{Max} (N)	Compliance in Extension	A (J)
0	5.14	119.96	0.043	170.0564
2	9.73	97.48	0.100	212.8536
4	13.18	84.98	0.155	258.7621
5	16.87	76.35	0.221	325.4685
6	19.45	65.12	0.299	512.5346
8	20.61	58.32	0.353	623.851

The ductility test showed that the addition of BRA reduced penetration, compliance, and power. That is, the toughness and deformation ability of the asphalt deteriorated, which resulted in the poor low-temperature performance of the BRA-modified asphalt. Compared with BRA ash asphalt mortar, the magnitude of the deterioration was small, indicating that the presence of pure asphalt tends to cause poor low-temperature performance. SBR is an unsaturated olefin polymer that allows the asphalt to form a more stable colloidal structure. Therefore, the low-temperature performance is improved, as shown in Figure 5.

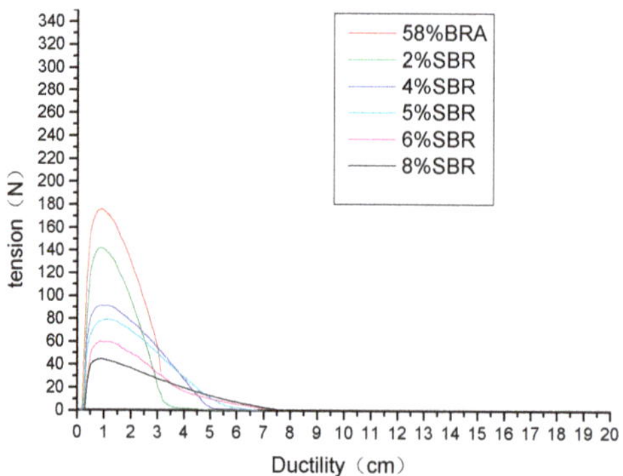

Figure 5. Ductility test of BRA–SBR composite modified asphalt.

3.3. Low-Temperature Bending Beam Rheological Test

Analysis of Tables 15 and 16 can be obtained:

Table 15. Beam bending rheometer (BBR) test results of BRA-modified asphalt.

T (°C)	Test Results	Amount of BRA (%)					
		0	19	39	58	77	97
−6	S (MPa)	74.9	115	186	249	328	437
	m	0.497	0.439	0.346	0.301	0.245	0.215
−12	S (MPa)	161	270	406	522	663	885
	m	0.427	0.379	0.284	0.253	0.215	0.204

Table 16. BBR test results of BRA ash asphalt mortar.

T (°C)	Test Results	Amount of BRA Ash (%)					
		0	14	29	43	57	72
−6	S (MPa)	74.9	94.2	141	179	208	264
	m	0.497	0.461	0.388	0.346	0.340	0.316
−12	S (MPa)	161	208	266	347	485	637
	m	0.427	0.404	0.356	0.306	0.278	0.258

At the same temperature, with the rise of BRA content, the stiffness modulus of BRA-modified asphalt increased, and the creep rate decreased. This shows that under constant load, the deformation of BRA-modified asphalt at the same temperature decreased with the increase of BRA content, and the stress relaxation performance of the material decreased and the low-temperature flexibility decreased.

BRA-modified asphalt could not meet the low-temperature performance requirement of −6 °C after the content of BRA was more than 58%, that is, the ash content was more than 39%. However, in the BRA ash asphalt mortar, the ash content greater than 43% met the low-temperature performance requirements of −6 °C. When the blending amount of BRA exceeded 19%, that is, the ash content was more than 14%, the BRA-modified asphalt could not meet the low-temperature performance requirement of −12 °C. In contrast, when the ash content was greater than 14% in the BRA ash asphalt mortar, the low-temperature performance requirement of −12 °C was satisfied. The reason is that when the blending amount of BRA is low, pure asphalt plays a major role in the modified asphalt, and when the amount of BRA is high, BRA ash plays a major role, as shown in Tables 17 and 18.

Table 17. Percentage of stiffness modulus of different components in BRA at −6 °C.

BRA Content (%)	BRA Ash Content (%)	Creep Rate		The Proportion of Ash and Pure Asphalt in the Difference between the Stiffness Modulus of BRA Modified Asphalt and Neat Asphalt	
		BRA Modified Asphalt	BRA Ash Asphalt Mortar	Ash (%)	Pure Asphalt (%)
0	0	0.497	0.497	0	0
19	14	0.439	0.461	62	38
39	29	0.346	0.388	72	28
58	43	0.301	0.346	77	23
77	57	0.245	0.340	62	38
97	72	0.215	0.316	64	36

Table 18. Percentage of stiffness modulus of different components in BRA at −12 °C.

BRA Content (%)	BRA Ash Content (%)	Creep Rate		The Proportion of Ash and Pure Asphalt in the Difference between the Stiffness Modulus of BRA Modified Asphalt and Neat Asphalt	
		BRA Modified Asphalt	BRA Ash Asphalt Mortar	Ash (%)	Pure Asphalt (%)
0	0	0.427	0.427	0	0
19	14	0.329	0.404	23	77
39	29	0.284	0.356	50	50
58	43	0.253	0.306	70	30
77	57	0.215	0.278	70	30
97	72	0.204	0.258	76	23

It can be seen in Table 19 that when the temperature was −6 °C, the stiffness modulus of the BRA-SBR composite modified asphalt with the SBR parameter of 0% was 1.16 times the parameter of 2%. The creep rate of the composite-modified asphalt with the SBR parameter of 0% was 1.09 times the parameter of 2%. When the temperature was −12 °C, the stiffness modulus of the composite-modified asphalt with the SBR parameter of 0% was 1.18 times the parameter of 2%. The creep rate of the composite-modified asphalt with the SBR parameter of 0% was 1.05 times the parameter of 2%. The creep rate of BRA-SBR composite modified asphalt increased with the increase of SBR content, which indicates that SBR can improve the flexibility of BRA-SBR composite modified asphalt. When the temperature decreased, the effect of SBR improved the low-temperature performance of BRA-SBR composite modified asphalt more obviously.

As the SBR latex increased, the stiffness modulus decreased. That is, the deformation ability of the asphalt at a low temperature increased. The stress caused by the shrinkage strain of the asphalt was small, and the low-temperature crack resistance was excellent. As the SBR content increased, the creep rate increased. This shows that the flexibility of the composite-modified asphalt increased and it was not easy to crack. When the SBR parameter was 0, the stiffness modulus was the largest, and the 43% BRA ash asphalt mortar was the second. The neat asphalt was the smallest. It shows that the addition of BRA caused the low-temperature performance of the neat asphalt to deteriorate, and the addition of SBR improved the low-temperature performance of BRA-modified asphalt.

Table 19. BBR test results of BRA-SBR compound modified asphalt.

Amount of SBR (%)		0	2	4	5	6	8
S (MPa)	−6 °C	249	213	180	142	128	116
	−12 °C	522	436	376	297	265	247
m	−6 °C	0.301	0.327	0.344	0.367	0.388	0.409
	−12 °C	0.253	0.265	0.290	0.327	0.333	0.346
m/S (MP^{-1})	−6	0.001209	0.001535	0.001911	0.002585	0.003031	0.003526
	−12	0.000485	0.000608	0.000771	0.001101	0.001257	0.001401

3.4. The Evaluation Index of BRA-SBR Composite Modified Asphalt under Low-Temperature

In this paper, the low-temperature performance of BRA-SBR composite modified asphalt was evaluated by penetration at 15 °C, equivalent brittle point, ductility at 10 °C, and stiffness modulus. Which index is more suitable to characterize the low-temperature performance of BRA-SBR composite modified asphalt is further studied.

The above studies show that penetration, ductility, creep rate, and equivalent brittleness point can reflect the low-temperature properties of asphalt. As the penetration increased, the asphalt became soft and the low-temperature performance was improved. As the ductility increased, the plastic deformation of the asphalt increased. The greater the creep rate, the stronger the low-temperature deformation ability of the asphalt and the better the low-temperature crack resistance of asphalt.

This paper fits the indicators of low-temperature index and SBR content. In order to analyze which indicator is more sensitive to low temperature performance, sensitivity was analyzed based on the slope. The greater the slope, the more sensitive it is to low-temperature performance. As can be seen from Figures 6–9, the slope of Figure 7 is the largest. With the increase of SBR content, the changing trend of the ductility is the most sensitive. It can be concluded that ductility is the most suitable for evaluating the performance at a low temperature.

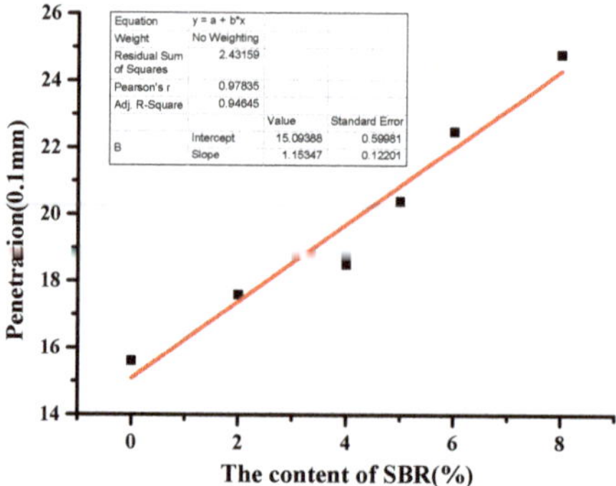

Figure 6. Fitting curves of SBR content and penetration.

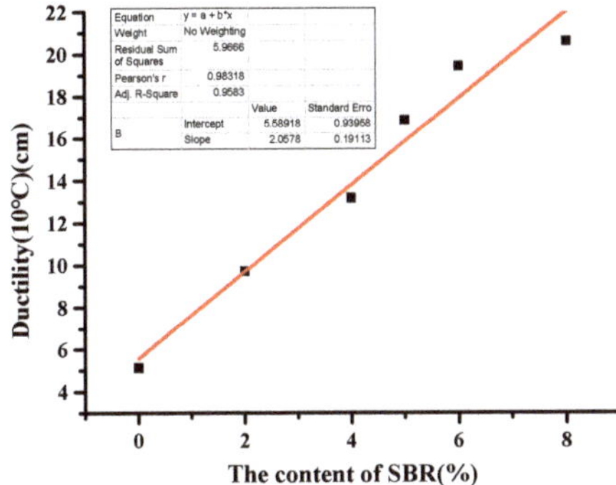

Figure 7. Fitting curves of SBR content and ductility.

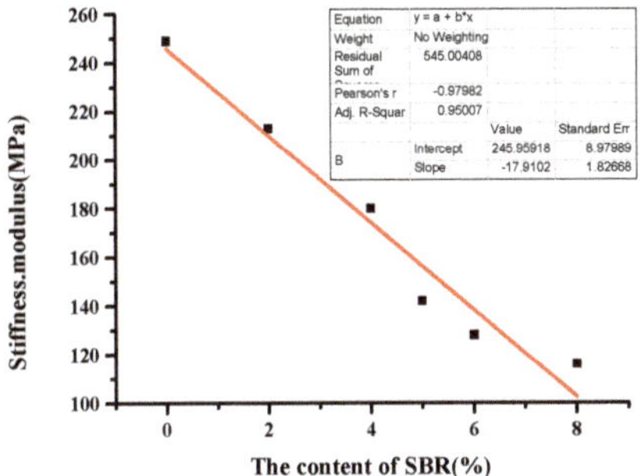

Figure 8. Fitting curves of SBR content and stiffness modulus.

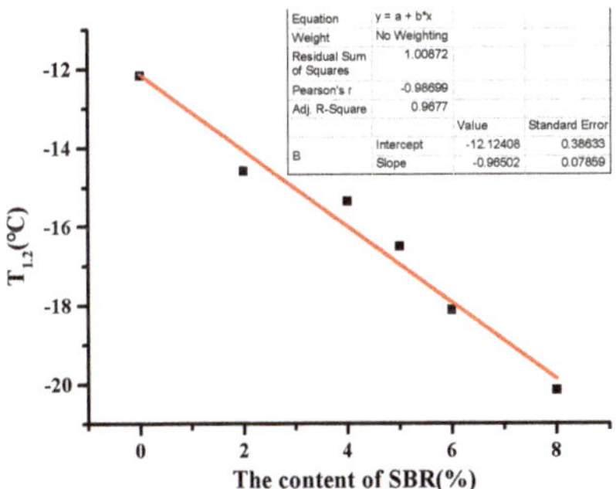

Figure 9. Fitting curves of SBR content and equivalent brittle point.

3.5. Low-Temperature Creep Bending Test of BRA-SBR Composite Modified Asphalt Mixture

The performance indexes of asphalt are shown in Table 20. The dense skeleton type gradation of aggregates was chosen according to the "Specifications for Design of Highway Asphalt Pavement" (Figure 10) [42]. The optimum asphalt ratio was determined using Marshall Tests (Table 21).

Table 20. Performance Index of BRA–SBR composite modified asphalt.

Type of Asphalt	Penetration 25 °C, 100 g, 5 s (0.1 mm)	Softening Point TR&B (°C)	Ductility 10 °C (cm)	Relative Density
neat asphalt	68.2	49.1	24.34	1.029
BRA modified asphalt	39.0	61.4	5.14	1.045
BRA-SBR compound modified asphalt	48.0	63.2	16.87	1.036

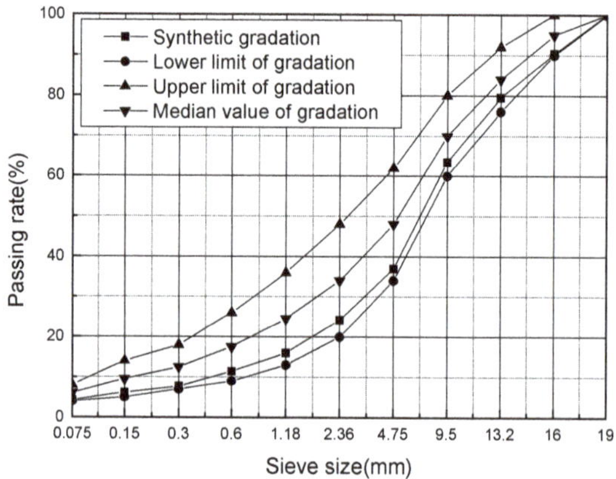

Figure 10. The aggregate gradation of AC-16C.

Table 21. Marshall test results at optimal asphalt content.

	Optimal Asphalt Content (%)	Bulk Specific Gravity (g·cm⁻³)	The Volume of Air Voids VV (%)	Voids Filled with Asphalt VFA (%)	Voids in Mineral Aggregate VMA/%	Marshall Stability (kN)	Flow Value (mm)
neat asphalt mixture	5.2	2.469	4.3	69.3	14	16.21	3.2
BRA modified asphalt mixture	4.7%	7.8	15.0	45.1	16.26	2.8	2.536
BRA-SBR compound modified asphalt mixture	4.7%	6.5	13.7	52.8	16.63	3.1	2.401

The low-temperature creep bending test results are as follows.

As shown in Table 22: Among the three asphalt mixtures, BRA-SBR composite modified asphalt mixture had the highest flexural-tensile failure strength, followed by BRA-modified asphalt mixture and neat asphalt mixture. SBR can improve the ability of BRA-modified asphalt to withstand damage at a low temperature.

The Technical Specification for Construction of Highway Asphalt Pavement (ITGF 40-2004) takes the maximum tensile strain of beams in low-temperature bending test of asphalt mixture as the evaluation index of low-temperature tensile performance of asphalt mixture. The maximum tensile strain of ordinary asphalt mixture was more than 2000 when it was fractured, while that of modified asphalt mixture was more than 2500 when it was fractured. Among the three asphalt mixtures, only BRA-modified asphalt failed to meet the requirements of specifications. This is mainly due to the addition of Buton rock asphalt, which makes the flexural strain smaller, the overall brittleness of asphalt mixtures, and slightly decreases the crack resistance of asphalt mixtures. The low-temperature failure strain of BRA-SBR composite modified asphalt mixture was significantly improved, which indicates that its low-temperature performance was significantly improved.

Table 22. Creep bending test results of BRA–SBR compound modified asphalt mixture.

Property	Specimen	Flexural Tensile Strength (MPa)	Average Value (MPa)	Failure Strain (με)	Average Value (με)	Stiffness Modulus (MPa)
neat asphalt mixture	1	7.12	7.27	2158	2164.3	3823.8
	2	7.17		2238		
	3	7.51		2097		
BRA modified asphalt	1	8.91	8.75	1544	1484.0	5898.5
	2	8.52		1413		
	3	8.83		1495		
BRA-SBR compound modified asphalt mixture	1	9.98	9.49	2675	2692.7	4014.6

In terms of stiffness modulus, the stiffness modulus of neat asphalt mixture was the lowest among the three asphalt mixtures, followed by BRA-SBR composite modified asphalt mixture, and BRA-modified asphalt mixture was the highest. This shows that BRA-modified asphalt caused the stiffness modulus of asphalt mixture to increase significantly, indicating that BRA weakens the low-temperature performance of asphalt mixture, and SBR improves the low-temperature performance of BRA-modified asphalt mixture.

3.6. High-Temperature Performance of BRA-SBR Composite Modified Asphalt

The test results from Table 23 show that the incorporation of SBR can reduce the Brookfield rotary viscosity. Moreover, when the SBR content was more than 5%, the Brookfield rotary viscosity of the BRA-SBR composite modified asphalt was smaller than that of the neat asphalt. That is, the workability of the BRA-SBR composite modified asphalt gradually became better as the amount of the SBR increased.

Table 23. Brookfield rotary viscosity of BRA-SBR compound modified asphalt (Pa·s).

T/°C	Amount of SBR (%)						Neat Asphalt
	0	2	4	5	6	8	
135	0.638	0.625	0.617	0.591	0.562	0.584	0.601
145	0.385	0.357	0.339	0.318	0.297	0.284	0.320
165	0.169	0.143	0.135	0.126	0.117	0.106	0.130
175	0.119	0.112	0.106	0.097	0.085	0.092	0.102

It can be seen from Figure 11 that the Rutting factor decreased with increasing temperature. The rutting factors of BRA-SBR composite modified asphalt are larger than BRA-modified asphalt. It means that the BRA-SBR composite modified asphalt is more elastic. When the content of SBR increased, the elastic of BRA-SBR composite modified asphalt increased. It is potentially because the high temperatures of BRA-SBR composite modified asphalt was improved by SBR additives.

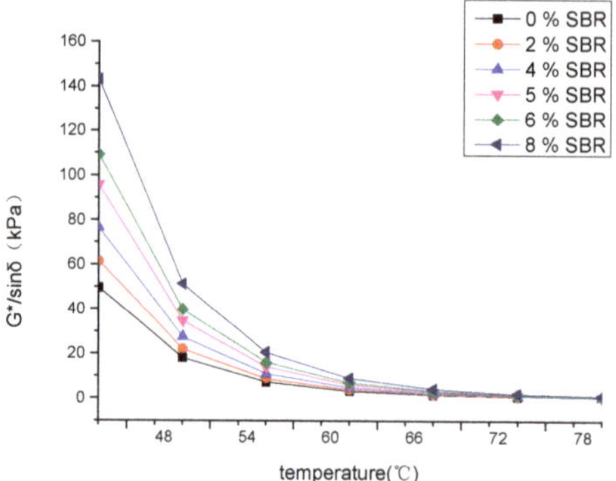

Figure 11. Rutting factor G*/sinδ with different temperatures.

It can be seen from Figure 12 that the phase angles δ of asphalt increased as the temperature increased. At the same temperature, the phase angle δ of BRA-SBR composite modified asphalt became smaller as the content of SBR increases. The phase angle δ of the BRA-SBR composite was smaller than BRA-modified asphalt, which means that the BRA-SBR composite modified asphalt is more elastic. When the content of SBR increased, the elasticity of BRA-SBR composite modified asphalt increased. There is potential that the high temperatures of asphalt binders and mixtures improved by SBR additives. The resistance of asphalt binders and mixtures to deformation is enhanced with improved durability.

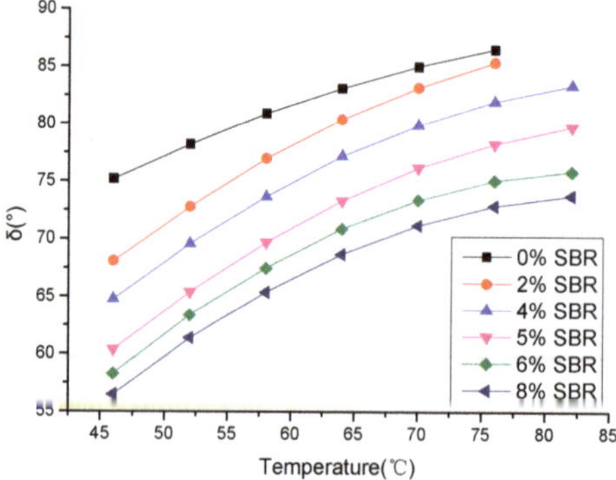

Figure 12. Phase angle δ with different temperatures.

In summary, when SBR is incorporated in the BRA-modified asphalt, the low-temperature performance can be remarkably improved on the basis of ensuring high-temperature performance. The reason is that SBR is an unsaturated olefin polymer, which can be dissolved in most of the solubility parameters and in the hydrocarbon solution close to styrene-butadiene rubber and the glass transition temperature is as low as −50 °C. BRA particles and neat asphalt have excellent compatibility. BRA particles can improve the poor compatibility of SBR with neat asphalt, enabling SBR and BRA particles as well as neat asphalt to form a more stable colloidal structure. When subjected to loads, micro-cracks appear, and SBR particles can play the role of toughening and plasticizing, offset some of the load effects, and hinder the further expansion of micro-cracks. Therefore, SBR can improve the low-temperature performance of BRA-modified asphalt, so that it can exhibit good flexibility, ductility, and crack resistance at lower temperatures.

4. Conclusions

This study aims to fill the knowledge gap on the effect of individual BRA component on the modification of asphalt binder and improve its low-temperature performance based on the obtained correlation. The performance difference between the BRA and BRA-ash modified asphalt was compared to determine the influence of a single component. Then, the SBR content was further applied to improve low-temperature performance of the BRA-modified asphalt. The main conclusion of this study was shown below.

1. The individual modification effect of BRA-binder and BRA-ash content was determined based on the characterization on BRA and BRA-ash modified asphalt, respectively. It was found that the asphalt was mainly affect by the BRA-binder content with a relatively low replacement ratio (within 20%), and the BRA-ash content played a main role in asphalt modification when the replacement ratio was relatively high (larger than 30%).
2. The addition of SBR can improve the low-temperature performance of BRA-modified asphalt. The ultimate failure strain and the failure strength were both enhanced with the added SBR content.
3. The correlation analysis indicated the ductility is more sensitive to the SBR content and hence, the test was recommended to evaluate the low-temperature performance of SBR-modified asphalt.

Author Contributions: Conceptualization, X.F. and S.L.; methodology, S.L. and W.L.; software, X.F. and S.L.; validation, X.F., S.L. and W.L.; formal analysis, F.H. and W.L.; investigation, X.F. and W.L.; resources, S.L.; data curation, X.F., S.L. and F.H.; writing—original draft preparation, X.F., S.L. and W.L.; writing—review and editing, X.F., S.L. and W.L.; visualization, X.F., S.L. and W.L.; supervision, S.L.; project administration, S.L.; funding acquisition, S.L.

Funding: This research was funded by the National Natural Science Foundation of China (51578081, 51608058); The Ministry of Transport Construction Projects of Science and Technology (2015318825120); Key Projects of Hunan Province-Technological Innovation Project in Industry (2016GK2096); The Inner Mongolia Autonomous Region Traffic and Transportation Department Transportation Projects of Science and Technology (NJ-2016-35, HMJSKJ-201801) and The Hunan Province- Transport Construction Projects of Science and Technology (201701). Hunan Graduate Research Innovation Project (CX2018B526).

Conflicts of Interest: The authors declare no conflict of interest.

References

1. Lv, S.; Xia, C.; You, L.; Wang, X.; Li, J.; Zheng, J. Unified fatigue characteristics model for cement-stabilized macadam under various loading modes. *Constr. Build. Mater.* **2019**, *223*, 775–783. [CrossRef]
2. Lv, S.; Liu, C.; Chen, D.; Zheng, J.; You, Z.; You, L. Normalization of fatigue characteristics for asphalt mixtures under different stress states. *Constr. Build. Mater.* **2018**, *177*, 33–42. [CrossRef]
3. Lv, S.; Xia, C.; Liu, H.; You, L.; Qu, F.; Zhong, W.; Yang, Y.; Washko, S. Strength and fatigue performance for cement-treated aggregate base materials. *Int. J. Pavement Eng.* **2019**, *1*, 1–10. [CrossRef]
4. Lv, S.; Yuan, J.; Liu, C.; Wang, J.; Li, J.; Zheng, J. Investigation of the fatigue modulus decay in cement stabilized base material by considering the difference between compressive and tensile modulus. *Constr. Build. Mater.* **2019**, *223*, 491–502. [CrossRef]
5. Lv, S.; Xia, C.; Liu, C.; Zheng, J.; Zhang, F. Fatigue equation for asphalt mixture under low temperature and low loading frequency conditions. *Constr. Build. Mater.* **2019**, *211*, 1085–1093. [CrossRef]
6. Karami, M.; Nega, A.; Mosadegh, A.; Nikraz, H. Evaluation of Permanent Deformation of BRA Modified Asphalt Paving Mixtures Based on Dynamic Creep Test Analysis. *Adv. Eng. Forum* **2016**, *16*, 69–81. [CrossRef]
7. Jing, L.; Guo, X.; Jiang, Y.; Bai, N.; Liu, Y.; Wu, C. A micro-analysis of Buton rock asphalt. *J. Comput. Theor. Nanosci.* **2015**, *12*, 2751–2756. [CrossRef]
8. Zeng, S.; Wang, H.M.; Ling, T.Q. Test on Road Performance of BRA Modified Asphalt and Mixture. *Appl. Mech. Mater.* **2014**, *584–586*, 972–979. [CrossRef]
9. Lv, S.; Fan, X.; Yao, H.; You, L.; You, Z.; Fan, G. Analysis of performance and mechanism of Buton rock asphalt modified asphalt. *Appl. Polym.* **2018**, *136*, 46903. [CrossRef]
10. Hadiwardoyo, S.P.; Sinaga, E.S.; Fikri, H. The influence of Buton asphalt additive on skid resistance based on penetration index and temperature. *Constr. Build. Mater.* **2013**, *42*, 5–10. [CrossRef]

11. Yang, J. Research on Application of Rock Natural Rock Asphalt Modifier in Highway. *Heilongjiang Trans. Sci. Technol.* **2016**, *39*, 46–47.
12. Ma, L. Research and Application of North American Natural Rock Asphalt Modification—Science and Technology Promotion Project of Ministry of Communications. *China Highw.* **2003**, *16*, 56–57.
13. Zamhari, K.A.; Hermadi, M.; Ali, M.H. Comparing the performance of granular and extracted binder from Buton Rock Asphalt. *Int. J. Pavement Res. Technol.* **2014**, *7*, 25–30.
14. Fu, Q. Research on application technology of rock modified asphalt in Leyi Expressway. Master's Thesis, Chongqing Jiaotong University, Chongqing, China, June 2013.
15. Liu, S.; Cao, W.; Li, X.; Li, Z.; Sun, C. Principle analysis of mix design and performance evaluation on Superpave mixture modified with Buton rock asphalt. *Constr. Build. Mater.* **2018**, *176*, 549–555. [CrossRef]
16. Li, R.; Karki, P.; Hao, P.; Bhasin, A. Rheological and low temperature properties of asphalt composites containing rock asphalts. *Constr. Build. Mater.* **2015**, *96*, 47–54. [CrossRef]
17. Kim, W.S.; Yi, J.; Lee, D.H.; Kim, I.J. Effect of 3-aminopropyltriethoxysilane and N, N-dimethyldodecylamine as modifiers of Na+-montmorillonite on SBR/organoclaynanocomposites. *J. Appl. Polym. Sci.* **2010**, *116*, 3373–3387.
18. Yang, S.; Liu, L.; Jia, Z.; Jia, D.; Luo, Y. Structure and mechanical properties of rare-earth complex La-GDTC modified silica/SBR composites. *Polymer* **2011**, *52*, 2701–2710. [CrossRef]
19. Zhang, F.; Yu, J. The research for high-performance SBR compound modified asphalt. *Constr. Build. Mater.* **2010**, *24*, 410–418. [CrossRef]
20. Chen, Z.; Hao, P. Rheological properties and mechanism analysis of SBR modified asphalt with RET. *J. Beijing Univ. Technol.* **2016**, *42*, 1691–1696.
21. Cong, Y.F.; Tang, D.; Huang, W.; Bai, S.F. Preparation and Performance Analysis of New Styrene-Butadiene Rubber Composite Modified Asphalt. *New Chem. Mater.* **2018**, *46*, 268–271.
22. Zhang, B.; Xi, M.; Zhang, D.; Zhang, H.; Zhang, B. The effect of styrene–butadiene–rubber/montmorillonite modification on the characteristics and properties of asphalt. *Constr. Build. Mater.* **2009**, *23*, 3112–3117. [CrossRef]
23. Khabaz, F.; Khare, R. Glass Transition and Molecular Mobility in Styrene-Butadiene Rubber Modified Asphalt. *J. Phys. Chem. B* **2015**, *119*, 14261–14269. [CrossRef] [PubMed]
24. Ming, L.Y.; Feng, C.P.; Siddig, E.A. Effect of phenolic resin on the performance of the styrene-butadiene rubber modified asphalt. *Constr. Build. Mater.* **2018**, *181*, 465–473. [CrossRef]
25. Ameri, M.; Reza Seif, M.; Abbasi, M.; Khavandi Khiavi, A. Viscoelastic fatigue resistance of asphalt binders modified with crumb rubber and styrene butadiene polymer. *Pet. Sci. Technol.* **2017**, *35*, 30–36. [CrossRef]
26. Li, R.; Hao, P.; Wang, C.; Zhang, Q. Modified Mechanism of Buton Rock Asphalt. *J. Highw. Transp. Res. Dev.* **2011**, *12*, 20–24.
27. Tang, B.; Jiang, S.; Liu, N. Preparation Process and Properties of Rock Bitumen-SBR Composite Modified Emulsified Asphalt. *China Foreign Highw.* **2018**, *38*, 285–289.
28. Zhou, L. Research on Technical Performance of BRA and SBR Composite Modified Asphalt and Its Mixture. *J. Highw. Eng.* **2014**, *6*, 277–282.
29. Jiang, X.; Chen, X. Preparation and Properties of Qingchuan Rock Asphalt and SBR Composite Modified Emulsified Asphalt. *J. Hunan City Univ.* **2018**, *27*, 32–35.
30. Du, S.W.; Liu, C.F. Performance Evaluation of High Modulus Asphalt Mixture with Button Rock Asphalt. *Adv. Mater. Res.* **2012**, *549*, 558–562. [CrossRef]
31. Zhang, J.; Wang, J.; Wu, Y.; Wang, Y.; Wang, Y. Preparation and properties of organic palygorskite SBR/organic palygorskite compound and asphalt modified with the compound. *Constr. Build. Mater.* **2008**, *22*, 1820–1830. [CrossRef]
32. Wang, M.; Lin, F.; Liu, L. Fatigue properties of rock asphalt modified asphalt based on simplified energy dissipation rate. *J. Build. Mater.* **2015**, *18*, 1024–1027.
33. Huang, W.-T.; Xu, G.-Y. Experimental Investigation into Pavement Performance of Buton Rock Asphalt Mixtures. *J. South China Univ. Technol.* **2012**, *208*, 122–135.
34. Yin, Y.; Zhang, X. Research on High-temperature Rheological Characteristics of Asphalt Mastics with Indonesian Buton Rock Asphalt (BRA). *J. Wuhan Univ. Technol.* **2010**, *32*, 85–89.

35. Zhang, Q.; Fan, W.; Wang, T.; Nan, G. Studies on the temperature performance of SBR modified asphalt emulsion. In Proceedings of the International Conference on Electric Technology & Civil Engineering, Lushan, China, 22–24 April 2011.
36. Wen, L.; Wang, X.J.; Liu, H.; Wu, H.X.; Cui, B.L.; Cui, X.Y. Performance Evaluation of BRA Modified Asphalt Mastic Based on Analysis of Microstructure and Rheological Mechanics Property. *Adv. Mater. Res.* **2011**, *243*, 4119–4124. [CrossRef]
37. Kang, A.H.; Zhang, W.H.; Sun, L.J. Preparation Method of Modified Asphalt Fluorescence Optical Microscopy Sample. *J. Sichuan Univ. Eng. Sci.* **2012**, *44*, 154–158.
38. Li, X.; Han, S.; Li, Y.; Li, B. Experimental study on low-temperature crack resistance index of asphalt binders. *J. Wuhan Univ. Technol.* **2010**, *32*, 81–84.
39. Li, J.; Zhang, J.; Qian, G.; Yao, Y.; Zheng, J. Three-dimensional simulation of aggregate and asphalt mixture using parameterized shape and size gradation. *J. Mater. Civil Eng.* **2019**, *31*, 04019004. [CrossRef]
40. Wang, Z.; Zhang, S.; Li, R.; Long, J. Study on Temperature Sensitivity of Domestic AH-90 Heavy Traffic Road Asphalt. *Petrol. Asphalt* **2002**, *16*, 6–9.
41. Fan, X.; Lv, S. Comparison of Road Performance of Budun Rock Asphalt Modified Asphalt Mixture. *J. Transp. Sci. Eng.* **2016**, *32*, 22–29.
42. Yue, X.J.; Huang, X.M.; Li, W.L.; Liu, J.H.; Li, Z.D. Research on Force Ductility Test and Toughness Ratio Evaluation Index. *J. Highw. Transp. Res. Dev.* **2007**, *24*, 33–36.

 *materials*

Article

Chemical and Rheological Evaluation of Aged Lignin-Modified Bitumen

Yi Zhang [1,2,*], **Xueyan Liu** [2,*], **Panos Apostolidis** [2], **Wolfgang Gard** [2], **Martin van de Ven** [2], **Sandra Erkens** [2] and **Ruxin Jing** [2]

1. School of Highway, Chang'an University, Xi'an 710064, China
2. Faculty of Civil Engineering and Geosciences, Delft University of Technology, Stevinweg 1, 2628 CN Delft, The Netherlands; P.Apostolidis@tudelft.nl (P.A.); W.F.Gard@tudelft.nl (W.G.); M.F.C.vandeVen@tudelft.nl (M.v.d.V.); S.M.J.G.Erkens@tudelft.nl (S.E.); R.Jing@tudelft.nl (R.J.)
* Correspondence: yizhang@chd.edu.cn (Y.Z.); X.Liu@tudelft.nl (X.L.); Tel.: +31-152787918 (X.L.)

Received: 24 October 2019; Accepted: 9 December 2019; Published: 12 December 2019

Abstract: As bitumen oxidizes, material stiffening and embrittlement occur, and bitumen eventually cracks. The use of anti-oxidants, such as lignin, could be used to delay oxidative aging and to extend the lifetime of asphalt pavements. In this study, the chemical and rheological effect of lignin on bitumen was evaluated by using a single dosage organsolv lignin (10 wt.% dosage). A pressure aging vessel (PAV) was used to simulate the long-term aging process after performing the standard short-term aging procedure, and the lignin-modified bituminous binders were characterized by an environmental scanning electron microscope (ESEM), Fourier-transform infrared (FTIR) spectroscopy, and a dynamic shear rheometer (DSR). From the ESEM results, the uniform microstructure was observed, indicating that the addition of lignin did not affect the worm structure of bitumen. Based on the FTIR test results, lignin-modified bitumen showed that a lower number of carbonyl and sulfoxide compounds were generated after aging than for neat bitumen. Based on the linear amplitude sweep (LAS) results, the addition of lignin slightly reduced the fatigue life of bitumen. From the frequency sweep results, it showed that lignin in bitumen acts as a modifier since the physical interaction between lignin and bitumen predominantly affects the material rheology. Overall, lignin could be a promising anti-oxidant due to its economic and environmental benefits.

Keywords: lignin; bitumen; aging; microstructure; chemistry; rheology

---

1. Introduction

Bitumen is a hydrocarbon residue produced from oil refining and comprises a plethora of different organic molecules causing its vulnerability to environmental conditions [1]. Currently, the rising cost of bitumen, together with the fact that asphalt production is one of the largest energy consumers, globally encourages the use of alternative systems to replace petroleum-based binders to enhance the quality of pavement materials. Therefore, the environmental concerns and the demand for developing long-lasting pavements drive the asphalt industry to assess the possible use of bio-based artificial made binders [2,3] or waste and easily available polymers in bitumen.

Lignin is the most abundant bio-based polymer that can be found in co-products of the wood industry making up about 20% to 25% of the dry mass of every plant [4]. The total amount of lignin present in the biosphere exceeds 300 billion tons and increases by approximately 20 billion tons every year [5]. Specifically, lignin is a type of complex organic polymers that contributes to forming the cell walls in plants. In addition, bitumen is composed of millions of different organic molecules, the utilization of lignin may be used to substitute partially petroleum-based binders that assist toward more sustainable development in the bitumen industry. Therefore, the utilization of lignin in bitumen, specially designed for pavements, attracted considerable attention in recent years [6–17].

One of the reasons that asphalt pavement cracks is the stiffening and embrittlement of the bitumen due to aging [1]. Based on previous studies, lignin shows oxidative aging resistance because of its radical scavenging activity and polyphenolic structure [18–20]. Recent research focused on the addition of lignin as a type of anti-oxidant to bitumen [7–16]; however, it is important to verify whether the lignin reacts with bitumen to improve the aging resistance of bitumen. Moreover, the microstructure of lignin–bitumen binders is not yet clear. Thus, an in-depth understanding of the effect of lignin on bitumen will help to optimize the technology of bio-based binders, leading to more environmentally friendly pavement materials. In this research, special emphasis was given on assessing the impact of wood lignin powder extracted by the organosolv method on the chemistry and rheology of bitumen after aging.

2. Objective and Approach

The objective of this study was to evaluate the compositional and rheological changes of lignin-modified bitumen due to aging. This study consists of three parts. In the first part, the microstructure of different materials was measured by an environmental scanning electron microscope (ESEM). In the second part, the chemical components of lignin and bitumen were characterized by Fourier-transform infrared (FTIR) spectroscopy, and, finally, the mechanical properties of various lignin-bitumen combinations were studied using a dynamic shear rheometer (DSR). Frequency sweep, linear amplitude sweep, and relaxation tests performed in DSR. A pressure aging vessel (PAV) was used to simulate the aging of bitumen.

Overall, this article was designed to achieve the following aims:

1. Examine the effect of lignin on bitumen performance. Also, few studies focused on the aging of lignin and, thus, the aged lignin was evaluated using ESEM and FTIR spectroscopy.
2. Assess the effect of aging on lignin-modified bitumen. ESEM, FTIR, and DSR tests were conducted on samples in order to explore the microstructural, chemical, compositional, and rheological changes of new binders.

3. Materials and Methods

3.1. Materials and Samples Preparation

A 70/100 pengrade bitumen was used in this study. The softening point of this bitumen was 47.5 °C. The wood lignin was a brown powder obtained from Chemical Point UG (Oberhaching, Germany). After extraction by the organosolv method, a content of 88% lignin was obtained. The density of the lignin was 1.0774 g/cm^3, which was measured by a helium pycnometer test, and the specific surface area was 147.0593 m^2/g, which was measured by a surface analyzing system (DVS Resolution). The physical properties of the lignin were measured after aging as well. The overall color of lignin particles became darker, the density was increased to 1.5029 g/cm^3, and the specific surface area was decreased to 65.0475 m^2/g. As mentioned in Reference [8], 10 wt.% lignin was added by substituting an equivalent amount of bitumen. An overview of the studied materials is provided in Table 1.

Table 1. Studied materials.

Studied Materials	Modification by Bitumen Mass	Explanation
Bref_F	0%	Fresh neat bitumen, as reference
Bref_A		Aged neat bitumen
BL10_F	10%	Fresh bitumen mixed with fresh lignin
BL10_A		Fresh bitumen mixed with fresh lignin, then aging
B(A) + L(F)		Aged bitumen mixed with fresh lignin
B(A) + L(A)		Bitumen and lignin aged separately, then mixing

The mixing time and temperature of lignin in bitumen were determined and described elsewhere [10]. Lignin (10 wt.% of bitumen) was gradually added to the bitumen, and then the two materials were mixed by a high shear mixing device at 163 °C and a mixing rate of 3000 rpm. The mixing was continued for about 30 min until the bubbles disappeared.

According to the standard testing procedure (ASTM D 6521-19 [21]), PAV was used to simulate the long-term aging process of bitumen and performed after the standard short-term aging procedure. In this study, 50 ± 0.5 g bitumen was poured into a PAV pan to form a film with 3.2-mm thickness. Then, the PAV test was performed at a temperature of 100 °C under pressurized air at 2.10 MPa for 20 h.

3.2. Microstructural Morphology

The microstructural observations were conducted at room temperature with a Philips XL30 environmental scanning electron microscope (ESEM, Eindhoven, Netherlands) under an acceleration voltage of 20 keV, similar to Reference [22], with a spot size of 3.5 in a chamber at 1.0 Torr pressure in low vacuum and secondary electron detector mode. The magnification was varied from ×125 to ×250, ×500, and ×1000. The scanning method was 1.68 ms × 484 lines, and the integration was 16 frames. The time of exposure was gradually increased from 0.5 to 1, 3, 5, and 10 min. As the exposure time increases, the energy absorbed by the surface of the material increases, and the light components in bitumen, such as saturates and aromatics, mostly evaporate; thus, the internal structure is much easier to be observed.

Through the sample preparation for ESEM analyses, lignin and lignin-modified bitumen were placed on special sample holders as in Reference [22]. In particular, lignin was placed in an oven at 150 °C for 24 h to ensure drying before scanning. After attaching a small black sheet with adhesive on both sides of the plate, approximately 20 mg of lignin was poured onto an 8-mm-diameter sample holder (Figure 1a). The plate was tapped and vibrated to prevent powder build-up and to distribute it as evenly as possible on the black sheet for scanning. For the lignin-modified bitumen, the diameter and height of the sample holder cylinder were 9.20 mm and 5.20 mm, respectively (Figure 1b). The bitumen was placed in the oven to become liquid and flow easily on the holder After evenly stirring, a small amount of bitumen was dropped on the cylinder holder, and then the sample was placed back in the oven to set flat and uniformly (Figure 1c). It is important to protect the samples from dust or other impurities before ESEM scanning.

(a)   (b)   (c)

Figure 1. (a,b) Environmental scanning electron microscope (ESEM) samples on the holder, and (c) samples ready to be analyzed in ESEM.

3.3. Chemical Characterization

FTIR is the most commonly used tool to detect the chemical compounds in bitumen [23] and lignin [18,24]. Different functional groups have a different light-absorption spectrum. Wavenumbers of typical bands of lignin and bitumen are listed in Table 2. In this study, attenuated total reflectance (ATR) FTIR was performed to collect spectral data of lignin and bitumen samples. The Spectrum 100 FTIR Perkin Elmer spectrometer with a single-point ATR fixture (Waltham, MA, USA) was used. The wavenumber ranged from 600 to 4000 cm^{-1} with a resolution of 4 cm^{-1}. Before scanning, the lignin samples were dried at 140 °C for 30 min to remove any volatiles from samples. For the bitumen samples, the prism was cleaned with methylene chloride after each scan. Nine replicates per material were analyzed.

Table 2. Main functional groups of lignin ($*^1$) and bitumen ($*^2$) in Fourier-transform infrared (FTIR) spectra.

Wavenumber (cm^{-1})	Function Groups
3500–3100	Stretching vibrations of –OH groups $*^1$
1753–1660	Stretching vibrations of C=O bonds $*^1$
1620–1555	Vibrations of aromatic ring $*^1$
1525–1480	Vibrations of aromatic ring $*^1$
1280–1245	Stretching vibrations of C–O bonds $*^1$
1170–1140	Deformation vibrations of C–H bonds in guaicyl rings $*^1$
1140–1100	Deformation vibrations of C–H bonds in syryngyl rings $*^1$
1095–1070	Deformation vibrations of C–O bonds in secondary alcohols and aliphatic ethers $*^1$
1070–995	Deformation vibrations of C–H bonds in the aromatic rings and C–O bonds in primary alcohols $*^1$
2990–2880	Stretching aromatic $*^2$
2880–2820	Stretching symmetric $*^2$
1753–1660	Oxygenated functional group (carbonyl) $*^2$
1670–1535	Aromatic structures $*^2$
1525–1395	Aliphatic structures $*^2$
1390–1350	Branched aliphatic structures $*^2$
1047–995	Oxygenated functional group (sulfoxide) $*^2$
912–838	Out of singlet $*^2$
838–783	Out of adjacent $*^2$
783–734	Out of adjacent $*^2$
734–710	Long chains $*^2$

The functional group absorbance index (*AI*) was used for the main absorption bands of lignin to compare the changes of functional groups with the changes in spectra, and it was determined as follows:

$$AI = \frac{A_{ab}}{\sum A},$$ (1)

where A_{ab} is the integral area of absorption band ab, and $\sum A$ is the sum of the integral areas of several characteristic functional group peaks. The range of chemical functional groups to be calculated and considered is summarized in Table 2.

Conventional aging indices of bitumen are the carbonyl (C=O) and sulfoxide (S=O) indices [25]. The effect of lignin as an anti-oxidant can be estimated by measuring the changes in the carbonyl and sulfoxide groups. Lignin is a combination of organic substances, and it contains carbonyl groups as well. The question of whether the aging of lignin has an impact on the aging index during the aging process should be strictly verified. Two aging indices were used to evaluate the anti-oxidation effect of lignin in bitumen, based on changes in carbonyl and sulfoxide groups, as follows:

$$I_{C=o} = \frac{A_{S=O}}{\sum A}, I_{S=o} = \frac{A_{S=O}}{\sum A},$$ (2)

where $A_{C=O}$ and $A_{S=O}$ are the integrated areas of carbonyl (C=O) and sulfoxide (S=O) groups, and $\sum A$ is the sum of the integrated areas of several characteristic functional group peaks as summarized in Table 2.

3.4. Rheological Characterization

3.4.1. Frequency Sweep Tests

Based on a standard testing procedure (AASHTO T 315-19 [26]), the complex shear modulus (G*) and phase angle (δ) were obtained over a wide range of temperatures and frequencies by means of DSR with oscillatory loading. In this study, DSR tests were performed using an 8-mm-diameter parallel plate with a 2-mm gap at temperatures from -10 to 30 °C and a 25-mm-diameter geometry with a 1-mm gap at temperatures from 30 to 70 °C (10 °C temperature step). The tests were carried out at a frequency sweep range from 100 to 0.1 rad/s (15.9 to 0.0159 Hz) and a strain load of 0.1%. The master curves of the complex shear modulus and phase angle at a reference temperature of 20 °C were constructed by applying the time–temperature superposition principle (TTSP). The TTSP-based master curves were used to evaluate the effect of lignin on bitumen performance.

3.4.2. Linear Amplitude Sweep Tests

According to the standard testing procedure (AASHTO TP 101-14 [27]), a cyclic loading with linearly increasing strain amplitudes was used in the LAS test to assess the fatigue behavior of different binders [28]. The 8-mm-diameter parallel plates with a 2-mm gap were used in LAS tests. The LAS test consisted of two steps; in the first step, the rheological properties of the sample were tested using a frequency sweep test, which was designed to obtain information about the rheological properties. The frequency sweep test was performed at 20 °C and applied oscillatory shear loading at constant amplitude over a range of loading frequencies, employing an applied load of 0.1% strain over a range of frequencies from 0.2–30 Hz, whereby data were sampled at the following 12 typical frequencies: 0.2, 0.4, 0.6, 0.8, 1.0, 2.0, 4.0, 6.0, 8.0, 10, 20, and 30 Hz. After that, the samples were tested by applying a strain sweep, in which the frequency was 10 Hz. Bitumen is more likely to experience cracking failure under cyclic loading in DSR at low values of intermediate temperature rather than at higher testing temperatures [29]. Bitumen is soft, and its response is dictated by instability flow at high temperatures. Thus, a temperature of 20 °C was chosen to perform LAS tests to evaluate the fatigue performance of studied materials. At the selected temperature, continuous oscillatory strain-controlled cycles with linearly increasing strain amplitudes from 0% to 30% were applied to accelerate the fatigue damage of bitumen.

The fatigue resistance was then calculated based on the frequency sweep and the amplitude sweep results, as shown in Equations (3)–(6). The damage accumulation, $D(t)$, of the studied binders with testing time, t, can be expressed as follows:

$$D(t) \cong \sum_{t=1}^{N} [\pi \gamma_0^2 (C_{i-1} - C_i)]^{\frac{a}{1+a}} (t_i - t_{i-1})^{\frac{1}{1+a}} \tag{3}$$

where $C(t) = \frac{|G^*|(t)}{|G^*|_{initial}}$, $|G^*|(t)$ is the complex modulus at time t (MPa), $|G^*|_{initial}$ is the initial state value, γ_0 is the applied strain for a given data point (%), $\alpha = m^{-1}$, in which m is the slope of an optimum-fit line in the logarithmic scale plot relating storage modulus to frequency, and i refers to the cycle number.

At any given time, the values of $C(t)$ and $D(t)$ can be obtained by fitting the relationship as follows:

$$C(t) = C_0 - C_1 D(t)^{C_2}, \tag{4}$$

where $C_0 = 1$, and C_1 and C_2 are curve-fitting coefficients.

The damage values at failure correspond to the peak stress as follows:

$$D_f = \left(\frac{C_0 - C_{at\ Peak\ Stress}}{C_1} \right)^{\frac{1}{C_2}}. \tag{5}$$

The fatigue parameter (N_f) can be calculated as follows:

$$N_f = A(\gamma_{max})^{-B},\tag{6}$$

where γ_{max} is the expected maximum strain (%), $A = \dfrac{f \times (D_f)^k}{k(\pi C_1 C_2)^\alpha}$, f is the loading frequency (10 Hz), $k = 1 + (1 - C_2)\alpha$, and $B = 2\alpha$.

3.4.3. Relaxation Tests

The stress relaxation demonstrates the ability of a material to relieve stress under a constant strain. The studied material is intended to be applied as the surfacing layer on a pavement structure (upper layer). The relaxation tests were performed in a DSR by using a parallel-plate configuration of 8-mm diameter and 2-mm gap under strain-controlled mode at 0 °C. The tests were conducted as follows: firstly, the strain was increased from 0% to 1% shear strain in 0.1 s, and then the 1% shear strain was kept constant during a relaxation period of 200 s, while the change of shear stress was measured [30]. Longer relaxation times imply that materials are more susceptible to stress accumulation. The relaxation time should be small enough to prevent high stress accumulation in the asphalt pavement, caused by the continuous traffic load. If the stress within the pavement material does not relax sufficiently, the load of the next vehicle would accumulate more stress in the pavement.

3.4.4. Glover–Rowe Parameter Tests

The location in black space diagrams (BSD) at low temperatures is an effective performance indicator to assess the cracking vulnerability of asphalt pavement materials [31]. The initial quality of bitumen as determined in the black space is an important performance indicator that can be successfully applied together with the complex modulus and phase angle to assess the aging effect on bitumen [32]. In addition, the black space diagrams could be useful for comparing the various proposed damage parameters. Based on results of the angle frequency (ω), the complex modulus (G^*), phase angle (δ), and the dynamic viscosity (η'), a damage curve in black space can be built as follows:

$$\frac{G'}{\frac{\eta'}{G'}} = G \times \left(\frac{\cos \delta^2}{\sin \delta}\right)\omega.\tag{7}$$

Given a black space function as defined by the Glover–Rowe (G–R) parameter, an aged sample can be tested to assess the degree of damage without imposing a rigid single test temperature and frequency. Material damage due to aging is initiated when the ductility is below 5 cm, and cracking is serious when ductility reaches 3 cm [33]. The parameters were measured at a temperature of 15 °C and a frequency of 0.005 rad/s. A failure curve in the black space represents the onset of cracking as follows:

$$G \times \left(\frac{\cos \delta^2}{\sin \delta}\right) = 180 \text{ kPa}.\tag{8}$$

Surface cracking is observed when the ductility falls to 3 cm, and the relative value of the Glover–Rowe parameter is represented by

$$G \times \left(\frac{\cos \delta^2}{\sin \delta}\right) = 450 \text{ kPa}.\tag{9}$$

These two equations provide a damage zone in black space diagrams.

4. Results and Discussion

4.1. Microstructural Observation

As shown in Figure 2, lignin contains smaller fractions of particles that seem to be crushed from larger ones. The size of lignin particles ranged from 10 to 200 μm. Moreover, the fresh lignin particles had some angularity, and the surface of them was rough. The specific surface area of fresh (unaged) lignin was two times more than that of aged lignin. Generally, a finer powder results in a more irregular particle shape, a rougher surface of the particle, a more complex particle structure, and a larger area. A larger powder area results in greater friction between the particles. After aging, the density of lignin increases, and its specific surface area decreases. In addition, the density of lignin increases and its surface becomes smoother, because it contacts oxygen at high temperatures during aging.

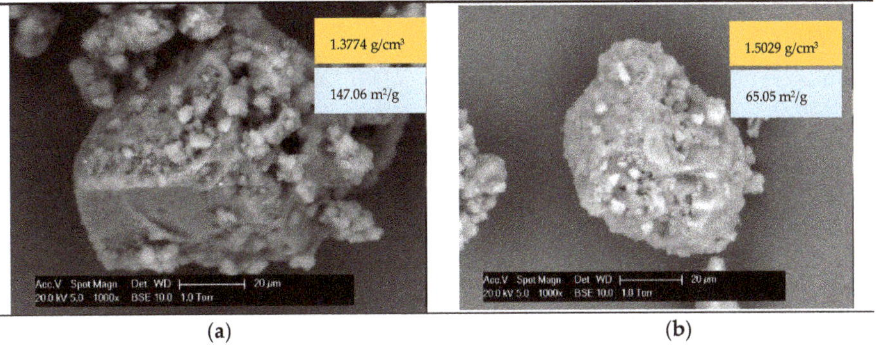

Figure 2. ESEM analysis results: ×1000 images of (**a**) fresh and (**b**) aged lignin.

Regarding the microstructural morphology of lignin-modified bitumen, the worm structures of fresh and aged binders (i.e., Bref_F, Bref_A, BL10, and BL10 A; see Table 1) were obtained by ESEM, as shown in Figure 3a,b. In particular, all fresh binders had a relatively clear slim worm structure. The structures of the binders changed significantly after PAV aging. The density of the worm structure increased, and the thickness of the worm structure substantially increased. For the long-term aged samples, it was difficult to observe the worm structure, and a longer exposure time was needed in the gaseous secondary electron detector (GSE) mode. Moreover, no significant changes were observed with the addition of lignin. In other words, the lignin particles were not embedded in the worm structure of bitumen. It may require a longer exposure time for the lignin-modified binders to display the same worm structures.

4.2. Chemical Characterization

In order to understand the aging of lignin itself, lignin samples were aged in different conditions: lignin in a fresh state without aging, and aged following one and two instances of PAV, for 20 h and 40 h, respectively, after conditioning lignin powder in the oven for 2 h at 163 °C. The FTIR spectral results of lignin in different aging conditions are shown in Figure 4a, demonstrating the functional groups of lignin. Each FTIR spectrum was the average result of the nine replications. The peak values of the curves were slightly different, but the peak positions of the curves were basically the same. This shows that the aged lignin did not produce new chemical functional group peaks. Additionally, the functional group absorbance index was used to measure the number of individual chemical components in bitumen.

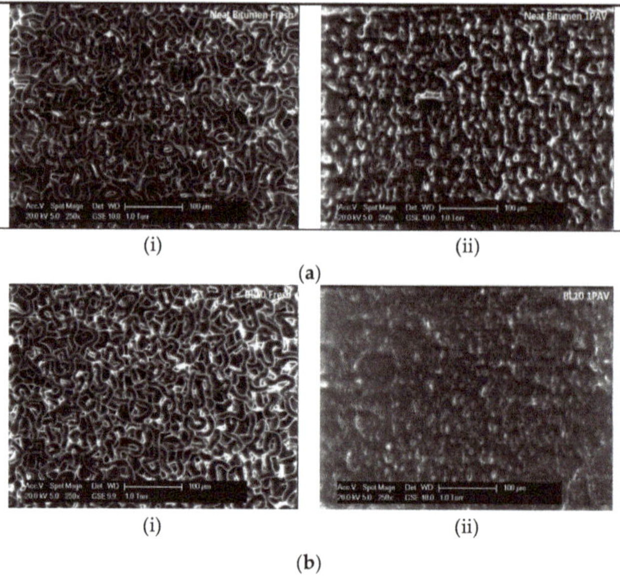

Figure 3. ESEM analysis results: ×1000 images of (**a**) neat bitumen and (**b**) lignin-modified binders ((i) fresh state and (ii) pressure aging vessel (PAV) aged state).

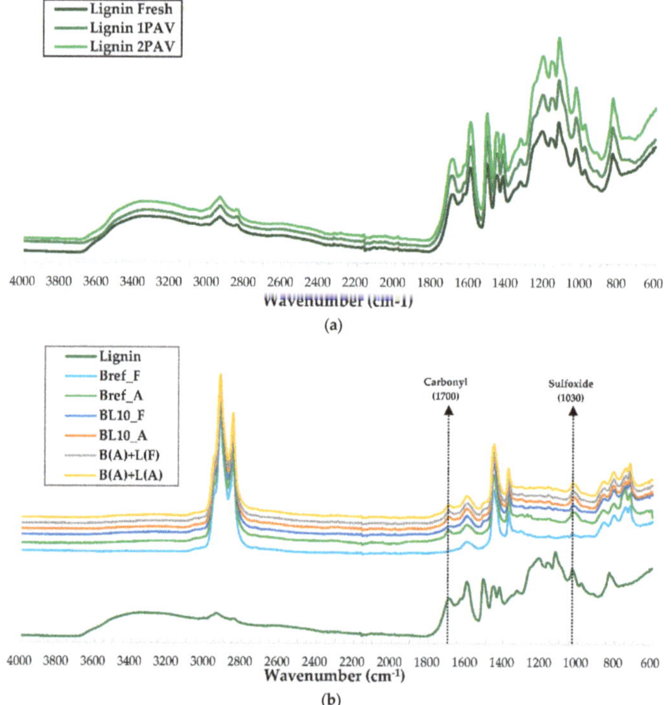

Figure 4. Fourier-transform infrared (FTIR) spectra of (**a**) lignin in different aging conditions, and (**b**) lignin, bitumen, and lignin-modified bitumen.

The typical absorbance indices are shown in Table 3. The values and the standard deviation of indices were the averages of the nine measurements. Most of the functional groups did not change or negligibly changed after the aging of lignin. This was illustrated by comparing the values of aging indices at different aging conditions. For example, the reduction of the hydroxyl group (3420 cm^{-1}) was mainly due to the volatilization of water in the short-term aging process and the reaction of hydroxide in the whole aging process. The carbonyl group (1708 cm^{-1}) increased due to oxidation. Most of the functional groups of lignin did not change with aging.

Table 3. The typical functional group absorbance index for lignin. PAV—pressure aging vessel.

Age State	Items	Absorbance Band Values (cm^{-1})								
		3420	1708	1597	1512	1269	1150	1125	1085	1032
Fresh	Value	0.2900	0.1640	0.1800	0.1772	0.0240	0.0095	0.0753	0.0054	0.0747
	SD	0.0189	0.0098	0.0033	0.0033	0.0022	0.0010	0.0114	0.0004	0.0068
1 PAV	Value	0.2736	0.1796	0.1833	0.1824	0.0252	0.0120	0.0652	0.0053	0.0734
	SD	0.0177	0.0050	0.0030	0.0040	0.0015	0.0011	0.0053	0.0007	0.0050
2 PAV	Value	0.2746	0.1770	0.1877	0.1813	0.0259	0.0123	0.0650	0.0056	0.0707
	SD	0.0191	0.0049	0.0035	0.0035	0.0024	0.0012	0.0065	0.0005	0.0056

The change in chemical composition of different aged lignin–bitumen systems is plotted in Figure 4b. The functional group peaks of lignin-modified bitumen were determined one by one through comparing the FTIR spectra of lignin, neat bitumen (Bref_F), and lignin-modified bitumen (BL10_F) in Figure 4b. It was determined that mixing lignin and bitumen does not cause a chemical reaction, because no new functional group peaks were produced in the FTIR spectra.

The aging indices of the studied materials were calculated, and they are provided in Figure 5. In Figure 5, the largest difference between the B(A) + L(F) and B(A) + L(A) binders was observed when aged lignin was added, indicating that the aging of lignin itself has little effect on the aging index of the whole system. The main reason was that lignin aged slightly, as shown in Table 3. In addition, the lignin content in lignin-modified bitumen was only 10% by mass of bitumen. The aging effect of bitumen was more obvious. Therefore, the difference between the fresh and aged lignin can be ignored. Carbonyl and sulfoxide aging indices can still be used to quantify the aging state of lignin-modified bitumen.

Both carbonyl and sulfoxide indices increased with aging. The fresh and aged virgin bitumen (Bref_F and Bref_A) were compared to indicate the aging extent of binders without lignin. Upon comparing Bref_A and BL10_A, it can be seen that the two materials (lignin and bitumen) were mixed firstly and then aged, which did reduce the rate of aging. Carbonyl and sulfoxide indices of neat bitumen increased from 0.0009 and 0.0098 to 0.0141 and 0.0231, respectively. However, for the BL10_F and BL10_A, the aging indices increased from 0.0076 and 0.0157 to 0.0138 and 0.0181 because lignin was added before aging. Comparing to BL10_A and B(A) + L(A), in which lignin and bitumen were aged separately and then mixed together, produced more aging functional groups.

Lignin was added and mixed uniformly in the bitumen. Lignin particles precipitate in bitumen. During the aging process of neat bitumen, the surface bitumen is exposed to the external environment, including temperature and air, and it would age first. As oxygen diffuses into the bitumen, the internal bitumen starts aging. For bitumen mixed with lignin particles, when oxygen enters the interior, it is necessary to bypass the obstacle formed by the lignin particles and the bituminous film. It would take a long time for oxygen in the air to enter the bitumen due to the entry path increasing. Therefore, the contact time between the bitumen and oxygen is reduced, and the oxidation effect of the bitumen is inhibited. Since the lignin particles precipitate in bitumen, the loss of light components in the bitumen is correspondingly reduced. This also delays the accumulation of asphaltenes. Overall, bitumen and lignin should be mixed together firstly and then aged to maximize the anti-oxidation effect of lignin in bitumen.

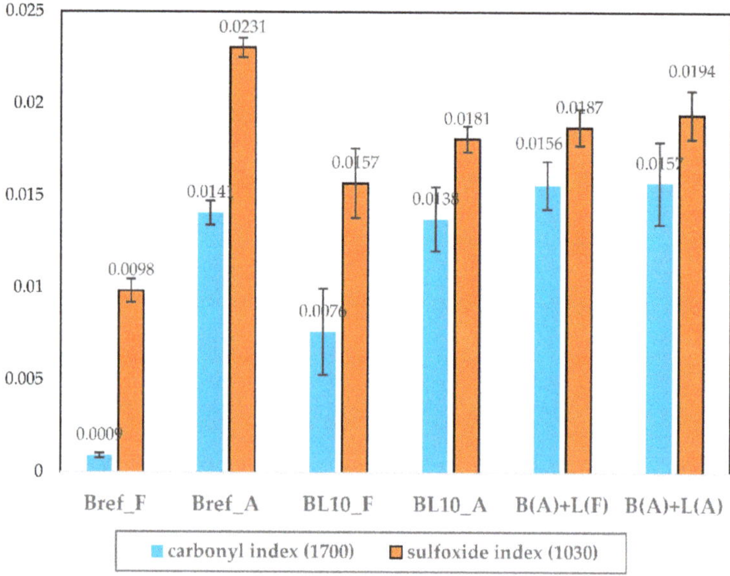

Figure 5. Carbonyl (1700 cm^{-1}) and sulfoxide (1030 cm^{-1}) indices of studied materials.

4.3. Rheological Characterization

4.3.1. Frequency Sweep Tests

The master curves of complex modulus and phase angle of different materials are shown in Figure 6. A higher modulus of the material indicates a stronger resistance to deformation. A lower phase angle means that the material is more elastic and that the delay in response between stress and strain is shorter. Each tested material had three replicates. Obviously, the complex modulus increased, and the phase angle decreased with aging. In the low-temperature region, the properties of the samples were very similar and stable in terms of modulus and phase angle. However, the effect of lignin on bitumen was mainly reflected at high temperatures.

The complex modulus and phase angle results of the three materials (i.e., BL10_A, B(A) + L(F), and B(A) + L(A)) were almost identical and overlapped each other, as determined by comparing them with fresh samples. The aging of lignin itself had little effect on the rheological properties of lignin-modified bitumen, as seen by comparing B(A) + L(A) and B(A) + L(F). In addition, there was little effect on the rheological properties of adding lignin before or after aging, as seen by comparing BL10_A and B(A) + L(A).

Additionally, to eliminate the influence of shift factor, the changes in bitumen rheology are depicted in black space diagrams [31] in Figure 7. After aging, the shape of the curve moved to a straighter curve. The decrease in phase angle and the increase in modulus denoted a tendency toward a more brittle material. It is clear from Figure 7 that, in addition to the fresh samples, the results of modified binders were quite close.

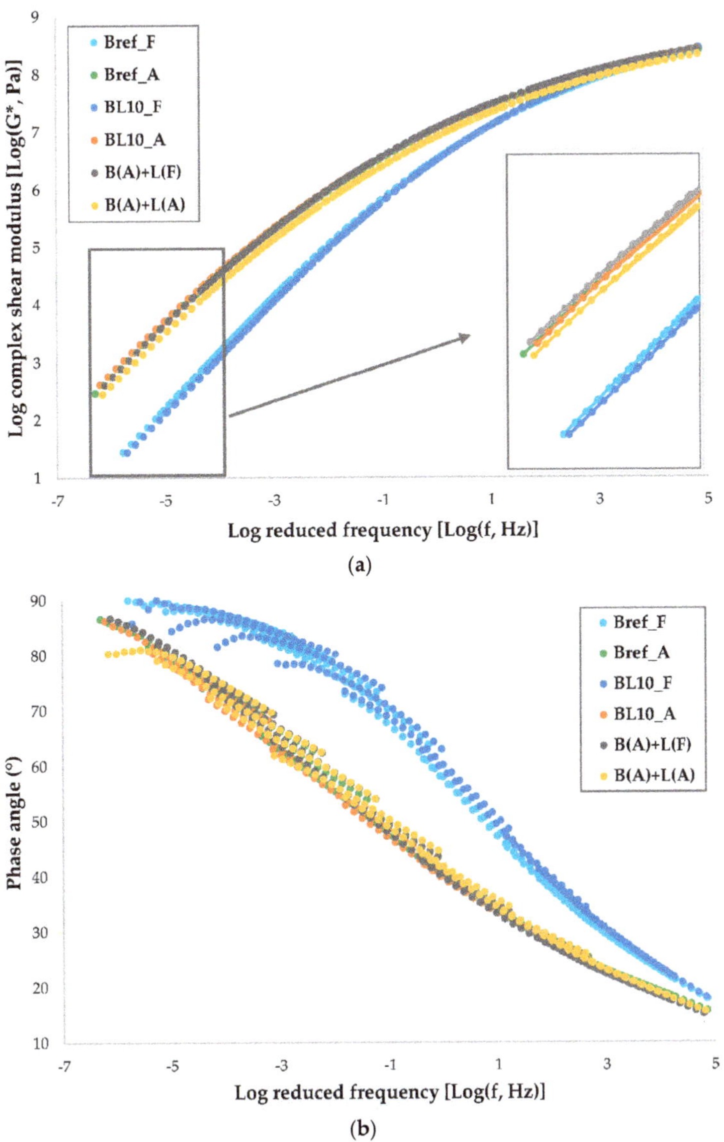

(a)

(b)

Figure 6. Master curves of (**a**) complex shear modulus and (**b**) phase angle.

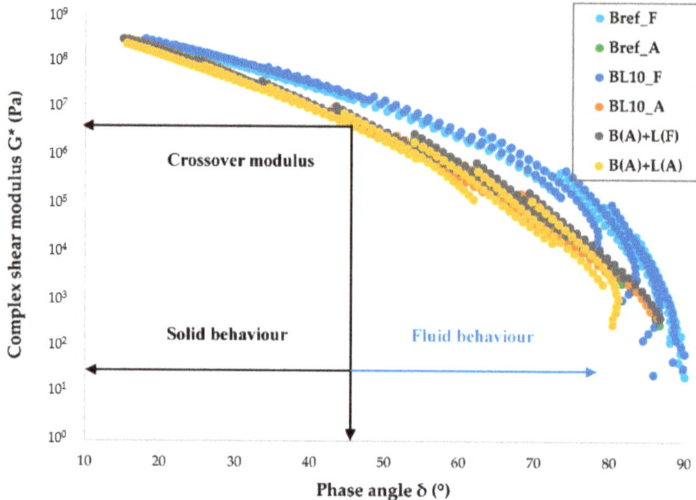

Figure 7. The rheological properties in a black diagram.

The most common method to characterize the viscoelastic fluid-to-solid transitional behavior is the crossover frequency (where the storage modulus and loss modulus are equal, i.e., the phase angle is 45°) of the storage modulus and the loss modulus [34]. The complex modulus corresponding to the crossover frequency is called the crossover modulus, which is shown in Figure 7. A lower crossover frequency reveals that bitumen has a higher molecular mass, longer relaxation time, and higher softening point, while a lower crossover modulus indicates wider molecular mass distribution and higher polydispersity [35,36]. The crossover modulus and frequencies of different samples are shown in Figure 8. It shows that aged bitumen had a lower crossover frequency and modulus. The results of aged lignin-modified bitumen were similar except for the unaged samples.

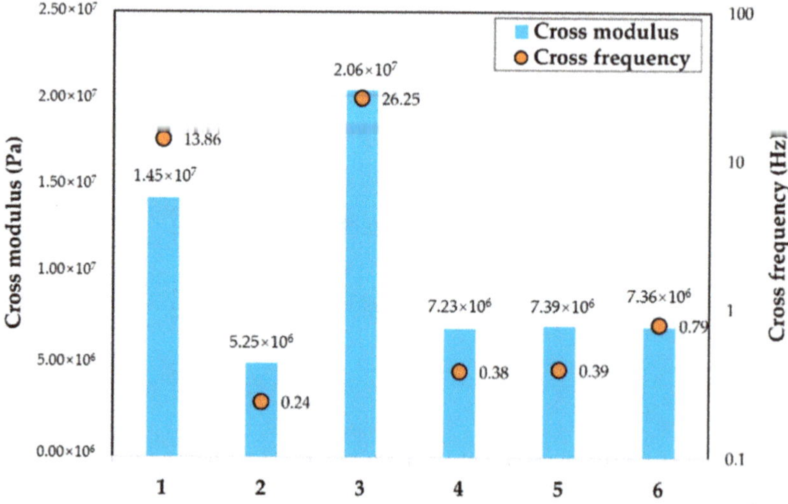

Figure 8. The rheological properties, crossover modulus, and crossover frequencies.

4.3.2. Linear Amplitude Sweep Tests

The fatigue resistance of studied materials was determined by LAS tests. Figure 9 shows the fatigue life (N_f) at different strain levels after performing the calculation, as mentioned in the previous section. The equation for the fatigue lines is listed in Table 4. With the increase in strain level, a significant decrease in N_f was observed. The strain level had a significant influence on the order of the fatigue life of the materials. In addition, the fatigue life performance was ranked as Bref_A to BL10_A, B(A) + L(A), B(A) + L(F), Bref_F, and BL10 at low strain levels (e.g., 2%). Bitumen becomes stiffer due to aging, and a stiffer material shows higher resistance to micro-deformation at low strain levels. In general, the addition of lignin lightly reduced the fatigue life of lignin-modified bitumen. On the other hand, the fatigue life performance at higher strain levels (e.g., 5%) was ranked as Bref_F to BL10, Bref_A, B(A) + L(A), BL10_A, and B(A) + L(F). Fresh bitumen showed a better fatigue life. Because the fresh bitumen has better viscous behavior than the aged one at high strain levels, damage is less likely to occur on the fresh bitumen. Moreover, the power of function for the fresh material was significantly larger than that of the other aged materials. Therefore, the aging process decreased the power of the fatigue function. Furthermore, the addition of lignin slightly reduced the fatigue life of bitumen, as seen by comparing the lines of Bref_F, BL10, Bref_A, and BL10_A samples. Another interesting point is that several straight lines had a similar fatigue life at 3% strain level. This point actually represents material strain sensitivity. A smaller point denotes more sensitivity. The performance at low strain levels should be emphasized. Compared to Bref_A, the other three aged samples (BL10_A, B(A) + L(A), and B(A) + L(F)) had very similar fatigue life. The aging of lignin itself and the moment that lignin was added to the bitumen had an inconspicuous effect on fatigue life. Physical interactions played a predominant role in fatigue life.

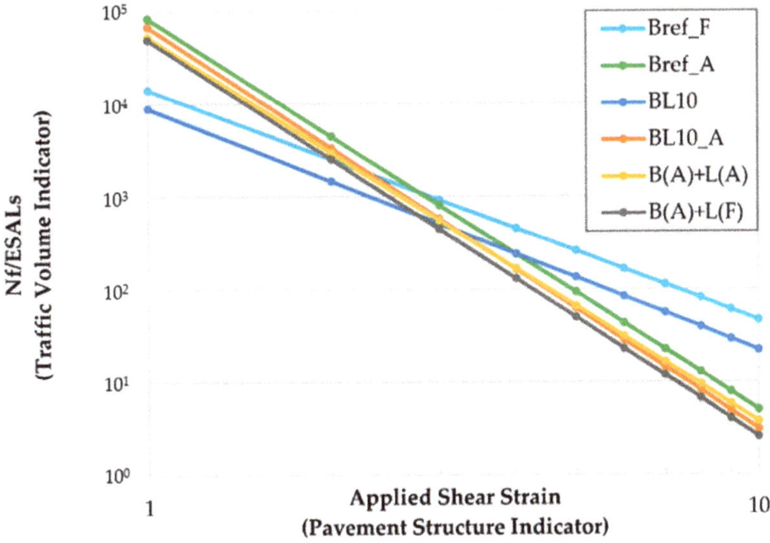

Figure 9. The plot of fatigue parameter N_f versus applied shear strain (20 °C).

Table 4. Fatigue lines of studied materials.

Studied Materials	$(N_f = A(\gamma)^{-B})$
Bref_F	$N_f = 14134(\gamma)^{-2.4832}$
Bref_A	$N_f = 83264(\gamma)^{-4.2192}$
BL10_F	$N_f = 8946(\gamma)^{-2.6073}$
BL10_A	$N_f = 67739(\gamma)^{-4.3408}$
B(A) + L(A)	$N_f = 52522(\gamma)^{-4.1466}$
B(A) + L(F)	$N_f = 48997(\gamma)^{-4.2816}$

4.3.3. Relaxation Tests

Figure 10 illustrates the change in shear stress of different studied materials with relaxation time. As materials age, the residual shear stress of aged materials increases after the same relaxation time period due to the increase in relaxation modulus [37]. Samples (BL10_A, B(A) + L(A), and B(A) + L(F)) produced using different preparation methods showed similar relaxation properties. To further evaluate the properties of these materials, the absolute value of shear stress at 0.1 s and 200 s and the ratio of residual shear stress (200 s) divided by the initial status (0.1 s) were plotted, as shown in Figure 11, depicting the stress at the initial and end times. Every sample had three replicates.

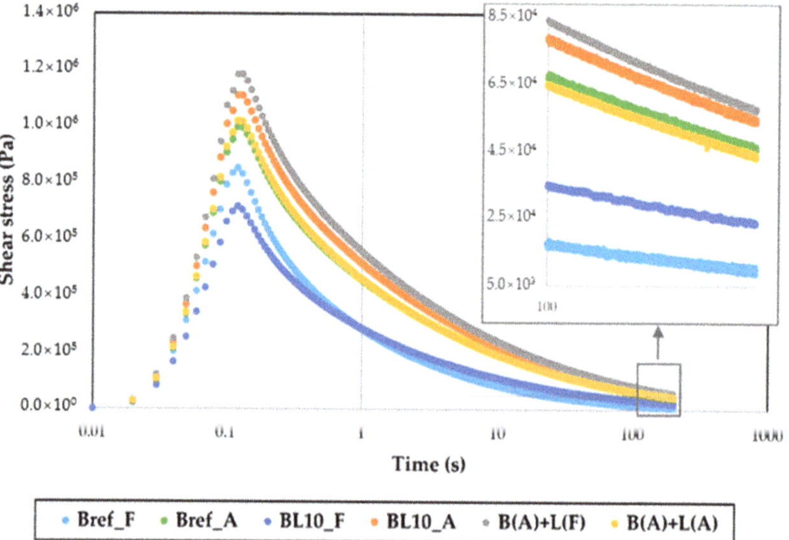

Figure 10. Relaxation results of the relationship between shear stress and relaxation time (at 1% shear strain and 0 °C).

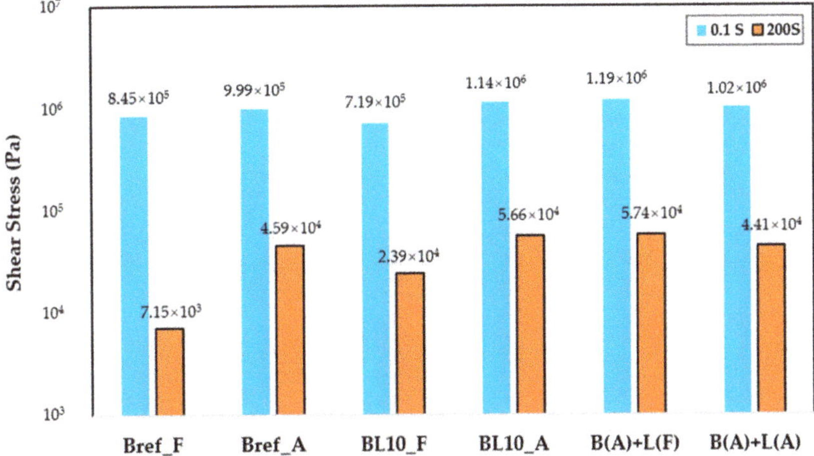

Figure 11. Shear stress at the initial (after 0.1 s) and end (200 s) times.

The initial and residual shear stresses of samples following different aging processes are depicted in Figure 11. Obviously, the aged samples had higher initial and residual shear stresses compared with fresh ones. The initial and residual shear stresses increased with the aging process. Upon reaching the same strain level in a shorter time (0.1 s), a higher initial shear stress means that the material had a higher modulus. Analyzing the data of Bref_A and BL10_A together, mixing with lignin had little effect on the initial and residual shear stress. The initial shear stress of B(A) + L(A), B(A) + L(F), and BL_A was similar, and it was about 1.4 times larger than that of BL_F. After a relaxation period of 200 s, the residual shear stress of these three aged materials was twice the value of the fresh samples. Comparing different aged materials, the order of initial shear stress was from B(A) + L(F) to BL_A and B(A) + L(A). Figure 12 shows the ratio of residual shear stress versus the initial shear stress of different samples.

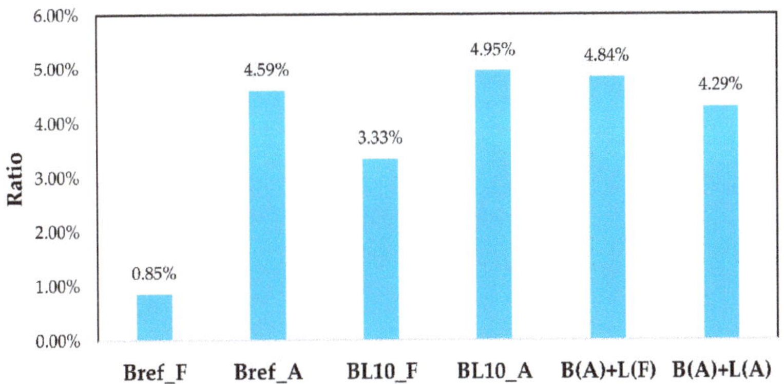

Figure 12. The ratio of residual shear stress (200 s) versus the initial shear stress (0.1 s).

The results show that the shear stress ratio increased with aging. For the neat bitumen, 0.85% of the initial shear stress remained after 200 s of relaxation; however, 4.59% shear stress remained in the aged sample Bref_A after relaxation. The same conclusion could be obtained from BL10_F and BL10_A. A lower ratio denotes a better relaxation property. Fresh specimens showed better elasticity than aging specimens. Therefore, they showed a better recovery ability and the ratio was smaller. The addition of lignin to fresh bitumen increased the stress ratio and reduced the relaxation property. In three different combinations, lignin and bitumen mixed after aging showed the minimum stress ratio in the aged samples. Due to the fact that traffic loading is usually continuous, the relaxation time of bitumen needs to be short enough to prevent stress accumulation in the pavement. The testing load should correspond to the traffic load frequencies or load time periods. Considering the fact that the bituminous materials are viscoelastic, then this time period should be linked to the recovery phase after loading, which subsequently affects the stress and, thus, the damage accumulation in the material. The relaxation time, as the shear stress was reduced to 50% and 25% of the initial stress, is depicted in Figure 13.

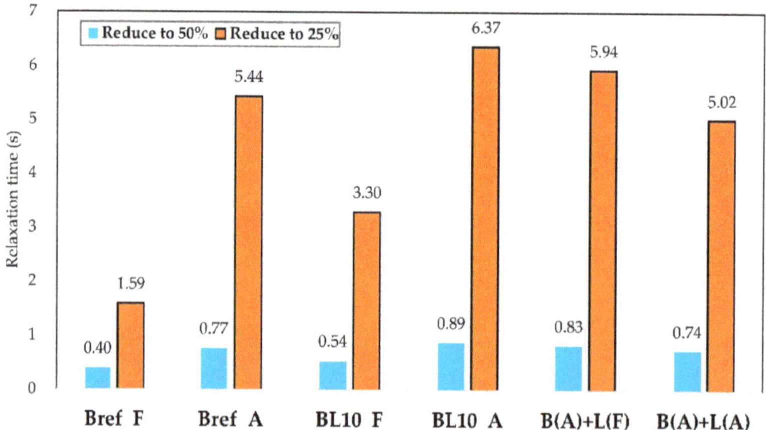

Figure 13. The relaxation time when the shear stress was reduced to 50% and 25% of the initial stress.

Figure 13 indicates that the relaxation time increased after aging when the shear stress was reduced to 50% and 25% of the initial stress. The viscosity of samples increased with aging, contributing to the relaxation time increase. For fresh materials, the shear stress of Bref_F reduction to 25% needed 1.59 s and that of BL10_F needed 3.30 s. However, for the other aged samples (Bref_A, BL_A, B(A) + L(F), and B(A) + L(A)) 5.44, 6.37, 5.94, and 5.02 s were required, respectively. When the lignin was added in advance, the relaxation time reduced to 50% and 25% when aging was increased (see BL_A with B(A) + L(A)). In summary, the aged sample had higher shear stress at initial and end times, a higher ratio of residual stress, and a longer relaxation time than the fresh sample. In addition, the relaxation properties of BL_A, B(A) + L(A), and B(A) + L(F) were similar compared to the fresh samples. The addition of lignin did not improve the relaxation properties dramatically. However, the aging of lignin itself and the moment that lignin was added to the bitumen had an effect on the relaxation properties, especially the relaxation ratio of residual and initial shear stress and the relaxation time when reducing to certain stress levels.

4.3.4. Glover–Rowe Parameter Tests

The shape of the Glover–Rowe (G–R) curve and the current Superpave fatigue parameter ($G^* \times \sin\delta$ = 5 MPa) were different. Based on the data of the samples, the curve shape of $G^* \times \sin\delta$ is not a logical damage indicator. The test conditions (15 °C, 0.005 rad/s) should be used for frequency sweeps to assess failure performance. Thus, the results of the Glover–Rowe damage parameter are shown in Figure 14. The two lines (G–R = 180 kPa; G–R = 450 kPa) provide the damage zone in the black space diagram. The red line shows the limit of the current Superpave fatigue parameter ($G^* \times \sin\delta$ = 5 MPa). The rhombuses, triangles, and squares with different colors denote different samples. Fresh specimens were in a very safe state. The properties approached the damage zone as aging proceeded. Based on the calculated data, the values of several aged samples were close to the damage onset line, but they did not enter the damage zone except for Bref_A. During the aging process, cracking began in the neat bitumen. However, with the addition of lignin, the time for the crack to appear was delayed. Lignin is, thus, beneficial for cracking resistance properties. Three samples (BL10_A, B(A) + L(A), and B(A) + L(F)) showed extremely close values and properties. This indicates that, independently of whether lignin ages or not, the adding process has little effect on the cracking resistance of the material.

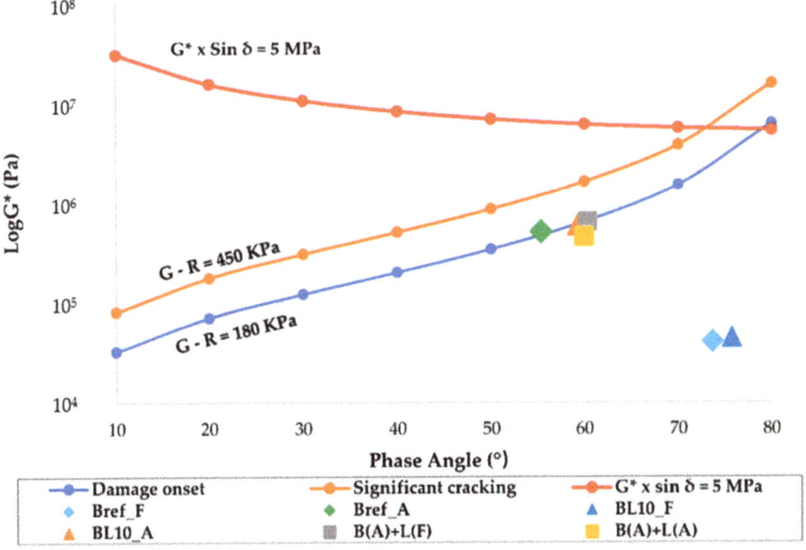

Figure 14. Glover–Rowe damage parameter in black space diagrams.

5. Conclusions

Lignin was added to bitumen to evaluate the interaction between the two materials after aging. Based on the current preliminary results, the main conclusions are as follows:

1. After aging, the specific surface area of lignin particles decreases. The microstructure of bitumen with and without lignin is almost the same. However, it becomes difficult to observe the worm structure of bitumen after aging and the addition of lignin.
2. The results from FTIR tests show that the various functional groups of lignin do not change remarkably during aging, and carbonyl and sulfoxide indices can still be used to assess the aging state of lignin-modified bitumen.
3. The effect of lignin added in advance or after aging has little effect on the viscoelastic characteristics of bitumen. The physical interaction between lignin and bitumen plays an important role, as shown by the changes in complex modulus, phase angle, crossover modulus, and crossover frequency.
4. The addition of lignin slightly reduces the fatigue life based on the results of LAS tests. As bitumen ages, its fatigue life increases at low strain levels and decreases at high strain levels due to the stiffening effect.
5. The addition of lignin does not improve the relaxation properties dramatically. However, the aging of lignin itself and the moment that lignin is added to the bitumen has an effect on the relaxation properties, especially the relaxation ratio of residual and initial shear stress and the relaxation time to certain stress levels.
6. Based on the results calculated by the Glover–Rowe parameter method, lignin provides an improvement in cracking resistance. Nevertheless, the aging of lignin itself and the moment that lignin is added to the bitumen have little impact on the cracking sensitivity of bitumen.

Overall, the addition of lignin has some positive effects as an anti-oxidant in bitumen. In this study, only a dosage of 10% by mass of bitumen, which was determined elsewhere, was used. Other lignin contents and bitumen types will be compared to verify the above conclusions. The compatibility between lignin and bitumen due to the differences in structure and density will be researched further, including the separation after blending and storage modulus.

Author Contributions: Conceptualization, Y.Z. and X.L.; methodology, Y.Z. and P.A.; validation, W.G., M.v.d.V., R.J., and S.E.; formal analysis, R.J.; investigation, P.A.; writing—original draft preparation, Y.Z. and P.A.; writing—review and editing, P.A., W.G., M.v.d.V., and R.J.; supervision, X.L.; project administration, S.E.

Funding: This research received no external funding.

Acknowledgments: The first author would like to acknowledge the scholarship from the China Scholarship Council and the financial support by Nedvang. Nedvang is also highly appreciated for providing base testing materials.

Conflicts of Interest: The authors declare no conflicts of interest.

References

1. Apostolidis, P.; Liu, X.; Kasbergen, C.; Scarpas, T. Synthesis of Asphalt Binder Aging and the State of the Art of Antiaging Technologies. *Transp. Res. Rec.* **2017**, *2633*, 147–153. [CrossRef]
2. Chailleux, E.M.; Audo, B.; Bujoli, C.; Queffelec, J.; Legrand; Lepine, O. Alternative Binder from Microalgae: Algoroute Project. In Proceedings of the Transportation Research E-Circular, Washington, DC, USA, 22 January 2012; pp. 7–14.
3. Audo, M.; Paraschiv, M.; Queffelec, C.; Louvet, I.; Hemez, J.; Fayon, F.; Lepine, O.; Legrand, J.; Tazerout, M.; Chailleux, E.; et al. Subcritical Hydrothermal Liquefaction of Microalgae Residues as a Green Route to Alternative Road Binders. *ACS Sustain. Chem. Eng.* **2015**, *3*, 583–590. [CrossRef]
4. Gosselink, R.J.A. Lignin as a Renewable Aromatic Resource for the Chemical Industry. Ph.D. Thesis, Wageningen University, Wageningen, The Netherlands, 2011.
5. Bruijnincx, P.; Gruter, G.-J.; Westenbroek, A.; Engelen-Smeets, E. *Lignin Valorisation: The Importance of a Full Value Chain Approach*; Utrecht University: Utrecht, The Netherlands, 2016.
6. Sundestrom, D.W.; Klel, H.E.; Daubenspeck, T.H. Use of Byproduct Lignins as Extenders in Asphalt. *Ind. Eng. Chem. Prod. Res. Dev.* **1983**, *22*, 496–500. [CrossRef]
7. McCready, N.S.; Williams, R.C. Utilization of Biofuels Coproducts as Performance Enhancers in Asphalt Binder. *Transp. Res. Rec.* **2008**, *2051*, 8–14. [CrossRef]
8. Xu, G.; Wang, H.; Zhu, H. Rheological Properties and Anti-aging Performance of Asphalt Binder modified with Wood Lignin. *Constr. Build. Mater.* **2017**, *151*, 801–808. [CrossRef]
9. Pan, T.A. First-principle based Chemophysical Environment for Studying Lignins as an Asphalt Antioxidant. *Constr. Build. Mater.* **2012**, *36*, 654–664. [CrossRef]
10. Pan, T.; Cheng, C. An Ab Initio Molecular Dynamics Analysis of Lignin as a Potential Antioxidant for Hydrocarbons. *J. Mol. Graph. Model.* **2015**, *62*, 325–341. [CrossRef]
11. Batista, K.B.; Padilha, R.P.L.; Castro, T.O.; Silva, C.F.S.C.; Araujo, M.F.A.S.; Leite, L.F.M.; Pasa, V.F.C.; Lins, V.M.D. High-temperature, Low-temperature and Weathering Aging Performance of Lignin modified Asphalt Binders. *Ind. Crop. Prod.* **2018**, *111*, 107–116. [CrossRef]
12. Arafat, S.; Kumar, N.; Wasiuddin, N.M.; Owhe, E.O. Lynam. Sustainable Lignin to Enhance Asphalt Binder Oxidative Aging Properties and Mix Properties. *J. Clean. Prod.* **2019**, *217*, 456–468. [CrossRef]
13. Perez, I.P.; Rodriguez Pasandin, A.M.; Pais, J.C.; Pereira, P.A.A. Use of Lignin Biopolymer from Industrial Waste as Bitumen Extender for Asphalt Mixtures. *J. Clean. Prod.* **2019**, *220*, 87–98. [CrossRef]
14. Van Vliet, D.; Slaghek, T.; Giezen, C.; Haaksman, I. Lignin as a Green Alternative for Bitumen. In Proceedings of the 6th Euroasphalt & Eurobitume Congress, Prague, Czech, 1–3 June 2016.
15. Yang, X.; Mills-Beale, J.; You, Z. Chemical Characterization and Oxidative Aging of Bio-asphalt and its Compatibility with Petroleum Asphalt. *J. Clean. Prod.* **2017**, *142*, 1837–1847. [CrossRef]
16. Kassem, E.; Khan, M.S.; Katukuri, S.; Sirin, O.; Muftah, A.; Bayomy, F. Retarding Aging of Asphalt Binders using Antioxidant Additives and Copolymers. *Int. J. Pavement Eng.* **2019**, *20*, 1154–1169. [CrossRef]
17. He, M.; Tu, C.; Cao, D.W.; Chen, Y.J. Comparative Analysis of Bio-binder Properties derived from Different Sources. *Int. J. Pavement Eng.* **2019**, *20*, 792–800. [CrossRef]

18. Boeriu, C.G.; Bravo, D.; Gosselink, R.J.; van Dam, J.E. Characterisation of Structure-dependent Functional Properties of Lignin with Infrared Spectroscopy. *Ind. Crop. Prod.* **2004**, *20*, 205–218. [CrossRef]

19. Dizhbite, T.; Telysheva, G.; Jurkjane, V.; Viesturs, U. Characterization of the Radical Scavenging Activity of Lignins-Natural Antioxidants. *Bioresour. Technol.* **2004**, *95*, 309–317. [CrossRef]

20. Vinardell, M.P.; Ugartondo, V.; Mitjans, M. Potential Applications of Antioxidant Lignins from Different Sources. *Ind. Crop. Prod.* **2008**, *27*, 220–223. [CrossRef]

21. ASTM D 6521-19. *Standard Practice for Accelerated Aging of Asphalt Binder Using a Pressurized Aging Vessel (PAV)*; ASTM: West Conshohocken, PA, USA, 2019.

22. Mikhailenko, P.; Kou, C.; Baaj, H.; Poulikakos, L.; Cannone-Falchetto, A.; Besamusca, J.; Hofko, B. Comparison of ESEM and Physical Properties of Virgin and Laboratory Aged Asphalt Binders. *Fuel* **2019**, *235*, 627–638. [CrossRef]

23. Lamontagne, J.; Dumas, P.; Mouillet, V.; Kister, J. Comparison by Fourier Transform Infrared (FTIR) Spectroscopy of Different Ageing Techniques: Application to Road Bitumens. *Fuel* **2001**, *80*, 483–488. [CrossRef]

24. Bykov, I. Characterization of Natural and Technical Lignins Using FTIR Spectroscopy. Master Thesis, Lulea University of Technology, Luleå, Sweden, 2008.

25. Petersen, J.C. A Review of the Fundamentals of Asphalt Oxidation: Chemical, Physicochemical, Physical Property, and Durability Relationships. In *Transportation Research E-Circular*; The National Acdemies of Sciences Engineering Medicine: Washington, DC, USA, 2009.

26. AASHTO T 315-19. *Standard Method of Test for Determining the Rheological Properties of Asphalt Binder Using a Dynamic Shear Rheometer (DSR)*; AASHTO: Washington, DC, USA, 2019.

27. AASHTO TP 101-14. *Standard Method of Test for Estimating Damage Tolerance of Asphalt Binders Using the Linear Amplitude Sweep*; AASHTO: Washington, DC, USA, 2014.

28. Hintz, C.; Bahia, H. Simplification of Linear Amplitude Sweep Test and Specification Parameter. *Transp. Res. Rec.* **2013**, *2370*, 10–16. [CrossRef]

29. Anderson, M.; Le Hir, Y.M.; Marasteanu, M.O.; Planche, J.-P.; Martin, D.; Gauthier, G. Evaluation of Fatigue Criteria for Asphalt Binders. *Transp. Res. Rec.* **2001**, *1766*, 48–56. [CrossRef]

30. Jing, R. Ageing of Bituminous Materials: Experimental and Numerical Characterization. Ph.D. Thesis, Delft University of Technology, Delft, The Netherlands, 2019.

31. Airey, G.D. Use of Black Diagrams to identify Inconsistencies in Rheological Data. *Road Mater. Pavement Des.* **2002**, *3*, 403–424. [CrossRef]

32. Mensching, D.J.; Rowe, G.M.; Daniel, J.S.; Bennert, T. Exploring Low-temperature Performance in Black Space. *Road Mater. Pavement Des.* **2015**, *16*, 230–253. [CrossRef]

33. King, G.; Anderson, M.; Hanson, D.; Blankenship, P. Using Black Space Diagrams to Predict Age-induced Cracking. In *7th RILEM International Conference on Cracking in Pavements*; Scarpas, A., Kringos, N., Al-QadiLoizos, I., Loizos, A., Eds.; Springer: Berlin, Germany, 2012; pp. 453–463.

34. Rahalkar, R.R. Correlation between the Crossover Modulus and the Molecular Weight Distribution using the Doi-Edwards Theory of Reptation and the Rouse Theory. *Rheol. Acta* **1989**, *28*, 166–175. [CrossRef]

35. Scarsella, M.; Mastrofini, D.; Barré, L.; Espinat, D.; Fenistein, D. Petroleum Heavy Ends Stability: Evolution of Residues Macrostructure by Aging. *Energy Fuel* **1999**, *13*, 739–747. [CrossRef]

36. Jing, R.; Varveri, A.; Liu, X.; Scarpas, A.; Erkens, S. Rheological, Fatigue and Relaxation Properties of Aged Bitumen. *Int. J. Pavement Eng.* **2019**, 1–10. [CrossRef]

37. Jing, R.; Varveri, A.; Xueyan, L.; Scarpas, T.; Erkens, S. Ageing Effect on the Relaxation Properties of Bitumen. In Proceedings of the Advances in Materials and Pavement Prediction: Papers from the International Conference on Advances in Materials and Pavement Performance Prediction (AM3P 2018), Doha, Qatar, 16–18 April 2018.

MDPI
St. Alban-Anlage 66
4052 Basel
Switzerland
Tel. +41 61 683 77 34
Fax +41 61 302 89 18
www.mdpi.com

Materials Editorial Office
E-mail: materials@mdpi.com
www.mdpi.com/journal/materials